Distillation

Distillation

Principles and Practices

Johann G. Stichlmair
Technische Universität München, München, Germany

James R. Fair
The University of Texas at Austin, Austin, TX

WILEY-VCH

NEW YORK / CHICHESTER / WEINHEIM / BRISBANE / SINGAPORE / TORONTO

This book is printed on acid-free paper. ∞

Published simultaneously in Canada.

Library of Congress Cataloging-in-Publication Data:

Stichlmair, Johann.
Distillation : principles and practice / by Johann Stichlmair and James R. Fair.
p. cm.
"A Wiley-Liss publication."
Includes index.
ISBN 0-471-25241-7 (cloth : alk. paper)
1. Distillation. I. Fair, James R., 1920– . II. Title.
TP156.D5S85 1998
660′.28425—dc21 97-48527
CIP

Printed in the United States of America.

10 9 8 7 6 5 4 3 2

Contents

	Preface	vii
	Nomenclature	ix
1.	Introduction	1
2.	Vapor-Liquid Equilibrium	6
3.	Single-Stage Distillation and Condensation	71
4.	Multistage Continuous Distillation (Rectification)	107
5.	Industrial Distillation Processes	192
6.	Multistage Batch Distillation	284
7.	Energy Economization in Distillation	328
8.	Design and Dimensioning Mass Transfer Equipment	361
9.	Control of Distillation Processes	474
	Index	519

Preface

Distillation is a widely used method for separating liquid mixtures into their components and had been called the workhorse separation of the petroleum, petrochemical, chemical, and related industries. It is old in the art, having been practiced in a crude form for several millennia. It represents a large portion of the capital investment in the above-noted industries, and is the largest consumer of energy in those industries. But distillation is also used elsewhere for such tasks as solvent purification and recovery, air separation, and environmental purity control. Wherever liquid–liquid extraction is used, there is likely to be a distillation column for solvent regeneration. It is quite safe to say that distillation is very much a part of today's industrial scheme—and will continue to be in the future. There appears to be no diminution in its dominance as a useful, reliable, and easily specified separation method to handle economically most needs for separating liquid mixtures.

Surveys have placed distillation in a preeminent position so far as technical understanding and convenience of use are concerned, higher on both counts than any other separation method. Yet, one may consider the method as still not completely developed, with much room for improvement in application and operation. And this older technique is still changing, as witness the use of related methods such as hybrid distillation (use of the method coupled with one or more different methods), reactive distillation (reaction and separation in situ), adsorptive distillation (combining the separation powers of distillation and selective adsorption), and membrane distillation (where an enhancement is made by a selective membrane). In essence, there is a large potential for investment economy and energy savings using "classical" distillation—and at the same time opportunities for combining our understanding of distillation principles with other separation disciplines.

The objective of this book is to provide the reader with a thorough background on the optimal use of distillation together with an understanding of the principles of distillation that in themselves can suggest new and improved applications of the technique to challenging separations. The book deals with all aspects of the subject: the thermodynamics of phase equilibria, the synthesis of distillation systems, the dimensioning and geometric aspects of distillation equipment, and the control distillation operations.

Special attention is given to the distillation of ternary mixtures, since they represent a more general case than binary mixtures, where most textbooks are focused. A thorough understanding of the principles of ternary mixture separations leads to a rational basis for dealing with multicomponent mixtures. New principles for separating azotropic mixtures are developed in a special section and are applied to a large collection of industrial processes. As indicated earlier, coverage has been extended to hybrid distillation and reactive distillation processes.

The authors decided to embark on this book project while working together at the University of Texas at Austin, where Dr. Stichlmair spent two leaves from the University of Germany. We felt that a collaborative effort could convey the essence of distillation technology from both European and American points of view. We expect the book to be useful to practitioners as well as academicians, and hope that it will stimulate newer generations of engineers to expand further the horizons of this important chemical engineering operation.

J. G. STICHLMAIR
Munich, Germany

J. R. FAIR
Austin, TX

Nomenclature

a	low boiler
a	specific area of a packing, m^{-1}
a	interfacial area per volume, m^{-1}
a_{eff}	effective interfacial area, m^{-1}
A	area, m^2
A_{ij}	Wilson parameter
A-1	absorber
b	intermediate boiler/high boiler
b	coefficient (equation of state), m^3/mol
B	amount of bottom product, kmol
B	second virial coefficient, kmol/m^3
$\dot{B}$	bottom stream (flow rate), kmol/s
c	intermediate boiler/high boiler
c	constant
C	constant
C	third virial coefficient, $(kmol/m^3)^2$
C_G	capacity factor, m/s
C_h	constant in Equation (8.78)
C-1	column
c.p.	critical point
cr	critical
d	high boiler
d	diameter, distance, m
D	diameter, m
D	diffusion coefficient, m^2/s
D	amount of overhead product, kmol (distillate)

$\dot{D}$	overhead stream (flow rate), kmol/s
D_E	dispersion coefficient, m^2/s
e	entrainer
E_{OG}	overall gasside point efficiency
E_{OGM}	overall gasside tray efficiency
E	exergy, kJ
E-1	heat exchanger, extractor
f	friction factor
f	fugacity, Pa
f	function
F	F factor (gas load), $Pa^{1/2}$
F	amount of feed, kmol
$\dot{F}$	feed stream (flow rate), kmol/s
Fr	Froude number
g	acceleration due to gravity, m/s^2
g	specific free enthalpy, kJ/kg or kJ/kmol
g_i	gas stream of component i, kmol/s
$\dot{G}$	gas or vapor stream (flow rate), kmol/s
G^E	free excess enthalpy, kJ
h	height, m
h_f	froth height, m
h_L	clear liquid height, m
h_L	holdup of liquid (packing)
h	specific enthalpy, kJ/kg or kJ/kmol
h_p	height of pressure drop, m
h_{Lo}	holdup below the loading point
h_w	weir height
H_{ij}	Henry coefficient for component i in solvent j, Pa
H	plate spacing, height of packing, m
$\dot{H}$	enthalpy stream, kJ/s or kW
HETP	height equivalent to a theoretical plate, m
H_{OG}	height of a transfer (vapor side), m
J	stripping factor, Figs. 4-4 and 4-6
k	binary interaction coefficient in cubic equations of state
k	number of components in a mixture
K	equilibrium ratio
K_R	reaction equilibrium constant
K, k	mass transfer coefficient, kmol $m^{-2}s^{-1}$
l	length, path length, m
l_i	liquid stream of component i, kmol/s
L	amount of liquid, kmol
$\dot{L}$	liquid stream (flow rate), kmol/s
m	coefficient (equation of state)
m	slope of equilibrium line, $m = dy^*/dx$
$\dot{M}$	mixing stream (flow rate), kmol/s

$M, \hat{M}_i$	molar mass of component i, kg/kmol
M	amount of mixture in the middle vessel
n	mole quantity, kmol
n	number of equilibrium stages; exponent in Equation (8A.1)
N	number of transfer units
$\dot{N}$	molar flow rate, kmol/s
p	pressure, Pa
p_i	partial pressure of component i, Pa
p_i^o	vapor pressure of pure component i, Pa
p_t	total pressure, Pa
q_F	thermal state of the feed
Q	heat, kJ
Q	dimensionless concentration change, Figs. 4-4 and 4-6
$\dot{Q}$	heat flow rate, kJ/s or kW
r	molar latent heat of vaporization, kJ/kmol
R-1	reactor
R	universal gas constant, R = 8.31451, J/(mol K)
$\dot{R}$	effluent of a reactor, kmol/s
R_G	external reboil ratio
R_L	external reflux ratio
s	plate thickness, m
$\dot{S}$	molar liquid flow after a decanter, side product stream, kmol/s
S-1	decanter
t	spacing, m
t	temperature, °C
t	time, s
T	absolute temperature, K
u	superficial velocity, m/s
v	specific volume, m^3/kg
V	volume, m^3
$\dot{V}$	volumetric flow, m^3/s
$\hat{v}_i$	molar volume of component i, m^3/kmol
W	work, kJ
x	mole fraction of liquid phase
x_g	weight fraction of liquid phase
X_i	transformed concentrations in reactive systems
y	mole fraction of gas phase
y_g	weight fraction of gas phase
z	mole fraction (gas-liquid mixture)
z	number of particles or channels
Z	compressibility factor

Greek Symbols

α	relative volatility, $\alpha_{ij} = p_i^o/p_j^o$
α	contraction number, discharge coefficient

α	coefficient in equation of state
β	mass transfer coefficient, m/s
ϵ	porosity, voidage
ζ	orifice coefficient, friction factor
ζ_E	friction factor in Ergun equation (Equation 8.75)
γ	activity coefficient
Δg^E	free excess enthalpy, kJ/kmol
Δg^{id}	free enthalpy change for mixing of ideal fluids, kJ/kmol
Δp	pressure drop, N/m^2 or Pa
$\Delta \rho$	density difference, kg/m^3
η	dynamic viscosity, kg/(m s)
ϑ	contact angle
ν	kinematic viscosity, m^2/s
ν_i	stoichiometric coefficient
λ_{ij}	Wilson parameter
π	pole on the enthalpy-concentration diagram, kJ/mol
π	circular constant, $\pi = 3.14159$
ρ	density, kg/m^3
σ	surface tension, N/m (= kg/s^2)
τ	contact time, s
τ_L	liquid residence time in the two-phase layer, s
ϕ	fugacity coefficient, caloric factor
ϕ	relative free area of a tray
ω	acentric factor

Subscripts and Superscripts

a	low boiler
ac	active
b	intermediate boiler/high boiler
B	bottom product
c	intermediate boiler/high boiler
c	column
c	continuous phase
C	condenser, cooling
cl	clearance under downcomer (skirt clearance)
cr	critical
d	high boiler
d	downcomer, dry
d	dispersed phase
D	overhead product
e	vena contracta
E	entrainment
eq	equivalent
f	froth layer, flooding
F	feed
G	gas

h	hole
H	heating
i	component i
id	ideal
irr	irrigated
j	component j
k	highest boiling component
l	component l
L	liquid
m	intermediate
max	maximum
min	minimum
n	number of plates or steps
n	nominal
o	orifice, single particle
OG	overall gas-phase state
OL	overall liquid-phase state
p, P	pinch point
R	reboiler
s	solid, particle swarm
sp	sphere
t	plate (tray)
T	transition point
V	valve
w	weir
o	initial state
∞	infinity
*	equilibrium state
∧	molar

Dimensionless Numbers

$Fr = \frac{u^2}{g\,d}$ Froude number

$Pe = \frac{l^2}{D\,\tau}$ Peclet number

$Re = \frac{\rho\,u\,d}{\eta}$ Reynolds number

$Sc = \frac{\eta}{\rho\,D}$ Schmidt number

$We = \frac{u^2\,\rho\,d}{\sigma}$ Weber number

Distillation

1

Introduction

1.1 Principle of Distillation Separation
1.2 Historical
Literature

1.1 PRINCIPLE OF DISTILLATION SEPARATION

The separation of a homogeneous liquid mixture into its individual components is an important step in many industrial processes. Of the various methods that might be used for effecting the separation, distillation is the most important one that is used commercially. Distillation utilizes a very simple separation principle: An intimate contact is created between the starting mixture and a second phase in order to enhance an effective mass transfer between these two phases. The thermodynamic conditions are chosen so that primarily the constituent to be separated from the mixture enters the second phase. The phases are subsequently separated into two single phases with different compositions.

Three steps are always involved in the implementation of this separation principle:

- Creation of a two-phase system
- Mass transfer between phases
- Separation of the phases

A large number of separation techniques utilize this very effective separation principle. Absorption, desorption, evaporation, condensation and distillation involve a gaseous and a liquid phase; solvent extraction involves two liquid phases. Separation techniques that utilize a fluid phase and a solid phase include adsorption, crystallization, drying and leaching. In most of these separation processes, the necessary two-phase system is created by adding an auxiliary phase to the feed mixture; the substances to be separated collect in diluted form in the auxiliary agent. In distillation, however, the second phase is generated by partial evaporization of the liquid feed. Hence, the use of an auxiliary substance (often called a *mass separating agent*), which usually requires costly recovery, can be avoided and the components to be separated can be recovered as relatively pure substances. Indeed, distillation requires energy only in the form of heat that can subsequently be easily removed from the system; this is an important advantage of distillation.

In practice, distillation requires intimate contacting of vapor and liquid under such conditions that the desired components of the liquid enter the vapor phase. Governing these conditions is thermodymanic equilibrium, and much of the art of distillation is devoted to the determination of how closely such equilibrium might be approached. In any case, it is necessary to separate the liquid and vapor phases for further processing.

The vapor and liquid are brought into intimate contact by countercurrent or cross-current flow, and mass transfer occurs because the two phases are not in thermodynamic equilibrium. The phases produced during distillation are formed by evaporation and condensation of the initial mixture. The separation process can be controlled solely by the heat supply.

The basis for planning a distillation process is a knowledge of the vapor-liquid equilibrium. As suggested earlier, the separation achieved depends primarily on the concentration of the individual substances in the vapor and liquid phases. In this book, principles of vapor-liquid equilibrium are discussed in Chapter 2, with special attention given to the equilibria of ternary and higher mixtures. Thermodynamic analysis of distillation and rectification is essential to establish the optimal conditions for mass transfer. The decisive factor is the driving force for mass transfer, i.e., the difference between the actual concentrations of the substances and their equilibrium concentrations. Operating conditions should ensure that this difference is sufficiently large. Appropriate relationships and methods for determining mass transfer are described in Chapters 3, 4, 5 and 6. Examples of industrially important separation processes and energy requirements are discussed in Chapters 5 and 7, respectively. Since separation is achieved by bringing the two phases into intimate contact, in practice, the problems created by multiphase flow and mass transfer between phases must be confronted. The mathematical interrelationships of multiphase flow are only beginning to be studied systematically from the viewpoint of fluid mechanics. As a result, only empirical approaches are usually available for practical equipment design, as described in Chapter 8. Chapter 9 deals with the control of single distillation columns and distillation processes.

1.2 HISTORICAL

Although several authors, e.g. [Krell 1958] support the view that the art of distillation has been well known to ancient Greece, this opinion has never been proven by historians. Ancient philosophers have been, however, very close by the correct understanding of the principles of distillation, for instance at the mere philosophical debate on the circulation of water in nature. Aristoteles (384–322 B.C.) writes in his Meteorologia ". . . . several authors support a similar view on the origin of the rivers. The water elevated by the sun and as rain condensated humidity collects . . . ". Later on in the same book he writes: "That evaporated sea water is drinkable and, after condensation, does not become sea water again, that can we state from experience." However, no practical application of these theoretical considerations has been reported and no device for performing the process of distillation has been described in ancient literature. Ancient Egypt and ancient China as well had probably no knowledge of the art of distillation. Forbes [Forbes 1948] agrees with several other authors e.g., [Underwood 1935] in the opinion that the art of distillation has been invented and pioneered in use in Alexandria, Egypt, in the first century A.D.

In the following centuries knowledge of distillation spread widely and was used around the eleventh century for the first time in northern Italy to produce alcoholic beverages. The development of distillation equipment has been influenced tremendously by this field of application. An interesting distillation equipment, described in the book "The Alchemy of Andreas Libavius," published in 1597, is illustrated in Fig. 1-1 [Libavius 1964]; it was used for the batch distillation of alcohol. Heat was supplied to the liquid contents of the boiler (a), built into the oven (b), and the vapor formed allowed to condense in two coolers (c). Cooling water was changed periodically. The only visible process was the dripping of the condensate into the receiver (d). This separation technique was named after the Latin word *destillare,* which means "dripping or trickling down". Even in early times, it was well known that an even higher alcohol content could be attained by using a second distillation step. In the apparatus shown in Fig. 1-1, two distillations could be carried out simultaneously. Condensate from the first distillation would be returned to the headpiece (e), the so-called *rectificatorium,* which was heated with vapor rising from the boiler.

Fig. 1-1. *Distillation and rectification equipment, taken from* The Alchemy of Andreas Libavius *[Libavius 1597]: (a) boiler, (b) oven, (c) coolers, (d) receiver, (e) headpiece, (f) receiver.*

The vapor thus produced was also condensed in the two coolers (c). A liquid with a higher alcohol content was then collected in a second receiver (f). The term *rectification* is derived from this process which, especially in Europe, is used to describe multistage distillation. The Latin words *recte facere* mean "to rectify or improve." Indeed, to this day, rectification refers to the process by which a further concentration change is achieved after the first evaporation step.

From such devices, distillation columns have been finally developed during the following centuries. Many authors, e.g., [Underwood 1935; Forbes 1970] agree in giving the credit of invention to the Frenchman Cellier-Blumenthal [Cellier-Blumenthal 1818]. Interesting are the circumstances that enhanced the development of improved distillation devices.

In 1807, Napoleon organized a blockade against England that was answered by a blockade against the European continent. Hence, goods from the overseas colonies no longer reached Europe, resulting in shortages of sugar-cane, among many other goods. It was well known that sugar could be produced from beets grown in Europe as well [Ullmann 1969]. However, the brown sugar from beets was much less attractive to the noblemen than the white sugar from canes. Napoleon opened a competition for producing white sugar from beets by setting a very high prize. A favorable process was extraction of sugar from the beets by alcohol instead of water—a process proposed again in recent years [Ullmann 1969]. The alcohol was recycled within this process; however, after longer periods of operation time, it needed purification. Cellier-Blumenthal developed the first distillation column (a tray column) for this process.

In the 19th and 20th century, the art of distillation developed rapidly, prompted by the oil and petrochemical industry [Deibele 1992] and by the chemical and pharmaceutical industry [Fair 1984]. The present importance of distillation is documented by the fact that approximately 40,000 distillation columns are under operation in the USA [Humphrey and Seibert 1992] that consume about 7% of the total energy demand of the USA [Gmehling et al. 1994].

LITERATURE

Cellier-Blumenthal, J. B.: *Brevet d'invention et de perfectionnement de quinze ans, pour des appareils destinés à la distillation continue et à l'évaporation,* French Patent No. 2266, from 1818.

Deibele, E.: "Die Entwicklung der Destillationstechnik im 19. Jahrhundert." Thesis, Technische Universität, München, 1992.

Fair, J. R.: "Historical Development of Distillation Equipment," *AIChE Symp. Ser. 79,* (1984) No. 235 pp. 1–14.

Forbes, R. J.: *Short History of the Art of Distillation.* E. J. Brill, Leiden, 1948

Gmehling, J.; Menke, J., Krafczyk, K., and Fischer, K.: *Azeotriopic Data, Part I and Part II.* VCH Publishers, Weiheim, New York, 1994.

Humphrey, J. L. and Seibert, A. F.: "New Horizons in Distillation," *Chemical Engineering* 1986 (December 1992).

Krell, E.: *Handbuch der Laboratoriumsdestillation*. VEB Deutscher Verlag der Wissenschaften, Berlin, 1958.

Libavius, A.: *Die Alchemie des Andreas Libavius, ein Lehrbuch der Chemie aus dem Jahre 1597*. Verlag Chemie, Weinheim, 1964.

Ullmann: *Ullmanns Enzyklopädie der Technischen Chemie*. Vol. 19, p. 201 and p. 217.

Underwood, A. J. V.: "The Historical Development of Distilling Plant," *Transactions—Institution of Chemical Engineers* 13 (1935) pp. 34–63.

Urban & Schwarzenberg, München, Berlin, Wien, 1969.

2

Vapor-Liquid Equilibrium

2.1 Dalton's Law, Raoult's Law, Henry's Law
    2.1.1 Boiling Point and Dew Point of Mixtures
    2.1.2 Special Cases of Vapor-Liquid Equilibria
2.2 Empirical Findings in Vapor-Liquid Equilibrium
    2.2.1 Binary Mixtures
    2.2.2 Ternary Mixtures
2.3 Formulation of Vapor-Liquid Equilibrium Relationships
    2.3.1 Equilibrium Ratio (K-value)
    2.3.2 Equilibrium Curve
    2.3.3 Distillation Line
2.4 Calculation of Phase Equilibrium from Excess Enthalpy
2.5 Calculation of Phase Equilibrium from Equations of State
2.6 Practical Determination of Phase Equilibria
Literature

If a vapor and a liquid are in intimate contact for a long period of time, equilibrium is attained between the two phases. This means that no net flow of heat, mass, or momentum occurs across the phase boundary. The complete formulation for a system at equilibrium is as follows:

$$\text{Thermal equilibrium} \qquad T'' = T' \tag{2.1}$$

$$\text{Mechanical equilibrium} \qquad p_t'' = p_t' \tag{2.2}$$

$$\text{Composition equilibrium} \qquad y_i^* = f(x_i \ldots) \tag{2.3}$$

where T is the absolute temperature, p_t the total pressure, y_i^* the mole fraction of component *i* in the vapor at phase equilibrium and x_i the mole fraction of *i* in the liquid. Single and double primes denote the liquid and vapor phases, respectively. As indicated in Chapter 1, the composition equilibrium between vapor and liquid is of primary importance in separation by distillation.

EXAMPLE 2-1: TRANSFORMATION OF WEIGHT FRACTIONS INTO MOLE FRACTIONS

Find the mole fractions x_i from the weight fractions x_{gi} for the ternary mixture i-butane, n-butane and n-pentane.

Definitions:

$$\text{Weight fraction} \qquad x_{gi} \equiv \frac{m_i}{\sum_{j=a}^{k} m_j}$$

with m_i = mass of constituents *a,b, . . . i,j, ... , k*

$$\text{Mole fraction} \qquad x_i \equiv \frac{n_i}{\sum_{j=a}^{k} n_j}$$

with n_i = moles of constituents *a,b, . . . i,j, ... , k*

Fundamental Relationship:

$$m_i = n_i \cdot \hat{M}_i \quad \text{with } \hat{M}_i = \text{formula weight of constituents}$$

$$x_{ga} = \frac{m_a}{m_a + m_b + m_c} = \frac{n_a \cdot \hat{M}_a}{n_a \cdot \hat{M}_a + n_b \cdot \hat{M}_b + n_c \cdot \hat{M}_c}$$

$$= \frac{\dfrac{n_a}{n_a + n_b + n_c} \cdot \hat{M}_a}{\dfrac{n_a}{n_a + n_b + n_c} \cdot \hat{M}_a + \dfrac{n_b}{n_a + n_b + n_c} \cdot \hat{M}_b + \dfrac{n_c}{n_a + n_b + n_c} \cdot \hat{M}_c}$$

$$\text{with } x_i \equiv \frac{n_i}{n_a + n_b + n_c};$$

$$x_{ga} = \frac{x_a \cdot \hat{M}_a}{x_a \cdot \hat{M}_a + x_b \cdot \hat{M}_b + x_c \cdot \hat{M}_c} \quad \text{or} \quad x_a = \frac{x_{ga}/\hat{M}_a}{x_{ga}/\hat{M}_a + x_{gb}/\hat{M}_b + x_{gc}/\hat{M}_c}$$

Multicomponent Mixture:

$$x_{gi} = \frac{x_i \cdot \hat{M}_i}{\sum_j^k x_j \cdot \hat{M}_j} \quad \text{or} \quad x_i = \frac{x_{gi}/\hat{M}_i}{\sum_j^k x_{gj}/\hat{M}_j}$$

Data:

Constituents i	Weight Fraction x_{gi}	Formular Weight $\hat{M}_i$
i-butane(= *a*)	0.17	58 kg/kmol
n-butane(= *b*)	0.39	58 kg/kmol
n-pentane(= *c*)	0.44	72 kg/kmol

Solution:

$$x_a = \frac{0.17/58}{0.17/58 + 0.39/58 + 0.44/72}$$

$$= \frac{2.9310 \cdot 10^{-3}}{2.9310 \cdot 10^{-3} + 6.7241 \cdot 10^{-3} + 6.1111 \cdot 10^{-3}}$$

$$x_a = \frac{2.931 \cdot 10^{-3}}{1.5766 \cdot 10^{-2}} = 0.186; \quad x_b = \frac{6.7241 \cdot 10^{-3}}{1.5766 \cdot 10^{-2}} = 0.426;$$

$$x_c = \frac{6.1111 \cdot 10^{-3}}{1.5766 \cdot 10^{-2}} = 0.388$$

2.1 DALTON'S LAW, RAOULT'S LAW, HENRY'S LAW

Vapor-liquid equilibria are best understood with the help of simple, empirical laws which may, however, only apply within certain limits. Dalton's law describes the vapor phase:

$$p_i''/p_t'' = n_i''/n_t'' \tag{2.4}$$

where p is the pressure and n the number of moles of substance. The subscript *i* denotes component *i*. Since $n_i''/n_t'' \equiv y_i$ (where y_i is the mole fraction of component *i* in the vapor), this law can be expressed as:

$$p_i'' = y_i \cdot p_t'' \tag{2.5}$$

Dalton's law generally holds at low pressure ($p_t < p_{cr}$, where p_{cr} is the critical pressure). At higher pressure, however, it may require modification. Instead of using pressure p, the equation is written in terms of fugacity f:

$$f_i'' = y_i \cdot f_t'' \tag{2.6}$$

The fugacity f can be determined from the fugacity coefficient $\phi \equiv f/p$ (Section 2.5).

The liquid phase is described by Raoult's law giving a relation for the partial pressure p_i' of component *i* over a liquid mixture:

$$p_i' = x_i \cdot p_i^o \tag{2.7}$$

The vapor pressure p^o of substance *i* can be evaluated by the Antoine equation. The Antoine parameters of some substances are listed in Table 2-1.

According to Equation (2.7) the partial pressure p_i' depends only on the vapor pressure p_i^o of pure component *i* and its liquid mole fraction x_i; it is not affected by the nature and concentration of other substances in the mixture. Raoult's law holds only as long as the forces of interaction between different types of molecules are equal (ideal mixture). Nonideal liquid-phase behavior (i.e., unequal interaction forces) is described by using the activity coefficient γ_i, which is defined as follows:

$$p_i' = \gamma_i \cdot x_i \cdot p_i^o \tag{2.8}$$

Equations (2.7) and (2.8) are valid only for mole fractions. The transformation of weight fractions x_{gi} into mole fractions x_i is demonstrated in Example 2-1.

The activity coefficient γ_i is often called a "Raoult's law correction factor" and is highly concentration-dependent. Methods used to determine activity coefficients are described in Section 2.4. Equations (2.7) and (2.8) apply only at temperatures below the critical temperature ($T < T_{cr}$) since a substance does not have a vapor pressure at temperatures $T > T_{cr}$. At higher temperatures, Henry's law is applied instead of Raoult's law:

$$p_i' = H_{ij} \cdot x_i \tag{2.9}$$

where H_{ij} is the Henry's coefficient for substance *i* in solvent *j*. Again, a correction must be made for nonideal mixtures by using the activity coefficient γ_i:

$$p_i' = H_{ij} \cdot \gamma_i \cdot x_i \tag{2.10}$$

This relationship is analogous to Raoult's law, except that the proportionality constant is not the vapor pressure p_i^o, but the Henry's coefficient H_{ij}.

Table 2-1. Normal boiling point (at p = 1.013 bar) and vapor pressure of some selected substances (selected from Gmehling et al. 1977: Reid et al. 1987)

Antoine equation for vapor pressure: $\frac{p_i^o}{bar} = \exp\left(A_i - \frac{B_i}{t/K + C_i}\right)$

Formula	Name	B.P. K	A_i	B_i	C_i	Range of Validity, K
Ar	argon	87.3	9.31039	832.778	2.361	73–133
CCl_4	tetrachloromethane	349.70	9.22001	2790.781	−46.741	287–350
$CHCl_3$	chloroform	334.40	9.39360	2696.249	−46.918	263–333
CH_2Cl_2	dichloromethane	313.20	10.44014	3053.085	−20.53	233–313
CH_2O	formaldehyde	254.00	9.94883	2234.878	−29.026	164–373
CH_2O_2	formic acid	373.70	9.37044	2982.446	−55.150	299–381
CH_3NO_2	nitromethane	374.30	10.14657	3331.696	−45.550	329–409
CH_4O	methanol	337.80	11.98705	3643.314	−33.424	288–357
CO	carbon monoxide	81.60	9.26679	769.935	1.637	52–121
CO_2	carbon dioxide	185.50	10.77151	1956.255	−2.112	146–285
C_2H_3N	acetonitrile	354.80	10.28058	3413.099	−22.627	246–355
C_2H_4O	acetaldehyde	293.70	9.97724	2532.406	−39.205	191–293
$C_2H_4O_2$	acetic acid	391.10	11.84896	4457.828	−14.699	291–391
C_2H_6	ethane	184.50	9.27428	1582.178	−13.762	145–284
C_2H_6O	ethanol	351.50	12.05896	3667.705	−46.966	293–366
C_2H_7N	ethylamine	289.70	10.38728	2618.730	−37.300	215–316
C_3H_6O	acetone	329.40	9.76775	2787.498	−43.486	260–328
C_3H_8	propane	231.10	9.10434	1872.824	−25.101	192–331
C_3H_8O	1-propanol	370.40	11.21152	3310.394	−74.687	333–378
C_3H_8O	2-propanol	355.40	13.82295	4628.956	−20.514	247–356
$C_3H_8O_2$	2-methoxy-ethanol	397.60	11.45476	4130.796	−36.273	290–447
C_4H_8O	2-butanone	352.80	9.64438	2904.340	−51.181	316–361
C_4H_8O	tetrahydrofuran	339.10	9.48686	2768.375	−46.896	296–373
$C_4H_8O_2$	1,4-dioxane	374.50	10.49171	3579.781	−32.813	293–378
$C_4H_8O_2$	butyric acid	436.40	13.43588	5602.222	−17.961	293–423
$C_4H_8O_2$	ethylacetate	350.30	9.73241	2866.606	−55.269	289–349
C_4H_9NO	morpholine	401.40	9.86713	3333.452	−63.150	317–443
C_4H_{10}	n-butane	272.70	9.05814	2154.897	−34.420	196–292
C_4H_{10}	isobutane	261.40	9.15169	2133.243	−28.162	165–345
$C_4H_{10}O$	diethylether	307.70	9.91763	2847.722	−20.110	308–432
$C_4H_{10}O_2$	2-ethoxyethanol	408.30	11.38407	4149.028	−43.150	336–407
C_5H_5N	pyridine	388.40	9.52860	3124.447	−60.495	254–388
C_5H_{12}	pentane	309.15	9.21312	2477.075	−39.945	223–331
C_6H_5Cl	chlorobenzene	404.90	9.77659	3485.354	−48.327	260–335
$C_6H_5NO_2$	nitrobenzene	484.00	9.79835	4032.655	−71.814	317–484
C_6H_6	benzene	353.20	9.22142	2755.642	−53.989	281–353
C_6H_6O	phenol	455.00	9.33802	3183.669	−113.657	336–455
C_6H_{12}	cyclohexane	353.90	9.15600	2778.000	−50.014	280–354
$C_6H_{12}O$	cyclohexanol	434.30	12.61197	5200.527	−21.526	317–434
$C_6H_{12}O_2$	butylacetate	399.20	9.79073	3293.659	−62.405	333–399
C_6H_{14}	hexane	341.90	9.29213	2738.418	−46.870	243–443
C_7H_8	toluene	383.80	9.38490	3090.783	−53.963	246–384
C_7H_{16}	heptane	371.60	9.25363	2911.320	−56.510	270–400
C_8H_{10}	ethylbenzene	409.30	9.41928	3291.661	−59.383	263–409
C_8H_{18}	octane	399.15	9.34011	3128.752	−63.295	259–399
H_2O	water	373.10	11.96481	3984.923	−39.724	274–373
NH_3	ammonia	239.80	10.88865	2363.237	−22.621	200–379
N_2	nitrogen	77.40	8.74393	648.592	−2.976	65–115
O_2	oxygen	90.20	8.43907	674.588	−10.093	60–125

2.1.1 Boiling Point and Dew Point of Mixtures

Raoult's and Dalton's laws can be used to find the boiling point of a liquid mixture. Boiling starts when the sum of the pressures p_i' of all components equals the total pressure p_t'' of the vapor phase:

$$\sum p_i' = p_t'' \quad \text{or} \quad \sum p_i/p_t'' = 1 \tag{2.11}$$

With Raoult's law, we obtain:

$$\sum x_i \cdot \gamma_i \cdot p_i^o = p_t'' \quad \text{or} \quad \sum x_i \cdot \gamma_i \cdot p_i^o/p_t'' = 1 \tag{2.12}$$

Since the vapor pressure p_i^o is a complex function of temperature, Equation (2.12) can be solved by iteration only. The procedure is demonstrated in Example 2-2 for an ideal ternary mixture.

By analogy to Equation (2.12), the dew point of a vapor mixture is found by:

$$\sum \frac{y_i \cdot p_t}{p_i^o \cdot \gamma_i} = 1 \quad \text{or} \quad \frac{1}{p_t} = \sum \frac{y_i}{p_i^o \cdot \gamma_i} \tag{2.13}$$

Applying Equations (2.12) and (2.13) to an ideal binary mixture ($\gamma_i = 1$) yields the result shown in Fig. 2-3A (second row). Both the boiling point line and the dew point line are curved over the range of molefraction x. For distillation, the difference $y - x$ is of major importance. This concentration difference can be seen more clearly in the $y - x$ plot (known as a "McCabe-Thiele diagram") in Fig. 2-3A (third row). The evaluation of the dew point of an ideal ternary mixture is demonstrated in Example 2-3.

EXAMPLE 2-2: BOILING POINT OF AN IDEAL TERNARY MIXTURE

Find the boiling point of a liquid mixture with 17 wt-% i-butane (*a*), 39 wt-% n-butane (*b*) and 44 wt-% n-pentane (*c*) at a pressure $p = 0.7$ bar.

Solution:

1. Transformation from weight fractions x_{gi} into mole fractions x_i (see Example 2-1): $x_a = 0.186$; $x_b = 0.426$; $x_c = 0.388$.
2. Solution with a vapor pressure/temperature diagram in Fig. 2-1—Boiling points of the constituents at $p = 0.7$ bar: $t_a = -20.5°C$; $t_b = -10°C$; $t_c = 25°C$.
   2.1. Estimation of the boiling temperature t: $t \approx \sum t_i \cdot x_i = -20.5 \cdot 0.186 + (-10) \cdot 0.426 + 25 \cdot 0.388 = 1.63°C$; we set: $t = 1°C$.

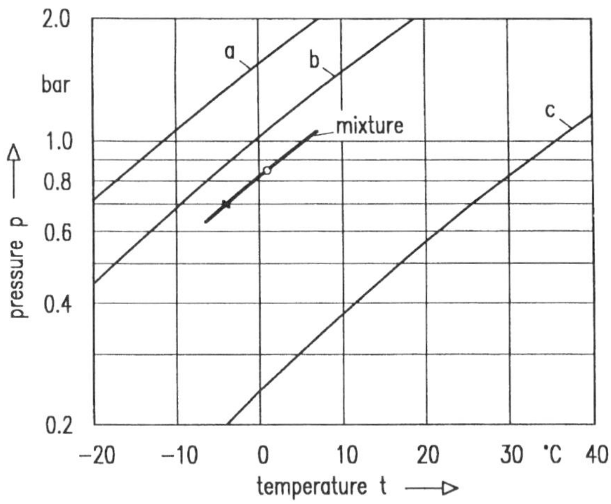

Fig. 2-1. Determination of the boiling point of the ideal ternary mixture i-butane/n-butane/n-pentane at a pressure of 0.7 bar (Example 2-2).

2.2. Vapor pressures p_i^o at t = 1°C: p_a^o = 1.60 bar; p_b^o = 1.06 bar; p_c^o = 0.25 bar.

2.3. Vapor pressure p_m of the mixture (Equation 2.12): $p_m = 0.186 \cdot 1.60 + 0.426 \cdot 1.06 + 0.388 \cdot 0.25 = 0.848$ bar; $p_m \neq 0.7$ bar, the estimation of t was wrong.

2.1a. Better estimation—Mark the point (t, p_m) in the diagram of Fig. 2-1 and draw through this point a parallel to the vapor pressure lines. The intersection with p = 0.7 bar gives a better value for the temperature: t = −4°C.

2.2a. Vapor pressure p_i^o at t = −4°C: p_a^o = 1.36 bar; p_b^o = 0.88 bar; p_c^o = 0.20 bar.

2.3a. Vapor pressure p_m of the mixture: $p_m = 0.186 \cdot 1.36 + 0.426 \cdot 0.88 + 0.388 \cdot 0.20 = 0.2511 + 0.3749 + 0.0776 = 0.7036$ bar; $p_m \approx$ 0.70 bar, the estimation was good enough: t = 4°C.

2.4. Composition of the first bubble: $y_a = 0.2511/0.7036 \approx 0.357$; $y_b = 0.3749/0.7036 \approx 0.533$; $y_c = 0.0776/0.7036 \approx 0.110$.

3. Solution with Antoine equation—Antoine equation for vapor pressure of pure components:

$$\frac{p_i^o}{\text{bar}} = \exp\left(A_i - \frac{B_i}{t/K + C_i}\right);$$

Antoine constants Table 2-1:

i	A_i	B_i	C_i
i-butane (*a*)	9.15169	2133.243	−28.162
n-butane (*b*)	9.05814	2154.897	−34.420
n-pentane (*c*)	9.21312	2477.075	−39.945

3.1. Estimation of the boiling point: $t = 1°C \triangleq 274.15$ K.

3.2. Vapor pressures p_i^o at $t = 1°C$:

$$\frac{p_a^o}{\text{bar}} = \exp\left(9.15169 - \frac{2133.243}{274.15 - 28.162}\right) = 1.6153;$$

$p_a^o = 1.6153$ bar; $p_b^o = 1.0717$ bar; $p_c^o = 0.2558$ bar

3.3. Vapor pressure p_m of the mixture at $t = 1°C \triangleq 274.15$ K: $p_m = 0.186 \cdot 1.6153 + 0.426 \cdot 1.0717 + 0.388 \cdot 0.2558 = 0.8563$ bar; $p_m \neq 0.7$ bar; the estimation was wrong.

3.1a. Better estimation—The temperature dependency of p_m can be described by the Antoine equation of substance *b* (= highest vapor concentration) when a correction factor Φ_b is used.

$$\frac{p_m}{\text{bar}} = \Phi_b \cdot \exp\left(A_b - \frac{B_b}{t/K - C_b}\right) \quad \text{with } \Phi_b \equiv \frac{p_m}{p_b^o};$$

$$\Phi_b \equiv \frac{0.8563}{1.0717} = 0.7990;$$

$$\frac{t_m}{K} = \frac{B_b}{A_b - \ln(p/\Phi_b)} - C_b$$

$$= \frac{2154.897}{9.05814 - \ln(0.7/0.7990)} + 34.420 = 268.892;$$

$$t = 268.892 \text{ K} \triangleq -4.25°C$$

3.2a. Vapor pressure p_i^o at $t = -4.25°C$: $p_a^o = 1.3370$ bar; $p_b^o = 0.8764$ bar; $p_c^o = 0.2007$ bar.

3.3a. Vapor pressure p_m of the mixture: $p_m = 0.186 \cdot 1.3370 + 0.426 \cdot 0.8764 + 0.388 \cdot 0.2007 = 0.2487 + 0.3733 + 0.0779 = 0.6999$ bar; $p \approx 0.70$ bar, the estimation was good: $t = -4.25°C$.

3.4. Composition of the first bubble: $y_a = 0.2487/0.6999 = 0.355$; $y_b = 0.3733/0.6999 = 0.533$; $y_c = 0.0779/0.6999 = 0.111$.

EXAMPLE 2-3: DEW POINT OF AN IDEAL TERNARY MIXTURE

Find the dew point of a vapor mixture with 17 wt-% i-butane (*a*), 39 wt-% n-butane (*b*) and 44 wt-% n-pentane (*c*) at a pressure $p = 0.7$ bar.

Solution:

1. Transformation from weight fraction y_{gi} into mole fraction y_i (see Example 2-1):

$$y_a = 0.186;\ y_b = 0.426;\ y_c = 0.388.$$

2. Solution with a vapor pressure-temperature diagram in Fig. 2-2—Dew points of the constituents at $p = 0.7$ bar: $t_a = -20.5°C$; $t_b = -10°C$; $t_c = 25°C$.
   - 2.1. Estimation of the dew point temperature t: $t \approx \Sigma t_i \cdot y_i = -20.5 \cdot 0.186 + (-10) \cdot 0.426 + 25 \cdot 0.388 = 1.63 \approx 1°C$.
   - 2.2. Vapor pressure p_i^o at $t = 1°C$: $p_a^o = 1.60$ bar; $p_b^o = 1.06$ bar; $p_c^o = 0.255$ bar.
   - 2.3. Dew point calculation (Equation 2.13): $\Sigma y_i/p_i^o \cdot p = 1$ or $\Sigma y_i/p_i^o = 1/p$; $0.186/1.60 + 0.426/1.06 + 0.388/0.255 = 2.0397 = 1/p$; $p_m = 0.4903$ bar; $p_m \neq 0.70$ bar. The estimation was wrong.
   - 2.1a. Better estimation—Mark the point (t, p_m) in the diagram of Fig. 2-2 and draw through this point a parallel to the vapor pressure lines. The intersection with $p = 0.7$ bar gives a better value for t: $t = 9.5°C$.
   - 2.2a. Vapor pressures p_i^o at $t = 9.5°C$: $p_a^o = 2.15$ bar; $p_b^o = 1.45$ bar; $p_c^o = 0.375$ bar
   - 2.3a. Dew point (Equation 2.13): $1/p_m = 0.186/2.15 + 0.426/1.45 + 0.388/0.375 = 1.4150$; $p_m = 0.7067$ bar; $p_m \approx 0.70$ bar. The estimation was good enough (deviation < 1%): $t = 9.5°C$.

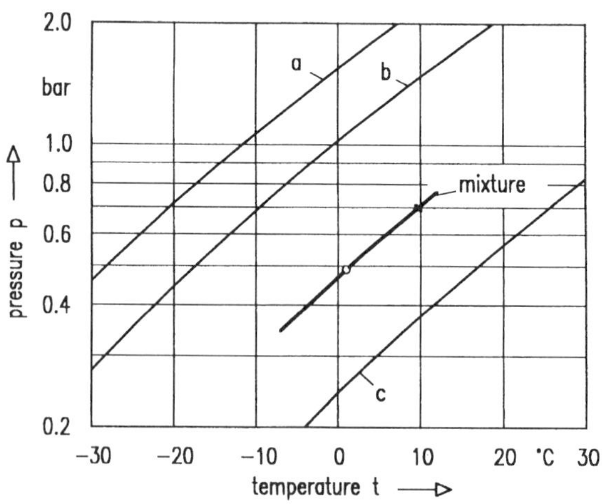

Fig. 2-2. Determination of the dew point of the ideal ternary mixture i-butane/n-butane/n-pentane at a pressure of 0.7 bar (Example 2-3).

2.4. Composition of the first liquid droplet—Equation (2.17): $x_i = y_i/p_i^o \cdot p$; $x_a = 0.186/2.15 \cdot 0.7067 = 0.0611$; $x_b = 0.426/1.45 \cdot 0.7067 = 0.2076$; $x_c = 0.388/0.375 \cdot 0.7067 = 0.7312$.

3. Solution with Antoine equation—Antoine equation for vapor pressure of pure components:

$$\frac{p_i^o}{\text{bar}} = \exp\left(A_i - \frac{B_i}{t/K + C_i}\right);$$

Antoine constants, Table 2-1:

i	A_i	B_i	C_i
i-butane (*a*)	9.15169	2133.243	−28.162
n-butane (*b*)	9.05814	2154.897	−34.420
n-pentane (*c*)	9.21312	24.77075	−39.945

3.1. Estimation of dew point temperature t: $t = 1°C \triangleq 274.15$ K.

3.2. Vapor pressures p_i^o at $t = 1°C$:

$$\frac{p_a^o}{\text{bar}} = \exp\left(9.15169 - \frac{2133.243}{274.15 - 28.162}\right) = 1.6153;$$

$p_a^o = 1.6153$ bar; $p_b^o = 1.0717$ bar; $p_c^o = 0.2558$ bar.

3.3. Dew point pressure p_m: $1/p_m = 0.186/1.6153 + 0.426/1.0717 + 0.388/0.2558 = 2.0295$; $p_m = 0.4927$ bar; $p_m \neq 0.7$ bar. The estimation was wrong.

3.1a. Better estimation—The temperature dependency of p_m can be described by the Antoine equation of substance *c* (= highest liquid concentration) when a correction factor Φ_c is used.

$$\frac{p_m}{\text{bar}} = \Phi_c \cdot \exp\left(A_c - \frac{B_c}{t/K + C_c}\right)$$

$$\text{with } \Phi_c \equiv \frac{p_m}{p_c^o} = \frac{0.4927}{0.2558} = 1.9261;$$

$$\frac{t}{K} = \frac{B_c}{A_c - \ln(p/\Phi_c)} - C_c$$

$$= \frac{2477.075}{9.2131 - \ln(0.7/1.9261)} + 39.945 = 282.20;$$

$$t = 282.20 \text{ K} \triangleq 9.05°C$$

3.2a. Vapor pressures p_i^o at $t = 9.05°C$: $p_a^o = 2.1262$ bar; $p_b^o = 1.4352$ bar; $p_c^o = 0.3635$ bar.

3.3a. Dew point pressure p_m: $1/p_m = 0.186/2.1262 + 0.426/1.4352 + 0.388/0.3635 = 1.4517$; $p_m = 0.6889$ bar; $p_m \neq 0.7$ bar. The estimation was wrong.

3.1b. Better estimation: $\Phi_c \equiv p_m/p_c^o = 0.6889/0.3635 = 1.8952$;

$$\frac{t}{K} = \frac{2477.075}{9.2131 - \ln(0.7/1.8952)} + 39.945 = 282.579;$$

$t = 282.579$ K $\triangleq$ 9.43°C.

3.2b. Vapor pressures p_i^o at $t = 9.43$°C: $p_a^o = 2.1530$ bar; $p_b^o = 1.4544$bar; $p_c^o = 0.3694$ bar.

3.3b. Dew point pressure p_m: $1/p_m = 0.186/2.1530 + 0.426/1.4544 + 0.388/0.3694 = 1.4298$; $p_m = 0.6994$ bar; $p_m \approx 0.7$ bar. The estimation was good enough: $t = 9.43$°C.

3.4. Composition of the first liquid droplet: $x_a = 0.186/2.1530 \cdot 0.6994 = 0.0604$; $x_b = 0.426/1.4544 \cdot 0.6994 = 0.2049$; $x_c = 0.388/0.3694 \cdot 0.6994 = 0.7346$.

2.1.2 Special Cases of Vapor-Liquid Equilibria

Dalton's and Raoult's law will now be applied to demonstrate the behavior of three special cases of a binary mixture. The first case is an ideal mixture where the intermolecular forces between the molecules of the same type (i.e., *a-a* or *b-b*) are the same as between the molecules of different types (i.e., *a-b*). Thus, a molecule does not realize whether it collides with a brother or a foreigner. In the second case, it is assumed that molecules of different types repel one another so heavily in the liquid phase that they cannot be mixed (total immiscibility). Accordingly, the liquid splits into two liquid phases, each phase containing just a single substance. The reverse situation is considered for the third case. The molecules of different types are supposed to attract each other so strongly that they undergo a chemical reaction to form a new substance *c* (i.e., irreversible chemical reaction). In all these cases, substance *a* is considered to be the light component (high vapor pressure, low normal boiling point) and substance *b* to be the heavy component (low vapor pressure, high normal boiling point).

The behavior of the ideal mixture is demonstrated in Fig. 2-3A. According to Raoult's law (Equation 2.7), the partial pressure of component *a*, p_a, is a straight line from p_a^o at $x_a = 1$ to $p_a = 0$ at $x_a = 0$. The same holds for the partial pressure p_b of component *b*. According to Dalton's law (Equation 2.5), the system pressure p_t is the sum of the partial pressures p_a and p_b. This pressure is also a straight line from p_a^o to p_b^o.

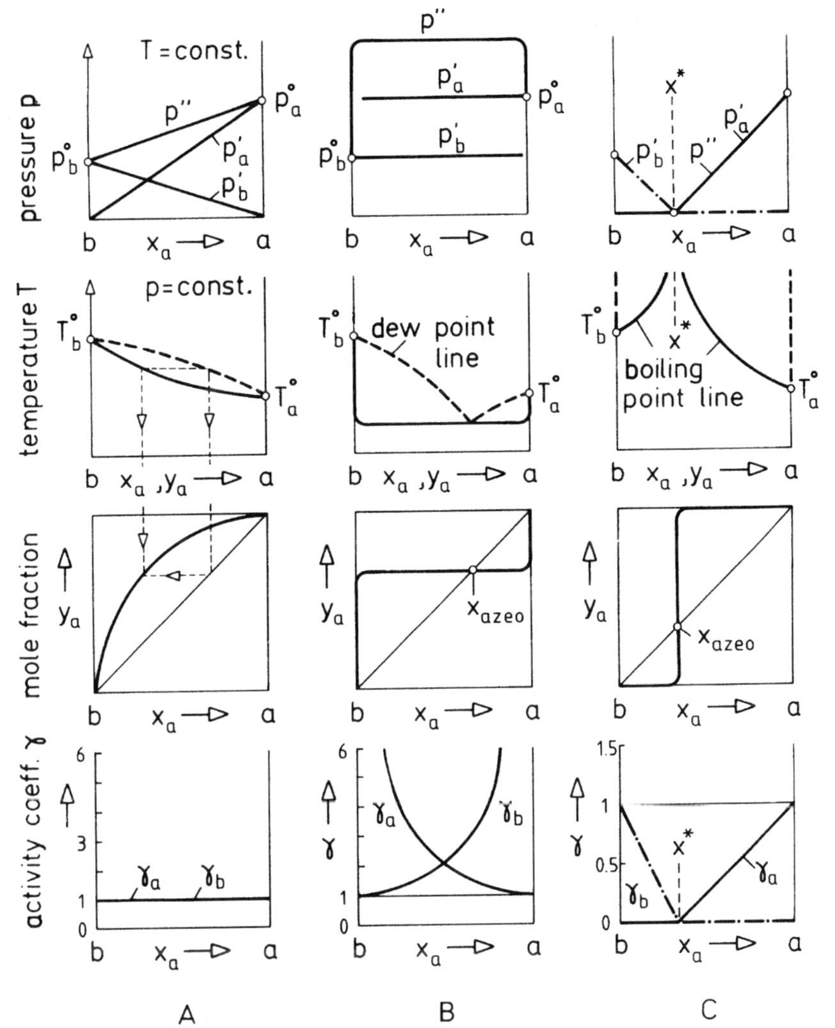

Fig. 2-3. *Gas-liquid equilibria of three extreme cases of binary mixtures: (A) ideal in gas and liquid phase; (B) ideal in gas phase, total immiscibiliy in liquid phase; (C) ideal in gas phase, chemical reaction in liquid phase.*

Case 2 is represented by Fig. 2-3B. The constituents *a* and *b* are supposed to behave ideally in the vapor phase, but are totally immiscible in the liquid phase. The partial pressure p_i' always equals the vapor pressure p_i^o because each substance exists as a pure liquid in the mixture. Therefore, in the p-x diagram of Fig. 2-3B the partial pressures p_a' and p_b' are horizontal lines. Following Dalton's law, the system pressure p_t equals $p_a^o + p_b^o$. The pressure of the mixture is higher than the vapor pressure of the pure constituents. The increased vapor pressure results in a decreased boiling point of the two-liquid phase mixture. The vapor composition always equals $y_a = p_a/(p_a + p_b)$; it no longer depends on liquid composition. This can be seen very clearly in the $y - x$ plot in Fig. 2-3B. The dew point line is a section of the vapor pressure line of pure constituents. The equilibrium line intersects the diagonal of the diagram. At the intersection point, gas concentration y and liquid concentration x become equal (see Example 2-4).

For totally immiscible liquids, an equation for the activity coefficient γ can easily be derived. As the partial pressure p_i always equals the vapor pressure p_i^o, Equation (2.8) becomes:

$$p_i^o = \gamma_i \cdot x_i \cdot p_i^o \quad \text{or} \quad \gamma_i = 1/x_i \tag{2.14}$$

The activity coefficients γ_i of all constituents are higher than 1 and range from $\gamma_i = 1$ at $x_i = 1$ to $\gamma_i = \infty$ at $x_i = 0$. In real mixtures there is always at least a small mutual solubility of the constituents *a* and *b*, and therefore, the activity coefficient γ_i^∞ at infinite dilution (at $x_i = 0$) is finite. The activity coefficient γ_i^∞ at infinite dilution strongly depends on the forces of interaction between the constituents of the mixture. Its value shows the degree of nonideality very clearly. Knowledge of the activity coefficients at infinite dilution is especially helpful in the initial steps of process design (see Section 5.4).

Figure 2-3C deals with case 3, where the constituents *a* and *b* are supposed to attract each other and undergo a chemical reaction. It is assumed that the product *c* of the chemical reaction has a very low vapor pressure ($p_c^o \approx 0$). If the mixture is rich in *a*, all of substance *b* will react with *a* to form *c*. The mixture consists then of *a* and *c* only; there is no free *b* in the liquid. Analogously, a *b*-rich mixture consists only of *b* and *c* (no free *a*). At the stoichiometric composition x^*, only *c* exists in the liquid phase. Assuming that Raoult's law holds for *a* and *c*, the partial pressure p_a' is linear from p_a^o at $x_a = 1$ to $p_b' = 0$ at x^*. As the vapor pressure p_c^o is supposed to be zero, the vapor in equilibrium with an *a*-rich liquid consists of substance *a* only. Analogously, the vapor in equilibrium with a *b*-rich liquid consists of substance *b* only. At the stoichiometric composition x^*, the boiling point is very high as a consequence of the very low vapor pressure of substance *c*. In the $y - x$ plot of Fig. 2-3C, the equilibrium curve intersects the diagonal of the diagram. At the point of intersection, y equals x (or $y - x = 0$). The activity coefficients γ_i are both lower than 1 in the mixture and they range between 0 and 1.

The three special cases described in this section are extreme cases, perhaps rarely encountered in real systems. However, they are very helpful for understanding the basic principles of vapor-liquid equilibrium.

EXAMPLE 2-4: BOILING POINT OF A BINARY LIQUID MIXTURE WITH TOTAL IMMISCIBILITY

Find the boiling point of the liquid mixture water (*a*) and toluene (*b*) at a pressure $p = 1$ bar.

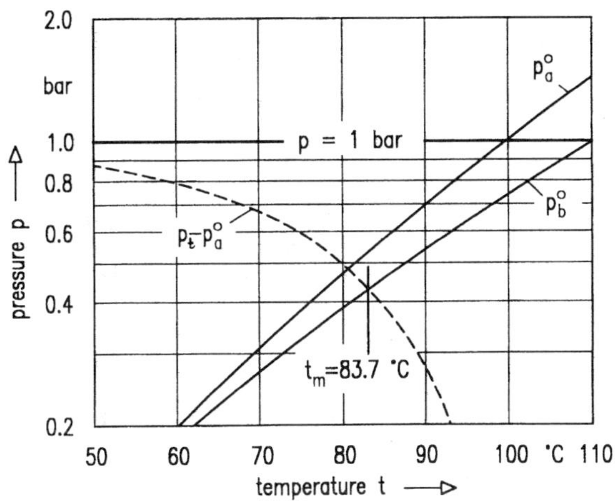

Fig. 2-4. Determination of the boiling point of a binary mixture with total immiscibility in the liquid phase.

Solution:

Equation (2.11): $\Sigma\, p_i'' = P$;
At total immiscibility: $p_a' = p_a^o$; $p_b' = p_b^o$; $p_a^o + p_b^o = P = 1$ bar
From the diagram in Fig. 2-4 follows: $t = 83.7°C$
Composition of the first bubble: $y_a = p_a/p = 0.555/1 = 0.555$ $y_b = 0.445/1 = 0.445$
At total immiscibility, the bubbles can be generated at the liquid-liquid interface only since both liquid phases have to contribute to the formation of the bubbles.

2.2 EMPIRICAL FINDINGS IN VAPOR-LIQUID EQUILIBRIUM

The most important empirical findings pertaining to the vapor-liquid equilibrium of binary and ternary mixtures are discussed below.

2.2.1 Binary Mixtures

Phase equilibria measured at moderate pressure for binary systems composed of substances *a* and *b* are shown in Fig. 2-5 [Mersmann 1980]. In the upper (first) row, total and partial pressures are plotted vs. the mole fraction x_a of substance *a* in the liquid at constant temperature. Substance *a* is more volatile than substance *b* and thus has a higher vapor pressure and a lower boiling point. The second row shows boiling point and dew point curves plotted vs. the liquid or vapor compositions at constant pressure. The third row of Fig. 2-5 presents the concentration y_a of *a* in the vapor phase plotted vs. its concentration in the liquid phase (x_a) at constant

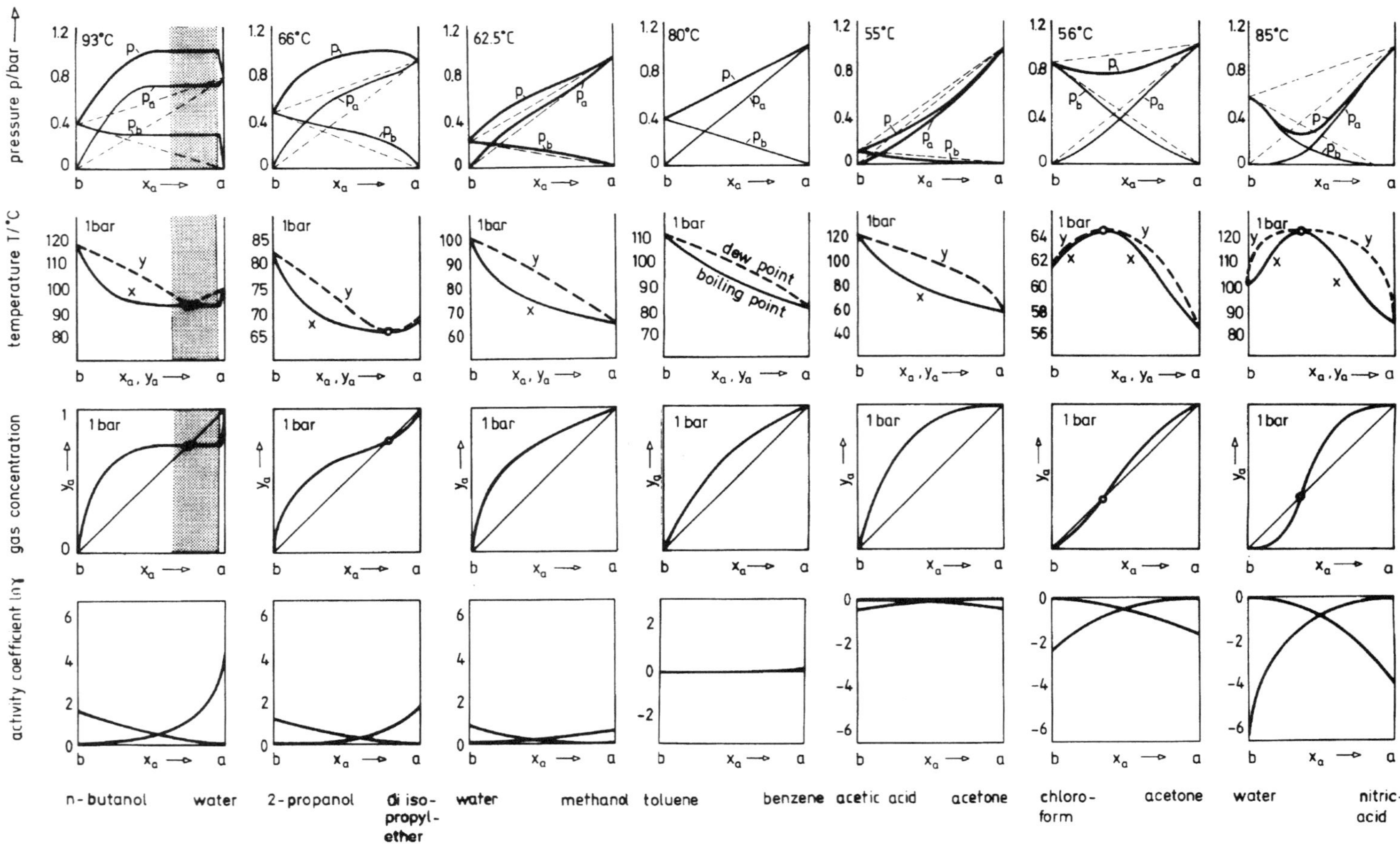
pressure p/bar
temperature T/°C
gas concentration
activity coefficient ln γ
93°C
66°C
62.5°C
80°C
55°C
56°C
85°C
1 bar
dew point
boiling point
n-butanol
water
2-propanol
di iso-propyl-ether
water
methanol
toluene
benzene
acetic acid
acetone
chloro-form
acetone
water
nitric-acid

pressure. The fourth row gives the activity coefficients γ_a and γ_b as functions of composition.

Significant differences in the behavior of these binary systems at equilibrium are due to the varying forces of interaction between the two molecular species *a* and *b*. Mixtures in which molecules *a* and *b* repel one another are shown on the left side of Fig. 2-5. Both substances undergo an increase in partial pressure p_i of the liquid phase. This increase in vapor pressure leads to a decrease in boiling point. In the 1-butanol-water system, repulsive forces between the molecules are so strong that splitting into two liquid phases occurs at intermediate concentration ranges.

Binary mixtures shown on the right side of Fig. 2-5 exhibit exactly the opposite behavior; molecules of substances *a* and *b* attract each other and tend to form complexes. This leads to a decrease in partial pressure p_i and thus to an increase in boiling point. In the binary mixture benzene-toluene, shown in the center of Fig. 2-5, forces of interaction between molecules *a* and *b* are the same as between molecules *a* and *a* or *b* and *b*, respectively. The molecule does not "realize" whether it is colliding with a molecule of the same or the other substance. Such mixtures are known as ideal mixtures; their phase equilibria can be represented by very simple equations (see Section 2.1).

The conditions at which the vapor and liquid compositions are equal (i.e., y = x) are very important. Such mixtures, known as azeotropic mixtures (Greek: $\tau\rho o\pi o\sigma$ = composition, $\zeta\epsilon\omega$ = to boil, α = contradiction, i.e., a mixture that does not change composition when boiled), have a minimum or a maximum boiling point. In a minimum-boiling azeotrope the molecular species repel each other, whereas in a maximum-boiling azeotrope they attract each other. If the azeotrope occurs in the composition range in which the two liquids are immiscible, phase splitting occurs and a heterogeneous azeotrope is formed. A heterogeneous azeotrope is always a minimum boiling one. Table 2-2 lists some examples of binary mixtures that form azeotropes.

A mixture of close-boiling components tends to form an azeotrope when the chemical structures of the components are dissimilar. Binary mixtures with two azeotropes (one minimum and one maximum) have been observed occasionally, e.g., perfluorobenzene/benzene.

Besides composition, pressure has a strong influence on the phase equilibrium of gas-liquid mixtures. The influence of pressure on the equilibrium of a mixture of nitrogen and methane is shown in Fig. 2-6A. At higher pressures, differences between concentrations of vapor y and liquid x decrease significantly. When the

---

Fig. 2-5. *Phase equilibria of seven binary mixtures composed of substances* a *and* b *[Mersmann 1980]. Substance* a *is more volatile than substance* b. *The dashed, thin lines in the upper row represent ideal behavior. Temperature and vapor concentration equilibria (middle and bottom rows) were measured at constant pressure (0.1 MPa). Shaded area represents the range of immiscibility in the liquid phase. Azeotropic points are indicated by open circles.* p_a = *partial pressure of* a; p_b = *partial pressure of* b; x_a = *mole fraction of a in liquid phase;* y_a = *mole fraction of a in vapor phase.*

Table 2-2. Examples of binary mixtures with an azeotrope (selected from Horsley 1973; Gmehling et al. 1977; Weast et al. 1986)

						Azeotropic Data		
Formula	Name	B.P.°C	Formula	Name	B.P.°C	x_a mol-%	B.P.°C	Type
H_2O	water	100	HBr	hydrobromic acid (*a*)	−67	16.7	126	max.
			HCl	hydrochloric acid (*a*)	−85	11.1	108.6	max.
			HNO_3	nitric acid (*a*)	86	38.3	121	max.
			$CHCl_3$	chloroform (*a*)	61.2	83.9	56.12	hetero
			CH_2Cl_2	dichloromethane (*a*)	40.1	93.3	38.1	hetero
			CH_3NO_2	nitromethane (*b*)	101.2	51.1	83.6	hetero
			C_2H_6O	ethanol (*a*)	78.3	90.4	78.2	min.
			$C_3H_6O_2$	methyl acetate (*a*)	57	87.2	56.4	min.
			C_3H_8O	n-propyl alcohol (*a*)	97.3	42.2	87.8	min.
			$C_4H_6O_2$	methyl acrylate (*a*)	80	72.9	71	min.
			C_4H_8O	tetrahydrofurane (*a*)	66	81.7	64	min.
			C_4H_8O	2-butanone (*a*)	79.6	66.2	73.4	min.
			C_4H_8O	butyraldehyde (*a*)	74	79.6	68	min.
			$C_4H_8O_2$	dioxane (*b*)	101.3	51.7	87.8	min.
			$C_4H_8O_2$	ethyl acetate (*a*)	77.1	68.2	70.4	hetero
			$C_4H_{10}O$	butyl alcohol (*b*)	117.4	75.2	92.7	hetero
			$C_4H_{10}O_2$	2-ethoxyethanol (*b*)	135.1	92.5	99.4	min.
			C_5H_5N	pyridine (*b*)	115	76.8	94	min.
			$C_5H_{12}O$	ethyl propyl ether (*a*)	63.6	83.1	59.5	hetero
			C_6H_6	benzene (*a*)	80.2	70.4	69.3	hetero
			C_7H_8	toluene (b)	110.7	55.6	84.1	hetero
$CHCl_3$	chloroform	61.2	CH_4O	methanol (*b*)	64.7	65.0	53.4	min.
			C_2H_6O	ethanol (*b*)	78.3	83.7	59.4	min.
			C_3H_6O	acetone (*a*)	56.1	36.7	64.4	max.

CH_4O	methanol	64.7	CH_2Cl_2	dichloromethane (*a*)	40.1	70.5	37.8	min.
			C_3H_6O	acetone (*a*)	56.2	80.2	55.5	min.
			C_4H_8O	2-butanone (*b*)	79.6	84	63.5	min.
			$C_4H_8O_2$	ethyl acetate (*b*)	77.1	69.9	62.3	min.
			C_6H_6	benzene (*b*)	80.2	61	57.5	min.
			C_7H_8	toluene (*b*)	110.7	86.5	63.8	min.
			C_7H_{16}	n-heptane (*b*)	98.5	76.5	59.1	min.
			C_8H_{18}	n-octane (b)	125.6	90.2	63	min.
C_2H_6O	ethanol	78.3	$C_4H_6O_2$	methylacrylate (*b*)	80	57.5	73.5	min.
			C_4H_8O	2-butanone (*b*)	79.6	50.0	74	min.
			$C_4H_8O_2$	ethylacetate (*a*)	77	53.8	71.8	min.
			$C_4H_{10}O_2$	eth.methoxymeth (*a*)	65.9	77	64	min.
			$C_5H_{12}O$	ethylpropylether (*a*)	63.6	61	61.2	min.
			C_6H_{14}	n-hexane (*a*)	68.7	66.5	58.4	min.
			C_7H_8	toluene (b)	110.7	81	76.7	min.
C_3H_6O	acetone	56.2	C_6H_{12}	cyclohexane (*b*)	80.8	74.6	53	min.
			C_6H_{14}	hexane (*b*)	68.9	68.1	49.8	min.
			C_7H_{16}	heptane (b)	98.4	93.6	55.9	min
$C_4H_{10}O_2$	2-ethoxyethanol	135.1	C_8H_{10}	ethylbenzene (b)	136.2	56.9	127.1	min.
			C_8H_{18}	octane (a)	125.8	61	116.1	min.
C_5H_5N	pyridine	115	C_7H_8	toluene (b)	110.7	75.3	110.1	min.
			C_8H_{18}	n-octane (*b*)	125.8	<92.9	112.8	min.
C_6H_6	benzene	80.2	C_3H_8O	2-propanol (b)	82.5	61	71.8	min.
			C_6H_{12}	cyclohexane (*b*)	80.8	53	77.6	min.

B.P. = boiling point, x_a = mol-% of component *a* (low boiler) at the azeotrope, p = 1.013 bar.

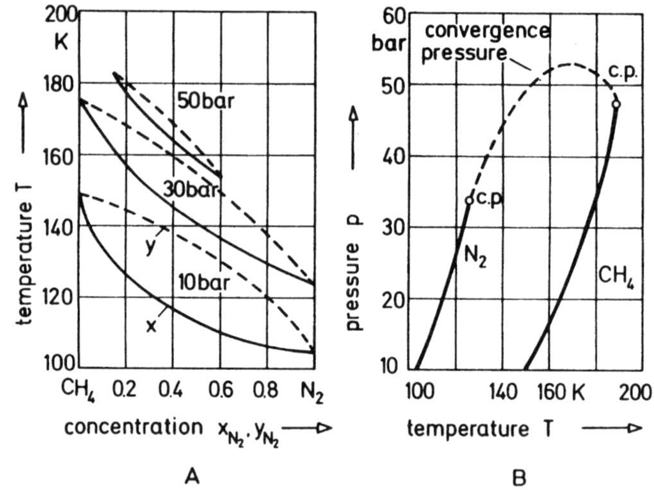

Fig. 2-6. *Influence of pressure p on the phase equilibrium of nitrogen and methane: (A) boiling point lines (—) and dew point lines (---) of nitrogen/methane mixtures. x_{N2} = mole fraction of nitrogen in the liquid phase; y_{N2} = mole fraction of nitrogen in the vapor phase. (B) Critical points of nitrogen (a), methane (b), and their mixtures (convergence pressure curve ---).*

critical pressure of the pure substances is exceeded, they can no longer be liquefied (supercritical region). However, mixtures of supercritical vapors can split into two phases, as shown in Fig. 2-6B. The region in which this can occur is restricted by the convergence pressure curve, which represents the locus of critical points of mixtures of the two substances. The convergence pressure curves typically pass through a maximum (see Section 2.3.1).

2.2.2 Ternary Mixtures

The phase equilibrium of ternary mixtures is depicted conveniently on a triangular diagram, in which the corners represent the pure substances *a, b* and *c;* the sides the three binary mixtures; and the area the ternary mixture. A possibile way to show the phase equilibrium compositions is to mark the liquid composition x and the vapor composition y in the grid of the triangle by different symbols and to connect pairs of data by a line (see Fig. 2-7A).

In Fig. 2-7B, the mole fraction of component *i* in the liquid, x_i, is represented by the basic grid, and the vapor mole fraction y_i^* in equilibrium with the liquid is shown as parametric curves. For ideal mixtures with constant relative volatilities α_{ik} (see Section 2.3.2), these curves are linear. However, for nonideal systems, such plots do not yield straight lines, and the entire representation becomes confusing. In addition, the azeotropic points, which are very important in distillation, are difficult to recognize.

For a better understanding of the vapor-liquid equilibria of ternary mixtures, a

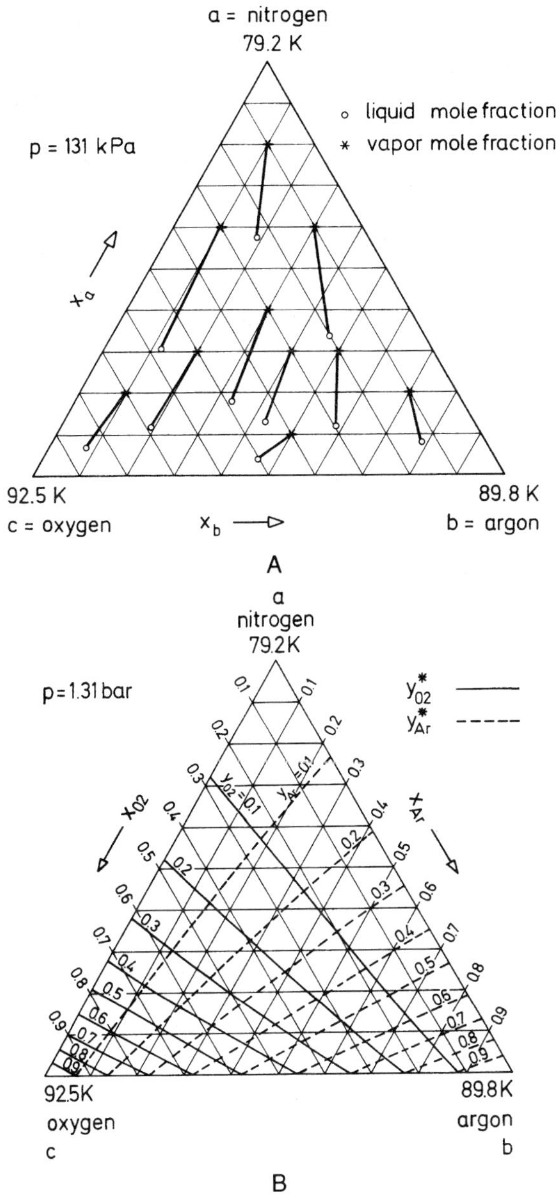

Fig. 2-7. *Representation of phase equilibria of a ternary mixture: (A) Gas and liquid composition are marked in a triangular diagram. Pairs are connected by straight lines. (B) The uniformly divided grid of the triangular diagram represents the liquid composition x_{Ar}, x_{O2}, and x_{N2}. The bold and dashed lines indicate equilibrium vapor mole fractions of oxygen y^*_{O2} and argon y^*_{Ar}, respectively. Temperatures indicate boiling points of the pure components.*

knowledge of the boiling point behavior is especially helpful. Figure 2-8 shows a three-dimensional plot of the boiling point surfaces of eight ternary mixtures. The boiling point temperatures at constant pressure are presented as functions of liquid composition. Most of the systems shown in Fig. 2-8 were calculated by the Wilson equation (see Section 2.4) with the parameters given by Gmehling [Gmehling et al. 1977].

Figure 2-8A depicts the boiling point surface of the nearly ideal mixture, nitrogen-argon-oxygen. At the sides of the triangle (representing the composition of the ternary mixture), the well-known behavior of a binary mixture can be seen. The ternary mixture exhibits a boiling point surface similar to the slope of a mountain. The highest point (peak) lies at the pure high boiling constituent c (here oxygen) and the lowest point at the low boiling constituent a (here nitrogen).

In Fig. 2-8B, a mixture is depicted where the light and the intermediate boiling components a and b form a low boiling azeotrope (minimum azeotrope). The boiling point curve of that binary mixture exhibits a temperature minimum. Consequently, there exists a valley in the boiling point surface of the ternary mixture that runs from the azeotrope upward (with indefinite ending). The slopes on both sides of that valley have the highest and lowest points in common. This fact is very important, as will be shown in Section 4.3.2.

A significantly different situation is encountered when the azeotrope is formed by the intermediate and the high boiler, b and c, respectively. This is shown in Fig. 2-8C. Here, a valley runs from the azeotrope to the low boiler a. The slopes on both sides of the valley have different peaks. This is an important difference from the system shown in Fig. 2-8B. A similar situation is met in Fig. 2-8D. This system forms two binary minimum azeotropes that are connected by a valley whose slopes have different peaks.

Figure 2-8E presents the boiling point surface of a ternary mixture with three binary minimum azeotropes. Additionally, a ternary azeotrope that has the lowest boiling point of the whole system is formed. The boiling point surface exhibits three valleys, each of them running from a binary azeotrope downhill to the ternary azeotrope. These valleys divide the mixture into three regions with different peaks.

A mixture that splits partially into two liquid phases is depicted in Fig. 2-8F. In addition to the heterogenous azeotrope in the two-liquid phase region, an a-b minimum azeotrope exists. A valley surrounded by two slopes with different peaks runs from one azeotrope to the next. The valley widens to a plain in the immiscibility region.

The mixture depicted in Fig. 2-8G forms an a-b maximum azeotrope. In the boiling point surface, a ridge runs downhill to the azeotrope. The ternary mixture is divided into two regions by this ridge. The two regions have a peak in common but different minima at the a and b corners, respectively.

A very complex mixture is shown in Fig. 2-8H. This mixture forms two binary minimum azeotropes and a binary maximum azeotrope. On the boiling point surface, a valley runs from one binary minimum azeotrope to the other. In addition, a ridge connects the high boiler with the maximum azeotrope. At the intercept of valley and ridge, a ternary azeotrope, is formed that is called a "saddle azeotrope."

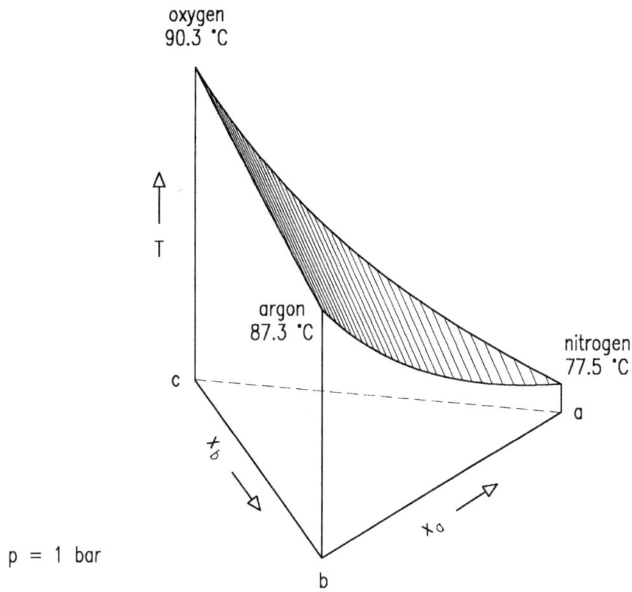

A

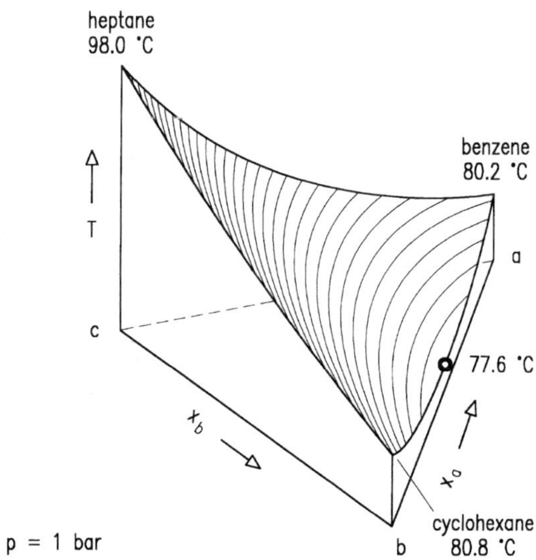

B

Fig. 2-8. *Boiling point surfaces of eight different ternary mixtures at constant pressure. Presented is the boiling point temperature as a function of the liquid compositon x_i (o azeotropes).*

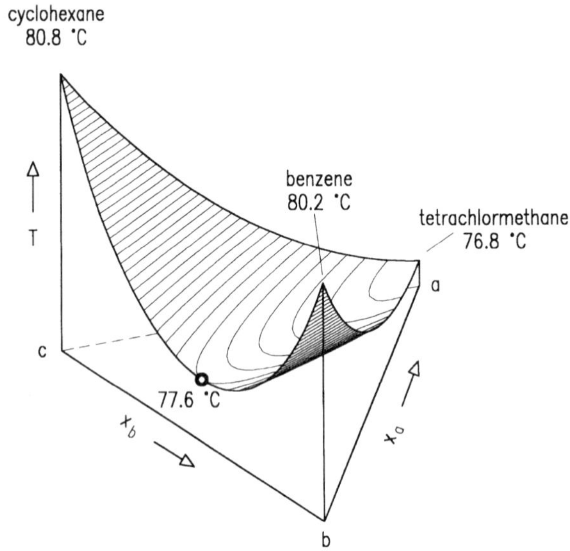

C

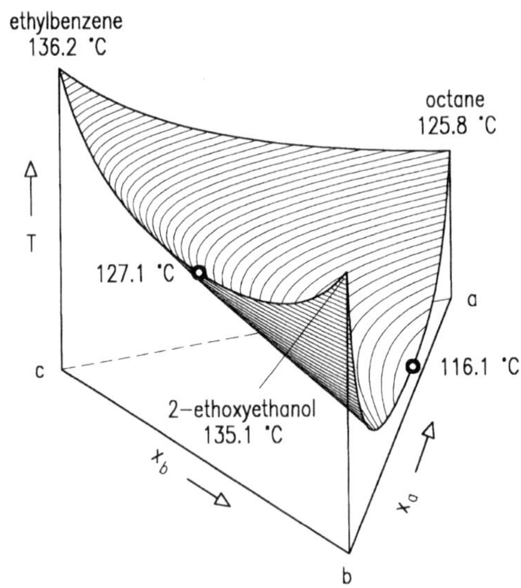

D

Fig. 2-8. *(continued)*

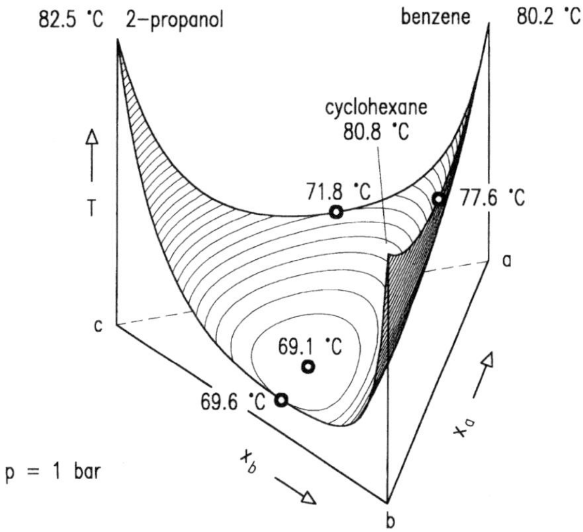

E

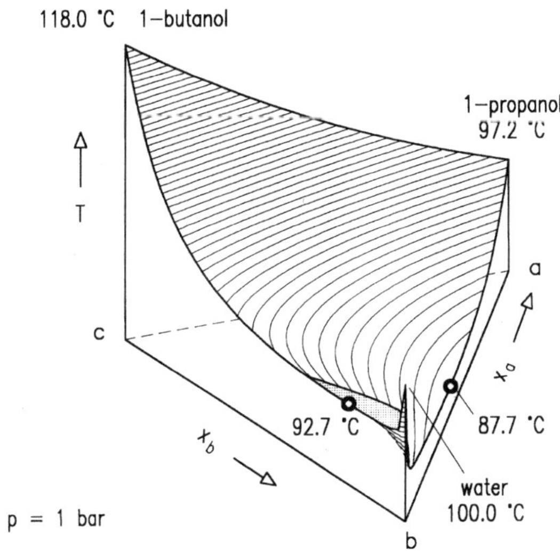

F

Fig. 2-8. *(continued)*

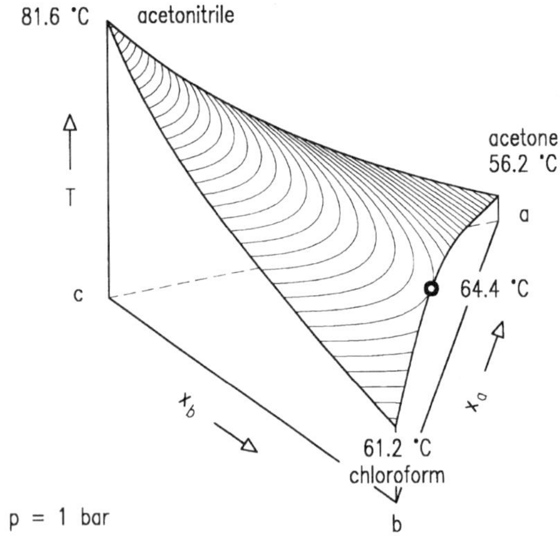

G

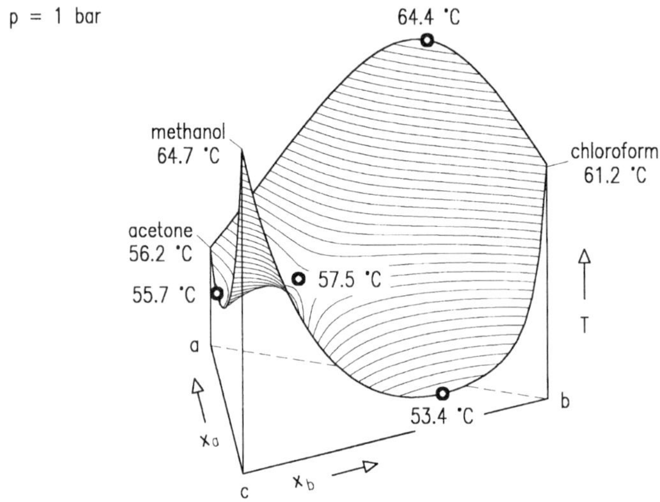

H

Fig. 2-8. *(continued)*

The dew point surface can be plotted in a similar way. Figure 2-9 depicts the dew point surface and the visible part of the underlying boiling point surface for the system acetone-chloroform-methanol. The dew point surface touches the boiling point surface at the pure constituents and the azeotropes. At the valleys and ridges, the distance between dew point and boiling point surfaces reaches a local minimum but does not become zero.

The perspective plots in Fig. 2-8 make it clear that the boiling point surfaces of

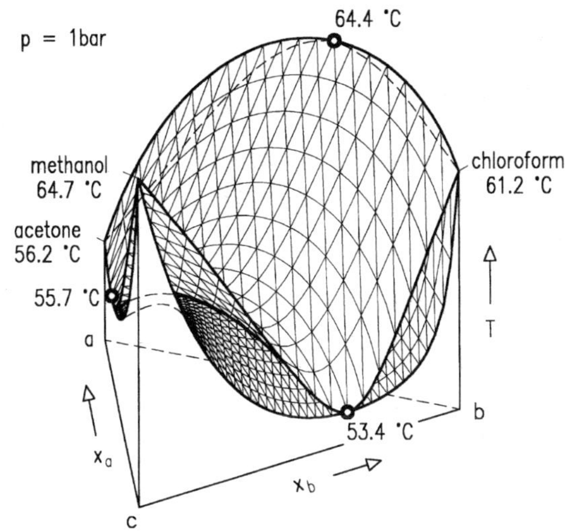

Fig. 2-9. Boiling and dew point surface of the mixture acetone/chloroform/methanol as a function of liquid composition x_i at p = 1.013 bar.

ternary mixtures differ very much. The existence of binary azeotropes always results in the formation of valleys or ridges in the boiling point surface. Those ridges or valleys are of major importance for distillations that are surrounded by slopes with different boiling point peaks or depressions (see Section 4.3.2).

2.3 FORMULATION OF VAPOR-LIQUID EQUILIBRIUM RELATIONSHIPS

There are several methods available to formulate vapor-liquid equilibrium (VLE) relationships. Equilibrium ratios (K-values) are often used, especially for multicomponent mixtures. For binary mixtures, the most convenient formulation is the equilibrium curve with temperature-independent relative volatility α. The distillation line is most appropriate for ternary mixtures. These approaches are discussed below.

2.3.1 Equilibrium Ratio (K-value)

The vapor-liquid equilibrium ratio value for component i, K_i, is commonly used to represent phase equilibrium data, especially for multicomponent mixtures. It is defined as:

$$K_i \equiv y_i^*/x_i \tag{2.15}$$

Since, at phase equilibrium

$$p_i'' = p_i' \tag{2.16}$$

then from Equations (2.5) and (2.6),

$$K_i = p_i^o/p_t \quad \text{or} \quad K_i = p_i^o \cdot \gamma_i/p_t \tag{2.17}$$

Similarly, at temperatures above the critical point,

$$K_i = H_{ij}/P_t \quad \text{or} \quad K_i = H_{ij} \cdot \gamma_i/p_t \tag{2.18}$$

Both the vapor pressure p_i^o and Henry's coefficient H_{ij} are strongly dependent on temperature. At high pressures the fugacity f must be used instead of the pressure p in Equations (2.17) and (2.18).

Hence, it is necessary to know the boiling point and the dew point for the conditions under consideration. The boiling point can be calculated by analogy to Equation (2.12) by:

$$\sum x_i \cdot K_i = 1 \tag{2.19}$$

For determining the dew point of a vapor mixture, the following equation holds:

$$\sum y_i/K_i = 1 \tag{2.20}$$

Figure 2-10 shows a plot of K-values vs. pressure for the binary mixture methane/ethane at two different temperatures [Foerg and Stichlmair 1969]. As long as Dalton's law holds, the K-values follow a straight line of the slope = −1 in the logarithmic plot. At higher pressures, the K-values depart from the linear relationship and finally converge all to a value of unity. The pressure at which this occurs is called the "convergence pressure." It is the critical pressure of the mixture that may be substantially higher than the critical pressures of the pure constituents.

The increase of the K-value of the heavy component in the mixture means that this substance becomes more volatile with increasing pressure (the pressure effect!), an important phenomenon that is encountered very often.

Figure 2-11 shows a plot of the K-values of the ternary mixture hydrogen/methane/ethylene. All three K-values converge to unity at the same pressure. This type of convergence holds for multicomponent mixtures, also. The convergence pressure depends on the nature and composition of the system. Some methods exist for precalculating or estimating the convergence pressure of a system and for making rough estimates of the K-values by interpolating between the ideal behavior (straight line on the K-p diagram) and the convergence pressure.

A highly developed method is available for natural gases [Gas Processors Suppliers Assn. 1972]. Figure 2-12 presents convergence pressure plots for a number of systems. In using this method, one replaces the multicomponent mixture by a binary mixture with the actual light component and a pseudo heavy component. The

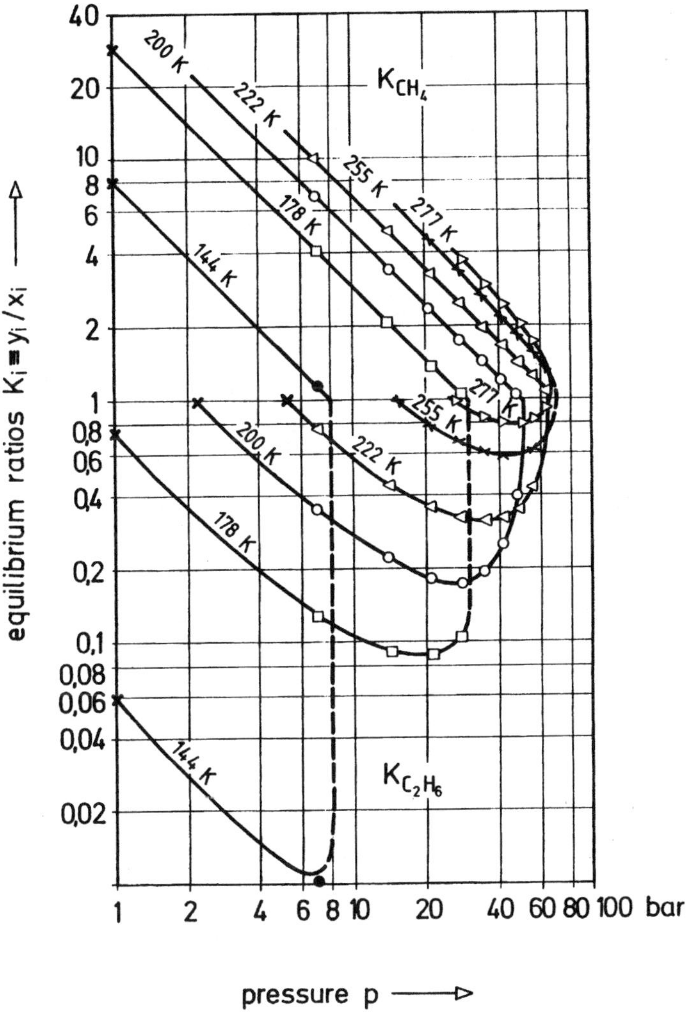

Fig. 2-10. K-values of the system methane/ethane as a function of system pressure [Foerg and Stichlmair 1969].

weight average critical temperature T_{crm} of the heavy components is used to represent the pseudo heavy component. The critical locus between T_{crm} and the critical point of the light component is traced, and the critical pressure is determined at the system temperature. The data collection contains K-charts for all substances labeled in Fig. 2-12 for critical pressures of 800, 1000, 1500, 2000, 3000, 5000 and 10 000 psia (i.e., 55.2, 69.0, 103.5, 138.0, 206.9, 344.9 and 689.8 bar). Quite good K-values for each critical pressure can be found by interpolation.

In this method for VLE estimation, the nature and composition of the mixture is

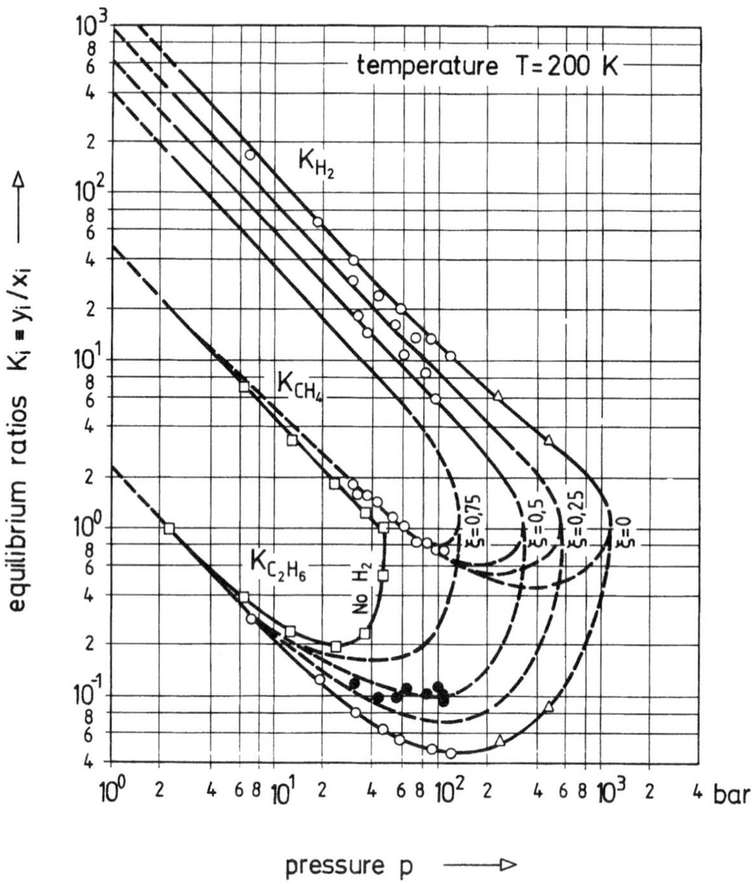

Fig. 2-11. K-values of the ternary system hydrogen/methane/ethane as a function of system pressure [Foerg and Stichlmair 1969] ($\xi = x_{CH4}/(x_{CH4} + x_{C2H6})$).

expressed by its critical pressure. At low pressure, the K-value of each component is independent of the other constituents. The higher the pressure, the more the K-value of each component is affected by the critical pressure of the system, i.e., by the other constituents in the mixture.

2.3.2 Equilibrium Curve

For binary systems consisting of components *a* and *b*, the mole fraction y_a^* is calculated by Dalton's and Raoult's laws:

$$y_a^* = \frac{p_a'}{p_a' + p_b'} = \frac{x_a \cdot p_a^o}{x_a \cdot p_a^o + x_b \cdot p_b^o} \tag{2.21}$$

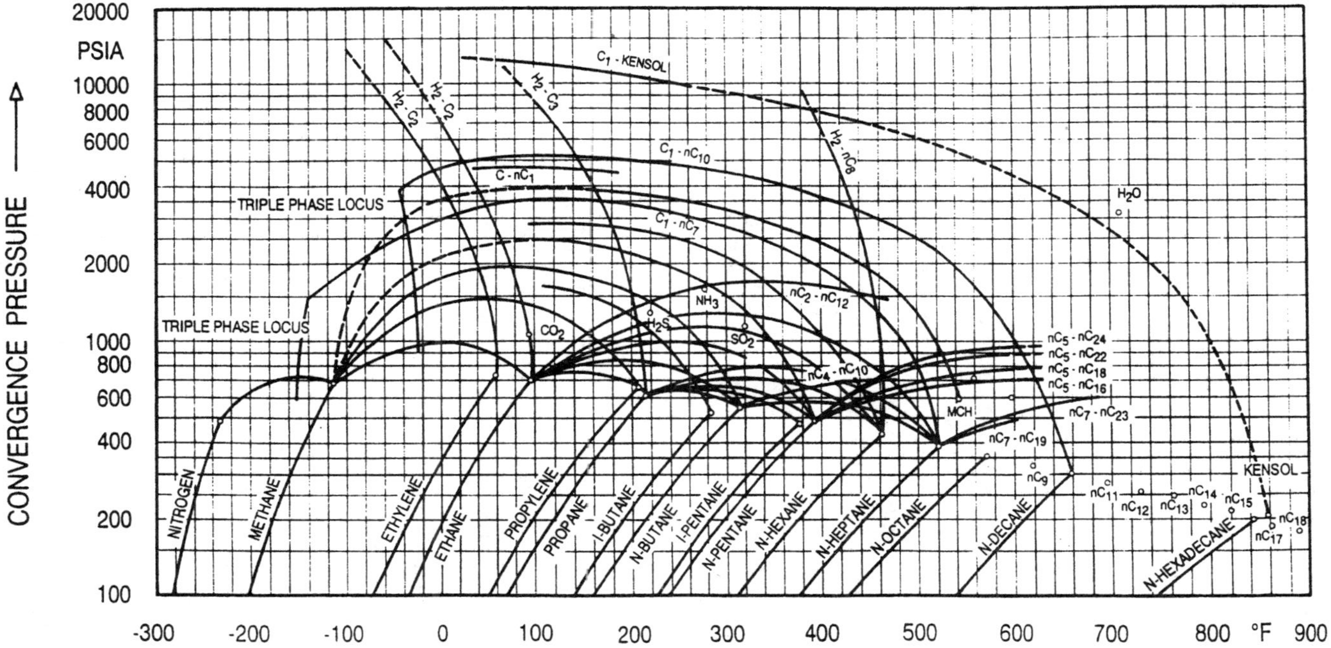

Fig. 2-12. *Convergence pressure chart for hydrocarbons (Gas Processors Suppliers,* Engineering Data Book*, 1972).*

Since the vapor pressures of the pure components p_a^o and p_b^o are highly temperature-dependent, this equation applies only at constant temperature. However, the ratio p_a^o/p_b^o is often independent of temperature; therefore, vapor-liquid equilibria are frequently expressed in this form as the relative volatility α:

$$\alpha_{ab} \equiv p_a^o/p_b^o \tag{2.22}$$

From the equation $x_b = 1 - x_a$,

$$y_a^* = \frac{\alpha_{ab} \cdot x_a}{1 + (\alpha_{ab} - 1) \cdot x_a} \quad \text{or} \quad x_a^* = \frac{\alpha_{ab}^{-1} \cdot y_a}{1 + (\alpha_{ab}^{-1} - 1) \cdot y_a} \tag{2.23}$$

This relationship can be represented by the so-called ''equilibrium curve'' and is valid when the temperature is constant. If α_{ab} does not vary with temperature, Equation (2.21) can also be employed at other temperatures or pressures; this is particularly important in distillation. Figure 2-3A (third row) shows a typical plot of Equation (2.23). For ternary systems consisting of components *a, b* and *c,* analogous relationships exist:

$$y_a^* = \frac{p_a'}{p_a' + p_b' + p_c'} = \frac{x_a \cdot p_a^o}{x_a \cdot p_a^o + x_b \cdot p_b^o + x_c \cdot p_c^o} \tag{2.24}$$

Substituting $\alpha_{ac} \equiv p_a^o/p_c^o$, $\alpha_{bc} \equiv p_b^o/p_c^o$, and $x_c = 1 - x_a - x_b$ gives

$$y_a^* = \frac{x_a \cdot \alpha_{ac}}{1 + (\alpha_{ac} - 1) \cdot x_a + (\alpha_{bc} - 1) \cdot x_b}$$

$$y_b^* = \frac{x_b \cdot \alpha_{bc}}{1 + (\alpha_{ac} - 1) \cdot x_a + (\alpha_{bc} - 1) \cdot x_b} \tag{2.25}$$

The general equation for multicomponent mixtures is:

$$y_i^* = \frac{\alpha_{ik} \cdot x_i}{1 + \sum_{j=1}^{k-1} (\alpha_{jk} - 1) \cdot x_j} \tag{2.26}$$

for component 1 to $k - 1$, where k refers to the highest boiling substance. In nonideal systems, the activity coefficient γ_i must be incorporated into the equations.

2.3.3 Distillation Line

If an initial liquid concentration of component *i*, x_{i0} is assumed, the equilibrium vapor concentration y_{i0}^* is determined. The vapor is then assumed to be totally condensed. Thus, a new liquid of composition $x_{i1} = y_{i0}^*$ is obtained and may be located on the triangular diagram. The vapor state y_{i1}^* in equilibrium with this liquid state

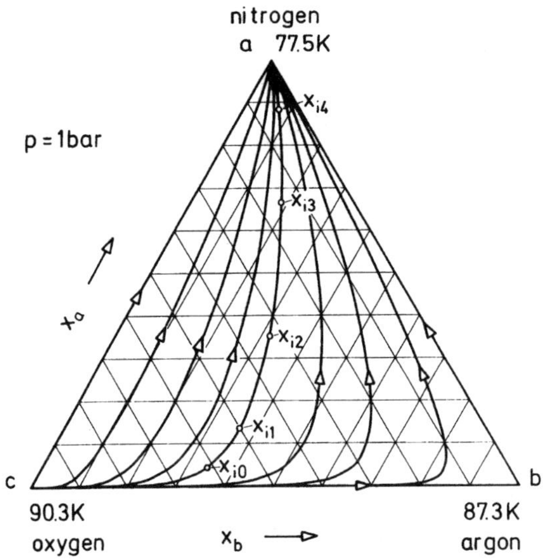

Fig. 2-13. Representation of the phase equilibrium of the ternary mixture nitrogen/argon/oxygen at p = 1.31 bar (131 kPa) by distillation lines. The distillation lines start at the component with the highest boiling point and end at the component with the lowest boiling point. The direction of decreasing temperatures is indicated by arrows. Distinct equilibrium states are shown on one line by dots.

x_{i1} is, in turn, determined and so on, successively. A series of liquid states x_{i1}, x_{i2}, x_{i3}, . . . is finally obtained; the composition difference in each case represents an equilibrium step. These liquid states can be joined together to give a distillation line. This process leads to a step-by-step enrichment of the more volatile component, which is accompanied by a decrease in boiling point (Fig. 2-13). The distillation lines of ideal mixtures can be expressed by a simple equation. Rearrangement of Equation (2.23) gives the equilibrium curve of ideal mixtures in the following form:

$$\frac{y_0}{1 - y_0} = \alpha \cdot \frac{x_0}{1 - x_0} \tag{2.27}$$

where the subscript 0 denotes the initial state. Total condensation of the vapor rising from the liquid is assumed in determining the paths of distillation lines. A new liquid with the concentration $x_1 = y_0$ is obtained. Thus:

$$\frac{x_1}{1 - x_1} = \alpha \cdot \frac{x_0}{1 - x_0} \tag{2.28}$$

Similarly, $x_2 = y_1$ is valid for the second step, giving:

$$\frac{x_2}{1 - x_2} = \alpha \cdot \frac{x_1}{1 - x_1} = \alpha^2 \cdot \frac{x_0}{1 - x_0} \tag{2.29}$$

For the nth step,

$$\frac{x_n}{1 - x_n} = \alpha^n \cdot \frac{x_0}{1 - x_0} \tag{2.30}$$

Rearrangement gives:

$$x_n = \frac{\alpha^n \cdot x_0}{1 + (\alpha^n - 1) \cdot x_0} \tag{2.31}$$

This equation is similar to that of the equilibrium curve (Equation 2.23), except that here the term α^n replaces α. Thus, for ternary mixtures,

$$\begin{aligned} x_{an} &= \frac{\alpha_{ac}^n \cdot x_{a0}}{1 + (\alpha_{ac}^n - 1) \cdot x_{a0} + (\alpha_{bc}^n - 1) \cdot x_{b0}} \\ x_{bn} &= \frac{\alpha_{bc}^n \cdot x_{b0}}{1 + (\alpha_{ac}^n - 1) \cdot x_{a0} + (\alpha_{bc}^n - 1) \cdot x_{b0}} \end{aligned} \tag{2.32}$$

Similarly, for multicomponent mixtures, the distillation line is represented by:

$$x_{in} = \frac{\alpha_{ik}^n \cdot x_{i0}}{1 + \sum_{j=1}^{k-1} (\alpha_{jk}^n - 1) \cdot x_{j0}} \tag{2.33}$$

The distillation lines of ideal mixtures with a constant relative volatility α_{ik} can easily be calculated by using these equations (see Example 2-5). For mixtures that are nonideal in the liquid phase, the activity coefficients γ (see Section 2.4) must be included in the equations.

Distillation lines describe, in a rough approximation, the path of the largest gradient on the boiling point surface (see Fig. 2-8). They mark the path of a rolling ball on the boiling point surface. Distillation lines always originate at a peak and terminate at a depression of the boiling point surface. The direction of decreasing boiling points is conveniently marked by an arrow.

EXAMPLE 2-5: DISTILLATION LINE OF AN IDEAL TERNARY MIXTURE

Find the distillation line for the mixture nitrogen ($=a$)/argon ($=b$)/oxygen ($=c$) through the point $x_{ao} = 0.05$, $x_{bo} = 0.36$. The mean relative volatilities are $\alpha_{ac} = 4.032$; $\alpha_{bc} = 1.339$.

Solution:

Equation (2.32):

$$x_{an} = \frac{\alpha_{ac}^{n} \cdot x_{ao}}{1 + (\alpha_{ac}^{n} - 1) \cdot x_{ao} + (\alpha_{bc}^{n} - 1) \cdot x_{bo}};$$

$$x_{bn} = \frac{\alpha_{bc}^{n} \cdot x_{bo}}{1 + (\alpha_{ac}^{n} - 1) \cdot x_{ao} + (\alpha_{bc}^{n} - 1) \cdot x_{bo}}$$

Evaluation for n = 1:

$$x_{a1} = \frac{4.032^{1} \cdot 0.05}{1 + (4.032^{1} - 1) \cdot 0.05 + (1.339^{1} - 1) \cdot 0.36} = \frac{0.2016}{1.2736} = 0.158;$$

$$x_{b1} = \frac{1.339^{1} \cdot 0.36}{1.2736} = 0.3785$$

Evaluation for n = 2:

$$x_{a2} = \frac{4.032^{2} \cdot 0.05}{1 + (4.032^{2} - 1) \cdot 0.05 + (1.339^{2} - 1) \cdot 0.36} = \frac{0.3968}{2.0483} = 0.397;$$

$$x_{b2} = \frac{1.339^{2} \cdot 0.36}{2.0483} = 0.3151$$

Evaluation for n = −1:

$$x_{a-1} = \frac{4.032^{-1} \cdot 0.05}{1 + (4.032^{-1} - 1) \cdot 0.05 + (1.339^{-1} - 1) \cdot 0.36} = \frac{0.0124}{0.8713} = 0.0142;$$

$$x_{b-1} = \frac{1.399^{-1}\ 0.36}{0.8713} = 0.3086$$

List of results:

n	Nitrogen x_a	Argon x_b	Oxygen x_c
4	0.0003	0.159	0.8407
−3	0.001	0.202	0.797
−2	0.004	0.253	0.743
−1	0.014	0.309	0.677
0	0.050	0.360	0.590
1	0.158	0.379	0.463
2	0.397	0.315	0.288
3	0.693	0.183	0.125
4	0.883	0.077	0.040
5	0.961	0.028	0.011

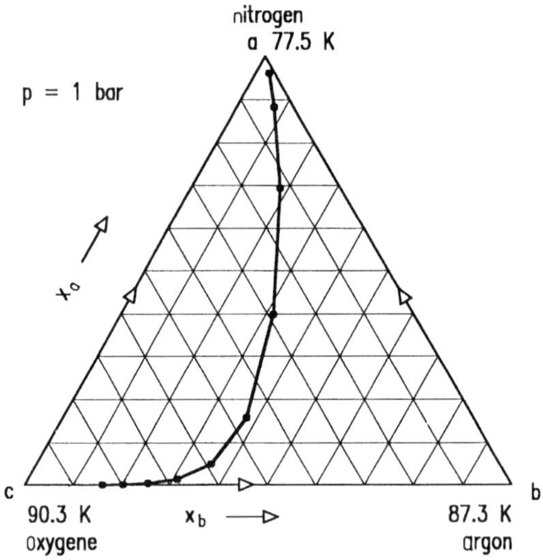

Fig. 2-14. *Determination of the course of the distillation line of an ideal ternary mixture (Example 2-5).*

The plot of the points is shown in the diagram of Fig. 2-14. The distillation line starts at the oxygen corner and ends at the nitrogen corner.

Distillation lines of eight different systems are shown in Fig. 2-15. In each case, the lowest boiling constituent *a* is at the top of the triangle, the intermediate boiling constituent *b* at the right bottom and the high boiling constituent *c* at the left bottom (clockwise sequence).

The mixture depicted in Fig. 2-15A exhibits an almost ideal behavior. All the distillation lines originate in the *c*-corner and terminate in the *a*-corner of the triangular diagram.

In Fig. 2-15B the components *a* and *b* form a minimum azeotrope that has the lowest boiling point of the system. Here, all distillation lines terminate at the *a*-*b* azeotrope. As in the case of an ideal system, just a single origin and a single terminus of distillation lines exists.

A different situation is met in the system depicted in Fig. 2-15C. Here, the intermediate and the high boiler *b* and *c* form a mimimum azeotrope. The distillation lines begin either at the *c*-corner or at the *b*-corner of the triangle and all terminate at the *a*-corner. A border distillation line splits the ternary mixture into two regions with different distillation line origins. Approximately, this border distillation line follows the course of the valley in the boiling point surface, as may be seen from Fig. 2-8C.

The system shown in Fig. 2-15D exhibits two binary minimum azeotropes. Here again a border distillation line runs from the *b*-*c* azeotrope to the *a*-*b* azeotrope, dividing the mixture in two regions with two different origins of distillation lines.

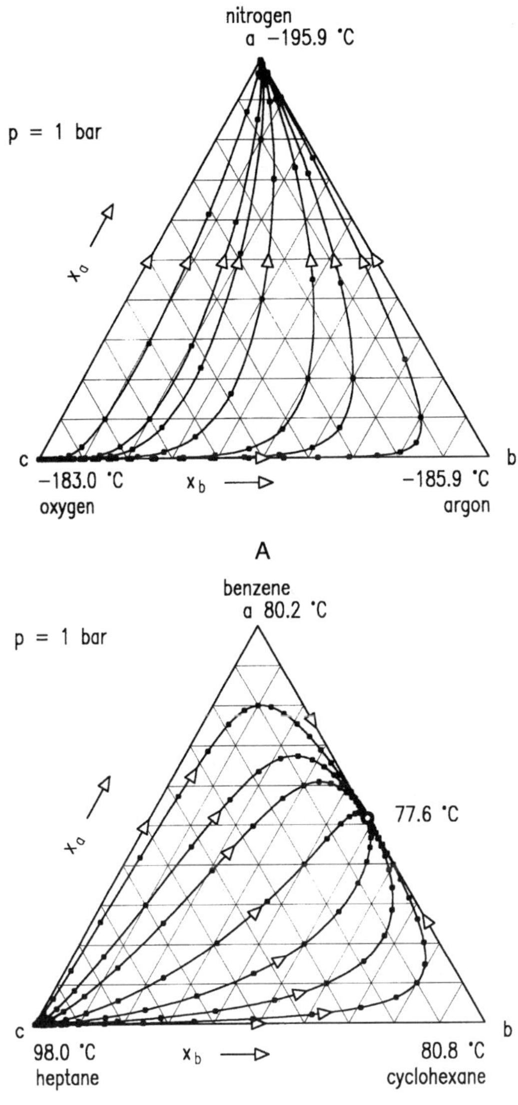

Fig. 2-15. *Phase equilibria of eight ternary mixtures at p = 1.013 bar (101.3 kPa). The distillation lines proceed from a temperature maximum to a temperature minimum. Distillation boundaries are represented by bold lines. Shaded areas indicate the presence of two liquid phases. Open circles denote azeotropic points.*

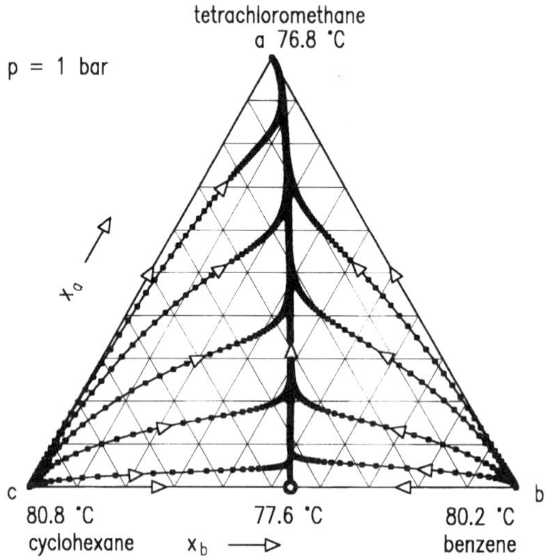

C

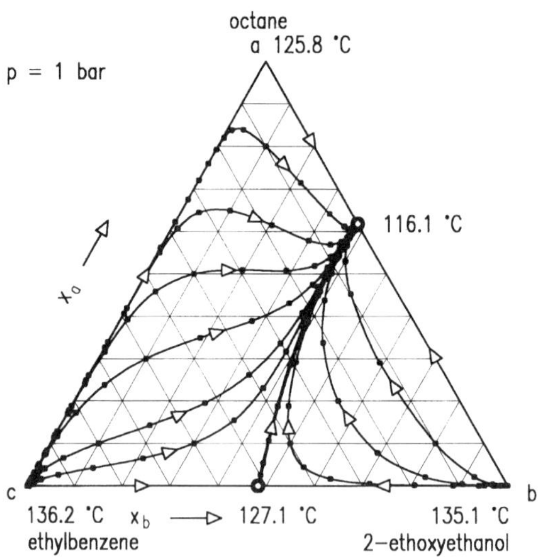

D

Fig. 2-15. *(continued)*

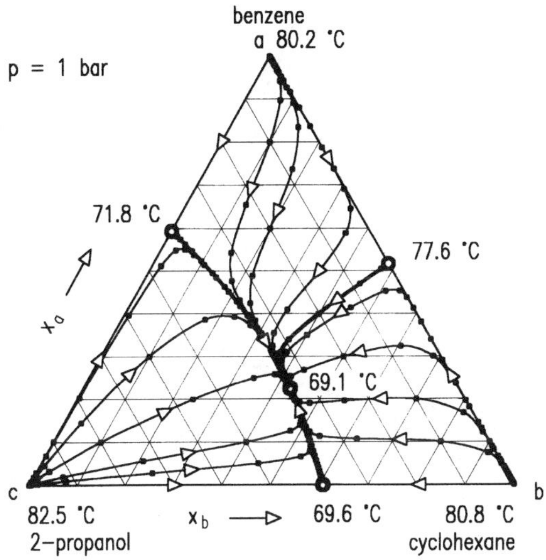

E

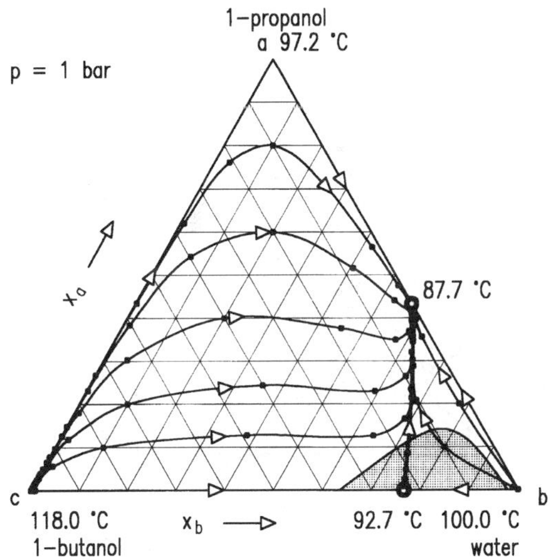

F

Fig. 2-15. *(continued)*

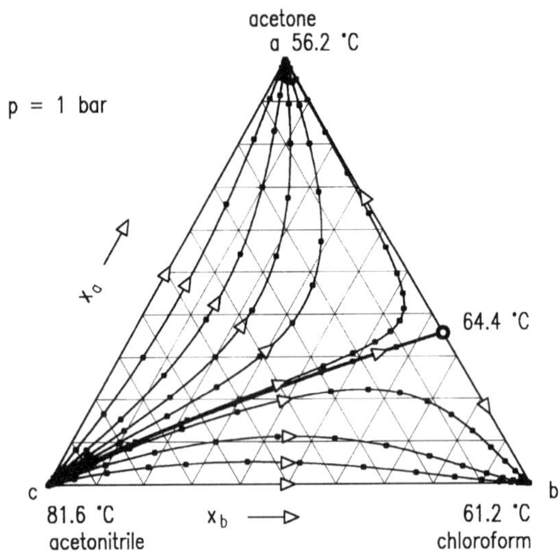

G

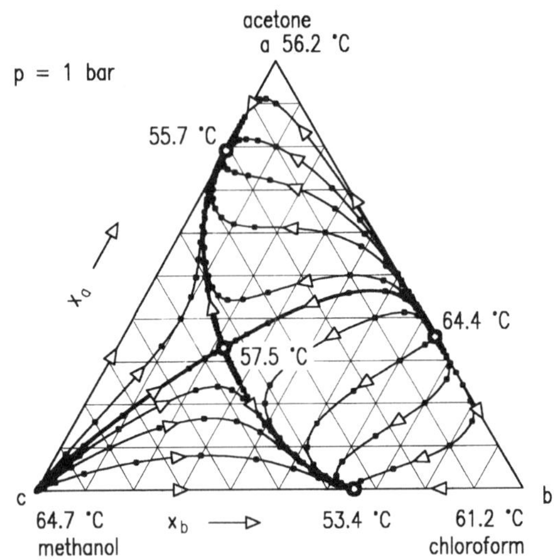

H

Fig. 2-15. *(continued)*

A mixture with three binary azeotropes and one ternary minimum azeotrope is shown in Fig. 2-15E. Each corner of the triangle is an origin of distillation lines, and they all terminate in the ternary azeotrope that has the lowest boiling point. Border distillation lines run from each binary azeotrope toward the ternary azeotrope, dividing the mixture into three regions with different origins.

A system with a heterogeneous *b-c* azeotrope (in a two-liquid phase region) and an *a-b* minimum azeotrope is depicted in Fig. 2-15F. As can be seen, the region of immiscibility affects the path of distillation lines very little.

The system acetone-chloroform-acetonitrile (Fig. 2-15G) forms an *a-b* maximum azeotrope. Under such conditions all distillation lines originate in the *c*-corner and terminate either in the *b*-corner or *a*-corner. A border distillation line runs from the high boiler *c* toward the maximum azeotrope. As may be noted from Fig. 2-5G, this border line approximately follows the course of the ridge on the boiling point surface. The very complex mixture acetone/chloroform/methanol is presented in Fig. 2-15H. It contains two binary minima, one binary maximum and one ternary saddle azeotrope. Two origins and two termini of distillation lines exist. The mixture is divided by border lines into four regions with different origins and termini, respectively.

The formulation of the VLE of ternary mixtures by distillation lines is very advantageous for process synthesis and design. In such diagrams, the existence of azeotropes can be realized very clearly. The origins or termini of distillation lines are—as will be demonstrated in Section 4.3.2.2—feasible top or bottom fractions of distillation columns. Border distillation lines are barriers that in most cases cannot be crossed by distillation. Border distillation lines are as important for ternary mixtures as azeotropes are for binary mixtures. It is essential to recognize that border distillation lines are either valleys or ridges on the boiling point surface. Thus, they run between local boiling point depressions or peaks only. Generally, border distillation lines are not straight, but are more or less curved.

2.4 CALCULATION OF PHASE EQUILIBRIUM FROM EXCESS ENTHALPY

Deviations from ideal behavior are more likely to occur in the liquid phase than in the vapor phase. As a result of smaller intermolecular distances, the forces of interaction between molecules in the liquid are considerably stronger. In contrast, the vapor phase can be assumed to behave ideally at moderate pressure. At higher pressures, the vapor phase must be described by equations of state (Section 2.5).

The phase behavior of real liquids is usually described by means of the activity coefficient γ (Equation 2.8). However, the activity coefficients of individual components i of a liquid mixture are not independent of each other, as formulated by the Gibbs-Duhem equation, e.g., [Prausnitz 1969].

$$\sum x_i \cdot d\ln\gamma_i = 0 \qquad (2.34)$$

For binary mixtures, the Gibbs-Duhem equation requires that both activity coefficients are either <1 or >1 and that they exhibit oppositely directed gradients when plotted as a function of the concentration.

Equation (2.34) is a fundamental thermodynamic equation which must always be satisfied. It has long been used to check the thermodynamic consistency of vapor-liquid equilibrium data. To obtain a relationship for the concentration dependency of the activity coefficient γ_i, it is useful to start with a value of the state of the entire mixture. In this way, a thermodynamically consistent formulation of activity coefficients is attained automatically. The mixing processes can be described most simply by using the free enthalpy g, e.g., [Prausnitz 1969]. The free enthalpy change for the mixing of ideal liquids, Δg^{id}, is:

$$\Delta g^{id} = R \cdot T \cdot \sum x_i \cdot \ln x_i \tag{2.35}$$

where R is the universal gas constant. The value of Δg^{id} is always negative. In an analogous fashion, for nonideal mixtures:

$$\Delta g = R \cdot T \cdot \sum x_i \cdot \ln x_i + R \cdot T \cdot \sum x_i \cdot \ln \gamma_i \tag{2.36}$$

The second term accounts for the nonideal behavior of the mixture and is called the free excess enthalpy Δg^E:

$$\Delta g^E \equiv R \cdot T \cdot \sum x_i \cdot \ln \gamma_i \tag{2.37}$$

Hence:

$$\Delta g = \Delta g^{id} + \Delta g^E \tag{2.38}$$

With $\Delta G^E \equiv n \cdot \Delta g^E$ and $x_i \equiv n_i/n$, we obtain:

$$\Delta G^E = R \cdot T \cdot \sum n_i \cdot \ln \gamma_i \tag{2.39}$$

Differentiation of the equation for the free excess enthalpy ΔG^E gives relationships for the activity coefficients γ_i of the individual components of a mixture at constant temperature. In general:

$$\ln \gamma_i = \frac{1}{R \cdot T} \left\{ \frac{\partial \Delta G^E}{\partial n_i} \right\}_{T,p,n_j \neq n_i} \tag{2.40}$$

This is equivalent to:

$$\ln \gamma_i = \frac{1}{R \cdot T} \left(\Delta g^E + \frac{\partial \Delta g^E}{\partial x_i} - \sum_{k=1}^{n} x_k \cdot \frac{\partial \Delta g^E}{\partial x_k} \right) \tag{2.41}$$

For a binary mixture of substances *a* and *b*:

$$\ln \gamma_a = \frac{1}{R \cdot T} \cdot \left[\Delta g^E + \frac{\partial \Delta g^E}{\partial x_a} - x_a \cdot \frac{\partial \Delta g^E}{\partial x_a} \right] \tag{2.42}$$

Using $x_b = 1 - x_a$ finally gives:

$$\ln \gamma_a = \frac{1}{R \cdot T} \cdot \left[\Delta g^E + x_b \cdot \frac{\partial \Delta g^E}{\partial x_a} \right] \tag{2.43}$$

Similarly:

$$\ln \gamma_b = \frac{1}{R \cdot T} \cdot \left[\Delta g^E - x_a \cdot \frac{\partial \Delta g^E}{\partial x_a} \right] \tag{2.44}$$

Figure 2-16 presents a graphical interpretation of Equations (2.43) or (2.44) [Gmehling and Kolbe 1992]. For a given concentration $x_a = 0.252$, the activity coefficient can be found by obtaining the tangent line to $g^E/(RT)$ at x_a, and from it determining the intersections at pure *a* and pure *b*.

Relationships for the free excess enthalpy Δg^E, which satisfy the condition $\Delta g^E = 0$ for $x_i = 1$, are required for the calculation of the activity coefficient

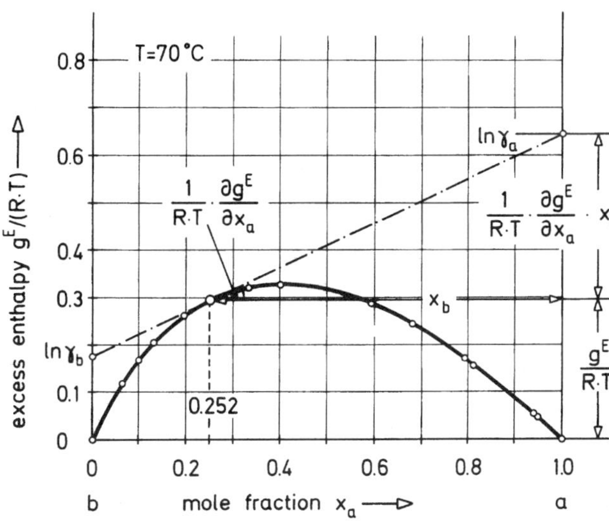

Fig. 2-16. *Excess enthalpy $g^E/(R \cdot T)$ for the system ethanol (=a) and water (=b) at 70°C [Gmehling and Kolbe 1992]. The determination of the activity coefficients $\ln\gamma_a$ and $\ln\gamma_b$ by Equations (2.43) or (2.44) is represented by a tangent to the curve.*

γ_i of a mixture from Equation (2.43). Empirical and semi-empirical correlations containing two or three correlating parameters are generally used for binary mixtures, e.g., [Prausnitz 1969]. Some important correlations are listed in Table 2-3. The application of the Margules equation is demonstrated in Example 2-6. However, improved theoretical formulations, usually based on the concept of local composition or on the increment method, are being employed increasingly.

Table 2-3. Some important equations for excess enthalpy g^E and activity coefficients γ_i of binary mixtures

Type of Equation	Adjustable Parameters	Equations
Margules	A_{ab}, A_{ba}	$\frac{g^E}{R \cdot T} = x_a \cdot x_b \cdot (A_{ba} \cdot x_a + A_{ab} \cdot x_b)$
		$\ln \gamma_a = x_b^2 \cdot (A_{ab} + 2 \cdot (A_{ba} - A_{ab}) \cdot x_a)$
		$\ln \gamma_b = x_a^2 \cdot (A_{ba} + 2 \cdot (A_{ab} - A_{ba}) \cdot x_b)$
van Laar	A_{ab}, A_{ba}	$\frac{g^E}{R \cdot T} = \frac{A_{ab} \cdot x_a \cdot x_b}{x_a \cdot (A_{ab}/A_{ba}) + x_b}$
		$\ln \gamma_a = A_{ab} \cdot \left(\frac{A_{ba} \cdot x_b}{A_{ab} \cdot x_a + A_{ba} \cdot x_b}\right)^2$
		$\ln \gamma_b = A_{ba} \cdot \left(\frac{A_{ab} \cdot x_a}{A_{ab} \cdot x_a + A_{ba} \cdot x_b}\right)^2$
Wilson[1]	$\lambda_{ab} - \lambda_{aa}$	$\frac{g^E}{R \cdot T} = -(x_a \cdot \ln(x_a + A_{ab} \cdot x_b) + x_b \cdot \ln(x_b + A_{ba} \cdot x_a))$
	$\lambda_{ba} - \lambda_{bb}$	$\ln \gamma_a = 1 - \ln(x_a + A_{ab} \cdot x_b) - \frac{x_a}{x_a + A_{ab} \cdot x_b} - \frac{A_{ba} \cdot x_b}{A_{ba} \cdot x_a + x_b}$
		$\ln \gamma_b = 1 - \ln(A_{ba} \cdot x_a + x_b) - \frac{A_{ab} \cdot x_a}{x_a + A_{ab} \cdot x_b} - \frac{x_b}{A_{ba} \cdot x_a + x_b}$
NRTL[2]	$g_{ab} - g_{bb}$	$\frac{g^E}{R \cdot T} = x_a \cdot x_b \cdot \left(\frac{\tau_{ba} \cdot G_{ba}}{x_a + x_b \cdot G_{ba}} + \frac{\tau_{ab} \cdot G_{ab}}{x_b + x_a \cdot G_{ab}}\right)$
	$g_{ba} - g_{aa}$	$\ln \gamma_a = x_b^2 \cdot \left(\tau_{ba} \cdot \left(\frac{G_{ba}}{x_a + x_b \cdot G_{ba}}\right)^2 + \frac{\tau_{ab} \cdot G_{ab}}{(x_b + x_a \cdot G_{ab})^2}\right)$
	α_{ab}	$\ln \gamma_b = x_a^2 \cdot \left(\tau_{ab} \cdot \left(\frac{G_{ab}}{x_b + x_a \cdot G_{ab}}\right)^2 + \frac{\tau_{ba} \cdot G_{ba}}{(x_a + x_b \cdot G_{ba})^2}\right)$

(continued)

Table 2-3. (*continued*)

Type of Equation	Adjustable Parameters	Equations
UNIQUAC[3]	$u_{ab} - u_{bb}$	$\frac{g^{EC}}{R \cdot T} = x_a \cdot \ln \frac{\varphi_a}{x_a} + x_b \cdot \ln \frac{\varphi_b}{x_b} + \frac{z}{2} \cdot \left(q_a \cdot x_a \cdot \ln \frac{\vartheta_a}{\varphi_a} + qb \cdot x_b \cdot \ln \frac{\vartheta_b}{\varphi_b}\right)$
	$u_{ba} - u_{aa}$	$\frac{g^{ER}}{R \cdot T} = -q_a \cdot x_a \cdot \ln(\vartheta_a + \vartheta_b \cdot \tau_{ba}) - q_b \cdot x_b \cdot \ln(\vartheta_b + \vartheta_a \cdot \tau_{ab})$ $\ln \gamma_a = \ln \gamma_a^C + \ln \gamma_a^R$ $\ln \gamma_a^C = \ln \frac{\varphi_a}{x_a} + \frac{z}{2} \cdot q_a \cdot \ln \frac{\vartheta_a}{\varphi_a} + \varphi_b \cdot \left(l_a - \frac{r_a}{r_b} \cdot l_b\right)$ $\ln \gamma^R_a = -q_a \cdot \ln(\vartheta_a + \vartheta_b \cdot \tau_{ba}) + \vartheta_b \cdot q_a \cdot \left(\frac{\tau_{ba}}{\vartheta_a + \vartheta_b \cdot \tau_{ba}} - \frac{\tau_{ab}}{\vartheta_a \cdot \tau_{ab} + \vartheta_b}\right)$ $\ln \gamma_b = \ln \gamma_b^C + \ln \gamma_b^R$ $\ln \gamma_b^C = \ln \frac{\varphi_b}{x_b} + \frac{z}{2} \cdot q_b \cdot \ln \frac{\vartheta_b}{\varphi_b} + \varphi_a \cdot \left(l_b - \frac{r_b}{r_a} \cdot l_a\right)$ $\ln \gamma_b^R = -q_b \cdot \ln(\vartheta_a \cdot \tau_{ab} + \vartheta_b) + \vartheta_a \cdot q_b \cdot \left(\frac{\tau_{ab}}{\vartheta_a \cdot \tau_{ab} + \vartheta_b} - \frac{\tau_{ba}}{\vartheta_a + \vartheta_b \cdot \tau_{ba}}\right)$ $l_i = \frac{z}{2} \cdot (r_i - q_i) - (r_i - 1)$; with $z \approx 10$

1
$$A_{ab} = \frac{\hat{v}_b}{\hat{v}_a} \cdot \exp\left(-\frac{\lambda_{ab} - \lambda_{aa}}{R \cdot T}\right) \qquad A_{ba} = \frac{\hat{v}_a}{\hat{v}_b} \cdot \exp\left(-\frac{\lambda_{ba} - \lambda_{bb}}{R \cdot T}\right)$$

$\hat{v}_i$ molar volume of pure liquid component i
λ_{ij} interaction energy between components i and j ($\lambda_{ij} = \lambda_{ji}$)

2
$$\tau_{ab} = \frac{g_{ab} - g_{bb}}{R \cdot T};\ \tau_{ba} = \frac{g_{ba} - g_{aa}}{R \cdot T};\ G_{ab} = \exp(-\alpha_{ab} \cdot \tau_{ab});\ G_{ba} = \exp(-\alpha_{ba} \cdot \tau_{ba})$$

g_{ij} parameter for interaction between components i and j ($g_{ij} = g_{ji}$)
α_{ij} nonrandomness parameter ($\alpha_{ij} = \alpha_{ji}$)

3 q_i area parameter
r_i volume parameter
u_{ij} parameter of interaction between components i and j ($u_{ij} = u_{ji}$)
γ^{iC} combinatorial part of activity coefficient
γ_i^R residual part of activity coefficientz coordination number

$$\vartheta_i = \frac{q_i \cdot x_i}{\sum_j^k q_i \cdot x_j}; \qquad \varphi_i = \frac{r_i \cdot x_i}{\sum_j^k r_i \cdot x_j}$$

$$\tau_{ab} = \exp\left(-\frac{u_{ab} - u_{bb}}{R \cdot T}\right); \qquad \tau_{ba} = \exp\left(-\frac{u_{ba} - u_{aa}}{R \cdot T}\right)$$

EXAMPLE 2-6: ESTIMATION OF VAPOR-LIQUID EQUILIBRIUM FROM KNOWLEDGE OF THE AZEOTROPE

Find the vapor-liquid equilibrium of the system chloroform (*a*), methanol (*b*) at $p = 1.013$ bar from the azeotropic data. Use the Margules equation.

Solution:

Azeotropic data (Table 2-2): $p = 1.013$ bar; $T = 53.4°C$; $x_a = 0.650$.

1. Evaluation of the activity coefficients at the azeotrope (Equations 2.5 and 2.8): $p_i = x_i \cdot p_i^o \cdot \gamma_i$ and $p_i = y_i \cdot p$; $\gamma_i = y_i \cdot p/x_i \cdot p_i^o$ with $y_i = x_i$ (azeotrope) $\gamma_i = p/p_i^o$. Vapor pressures p_i^o from Antoine equation (A_i, B_i, C_i from Table 2-1) at 53.4°C ≙ 326.55 K:

$$p_a^o = \exp\left(9.39360 - \frac{2696.249}{326.55 - 46.918}\right) = 0.7799 \text{ bar};$$

$$p_b^o = \exp\left(11.98705 - \frac{3643.314}{326.55 - 33.424}\right) = 0.6427 \text{ bar};$$

$$\gamma_a = 1.013/0.7799 = 1.2988; \; \gamma_b = 1.013/0.6427 = 1.5762.$$

2. Margules equation (see Table 2-3): $\ln \gamma_a = x_b^2 \cdot (A_{ab} + 2 \cdot (A_{ba} - A_{ab}) \cdot x_a)$; $\ln \gamma_b = x_a^2 \cdot (A_{ba} + 2 \cdot (A_{ab} - A_{ba}) \cdot x_b)$. Transformation:

$$A_{ab} = \frac{x_b - x_a}{x_b^2} \cdot \ln \gamma_a + \frac{2}{x_a} \cdot \ln \gamma_b;$$

$$A_{ba} = \frac{x_a - x_b}{x_a^2} \cdot \ln \gamma_b + \frac{2}{x_b} \cdot \ln \gamma_a.$$

   With the values at the azeotrope:

$$A_{ab} = \frac{0.35 - 0.65}{0.35^2} \cdot \ln 1.2988 + \frac{2}{0.65} \cdot \ln 1.5762 = 0.7598;$$

$$A_{ba} = \frac{0.65 - 0.35}{0.65^2} \cdot \ln 1.5762 + \frac{2}{0.35} \cdot \ln 1.2988 = 1.8171.$$

3. Activity coefficients: $\ln \gamma_a = x_b^2 \cdot (0.7498 + 2.1145 \cdot x_a)$; $\ln \gamma_b = x_a^2 \cdot (1.8171 - 2.1145 \cdot x_b)$. Evaluation gives:

x_a	0	0.1	0.2	0.3	0.4	0.5	0.6	0.7	0.8	0.9	1.0
γ_a	2.1379	2.1962	2.1318	1.9801	1.7825	1.5750	1.3834	1.2234	1.1030	1.0270	1.0
γ_b	1.0	0.9991	1.0050	1.0308	1.0917	1.2092	1.4186	1.7852	2.4407	3.6714	6.1540

4. Vapor-liquid equilibrium—We set: $x_a = 0.1$; $\rightarrow \gamma_a = 2.1962$; $\gamma_b = 0.9991$.

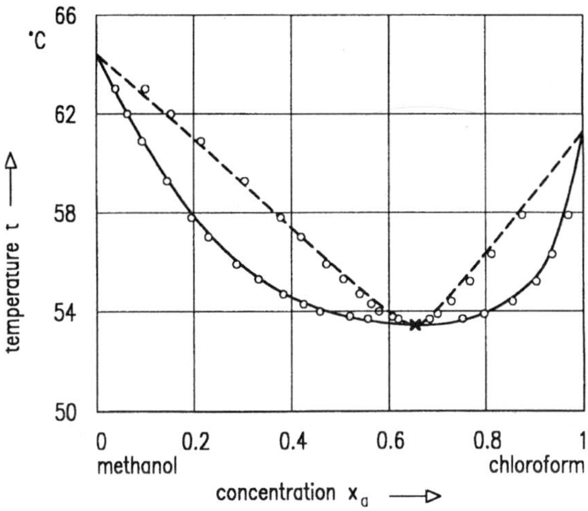

Fig. 2-17. *Vapor-liquid equilibrium of the binary system chloroform/methanol calculated from azeotropic data with the Margules equation (Example 2-6).*

Estimation: T = 61°C ≙ 334.15 K. From Antoine equation: $p_a^o = 1.0066$ bar; $p_b^o = 0.8798$ bar. Boiling point (Equation 2.12): $\Sigma x_i \cdot p_i^o \cdot \gamma_i = p_m$; $p_m = 0.1 \cdot 1.0066 \cdot 2.1962 + 0.9 \cdot 0.8798 \cdot 0.9991 = 0.2211 + 0.7912 = 1.0123$ bar; $p_m \approx 1.013$ bar. The estimation was good (if not: see Example 2-1); $y_a = 0.2211/1.0123 = 0.2184$; $y_b = 0.7816$. Listing of results:

x_a	0	0.1	0.2	0.3	0.4	0.5	0.6	0.7	0.8	0.9	1.0
y_a	0	0.2184	0.3831	0.4952	0.5673	0.6120	0.6396	0.6599	0.6864	0.7511	1.0000
T	64.54	61.0	58.0	55.7	54.4	53.7	53.4	53.4	53.7	55.0	61.19

Plot of the results and comparison with experimental data (Nagata 1962, see Fig. 2-19) in Fig. 2-17.

An example is the Wilson equation for a binary mixture [Wilson 1964]:

$$\frac{\Delta g^E}{R \cdot T} = -[x_a \cdot \ln(x_a + A_{ab} \cdot x_b) + x_b \cdot \ln(A_{ba} \cdot x_a + x_b)] \qquad (2.45)$$

where A_{ab} and A_{ba} are the Wilson coefficients for the binary pair *a-b*. Hence, the activity coefficients can be represented by:

$$\begin{aligned} \ln\gamma_a &= 1 - \ln(x_a + A_{ab} \cdot x_b) - \frac{x_a}{x_a + A_{ab} \cdot x_b} - \frac{A_{ba} \cdot x_b}{A_{ba} \cdot x_a + x_b} \\ \ln\gamma_b &= 1 - \ln(A_{ba} \cdot x_a + x_b) - \frac{A_{ab} \cdot x_a}{x_a + A_{ab} \cdot x_b} - \frac{x_b}{A_{ba} \cdot x_a + x_b} \end{aligned} \qquad (2.46)$$

Table 2-4. Wilson parameters for some binary mixtures (selected from Gmehling et al. 1997)

$$\mathbf{A}_{ab} = \frac{\hat{\mathbf{v}}_b}{\hat{\mathbf{v}}_a} \cdot \mathbf{exp}\left(-\frac{\lambda_{ab} - \lambda_{aa}}{\mathbf{R} \cdot \mathbf{T}}\right) \qquad \mathbf{A}_{ba} = \frac{\hat{\mathbf{v}}_a}{\hat{\mathbf{v}}_b} \cdot \mathbf{exp}\left(-\frac{\lambda_{ba} - \lambda_{bb}}{\mathbf{R} \cdot \mathbf{T}}\right) \quad \text{with } \mathbf{R} = 1.98721 \text{ cal/(mol} \cdot \text{K)}$$

Formula	Name	$\hat{v}$ cm³/ml	Formula	Name	$\hat{v}$ cm³/mol	$\lambda_{ab} - \lambda_{aa}$ cal/mol	$\lambda_{ba} - \lambda_{bb}$ cal/mol
H_2O	water	18.07	$CHCl_3$	chloroform (*a*)	80.67	8558.5400	16136.578
			CH_2Cl_2	dichloromethane (*a*)	64.50	2630.8762	4642.7827
			CH_2O_2	formic acid (*b*)	37.91	−278.0478	1037.5049
			CH_4O	methanol (*a*)	40.73	243.9483	435.1416
			C_2H_3N	acetonitrile (*a*)	52.86	579.4404	1510.4348
			C_2H_4O	acetaldehyde (*a*)	56.62	−592.3239	1567.4768
			$C_2H_4O_2$	acetic acid (*b*)	57.54	705.5876	111.6579
			C_2H_5ClO	2-chloroethanol (*b*)	67.29	896.5587	918.2200
			C_3H_6O	acetone (*a*)	74.05	462.4153	1464.1566
			C_3H_8O	1-propanol (*a*)	75.14	1618.7879	1270.2305
			C_3H_8O	2-propanol (*a*)	76.92	763.3100	−1138.227
			C_4H_8O	2-butanone (*a*)	90.17	1269.0280	1919.5114
			C_4H_8O	tetrahydrofuran (*a*)	81.55	1140.7179	1819.4033
			$C_4H_8O_2$	1,4-dioxane (*b*)	85.71	1967.0868	397.5832
			$C_4H_8O_2$	ethylacetate (*a*)	98.49	1147.4297	2084.0256
			C_4H_9NO	morpholine (*b*)	87.52	2578.6032	−1716.484
			$C_4H_{10}O$	1-butanol (*b*)	91.97	1536.8544	1679.4717
			$C_4H_{10}O_2$	2-ethoxyethanol (*b*)	97.41	1192.8085	235.8588
			C_5H_5N	pyridine (*b*)	80.86	1098.8670	1160.6292
			C_6H_6	benzene (*a*)	89.41	9954.9130	11848.842
			C_6H_6O	phenol (*b*)	83.14	1449.5866	1434.8670
			C_6H_{12}	cyclohexane (*a*)	108.75	35632.3070	18155.835
			C_7H_8	toluene (*b*)	106.85	2164.8740	12682.718
$CHCl_3$	chloroform	80.67	CCl_4	tetrachloromethane (*b*)	97.09	−409.8965	831.8887
			CH_4O	methanol (*b*)	40.73	−351.3486	1748.1927
			C_2H_3N	acetonitrile (*b*)	52.86	308.9227	25.7094
			C_2H_6O	ethanol (*b*)	58.68	−304.1230	1549.0997
			C_3H_6O	acetone (*a*)	74.05	28.8819	−484.3856
			$C_4H_{10}O$	1-butanol (*b*)	91.97	−152.1757	847.9199
			C_6H_6	benzene (*b*)	89.41	−161.8065	49.6010
			C_7H_8	toluene (*b*)	106.85	−365.8311	552.1459

CH_4O	methanol	40.73	CCl_4	tetrachloromethane (*b*)	97.09	1955.2659	189.0939
			CH_2Cl_2	dichloromethane (*a*)	64.50	−243.9874	1902.6643
			C_2H_6O	ethanol (*b*)	58.68	135.8113	−132.0576
			C_3H_6O	acetone (*a*)	74.05	−170.1753	594.1757
			C_3H_8O	1-propanol (*b*)	75.14	272.4310	−207.7119
			C_3H_8O	2-propanol (*b*)	76.92	140.1672	−132.7052
			C_4H_8O	2-butanone (*b*)	90.17	671.3573	−150.8649
			$C_4H_{10}O$	1-butanol (*b*)	91.97	−121.3194	9543.2503
			C_6H_6	benzene (*b*)	89.41	1853.1108	113.5712
			C_6H_{14}	hexane (*b*)	131.61	−444.5442	1494.9731
$C_2H_4Cl_2$	1,2 dichloroethane	78.87	$C_4H_{10}O$	1-butanol (*b*)	91.97	305.3873	583.0877
			C_6H_6	benzene (*a*)	89.41	121.9074	−83.4466
			C_7H_8	toluene (*b*)	106.85	−187.5366	422.5332
C_2H_6O	ethanol	58.68	C_2H_4O	acetaldehyde (*a*)	56.62	162.5974	−150.9298
			C_3H_6O	acetone (*a*)	74.05	252.2864	180.4365
			C_4H_8O	2-butanone (*b*)	90.17	477.6356	−40.1590
			C_4H_9NO	morpholine (*b*)	87.52	−573.3858	993.4870
			C_5H_5N	pyridine (*b*)	80.86	89.4230	−89.7636
			C_6H_6	benzene (*b*)	89.41	1264.4318	266.6118
			C_7H_8	toluene (*b*)	106.85	1772.9967	132.4691
C_3H_6O	acetone	74.05	C_2H_4O	acetonitrile (*b*)	52.86	−371.1771	422.1673
			C_5H_5N	pyridine (*b*)	80.86	−294.8147	855.7556
			C_6H_6	benzene (*b*)	89.41	543.9352	−182.5230
			C_6H_{12}	cyclohexane (*b*)	108.75	1016.1121	305.4414
$C_4H_{10}O$	1-butanol	91.97	C_5H_5N	pyridine (*a*)	80.86	−799.3720	2216.4214
			C_6H_6	benzene (*a*)	89.41	881.6972	137.6252
			C_7H_8	toluene (*a*)	106.85	197.0940	861.0298
C_6H_6	benzene	89.41	CCl_4	tetrachloromethane (*a*)	97.09	−223.5669	331.0329
			C_2H_4O	acetonitrile (*b*)	52.86	−112.5782	945.7542
			C_3H_8O	1-propanol (*b*)	75.14	210.4437	1076.8664
			C_5H_5N	pyridine (*b*)	80.86	−227.3437	423.2812
			C_7H_8	toluene (*b*)	106.85	794.4068	−550.8902
$C_8H_{10}O_2$	2-ethoxyethanol	97.41	C_8H_{10}	ethylenbenzene (*b*)	123.07	744.668	67.122
C_8H_{18}	octane	163.45	C_8H_{10}	ethylbenzene (*b*)	123.07	256.167	−9.290
			$C_8H_{10}O_2$	2-ethoxyethanol (*b*)	97.41	401.746	1221.125

The forces of interaction between molecules of *a* and *b* can be determined from a few measured values and are described by the coefficients A_{ab} and A_{ba}; these coefficients depend only slightly on pressure.

The Wilson equation for a ternary mixture is:

$$\frac{\Delta g^E}{R \cdot T} = -x_a \cdot \ln(A_{aa} \cdot x_a + A_{ab} \cdot x_b + A_{ac} \cdot x_c) - x_b \cdot \ln(A_{ba} \cdot x_a + A_{bb} \cdot x_b + A_{bc} \cdot x_c) - x_c \cdot \ln(A_{ca} \cdot x_a + A_{cb} \cdot x_b + A_{cc} \cdot x_c) \quad \text{with } A_{aa} = A_{bb} = A_{cc} = 1 \tag{2.47}$$

Therefore, the activity coefficient of substance *i* of a ternary mixture is:

$$\ln\gamma_i = 1 - \ln(A_{ia} \cdot x_a + A_{ib} \cdot x_b + A_{ic} \cdot x_c) - \frac{A_{ai} \cdot x_a}{A_{aa} \cdot x_a + A_{ab} \cdot x_b + A_{ac} \cdot x_c} - \frac{A_{bi} \cdot x_b}{A_{ba} \cdot x_a + A_{bb} \cdot x_b + A_{bc} \cdot x_c} - \frac{A_{ci} \cdot x_c}{A_{ca} \cdot x_a + A_{cb} \cdot x_b + A_{cc} \cdot x_c} \tag{2.48}$$

with $A_{aa} = A_{bb} = A_{cc} = 1$

The Wilson equation for a multicomponent mixture consisting of n components is:

$$\frac{\Delta g^E}{R \cdot T} = -\sum_{i=1}^{n} x_i \cdot \ln\left[\sum_{j=1}^{n} A_{ij} \cdot x_j\right] \quad \text{with } A_{ii} = 1 \tag{2.49}$$

Hence, the activity coefficients are given by:

$$\ln\gamma_i = 1 - \ln\left[\sum_{j=1}^{n} (A_{ij} \cdot x_j)\right] - \sum_{l=1}^{n} \frac{A_{li} \cdot x_l}{\sum_{j}^{n} A_{lj} \cdot x_j} \tag{2.50}$$

Table 2-4 lists the Wilson parameters A_{ij} for some binary mixtures. Even in ternary and higher systems, the Wilson equation contains only the interaction parameters A_{ij} for all binary mixtures involved. Hence, prediction of the phase equilibria of multicomponent systems from binary data is possible. The procedure is demonstrated in Example 2-6. The Wilson equation has proved its worth in practice; the majority of distillation lines presented in Fig. 2-15 were calculated by using this equation. However, unlike the Uniquac (UNIversal QUAsiChemical) equation [Prausnitz et al. 1980], it is not applicable to systems that undergo phase splitting in the liquid phase.

EXAMPLE 2-7: ACTIVITY COEFFICIENTS OF A TERNARY MIXTURE

Find the activity coefficients for the mixture acetone (= *a*) chloroform (= *b*), methanol (= *c*) at the temperature t = 50°C. Mole fractions $x_a = 0.25$, $x_b = 0.35$, $x_c = 0.40$. Use the Wilson equation.

Solution:

1. Wilson parameters from Table 2-4:

Acetone (= *a*)	$\hat{v}_a = 74.05$	$\lambda_{ab} - \lambda_{aa} = 28.8819$	$\lambda_{ac} - \lambda_{aa} = -170.1753$
Chloroform (= *b*)	$\hat{v}_b = 80.67$	$\lambda_{ba} - \lambda_{bb} = -484.385$	$\lambda_{bc} - \lambda_{bb} = -351.3486$
Methanol (= *c*)	$\hat{v}_c = 40.73$	$\lambda_{ca} - \lambda_{cc} = 594.1757$	$\lambda_{cb} - \lambda_{cc} = 1748.1927$

2. Evaluation of the parameters A_{ij} (Table 2-4) at T = 323 K:

$$A_{ab} = \frac{\hat{v}_b}{\hat{v}_a}\exp\left(-\frac{\lambda_{ab} - \lambda_{aa}}{R \cdot T}\right) = \frac{80.67}{74.05}\exp\left(-\frac{28.8819}{1.98721 \cdot 323}\right) = 1.0415$$

$$A_{ac} = \frac{\hat{v}_c}{\hat{v}_a}\exp\left(-\frac{\lambda_{ac} - \lambda_{aa}}{R \cdot T}\right) = \frac{40.73}{74.05}\exp\left(-\frac{-170.1753}{641.869}\right) = 0.7170$$

$$A_{ba} = \frac{\hat{v}_a}{\hat{v}_b}\exp\left(-\frac{\lambda_{ba} - \lambda_{bb}}{R \cdot T}\right) = \frac{74.05}{80.67}\exp\left(-\frac{-484.385}{641.869}\right) = 1.9523$$

$$A_{bc} = \frac{\hat{v}_c}{\hat{v}_b}\exp\left(-\frac{\lambda_{bc} - \lambda_{bb}}{R \cdot T}\right) = \frac{40.73}{80.67}\exp\left(-\frac{-351.3486}{641.869}\right) = 0.8728$$

$$A_{ca} = \frac{\hat{v}_a}{\hat{v}_c}\exp\left(-\frac{\lambda_{ca} - \lambda_{cc}}{R \cdot T}\right) = \frac{74.05}{40.73}\exp\left(-\frac{594.1757}{641.869}\right) = 0.7204$$

$$A_{cb} = \frac{\hat{v}_b}{\hat{v}_c}\exp\left(-\frac{\lambda_{cb} - \lambda_{cc}}{R \cdot T}\right) = \frac{80.67}{40.73}\exp\left(-\frac{1748.1927}{641.869}\right) = 0.1300$$

3. Evaluation of the activity coefficients from Equation (2.48):

$$\ln\gamma_a = 1 - \ln(x_a + A_{ab}\cdot x_b + A_{ac}\cdot x_c) - \frac{x_a}{x_a + A_{ab}\cdot x_b + A_{ac}\cdot x_c}$$

$$-\frac{A_{ba}\cdot x_b}{A_{ba}\cdot x_a + x_b + A_{bc}\cdot x_c} - \frac{A_{ca}\cdot x_c}{A_{ca}\cdot x_a + A_{cb}\cdot x_b + x_c}$$

$$= 1 - \ln(0.25 + 1.0415\cdot 0.35 + 0.7170\cdot 0.4)$$

$$-\frac{0.25}{0.25 + 1.0415\cdot 0.35 + 0.7170\cdot 0.4}$$

$$= 1 - \ln(A_{ba}\cdot x_a + x_b + A_{bc}\cdot x_c) - \frac{A_{ab}\cdot x_a}{x_a + A_{ab}\cdot x_b + A_{ac} x_c}$$

$$-\frac{x_b}{A_{ba}\cdot x_a + x_b + A_{bc}\cdot x_c} - \frac{A_{cb}\cdot x_c}{A_{ca}\cdot x_a + A_{cb}\cdot x_b + x_c}$$

$$= 1 - 0.1716 - 0.2889 - 0.2948 - 0.0831 = 0.1616;$$

$$\gamma_b = \exp(0.1616) = 1.1754;$$

$$\ln\gamma_c = 1 - \ln(A_{ca}\cdot x_a + A_{cb}\cdot x_b + x_c) - \frac{A_{ac}\cdot x_a}{x_a + A_{ab}\cdot x_b + A_a\cdot x_c}$$

$$- \frac{A_{bc}\cdot x_b}{A_{ba}\cdot x_a + x_b + A_{bc}\cdot x_c} - \frac{x_c}{A_{ca}\cdot x_a + A_{cb}\cdot x_b + x_c}$$

$$= 1 + 0.4690 - 0.1989 - 0.2573 - 0.6394 = 0.3734;$$

$$\gamma_c = \exp(0.3734) = 1.4527$$

4. Border activity coefficients for acetone
   4.1. Acetone in methanol: $x_a \rightarrow 0$; $x_b = 0$: $\ln(\gamma_a^\infty)_c = 1 - \ln A_{ac} - A_{ca} = 1 - \ln 0.717 - 0.7204 = 0.6123$;
   $(\gamma_a^\infty)_c = 1.8446$.
   4.2. Acetone in chloroform: $x_a \rightarrow 0$;

$$x - \frac{1.9523\cdot 0.35}{1.9523\cdot 0.25 + 0.35 + 0.8728\cdot 0.4} - \frac{0.7204\cdot 0.4}{0.7204\cdot 0.25 + 0.130\cdot 0.35 + 0.4}$$

$$= 1 + 0.1039 - 0.2774 - 0.5756 - 0.4606 = -0.2097;$$

$$\gamma_a = \exp(-0.2097) = 0.8108;$$

$$\ln\gamma_{bc} = 0\text{: } \ln(\gamma_a^\infty)_b = 1 - \ln A_{ab} - A_{ba} = 1 - \ln 1.0415$$

$$- 1.9523 = -0.9930;$$

$$(\gamma_a^\infty)_b = 0.3705.$$

   4.3. Acetone in an equimolar mixture of chloroform and methanol: $x_a \rightarrow 0$; $x_b = 0.5$, $x_c = 0.5$:

$$\ln(\gamma_a^*)_{b+c} = 1 - \ln(0.5(A_{ab} + A_{ac})) - \frac{A_{ba}}{A_{bc} + 1} - \frac{A_{ca}}{A_{cb} + 1}$$

$$= 1 - \ln(0.5(1.0415 + 0.717)) - \frac{1.9523}{0.8728 + 1} - \frac{0.7204}{0.13 + 1}$$

$$= -0.5513;$$

$$(\gamma_a^\infty)_{b+c} = 0.5762.$$

EXAMPLE 2-8: BOILING POINT OF A NONIDEAL TERNARY MIXTURE

Calculate the boiling point of the mixture acetone $(= a)$/chloroform $(= b)$/methanol $(= c)$ at a pressure $p = 1$ bar. The mole fractions of the liquid are $x_a = 0.25$; $x_b = 0.35$, $x_c = 0.40$. Use the Wilson equation. Give the composition of the vapor.

Solution:

Antoine constants from Table 2-1:

i	A_i	B_i	C_i
Acetone (= *a*)	9.76775	2787.498	−43.486
Chloroform (= *b*)	9.3936	2696.249	−46.918
Methanol (= *c*)	11.98705	3643.314	−33.424

1. Estimation of the boiling point temperature, t = 50°C (see Fig. 2-8H).
2. Vapor pressures p_i^o at t = 50°C: p_a^o = 0.8191 bar; p_b^o = 0.6926 bar; p_c^o = 0.5554 bar.
3. Activity coefficients γ_i from the Wilson equation (see Example 2-7): γ_a = 0.8108; γ_b = 1.1754; γ_c = 1.4527.
4. Vapor pressure p_m of the mixture (Equation 2.12): $p_m = \sum x_i \cdot p_i^o \cdot \gamma_i = 0.25 \cdot 0.8191 \cdot 0.8108 + 0.35 \cdot 0.6926 \cdot 1.1754 + 0.40 \cdot 0.5554 \cdot 1.4527 = 0.7736$ bar; $p_m \neq 1$ bar. The estimation was wrong.

1a. Better estimation—The temperature dependency of p_m'' can be roughly described by the Antoine equation when a correction factor ϕ is used (see Example 2-2):

$$\phi_b \equiv \frac{p_m}{p_b^o} = \frac{0.7736}{0.6926} = 1.1170;$$

$$t = \frac{B_b}{A_b - \ln(p/\phi_b)} - C_i = \frac{2696.249}{9.3936 - \ln(1/1.117)} + 46.918$$

$$= 330.61 \text{ K} \triangleq 57.46°\text{C};$$

$$t \approx 57°\text{C}.$$

2a. Vapor pressures p_i^o at t = 57°C: p_a^o = 1.0448 bar; p_b^o = 0.8816 bar; p_c^o = 0.7473 bar.

3a. Activity coefficients (see Example 2-7)—Evaluation of the equations with t = 57°C gives: γ_a = 0.811; γ_b = 1.178; γ_c = 1.449.

4b. Vapor pressures p_m of the mixture: $p_m = 0.25 \cdot 1.0448 \cdot 0.811 + 0.35 \cdot 0.8816 \cdot 1.178 + 0.40 \cdot 0.7473 \cdot 1.449 = 0.2118 + 0.3635 + 0.4331 = 1.0085$ bar; $p_m \approx 1$ bar. The estimation was good enough: t = 57°C.

5. Composition of the first bubble: $y_a = \frac{p_a}{p} = 0.2113/1.0085 = 0.210$; $y_b = 0.3635/1.0085 = 0.3605$; $y_c = 0.4331/1.0085 = 0.4295$.

EXAMPLE 2-9: BORDER DISTILLATION LINE OF A NONIDEAL TERNARY MIXTURE

Find the border distillation line for the mixture octane (= *a*)/2-ethoxyethanol (= *b*)/ethylbenzene (= *c*) at a pressure of p = 1.013 bar.

Solution

1. Rough estimation of the border distillation line—A border distillation line exists when there is more than one origin or terminus of distillation lines. As can be seen from Fig. 2-15D there exist two origins of distillation lines in the system. The border distillation line has to run from one local boiling point minimum (here $b = c$ azeotrope) to a lower boiling one (here the a-b azeotrope).
2. Calculation of the course of the border distillation line—Data:

	Octane (= *a*)	2-ethoxyethanol (= *b*)	Ethylbenzene (= *c*)
Antoine constants:	$A_a = 9.34011$	$A_b = 11.38407$	$A_c = 9.41928$
	$B_a = 3128.752$	$B_b = 4149.028$	$B_c = 3291.661$
	$C_a = -63.295$	$C_b = -43.150$	$C_c = -59.383$
Molar volume:	$\hat{v}_a = 163.54$	$\hat{v}_b = 97.41$	$\hat{v}_c = 123.07$
Wilson parameters:	$\lambda_{ab} - \lambda_{aa} = 401.746$	$\lambda_{ba} - \lambda_{bb} = 1221.125$	$\lambda_{ca} - \lambda_{cc} = -9.290$
	$\lambda_{ac} - \lambda_{aa} = 256.167$	$\lambda_{bc} - \lambda_{bb} = 744.668$	$\lambda_{cb} - \lambda_{cc} = 67.122$

Startpoint is any ternary mixture very close to the b-c azeotrope ($x_b = 0.483$, $x_c = 0.517$, $T = 127.1°C$). We set: $x_a = 0.024$; $x_b = 0.469$; $x_c = 0.507$.

2.1. Evaluation of the boiling point (see Example 2-8): Estimation $t = 127.2°C \triangleq 400.35$ K.

2.2. Vapor pressures p_i^o at $t = 127.2°C$—Antoine equation:

$$\frac{p_i^o}{\text{bar}} = \exp\left(A_i - \frac{B_i}{t/K + C_i}\right);$$

$p_a^o = 1.0592$ bar; $p_b^o = 0.7935$ bar; $p_c^o = 0.7909$ bar.

2.3. Activity coefficients γ_i from the Wilson equation (see Example 2-7): $\gamma_a = 2.001$; $\gamma_b = 1.266$; $\gamma_c = 1.227$.

2.4. Vapor pressure of the mixture (Equation 2.12): $p_m = \sum x_i \cdot p_i^o \cdot \gamma_i = 0.024 \cdot 1.0592 \cdot 2.001 + 0.469 \cdot 0.7935 \cdot 1.266 + 0.507 \cdot 0.7909 \cdot 1.227 = 0.0509 + 0.4711 + 0.4920 = 1.0140$ bar; $p_m \approx 1$ bar. The estimation of t was good (if not see Example 2-2).

2.5. Composition of the first bubble: $y_a = 0.0509/1.0140 = 0.0500$; $y_b = 0.4711/1.0140 = 0.4646$; $y_c = 0.4920/1.0140 = 0.4852$.

2.6. Composition of the new liquid mixture (after full condensation of the vapor): $x_a = 0.050$; $x_b = 0.465$; $x_c = 0.485$. This is the new start point. The steps 2.1 to 2.6 have to be repeated. List of data:

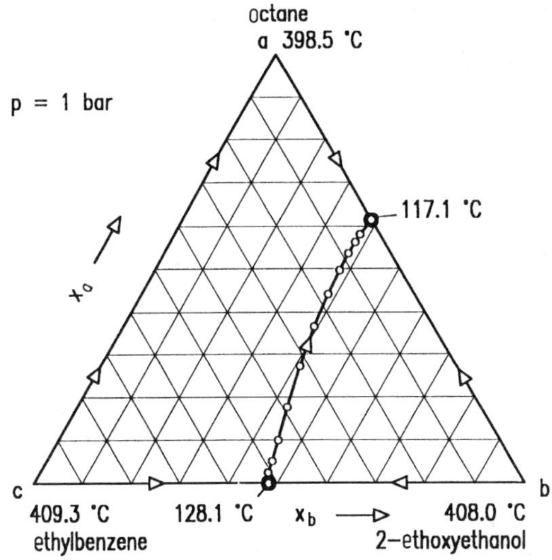

Fig. 2-18. Border distillation line of the system octane/2-ethoxyethanol/ethylbenzene calculated with the Wilson equation (Example 2-9).

Step	x_a	x_b	x_c	t/°C	γ_a	γ_b	γ_c
0	0.024	0.469	0.507	127.2	2.011	1.266	1.227
1	0.050	0.465	0.485	126.3	1.939	1.282	1.216
2	0.099	0.452	0.449	124.8	1.813	1.320	1.197
3	0.176	0.432	0.392	123.1	1.653	1.389	1.172
4	0.271	0.410	0.319	121.4	1.509	1.482	1.512
5	0.364	0.393	0.243	120.1	1.409	1.576	1.143
6	0.440	0.384	0.176	119.2	1.351	1.651	1.141
7	0.496	0.380	0.124	118.5	1.318	1.703	1.143
8	0.535	0.379	0.086	118.1	1.300	1.736	1.146
9	0.561	0.380	0.059	117.8	1.290	1.754	1.149
10	0.579	0.381	0.040	117.6	1.284	1.766	1.151

The points result in a curved line from the *b*-*c* azeotrope to the *a*-*b* azeotrope in the triangular diagram of Fig. 2-18. Knowledge of the curvature of the border distillation line is very important (see Section 5.6.2).

2.5 CALCULATION OF PHASE EQUILIBRIUM FROM EQUATIONS OF STATE

The method for calculating phase equilibrium from the excess enthalpy, described in Section 2.4, can be applied to systems that are nonideal in the liquid phase only. However, because Dalton's law does not hold at higher pressures, the vapor phase may also deviate from ideal behavior. An effective way of calculating phase equilibria at high pressure is based on equations of state which describe both the vapor and the liquid phases with sufficient accuracy. Partial fugacities f_i are used instead of partial pressures p_i. At equilibrium,

$$f_i'' = f_i' \tag{2.51}$$

where f_i'' and f_i' are the fugacities of component *i* in the vapor and liquid phases, respectively. By analogy with Equation (2.6) and using the fugacity coefficients $\phi_i \equiv f_i/p_i$, the following equation is obtained:

$$\phi_i'' \cdot y_i \cdot p_t = \phi_i' \cdot x_i \cdot p_t \tag{2.52}$$

Hence, the equilibrium ratio $K_i \equiv y_i/x_i$ is:

$$K_i = \phi_i'/\phi_i'' \tag{2.53}$$

Thus, the equilibrium ratio can be calculated from fugacity coefficients determined in the two phases. The following rigorous relationship is obtained by using the laws of thermodynamics, e.g., [Prausnitz 1969].

$$R \cdot T \cdot \ln \phi_i = \int_o^p (\hat{v} - R \cdot T/p) \cdot dp \tag{2.54}$$

with the partial molar volume $\hat{v}_i = (\partial V/\partial n_i)_{p,T,n,j \neq i}$ and the total volume $V = \Sigma\, \hat{v}_i \cdot n_i$.

The molar volume $\hat{v}$ and the derivative $\partial V/\partial n_i$ are determined by using a suitable equation of state. The real behavior of pure fluids can be described by the compressibility factor Z:

$$Z = p \cdot \hat{v}/(R \cdot T) \tag{2.55}$$

For ideal vapors, $Z = 1$. Deviations from this value are directly proportional to the degree of nonideal behavior. Two different types of equations of state describe the real behavior of pure fluids. One is based on the van der Waals equation (cubic equation of state):

$$Z = \frac{\hat{v}}{\hat{v} - b} - \frac{a}{R \cdot T \cdot \hat{v}} \tag{2.56}$$

The coefficients a and b can be interpreted as representing interaction forces and molecular volume, respectively. The second type of expression is based on the virial equation:

$$Z = 1 + B/\hat{v} + C/\hat{v}^2 + \cdots \tag{2.57}$$

Intermolecular forces are described by the virial coefficients B, C, etc., which for

Table 2-5: Some important cubic equations of state

Type	Equations
van der Waals	$Z = \frac{v}{v - b} - \frac{a}{R \cdot T \cdot v}$
	$a = \frac{27 \cdot R^2 \cdot T_{cr}^2}{64 \cdot p_{cr}}; \quad b = \frac{R \cdot T_{cr}}{8 \cdot p_{cr}}$
Redlich-Kwong	$Z = \frac{v}{v - b} - \frac{a}{R \cdot T^{3/2} \cdot (v + b)}$
	$a = 0.42748 \cdot R^2 \cdot T_{cr}^{2.5}/P_{cr}; \quad b = 0.08664 \cdot R \cdot T_{cr}/p_{cr}$
Redlich-Kwong-Soave	$Z = \frac{v}{v - b} - \frac{a(T)}{R \cdot T \cdot (v + b)}$
	$a(T) = 0.42748 \cdot \frac{R^2 \cdot T_{cr}^2}{p_{cr}} \cdot \alpha(T)$
	$\alpha(T) = (1 + (0.48 + 1.574 \cdot \omega - 0.176 \cdot \omega^2) \cdot (1 - \sqrt{T/T_{cr}}))^2$
	$b = 0.08664 \cdot R \cdot T_{cr}/p_{cr}; \quad (\omega = \text{azentric factor})$
Peng-Robinson	$Z = \frac{v}{v - b} - \frac{a(T) \cdot v}{R \cdot T \cdot (v \cdot (v + b) + b \cdot (v - b))}$
	$a(T) = 0.45724 \cdot \frac{R^2 \cdot T_{cr}^2}{p_{cr}} \cdot \alpha(T)$
	$\alpha(T) = (1 + (0.37464 + 1.54226 \cdot \omega - 0.26992 \cdot \omega^2) \cdot (1 - \sqrt{T/T_{cr}}))^2$
	$b = 0.0778 \cdot R \cdot T_{cr}/p_{cr}; \quad (\omega = \text{azentric factor})$
De Santis	$Z = \frac{1 + \xi + \xi^2 - \xi^3}{(1 - \xi)^3} - \frac{a(T)}{R \cdot T \cdot (v + b(T))}$
	$\ln a(T) = \ln a(T_{cr}) + \alpha \cdot (1 - T/T_{cr})$
	$b(T) = b(T_{cr}) \cdot \frac{\beta - T/T_{cr}}{\beta - 1}; \quad \xi = b(T)/(4 \cdot v)$
	α constant for temperature dependence of parameter a
	β constant for temperature dependence of parameter b

pure fluids are a function only of temperature. Both types of equations have been improved by numerous modifications to obtain a pressure-volume-temperature equation that is applicable to both vapor and liquid phases. Table 2-5 lists some important examples. Equations of state used to determine phase equilibria must correctly represent mixtures of substances and must include equations for the pure components as special cases. Equations of state for mixtures are generally derived from equations of state for the pure substances by using concentration-dependent coefficients. Unfortunately, theoretical methods are not available, and empirical rules for combination must be used. In principle, equations for excess enthalpy (see Section 2.3) can be applied successfully to the formulation of mixing rules [Wagner 1985]. The procedure can be illustrated by using the Soave modification of the Redlich-Kwong equation [Soave 1972].

$$Z = \frac{\hat{v}}{\hat{v} - b} - \frac{a}{R \cdot T \cdot (\hat{v} + b)} \tag{2.58}$$

The coefficients a and b can be determined from the critical values P_{cr}, T_{cr} and the azentric factor ω of the pure substances.

$$a = 0.42747 \cdot \alpha \cdot R^2 \cdot T_{cr}^2/p_{cr}$$

where $\alpha = [1 + (0.48 + 1.574 \cdot \omega - 0.176 \cdot \omega^2) \cdot (1 - \sqrt{T/T_{cr}})]$

and $b = 0.08664 \cdot R \cdot T_{cr}/p_{cr}$ (2.59)

The following mixing rules apply to mixtures of components i and j:

$$a = \sum_i \sum_j x_i \cdot x_j \cdot a_{ij}$$

Here,

$$a_{ij} = \sqrt{a_{ii} \cdot a_{jj}} \cdot (1 - k_{ij}), \quad \text{where } k_{ii} = 0 \tag{2.60}$$

Values for k_{ij} for $i \neq j$ can be found in tables (see Table 2-6):

$$b = \sum_i x_i \cdot b_i$$

Substituting Equation (2.58) into Equation (2.54) gives:

$$\ln \phi_i = \frac{b_i}{b \cdot R \cdot T} \cdot (p \cdot \hat{v} - R \cdot T) - \ln\left(\frac{p}{R \cdot T} \cdot (\hat{v} - b)\right.$$
$$\left. - \frac{a}{b \cdot R \cdot T}\right) \cdot \left(\frac{2}{a} \cdot \sum (x_j \cdot a_{ji}) - \frac{b_i}{b}\right) \cdot \ln\left(\frac{\hat{v} + b}{\hat{v}}\right) \tag{2.61}$$

Trial values for the concentration and molar volume of each phase must be estimated

Table 2-6A: Parameters of the Redlich-Kwong-Soave equation of state for some substances (selected from Knapp et al. 1982)

		Critical Data:				
Formula	Name	$\hat{M}$ g/mol	T_{cr} K	p_{cr} bar	$\hat{v}_{cr}$ cm³/mol	ω –
Ar	argon	39.948	150.71	48.64	74.848	0.000
CH_4	methane	16.040	190.63	46.17	99.418	0.010
CO	carbon monoxide	28.011	132.95	34.99	89.288	0.093
CO_2	carbon dioxide	44.01	304.15	73.76	92.863	0.231
C_2H_2	acetylene	26.038	309.45	62.42	113.436	0.180
C_2H_4	ethylene	28.054	282.41	50.50	131.636	0.087
C_2H_6	ethane	30.070	305.43	48.84	146.699	0.099
C_3H_6	propylene	42.080	365.05	46.00	183.799	0.140
C_3H_8	propane	44.090	369.82	42.50	200.830	0.152
H_2	hydrogen	2.016	43.60	20.47	51.453	0.000
N_2	nitrogen	28.013	126.15	33.94	88.586	0.045

Table 2-6B: Parameters K_{ab} for the Redlich-Kwong-Soave equation of state for some substances (selected from Knapp et al. 1982)

K_{ab}	Ar	CH_4	CO	CO_2	C_2H_2	C_2H_4	C_2H_6	C_3H_6	C_3H_8	H_2	N_2
Ar	0										
CH_4	0.0252	0									
CO		0.0322	0								
CO_2		0.0933		0							
C_2H_2					0						
C_2H_4		0.0189		0.0533	0.0596	0					
C_2H_6		−0.0078	−0.0278	0.1363		0.0089	0				
C_3H_6		0.0289		0.0944	−0.4200		0.0026	0			
C_3H_8		0.0090	0.0156	0.1289			−0.0022	0.0033	0		
H_2		−0.0222	0.0804	−0.3426		−0.0681	−0.1667	−0.3233	−0.2359	0	
N_2	0.000	0.0278	0.0374	−0.0315		0.0798	0.0407	0.0900	0.0763	0.0978	0

to solve this equation for each phase. Equation (2.58) is used to improve the values during iteration. The equilibrium ratio K for each component can then be calculated by using Equation (2.53) to check the assumed phase concentrations. This iteration method often converges to trivial solutions. However, other available procedures require good starting values close to those of the solution [Asselineau et al. 1979]. The application of the Redlich-Kwong-Soave equation to a ternary mixture is demonstrated in Example 2-10.

EXAMPLE 2-10: EVALUATION OF THE GAS LIQUID EQUILIBRIUM OF A TERNARY MIXTURE BY AN EQUATION OF STATE

Find the system pressure and vapor composition of a boiling liquid mixture at 200 K with 5 mol-% hydrogen, 70 mol-% methane and 25 mol-% ethane. Use the Redlich-Kwong-Soave equation.

Data and Redlich-Kwong-Soave parameters (from Table 2-6):

	Hydrogen (*a*)	Methane (*b*)	Ethane (*c*)
Liquid composition x_i	0.050	0.700	0.250
Critical temperature (K)	43.6	190.63	305.43
Critical pressure (bar)	20.47	46.17	48.84
Azentric factor ω	0.00	0.01	0.099
k_{ij} = factor	$k_{ab} = -0.0222$	$k_{bc} = -0.0078$	$k_{ac} = -0.1667$

Solution:

Equation (2.53): $K_i \equiv y_i/x_i = \varphi_i'/\varphi_i''$ with φ_i from Equation (2.61):

1. Redlich-Kwong-Soave parameters

   1.1. Pure constituents; Equation (2.59): $\alpha_i = (1 + (0.48 + 1.574 \cdot \omega_i - 0.176 \cdot \omega_i^2) \cdot (1 - \sqrt{T/T_{cr}}))^2$; $\alpha_a = (1 + 0.48 \cdot (1 - \sqrt{200/43.6}))^2 = 0.204$; $a_{ii} = 0.42747 \cdot \alpha \cdot R^2 \cdot T_{cr}^2/p_{cr}$; $a_{aa} = 0.42747 \cdot 0.204 \cdot 8314^2 \cdot 43.6^2/2.047 \cdot 10^6 = 5.605\ \text{Pa} \cdot \text{m}^6/\text{kmol}^2$; $b_{ii} = 0.08664 \cdot R \cdot T_{cr}/p_{cr}$; $b_{aa} = 0.08664 \cdot 8314 \cdot 43.6/2.047 \cdot 10^6 = 0.0153\ \text{m}^3/\text{kmol}$.

Listing of the results of all constituents:

	m_i	α_i	a_{ii}	b_{ii}
Hydrogen (*a*)	0.480	0.204	$5.605 \cdot 10^3$	$1.53 \cdot 10^{-2}$
Methane (*b*)	0.496	0.976	$2.270 \cdot 10^5$	$2.97 \cdot 10^{-2}$
Ethane (*c*)	0.634	1.257	$7.093 \cdot 10^5$	$4.50 \cdot 10^{-2}$

   1.2. Interaction parameters—Equation (2.60): $a_{ij} = \sqrt{a_{ii} \cdot a_{jj}} \cdot (1 - k_{ij})$;

$a_{ab} = \sqrt{5.605 \cdot 10^3 \cdot 2.27 \cdot 10^5} \cdot (1 - (-0.0222)) = 3.647 \cdot 10^4$ Pa · $m^6/kmol^2$.

Listing of the results of all constituents:

Hydrogen (*a*)	$a_{aa} = 5.605 \cdot 10^3$	$a_{ab} = 3.647 \cdot 10^4$	$a_{ac} = 7.357 \cdot 10^4$
Methane (*b*)	$a_{ba} = 3.647 \cdot 10^4$	$a_{bb} = 2.270 \cdot 10^5$	$a_{bc} = 4.044 \cdot 10^5$
Ethane (*c*)	$a_{ca} = 7.357 \cdot 10^4$	$a_{cb} = 4.044 \cdot 10^5$	$a_{cc} = 7.093 \cdot 10^5$

2. Estimation of the system pressure: $p = 50$ bar $\triangleq 5 \cdot 10^6$ Pa.
3. Liquid phase
   3.1. Mixing rules—Equation (2.60): $a = \sum_i^k \cdot \sum_j^k \cdot x_i \cdot x_j \cdot a_{ij}$; $a = x_a \cdot x_a \cdot a_{aa} + x_a \cdot a_b \cdot a_{ab} + x_a \cdot x_c \cdot a_{ac} + x_b \cdot x_a \cdot a_{ba} + x_b \cdot x_b \cdot a_{bb} + x_b \cdot x_c \cdot a_{bc} + x_c \cdot x_a \cdot a_{ca} + x_c \cdot x_b \cdot a_{cb} + x_c \cdot x_c \cdot a_{cc}$;
   $a = 3.015 \cdot 10^5$ Pa · $m^6/kmol^2$.

Equation (2.60): $b = \sum x_i \cdot b_i = x_a \cdot b_a + x_b \cdot b_b + x_c \cdot b_c$; $b = 0.05 \cdot 0.0153 + 0.70 \cdot 0.0297 + 0.25 \cdot 0.0450 = 0.0328$ $m^3/kmol$.

   3.2. Molar volume of the liquid (Equation (2.58):

$$p = \frac{R \cdot T}{\hat{v} - b} - \frac{a}{\hat{v} \cdot (\hat{v} + b)}; \qquad 5 \cdot 10^6 = \frac{8314 \cdot 200}{\hat{v} - 0.0328} - \frac{3.015 \cdot 10^5}{\hat{v} \cdot (\hat{v} + 0.0328)}.$$

   This cubic equation has three solutions. The lowest value is the molar volume of the liquid: $\hat{v}' = 0.0634$ $m^3/kmol$.

   3.3. Fugacity coefficient φ_i' (Equation (2.61):

$$\ln\varphi_i' = \frac{b_i}{b \cdot R \cdot T} \cdot (p \cdot \hat{v} - R \cdot T) - \ln\left(\frac{p}{R \cdot T}(\hat{v} - b) - \frac{a}{b \cdot R \cdot T}\right) \cdot \left(\frac{2}{a} \cdot \sum (x_j \cdot a_{ji}) - \frac{b_i}{b}\right) \cdot \ln\left(\frac{\hat{v} + b}{\hat{v}}\right);$$

$R \cdot T = 8314 \cdot 200 = 1.663 \cdot 10^6$; $\sum_{j=1}^{k} x_j \cdot a_{ja} = 0.05 \cdot 5.605 \cdot 10^3 + 0.70 \cdot 3.647 \cdot 10^4 + 0.25 \cdot 7.357 \cdot 10^4 = 4.4 \cdot 10^4$ Pa · $m^6/kmol^2$;

$$\ln\varphi_a' = \frac{0.0153}{0.0328 \cdot 1.663 \cdot 10^6} \cdot (5 \cdot 10^6 \cdot 0.0634 - 1.663 \cdot 10^6) - \ln\left(\frac{5 \cdot 10^6}{1.663 \cdot 10^6} \cdot (0.0634 - 0.0328) - \frac{3.015 \cdot 10^5}{0.0328 \cdot 1.663 \cdot 10^6}\right) \cdot \left(\frac{2}{3.015 \cdot 10^5} \cdot 4.4 \cdot 10^4 - \frac{0.0153}{0.0328}\right) \cdot \ln\frac{0.0634 + 0.0328}{0.0634} = 2.4088;$$

$\varphi_a' = \exp(2.4088) = 11.121$; $\varphi_b' = 0.7700$; $\varphi_c' = 0.07030$.

4. Vapor phase—Estimation of the vapor composition: $y_i = K_i \cdot x_i$ with $K_i = \varphi_i'/\varphi_i''$.

Assumption: ideal behavior of the vapor: $\varphi_i'' = 1$; $K_a = 11.121$; $K_b = 0.7700$; $K_c = 0.07030$; $y_a = 11.121 \cdot 0.05 = 0.556$; $y_b = 0.7700 \cdot 0.70 = 0.539$; $y_c = 0.07030 \cdot 0.25 = 0.0176$; $y_a + y_b + y_c = 1.113$.

Correction: $y_i = y_i/\sum y_i$; $y_a = 0.500$; $y_b = 0.484$; $y_c = 0.016$.

4.1. Mixing rules (see Step 2.1): $a = \sum_i^k \cdot \sum_j^k (y_i \cdot y_j \cdot a_{ij}) = 7.99 \cdot 10^4$ Pa $\cdot$ m^6/kmol^2; $b = \sum_i^k y_i \cdot b_i = 0.0228$ m^3/kmol.

4.2. Molar volume of the vapor:

$$p = \frac{R \cdot T}{\hat{v} - b} - \frac{a}{\hat{v} \cdot (\hat{v} + b)}$$

$$10^6 = \frac{8314 \cdot 200}{\hat{v} - 0.0229} - \frac{7.99 \cdot 10^4}{\hat{v} \cdot (\hat{v} + 0.0229)}.$$

This cubic equation has three solutions. The highest value is the molar volume of the vapor: $\hat{v}'' = 0.3110$ m^3/kmol.

4.3. Fugacity coefficient φ_i''—Equation (2.61):

$$\ln\varphi_i'' = \frac{b_i}{b \cdot R \cdot T} \cdot (p \cdot \hat{v} - R \cdot T) - \ln\left(\frac{p}{R \cdot T}(\hat{v} - b)\right.$$
$$\left. - \frac{a}{b \cdot R \cdot T}\right) \cdot \left(\frac{2}{a} \cdot \sum (x_j \cdot a_{ji}) - \frac{b_i}{b}\right) \cdot \ln\left(\frac{\hat{v} + b}{\hat{v}}\right);$$

$\varphi_a'' = 1.126$; $\varphi_b'' = 0.7794$; $\varphi_c'' = 0.549$.

5. Vapor-liquid equilibrium and system pressure

5.1. Equilibrium constants—Equation (2.53): $K_i = \varphi_i'/\varphi_i''$; $K_a = 11.121/1.126 = 9.8766$; $K_b = 0.7700/0.7794 = 0.9879$; $K_c = 0.07030/0.549 = 0.1281$.

Vapor composition: $y_i = K_i \cdot x_i$; $y_a = 9.8766 \cdot 0.05 = 0.4938$; $y_b = 0.9879 \cdot 0.70 = 0.6916$; $y_c = 0.1281 \cdot 0.25 = 0.0320$.

Partial pressures: $p_a = 0.4938 \cdot 5.10^6 = 2.469 \cdot 10^6$ bar; $p_b = 0.6916 \cdot 5 \cdot 10^6 = 3.458 \cdot 10^6$ bar; $p_c = 0.1600 \cdot 10^6$ bar.

5.2. System pressure: $p = 6.086 \cdot 10^6$ Pa $\triangleq$ 60.86 bar; the estimation p = 50 bar was wrong.

5.3. Better estimation by recursion: p = 60.86 bar. Steps 3.2, 3.3, 4.1, 4.2, 4.3, 5.1 and 5.2 have to be repeated several times.

List of the results of the iteration:

iteration	1	2	3	4	5	6	7	8	9	10
system pressure	60.86	66.85	70.18	72.08	73.18	73.83	74.21	74.44	74.58	74.66

Final result: p = 74.77 bar; $K_a = 6.202$; $K_b = 0.9234$; $K_c = 0.1741$; $y_a = 0.310$; $y_b = 0.646$; $y_c = 0.044$.

Comparison with experimental data (see Fig. 2-11):
$\xi = x_b/x_b + x_c = 0.737$; $K_a = 5.8$; $K_b = 0.96$; $K_c = 0.16$;
$y_a = 0.290$; $y_b = 0.672$; $y_c = 0.04$.

2.6 PRACTICAL DETERMINATION OF PHASE EQUILIBRIA

Reliable phase equilibrium data are essential for the design of distillation processes. Despite the highly advanced state of mixed-phase thermodynamics, good experimental data should be used whenever possible. Several thousand publications on vapor-liquid equilibria are available, which sometimes include complete descriptions of the p, v, T, x behavior of a system. However, the majority of the publications provide equilibrium concentrations only over a certain range of temperature and pressure. Useful data on azeotropic compositions are readily available in books [Horsley 1973, Gmehling et al. 1994]. The latter data collection contains more than 18 800 systems involving approximately 1700 compounds. Numerous vapor-liquid equilibrium data have been published, mainly in the form of data collections [Knapp 1982; Landolt-Boernstein 1962; Perry et al. 1984, Hala et al. 1967, Hirata et al. 1975; Reid et al. 1987; Fredenslund et al. 1977; Gmehling et al. 1977]. Older data collections consist merely of compilations of measured values; however, the newer data have been subjected to a thermodynamic consistency test. In addition, coefficients for several equations of state or for equations for excess enthalpy are presented in [Knapp et al. 1982 and Gmehling et al. 1977] Figure 2-19 shows a page of the Dechema Chemistry Data series [Gmehling et al. 1977].

Data for many binary mixtures can be found in the above collections. Available data for ternary or higher mixtures over the concentration range of interest are seldom adequate. At the very best, data are provided for binary mixtures present in the system. Usually, equilibrium data are far from complete. In all these cases, the methods of thermodynamic representation of phase equilibria described in Sections 2.3 and 2.4 must be used. Equations for excess enthalpy are useful at low pressure and when nonideal behavior occurs mainly in the liquid phase. Equations that use binary interaction parameters to calculate the phase equilibrium of multicomponent mixtures are especially useful. These are primarily equations based on the concept of local composition. Furthermore, the incremental method permits approximation of the interaction parameters from the chemical structure of the substances involved [Gmehling et al. 1982].

In high-pressure systems, the liquid phase, and especially the vapor phase, exhibit nonideal behavior. The convergence pressure method has often been used in the past [Edmister 1961] but now has been superseded by the more effective equations of state. Equations of state based on the virial relationship have the advantage of being theoretically based for mixtures. However, these equations are valid only at reduced pressure, e.g., $p/p_{cr} < 0.5$. Equations of the van der Waals type are preferred at higher pressure. Apart from empirical mixing rules, almost all equations of state contain a further empirical correlating parameter (k_{ij} in Equation 2.60), which greatly influences phase equilibrium.

(1) CHLOROFORM CHCL3

(2) METHANOL CH4O

+++++ ANTOINE CONSTANTS			REGION +++++			CONSISTENCY
(1)	6.95465	1170.966	226.232	-10- 60 C	METHCD 1	+
(2)	8.08097	1582.271	239.726	15- 84 C	METHOD 2	+

PRESSURE= 760.00 MM HG (1.013 BAR)

LIT: NAGATA I.,J.CHEM.ENG.DATA 7,367(1962).

CONSTANTS:	A12	A21	ALPHA12
MARGULES	0.8320	1.7365	
VAN LAAR	0.9356	1.8860	
WILSON	-361.7944	1694.0241	
NRTL	1364.6529	-134.3577	0.2932
UNIQUAC	1265.5380	-274.2024	

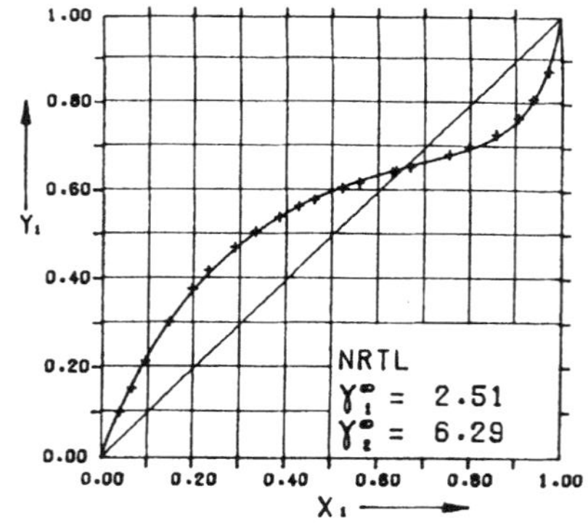

EXPERIMENTAL DATA			MARGULES		VAN LAAR		WILSON		NRTL		UNIQUAC	
T DEG C	X1	Y1	DIFF T	DIFF Y1	DIFF T	DIFF Y1	DIFF T	DIFF Y1	DIFF T	DIFF Y1	DIFF T	DIFF Y1
63.00	0.0400	0.1020	0.01	0.0045	0.20	-0.0014	0.30	-0.0048	0.17	-0.0003	0.22	-0.0022
62.00	0.0650	0.1540	-0.05	0.0008	0.17	-0.0050	0.30	-0.0088	0.13	-0.0037	0.20	-0.0058
60.90	0.0950	0.2150	-0.07	0.0005	0.14	-0.0031	0.27	-0.0064	0.09	-0.0019	0.16	-0.0038
59.30	0.1460	0.3040	0.0	-0.0011	0.10	0.0013	0.21	0.0002	0.07	0.0018	0.12	0.0009
57.80	0.1960	0.3780	-0.09	-0.0003	-0.13	0.0076	-0.07	0.0092	-0.14	0.0073	-0.11	0.0077
57.00	0.2300	0.4200	-0.06	-0.0003	-0.19	0.0105	-0.17	0.0138	-0.19	0.0097	-0.19	0.0109
55.90	0.2870	0.4720	-0.02	-0.0065	-0.26	0.0070	-0.31	0.0126	-0.23	0.0055	-0.27	0.0079
55.30	0.3320	0.5070	0.09	-0.0083	-0.21	0.0055	-0.30	0.0123	-0.17	0.0036	-0.23	0.0066
54.70	0.3830	0.5400	0.09	-0.0091	-0.22	0.0033	-0.35	0.0108	-0.17	0.0012	-0.25	0.0045
54.30	0.4250	0.5640	0.06	-0.0076	-0.25	0.0025	-0.40	0.0100	-0.18	0.0005	-0.28	0.0038
54.00	0.4590	0.5800	-0.01	-0.0069	-0.31	0.0007	-0.46	0.0078	-0.23	-0.0012	-0.34	0.0019
53.80	0.5200	0.6070	0.05	-0.0023	-0.20	0.0001	-0.36	0.0057	-0.12	-0.0013	-0.23	0.0010
53.70	0.5570	0.6190	0.05	-0.0014	-0.17	-0.0024	-0.33	0.0018	-0.10	-0.0033	-0.20	-0.0018
53.50	0.6280	0.6430	-0.07	0.0045	-0.26	-0.0026	-0.41	-0.0020	-0.18	-0.0026	-0.28	-0.0029
53.50	0.6360	0.6460	-0.07	0.0056	-0.25	-0.0021	-0.40	-0.0019	-0.18	-0.0019	-0.28	-0.0026
53.50	0.6670	0.6550	-0.08	0.0074	-0.25	-0.0025	-0.40	-0.0042	-0.18	-0.0019	-0.27	-0.0035
53.70	0.7530	0.6840	-0.02	0.0140	-0.18	0.0006	-0.34	-0.0063	-0.11	0.0024	-0.21	-0.0018
53.90	0.7970	0.7010	-0.02	0.0151	-0.15	0.0025	-0.31	-0.0065	-0.08	0.0046	-0.18	-0.0005
54.40	0.8550	0.7300	-0.06	0.0125	-0.08	0.0043	-0.23	-0.0053	-0.02	0.0064	-0.10	0.0015
55.20	0.9040	0.7680	-0.21	0.0056	-0.06	0.0047	-0.12	-0.0017	-0.03	0.0058	-0.05	0.0031
56.30	0.9370	0.8120	-0.23	0.0014	0.05	0.0064	0.11	0.0049	0.02	0.0062	0.08	0.0063
57.90	0.9700	0.8750	-0.48	-0.0110	-0.16	-0.0021	0.03	0.0022	-0.22	-0.0040	-0.11	-0.0009
MEAN DEVIATION:			0.09	0.0058	0.18	0.0036	0.28	0.0063	0.14	0.0035	0.20	0.0037
MAX. DEVIATION:			0.48	0.0151	0.31	0.0105	0.46	0.0138	0.23	0.0097	0.34	0.0109

Fig. 2-19. *Page of the Dechema Chemistry Data Series [Gmehling et al. 1977].*

Equations of state are preferred to equations of excess enthalpy, because the former permit calculation not only of phase equilibria but also of other values of state, e.g., density and enthalpy. Empirical mixing rules are not completely satisfactory because they are usually applied to both phases in the same way; this does not correspond fully with real conditions in multicomponent mixtures. The vapor-liquid equilibria cannot yet be reliably predicted. Progress in mixed-phase thermodynamics permits only a thermodynamically consistent representation of equilibria based on relatively few experimental data.

LITERATURE

Asselineau, L.; Bogdanic, G.; and Vidal, J.: "A Versatile Algorithm for Calculating Vapour-Liquid Equilibria," *Fluid Phase Equilibria* 3 (1979). No. 1, pp. 273/290.

Edmister, W.C.: *Applied Hydrocarbon Thermodynamics.* Gulf Publ. Co., Houston, 1961.

Foerg, W. and Stichlmair, J.: "Die Gewinnung von rein Wasserstoff aus Gemischen mit leichten aliphatischen Kohlenwasserstoffen mit Hilfe tierfer Temperaturen," *Kältetechnik—Klimatisierung* 21 (1969) No. 4, pp. 104–109.

Fredenslund, A.; Gmehling, J.; and Rasmussen, P.: *Vapor-Liquid Equilibria Using UNIFAC.* Elsevier, Amsterdam 1977.

Gas Processors Suppliers Association: *Engineering Data Book,* Tulsa, Oklahoma, 1972.

Gmehling, J.; Onken, U.; and Arlt, W.: *Vapor-Liquid Equilibrium Data Collection.* Dechema Chemistry Data Series, Frankfurt/Main, 1977 (continuing series with more than 40 volumes).

Gmehling, J.; Rasmussen, P.; and Fredenslund, A.: "Vapor-Liquid Equilibria by UNIFAC Group Contribution, Revision and Extension 2," *Ind. Eng. Chem. Process Des. Dev.* 21 (1982) no. 1, pp. 118/127.

Gmehling, J. and Kolbe, B.: *Thermodynamik.* VCH Publishers, Weinheim, New York, 1992.

Gmehling, J.; Menke, J.; Krafczyk, J.; and Fischer, K.: *Azeotropic Data, Part I and Part II.* VCH-Publishers, Weinheim, New York, 1994.

Hala, E.; Pick, J.; and Vilim, V.: *Vapor-Liquid Equilibrium.* Pergamon Press, Oxford, 1967.

Hirata, M.; Ohe, S.; and Nagahama, K.: *Computer Aided Data Book of Vapor-Liquid Equilibria.* Kodanska/Elsevier, Tokyo, 1975.

Horsley, L. H.: *Azeotropic Data III.* American Chemical Society, Washington, D.C., 1973.

Knapp, H.; Doering, R.; Oelbrich, L.; Ploecker, V.; and Prausnitz, J. M.: "Vapor-Liquid Equilibria for Mixtures of Low Boiling Substances," *Chemistry Data Series Vol. VI,* Dechema, Frankfurt/Main, 1982.

Landolt-Boernstein: *Zahlenwerte und Funktionen, Loesungsgleichgewichte I, Vol. II2b.* Springer Verlag, Berlin, 1962.

Mersmann, A.: *Thermische Verfahrenstechnik.* Springer Verlag, Berlin, 1980.

Nagata, I. in Gmehling, J.; Onken, U.; and Arlt, W.: *Vapor-Liquid Equilibrium Data Collection.* Dechema Chemistry Data Series, Frankfurt/Main, 1977, Vol. I, Part 2a, p. 19.

Perry, R. H.; Green, D. W.; and Maloney, J. O.: *Perry's Chemical Engineer Handbook*. McGraw-Hill, New York, 1984.

Prausnitz, J. M.: *Molecular Thermodynamics of Fluid-Phase Equilibria*. Prentice-Hall, Englewood Cliffs, N.J., 1969.

Prausnitz, J. M.; Anderson, T. F.; Grens, E. A.; Eckert, C. A.; R. Hsieh, R.; and O'Connel, J. P.: *Computer Calculations for Multicomponent Vapor-Liquid and Liquid-Liquid Equilibria*. Prentice-Hall, Englewood Cliffs, N. J., 1980.

Reid, R. C.; Prausnitz, J. M.; and Poling, B. E.: *The Properties of Gases and Liquids*. McGraw-Hill, New York, 1987.

Soave, G.: "Equilibrium Constants from a Modified Redlich-Kwong Equation of State," *Chem. Eng. Sci.* 27 (1972) No. 6, pp. 1197/1203.

Wagner, W.: "Berechnung von Hochdruckphasengleichgewichten azeotroper Gemische." Dissertation, Universitaet Stuttgart, 1985.

Weast, R. C.; Astle, M. J.; and Beyer, W. H.: *CRC Handbook of Chemistry and Physics*. CRC Press, Boca Raton, 1986.

Wilson, G. M.: "Vapor-Liquid Equilibrium—A new expression for the excess free energy of mixing," *Journal of the American Chemical Society* 86 (1964) pp. 127–130.

3

Single-Stage Distillation and Condensation

3.1 Continuous Closed Distillation and Condensation
  3.1.1 Closed Distillation of Binary Mixtures
  3.1.2 Closed Distillation of Multicomponent Mixtures (Flash)
    3.1.2.1 Relative Amount of Vapor in a Vapor-Liquid Mixture
    3.1.2.2 Partial Evaporation of a Multicomponent Liquid Mixture
    3.1.2.3 Adiabatic Flash
  3.1.3 Application of Flash Equations to a Separation Process
3.2 Batchwise Open Distillation and Open Condensation
  3.2.1 Binary Mixtures
    3.2.1.1 Open Distillation (Batch Distillation)
    3.2.1.2 Open Condensation
  3.2.2 Ternary Mixtures
    3.2.2.1 Open Distillation
    3.2.2.2 Open Condensation
  3.2.3 Multicomponent Mixtures
    3.2.3.1 Open Distillation
    3.2.3.2 Open Condensation
3.3 Semicontinuous Single-Stage Distillation
  3.3.1 Semicontinuous Single-Stage Distillation of Binary Mixtures
Literature

As discussed in Chapter 2, the composition of the vapor usually differs from that of the liquid from which it originates. This concentration difference is the basis of distillation, which is a simple method of separating liquid mixtures. The method involves partial evaporation of the liquid followed by total condensation of the vapor. The resulting condensate is enriched in the more volatile components, in accordance

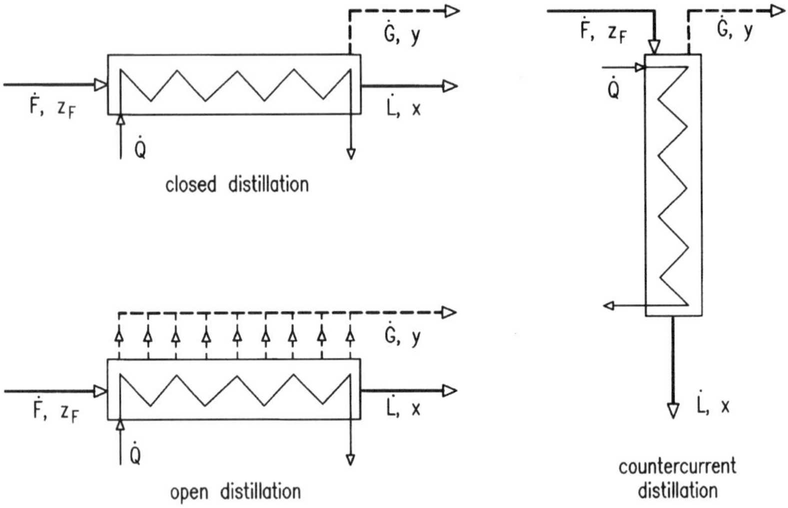

Fig. 3-1. Modes of single-stage distillation.

with the vapor liquid equilibrium for the system at hand. Thus, the starting mixture or feed is separated into two fractions with different compositions.

There exist three different modes of single-stage distillation, see Fig. 3-1. The most important one is the cocurrent or closed distillation. Here, the leaving gas stream is in equilibrium with the leaving liquid stream. In the second mode, called

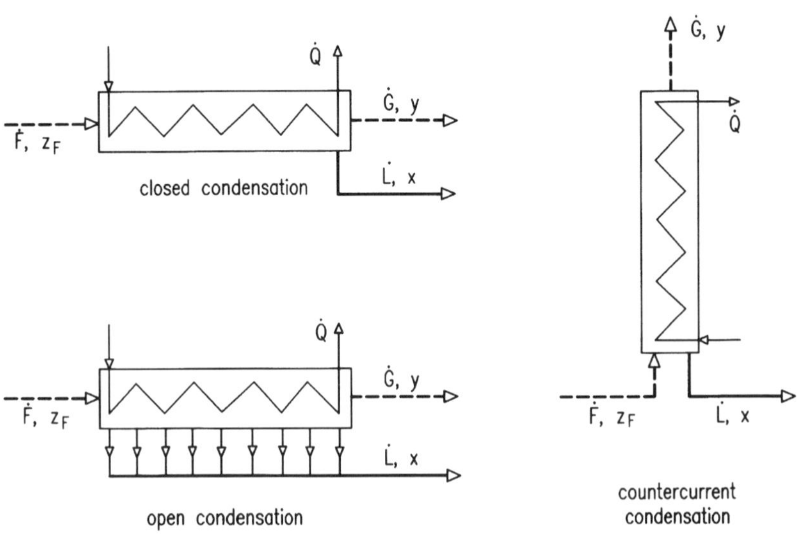

Fig. 3-2. Modes of single-stage partial condensation.

open distillation, the gas is removed from the liquid immediately after its generation. This mode of operation is simplest performed by batch distillation. The third mode is the countercurrent distillation. The leaving gas stream is supposed to be in equilibrium state with the liquid feed stream. Countercurrent distillation (in a boiler) is very seldom applied to industrial processes. Direct contact beween vapor and liquid in a column is preferred here (see Chapter 4).

All three modes of distillation have analoga in condensation, see Fig. 3-2. All the equations derived for distillation are after small modifications valid for partial condensation, too.

3.1 CONTINUOUS CLOSED DISTILLATION AND CONDENSATION

Figure 3-3 presents the schematic of continuous closed distillation of a liquid mixture. The feed (molar flow rate $\dot{F}$) is separated into an overhead product or distillate (molar flow rate $\dot{D}$) and a bottom product (molar flow rate $\dot{B}$). The composition difference between these two products depends on the vapor-liquid equilibrium.

3.1.1 Closed Stage Distillation of Binary Mixtures

A material balance around the boiler (see Fig. 3-3) yields:

$$\dot{F} = \dot{G} + \dot{L} \tag{3.1}$$

where $\dot{F}$, $\dot{G}$, and $\dot{L}$ are the molar flow rates of the feed, vapor and bottom liquid,

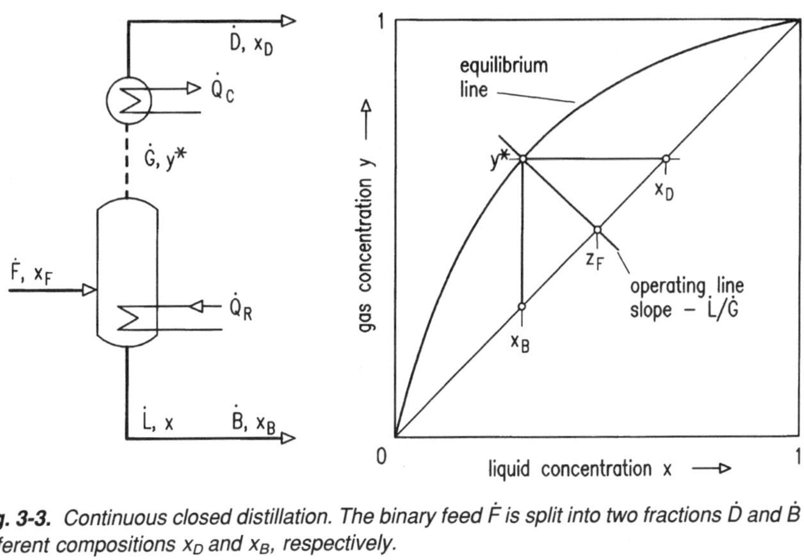

Fig. 3-3. Continuous closed distillation. The binary feed $\dot{F}$ is split into two fractions $\dot{D}$ and $\dot{B}$ with different compositions x_D and x_B, respectively.

respectively. Similarly, a material balance for the more volatile component *a* of a binary mixture *a*-*b* gives:

$$\dot{F}\, z_F = \dot{G}\, y + \dot{L}\, x \tag{3.2}$$

where z_F is the mole fraction of *a* in the feed, y is the mole fraction of *a* in the vapor, and x is the mole fraction of *a* in the bottom liquid. From both equation follows:

$$y = -\dot{L}/\dot{G}\, x + (1 + \dot{L}/\dot{G})\, z_F \tag{3.3}$$

This is the equation for a straight line with slope $-\dot{L}/\dot{G}$. When plotted on the $y = x$ diagram, it furnishes the operating line. The operating line intersects the diagonal at $x = z_F$ (Fig. 3-3). With the closely-met assumption that equilibrium exists between the vapor $\dot{G}$ and the liquid $\dot{L}$ in the boiler, (i.e., $y = y^*$), the following equation holds for ideal mixtures (see Section 2.3.2):

$$y^* = \frac{\alpha \cdot x}{1 + (\alpha - 1) \cdot x} \tag{3.4}$$

where α is the relative volatility. The concentrations x and y are determined by equating relationships (3.3) and (3.4) to obtain a quadratic equation which can be solved easily. With $\dot{L}/\dot{G} = \dot{F}/\dot{G} - 1$ we get:

$$x = \frac{1}{2 \cdot (1 - \dot{G}/\dot{F})} \cdot \Big[-(1 + (\alpha - 1) \cdot (\dot{G}/\dot{F} - z_F)) + \sqrt{(1 + (\alpha - 1) \cdot (\dot{G}/\dot{F} - z_F))^2 + 4 \cdot (1 - \dot{G}/\dot{F}) \cdot z_F} \Big] \tag{3.5}$$

The corresponding vapor or distillate concentration follows from Equation (3.4).

In most cases, a graphical solution based on the $y - x$ diagram is preferred to Equation (3.5), because it is a simple method that can also be easily applied to nonideal mixtures. The desired concentrations of vapor y and liquid x are determined from the point of intersection of the operating line with the equilibrium line. Subsequent total condensation of the vapor is not accompanied by a change in concentration; i.e., the distillate concentration x_D is the same as the vapor concentration y. This is represented graphically in the $y - x$ diagram by the intercept x_D of the horizontal line $y^* =$ constant with the diagonal. Figure 3-3 clearly shows the concentration difference between the distillate x_D and the bottoms x_B. The decisive factor here is the course of the equilibrium curve, i.e., for ideal systems, the relative volatility α. Generally, only a limited difference between the concentrations of the two fractions can be obtained by single-stage distillation.

3.1.2 Closed Distillation of Multicomponent Mixtures (Flash)

Boiling point and dew point temperatures are identical for a single-component system. The vapor pressure line describes both states. In a multicomponent mixture the

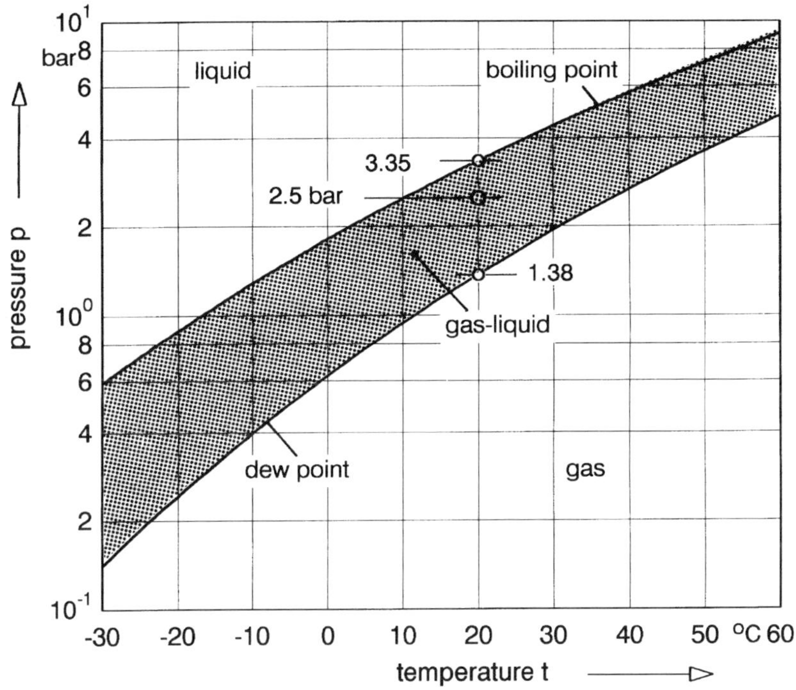

Fig. 3-4. Dew point and boiling point lines of the mixture propane/i-butane/n-butane/pentane. A two-phase vapor-liquid mixture exists in the shaded area.

vapor pressure line splits into a boiling point line and a dew point line with an intermediate two-phase region, see Fig. 3-4.

The boiling point line of a mixture is calculated by Equation (2.12) derived in Section 2.1.

$$p' = \sum z_{Fi} \cdot \gamma_i \cdot p_i^o \quad \text{with } p_i^o = f(t) \tag{3.6}$$

The dew point line of a mixture follows from Equation (2.13):

$$p'' = \frac{1}{\sum z_{Fi}/(\gamma_i \cdot p_i^o)} \tag{3.7}$$

The intermediate two-phase region is characterized by the relative content of vapor $\dot{G}/\dot{F}$. Under the assumption that gas and liquid are in equilibrium state, the following set of equations holds:

$$\text{Material:} \quad \dot{F} = \dot{G} + \dot{L} \qquad \dot{F} \cdot z_{Fi} = \dot{G} \cdot y_i + \dot{L} \cdot x_i \tag{3.8}$$

$$\text{Equilibrium:} \quad y_i = K_i \cdot x_i \quad \text{with } K_i = x_i \cdot \gamma_i \cdot p_i^o/p \tag{3.9}$$

$$\text{Energy:} \quad \dot{F} \cdot h_F + \dot{Q} = \dot{G} \cdot h_G + \dot{L} \cdot h_L$$

$$\text{with } h_L = c_{Lm} \cdot t \text{ and } h_G = c_{Lm} \cdot t + \sum (y_i \, r_i) \tag{3.10}$$

Here, the latent heat of vaporization r_i depends heavily on temperature, see Fig. 3-5.

Combining Equations (3.8) and (3.9) yields:

$$y_i = \frac{z_{Fi} \cdot K_i}{1 + \dot{G}/\dot{F} \cdot (K_i - 1)} \tag{3.11}$$

With the molefraction constraint $\sum y_i = 1$:

$$\sum \frac{z_{Fi} \cdot K_i}{1 + \dot{G}/\dot{F} \cdot (K_i - 1)} = 1 \tag{3.12}$$

Equation (3.12) is used to determine the relative gas content $\dot{G}/\dot{F}$ in the two-phase

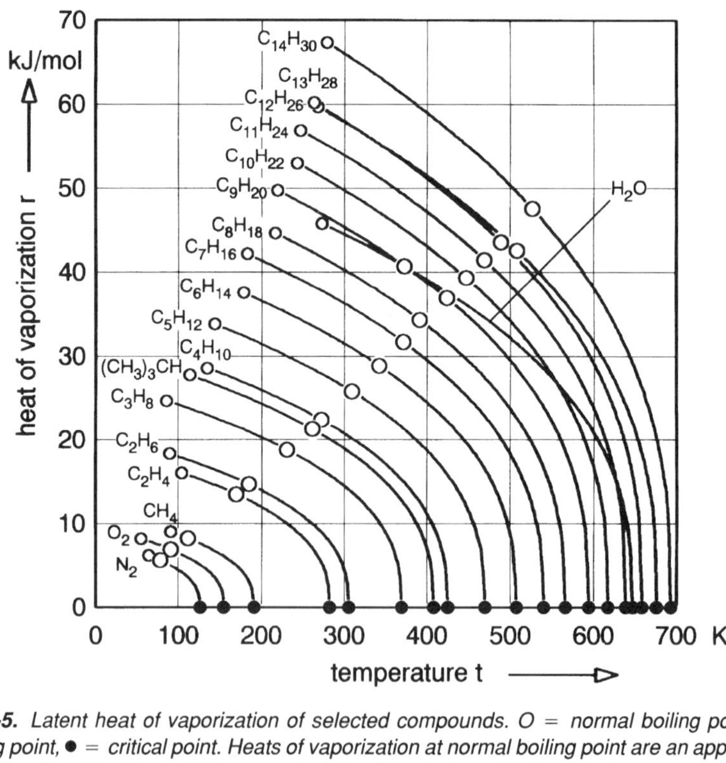

Fig. 3-5. *Latent heat of vaporization of selected compounds. O = normal boiling point, o = melting point, ● = critical point. Heats of vaporization at normal boiling point are an approximate linear function of boiling point temperature T_{BP} (Trouton's rule). Temperature dependence of r_i is described by: $r_i = r_{i,BP} \cdot ((T_{cr,i} - T)/(T_{cr,i} - T_{BP}))^{0.38}$*

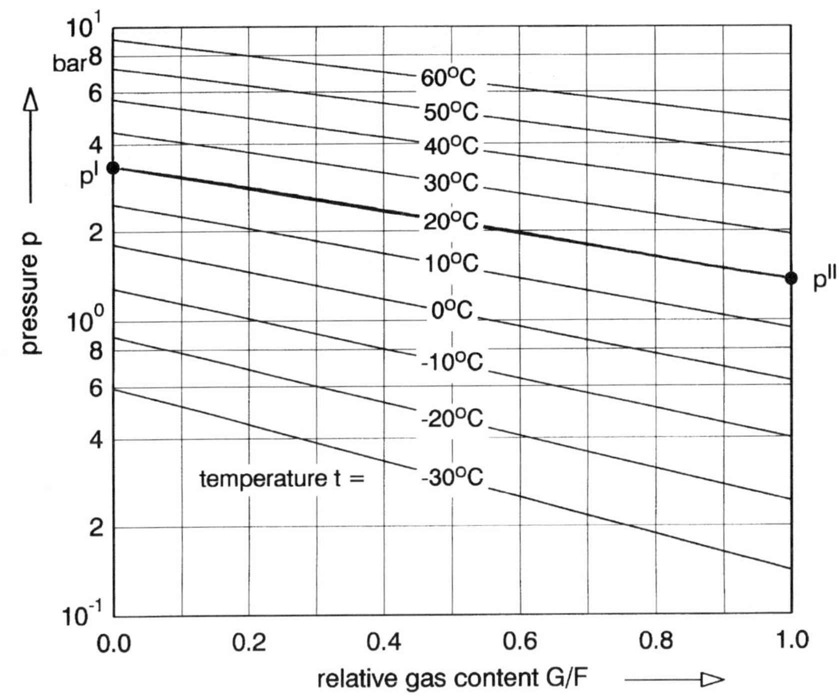

Fig. 3-6. Relation between relative gas content, $\dot{G}/\dot{F}$, and total pressure, p, for the system propane/i-butane/n-butane/pentane. The logarithm of the pressure, lnp, is approximately a linear function of the relative gas content $\dot{G}/\dot{F}$. The slightly curved bold line is determined by evaluating Equation (3.12) for t = 20°C. The thin lines are linear interpolations between boiling point pressure p′ and dew point pressure p″ at constant temperatures.

region when total composition z_{Fi}, pressure p and temperature t are known. Figure 3-6 displays the Equation (3.12) for an ideal quaternary mixture at a temperature of 20°C. The value at $\dot{G}/\dot{F} = 0$ is the boiling point pressure p′, the value at $\dot{G}/\dot{F} = 1$ is the dew point pressure p″. In the two-phase region, the logarithm of the pressure, lnp, is approximately a linear function of the relative vapor content $\dot{G}/\dot{F}$.

The solution procedure for the flash Equations (3.10) and (3.12) depends on which parameters are the unknowns. Some possibilities are treated in the next chapters.

3.1.2.1 Relative Amount of Vapor in a Vapor-Liquid Mixture

When the total concentration z_{Fi} of the mixture, the pressure p and the temperature t are set, the relative amount of vapor $\dot{G}/\dot{F}$ follows from:

$$\sum \frac{z_{Fi} \cdot K_i}{1 + \dot{G}/\dot{F} \cdot (K_i - 1)} = 1 \tag{3.13}$$

or with $K_i = \gamma_i \cdot p_i^o/p_t$, i.e., ideal behavior in the vapor and nonideal behavior in the liquid phase:

$$\sum \frac{z_{Fi} \cdot \gamma_i \cdot p_i^o}{p_t \cdot (1 - \dot{G}/\dot{F}) + \dot{G}/\dot{F} \cdot \gamma_i \cdot p_i^o} = 1 \tag{3.14}$$

For constant temperature, the logarithm of the total pressure, lnp, is approximately a linear function of the relative gas content $\dot{G}/\dot{F}$ of the mixture. Hence, a first estimation of $\dot{G}/\dot{F}$ can be determined by:

$$(\dot{G}/\dot{F})^o \approx \frac{\ln(p/p')}{\ln(p''/p')} \tag{3.15}$$

The exact solution of Equation (3.13) is best performed by the Newton algorithm.

$$(\dot{G}/\dot{F})^1 = (\dot{G}/\dot{F})^o - \frac{f((\dot{G}/\dot{F})^o)}{f'((\dot{G}/\dot{F})^o)} \tag{3.16}$$

with

$$f(\dot{G}/\dot{F}) = \sum \frac{z_{Fi} \cdot K_i}{1 + \dot{G}/\dot{F} \cdot (K_i - 1)} - 1 = 0$$

and

$$f'(\dot{G}/\dot{F}) = -\sum \frac{z_{Fi} \cdot K_i \cdot (K_i - 1)}{(1 + \dot{G}/\dot{F} \cdot (K_i - 1))^2}$$

Example 3-1 demonstrates the application of the calculation procedure to a quaternary mixture.

EXAMPLE 3-1: STATE OF A MULTICOMPONENT VAPOR-LIQUID MIXTURE

Find the state of the mixture propane (=*a*), i-butane (=*b*), n-butane (=*c*) and pentane (=*d*) at a pressure p = 2.5 bar and a temperature t = 20°C. The total concentrations are $z_{Fa} = 0.25$, $z_{Fb} = 0.15$, $z_{Fc} = 0.32$, $z_{Fd} = 0.28$.

Solution:

1. Vapor pressure of pure components from Antoine equation:

$$\frac{p_i^o}{\text{bar}} = \exp\left(A_i - \frac{B_i}{t/K + C_i}\right) \quad \text{with } A_i, B_i, C_i \text{ from Table 2.1;}$$

$p_a^o = 8.31$ bar, $p_b^o = 3.01$ bar, $p_c^o = 2.07$ bar, $p_d^o = 0.57$ bar.
Equilibrium ratios: $K_a = 3.33$; $K_b = 1.20$; $K_c = 0.83$; $K_d = 0.23$.

2. Boiling point pressure of the mixture (Equation 3.6):
$p' = 0.25 \cdot 8.31 + 0.15 \cdot 3.01 + 0.32 \cdot 2.07 + 0.28 \cdot 0.57 = 3.35$ bar.
3. Dew point pressure of the mixture (Equation 3.7):
$1/p'' = 0.25/8.31 + 0.15/3.01 + 0.32/2.07 + 0.28/0.57 = 0.725$; $p'' = 1.38$ bar.
The pressure of the system is between boiling point and dew point pressure → two phases exist. See Fig. 3-6.
4. Relative content of vapor $\dot{G}/\dot{F}$—Estimation by Equation (3.15):

$$\dot{G}/\dot{F} \approx \frac{\ln(2.5/3.35)}{\ln(1.38/3.35)} \approx 0.33.$$

Evaluation of Equation (3.13):

$$\frac{0.25 \cdot 3.33}{1 + 0.33 \cdot (3.33 - 1)} + \frac{0.15 \cdot 1.20}{1 + 0.33 \cdot (1.20 - 1)} + \frac{0.32 \cdot 0.83}{1 + 0.33 \cdot (0.83 - 1)} + \frac{0.28 \cdot 0.23}{1 + 0.33 \cdot (0.23 - 1)}$$
$$= 0.471 + 0.169 + 0.281 + 0.086 = 1.007 \neq 1.$$

5. Better value for $\dot{G}/\dot{F}$ by Newton algorithm (Equation 3.16):

$$f'(\dot{G}/\dot{F}) = -\left(\frac{0.25 \cdot 3.33 \cdot (3.33 - 1)}{(1 + 0.33 \cdot (3.33 - 1))^2} + \frac{0.15 \cdot 1.20 \cdot (1.20 - 1)}{(1 + 0.33 \cdot (1.20 - 1))^2} + \frac{0.32 \cdot 0.83 \cdot (0.83 - 1)}{(1 + 0.33 \cdot (0.83 - 1))^2} + \frac{0.28 \cdot 0.23 \cdot (0.23 - 1)}{(1 + 0.33 \cdot (0.23 - 1))^2}\right)$$
$$= -(0.6199 + 0.0317 - 0.0508 - 0.0891) = -0.5117;$$

$(\dot{G}/\dot{F})^1 = 0.33 - 0.0007/-0.5117 = 0.3437 \approx 0.345$. Equation (3.13):

$$\frac{0.25 \cdot 3.33}{1 + 0.345 \cdot (3.33 - 1)} + \frac{0.15 \cdot 1.20}{1 + 0.345 \cdot (1.20 - 1)} + \frac{0.32 \cdot 0.83}{1 + 0.345 \cdot (0.83 - 1)} + \frac{0.28 \cdot 0.23}{1 + 0.345 \cdot (0.23 - 1)}$$
$$= 0.4615 + 0.1684 + 0.2821 + 0.0877 = 0.9997 \text{ o.k.}$$

Concentrations of the vapor are given by the additive terms of the previous equation: $y_a = 0.4615$: $y_b = 0.1684$; $y_c = 0.2821$; $y_d = 0.0877$.
Concentration of the liquid with $x_i = y_i/K_i$: $x_a = 0.1386$; $x_b = 0.1403$; $x_c = 0.3399$; $x_d = 0.3813$.

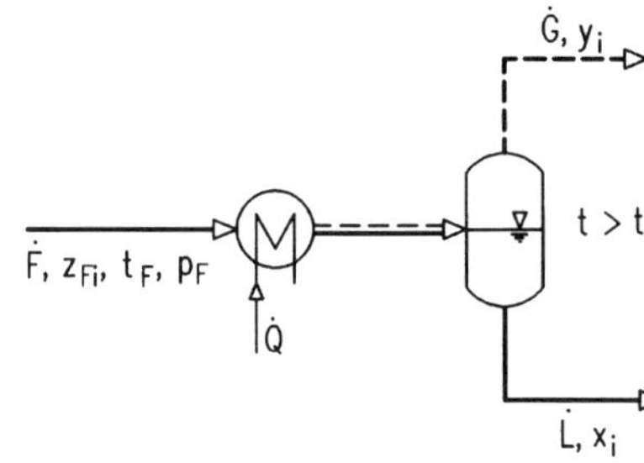

Fig. 3-7. Schematic of closed distillation of a multicomponent liquid mixture.

3.1.2.2 Partial Evaporation of a Multicomponent Liquid Mixture The principal process of single-stage distillation is the partial evaporation of a liquid mixture, see Fig. 3-7. For a multicomponent system the relevant equations are:

$$\sum \frac{z_{Fi} \cdot K_i}{1 + \dot{G}/\dot{F} \cdot (K_i - 1)} = 1 \tag{3.17}$$

and

$$\dot{Q}/\dot{F} = c_{Lm} \cdot (t - t_F) + \dot{G}/\dot{F} \cdot \sum (y_i r_i) \tag{3.18}$$

Often, the relative amount of distillate $\dot{D}/\dot{F}$ $(=\dot{G}/\dot{F})$ is given. A first approximation of the temperature in the still is taken from a graph analogous to Fig. 3-6. Then Equation (3.17) is solved iteratively giving the value $\dot{G}/\dot{F}$ for this temperature. If $\dot{G}/\dot{F}$ is too far from the desired amount of distillate, the procedure has to be repeated with a better estimation of the temperature. The procedure of solution is explained in detail in Example 3-2.

EXAMPLE 3-2: DISTILLATION OF A QUATERNARY IDEAL MIXTURE:

A mixture of propane ($x_{Fa} = 0.25$), i-butane ($x_{Fb} = 0.15$), n-butane ($x_{Fc} = 0.32$) and pentane ($x_{Fd} = 0.28$) is distilled at a pressure of 4 bar. The relative amount of distillate, $\dot{D}/\dot{F}$, is approximately 0.6. Find the temperature in the evaporator, the concentrations of distillate and the heat required.

Solution:

1. Estimation of the temperature in the evaporator—From Fig. 3-6: $t \approx 43°C$.
2. Vapor pressures and equilibrium ratios—From Antoine equation (see Table 2.1) follows for $t = 43°C$: $p_a^o = 14.43$ bar; $p_b^o = 5.72$ bar; $p_c^o = 4.09$ bar; $p_d^o = 1.28$ bar; $K_a = p_a^o/p = 3.61$; $K_b = 1.43$; $K_c = 1.02$; $K_d = 0.319$.

3. Determination of $\dot{G}/\dot{F}$ at 43°C with Equation (3.17):

$$\sum \frac{z_{Fi} \cdot K_i}{1 + \dot{G}/\dot{F} \cdot (K_i - 1)} = 1;$$

0.3517 + 0.1705 + 0.3229 + 0.1511 = 0.9963 ≠ 1. Iteration with Newton (see Example 3-1) yields: $\dot{G}/\dot{F} = 0.585$;

$$\sum \frac{z_{Fi} \cdot K_i}{1 + \dot{G}/\dot{F} \cdot (K_i - 1)} = 0.3571 + 0.1714 + 0.3231 + 0.1486$$
$$= 1.0001 \text{ o.k.}$$

4. Concentration of distillate (given by the additive terms of the previous equation) and bottoms: $x_{Da} = 0.3571$; $x_{Db} = 0.1714$; $x_{Dc} = 0.3231$; $x_{Dd} = 0.1486$. Liquid concentration from $x_i = y_i/K_i$: $x_a = 0.3571/3.61 = 0.0989$; $x_b = 0.1714/1.43 = 0.1198$; $x_d = 0.3231/1.02 = 0.3157$; $x_d = 0.1486/0.319 = 0.4652$.
5. Heat requirement from Equation (3.18): $\dot{Q}/\dot{F} = c_{Lm} \cdot (t - t_F) + \dot{G}/\dot{F} \cdot \sum y_i \cdot r_i(t)$. Latent heat of vaporization at 43°C from Fig. 3-5: $r_a = 13.1$ kJ/mol; $r_b = 17.7$ kJ/mol; $r_c = 21.0$ kJ/mol; $r_d = 26.1$ kJ/mol. Average heat capacity of liquid: $c_{Lm} \approx 0.13$ kJ/(mol°C); $r_m = 0.3571 \cdot 13.1 + 0.1714 \cdot 17.7 + 0.3231 \cdot 21.0 + 0.1486 \times 26.1 = 18.4$ kJ/mol; $\dot{Q}/\dot{F} = 0.13 \cdot (43 - 20) + 0.585 \cdot 18.4 = 13.75$ kJ/mol.

3.1.2.3 *Adiabatic Flash* Pressure reduction in a nozzle causes partial evaporation of a liquid near boiling point, see Fig. 3-8. Since the enthalpy remains constant in a nozzle, the relevant set of equations for an adiabatic flash is:

$$\sum \frac{z_{Fi} \cdot K_i}{1 + \dot{G}/\dot{F} \cdot (K_i - 1)} = 1 \qquad (3.19)$$

and

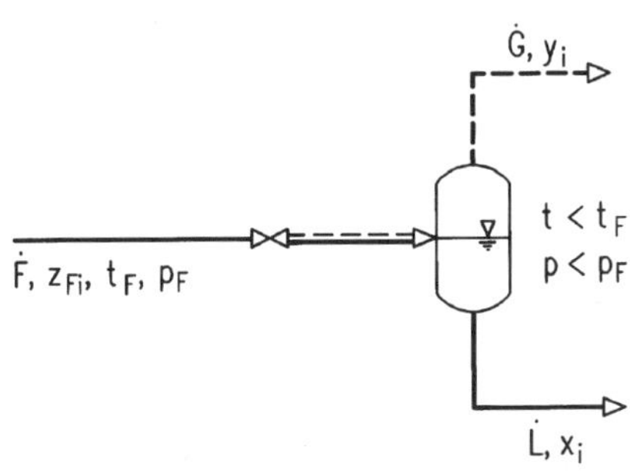

Fig. 3-8. Schematic of an adiabatic flash. Some vapor is generated by expanding a hot liquid in a nozzle and, hence, temperature decreases.

Table 3-1. Carbon Dioxide Outburst at Lake Nyos, Camerun, on August 21, 1986.

Lake Nyos, Camerun, is a 200 m deep volcanic crater filled with water today (see Fig. 3-9). Carbon dioxide leaks from the volcanic rock into the water and accumulates in the lowest layer to a relatively high CO_2 content (up to 2 mol.-%).

On August 21, 1986 the instable layers in the lake suddenly turned over for any reason (e.g., falling rock, light earthquake). As the CO_2-rich water from the ground approached the surface some CO_2 degassed due to the pressure loss (= adiabatic flash). The bubbles enhanced the upwards flow of the water and, in turn, more and more CO_2 degassed. A carbon dioxide cloud with a density higher than air emerged from the lake and floated down the hill in approximately a 20 m thick layer. At Nyos and a neighbor village, 1700 people and all their livestock suffocated on this night (National Geographic Society, No. 3, 1987).

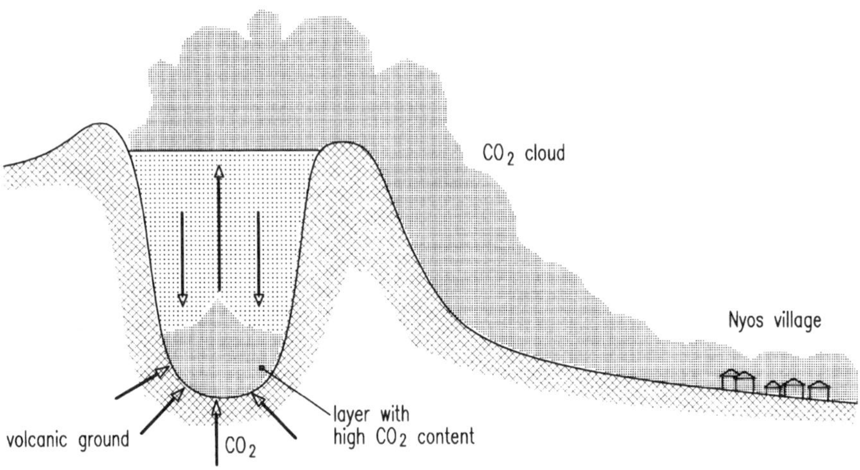

Fig. 3-9. Carbon dioxide outburst at Lake Nyos on August 21, 1986. See Table 3-1.

$$t = t_F - \frac{1}{c_{Lm}} \cdot \dot{G}/\dot{F} \cdot \sum (y_i \cdot r_i(t)) \tag{3.20}$$

Since the vapor has a higher heat content than the liquid, the temperature of the system falls.

The solution is easiest found with the help of a graph analogous to Fig. 3-6. First, the temperature t after the nozzle is estimated, and the relative amount of vapor, $\dot{G}/\dot{F}$, is approximately determined from the graph. With Equation (3.20) the estimated temperature is checked using a rough approximation for the heat of evaporation $r_m = \sum (z_{Fi} \cdot r_i)$. The exact solution is then found by the Newton iteration procedure as outlined in Example 3-3.

Adiabatic flash caused by pressure drop is a very essential element of many industrial separation processes, e.g., desalination of sea water (Chapter 7). However, it is also encountered in nature. A very impressive example is the carbon/dioxide outburst from Lake Nyos described in Table 3-1. Divers' sickness caused by degassing air from blood is another important example of a natural adiabatic flash.

EXAMPLE 3-3: ADIABATIC FLASH OF AN IDEAL QUATERNARY MIXTURE

A liquid mixture of propane (x_{Fa} = 0.25), i-butane (x_{Fb} = 0.15), n-butane (x_{Fc} = 0.32) and pentane (x_{Fd} = 0.28) at a temperature t_F = 20°C is expanded from 3.35 bar (= boiling pressure) to 1.2 bar in a nozzle. Find the end temperature t, the relative vapor content $\dot{G}/\dot{F}$ and the concentrations of vapor and liquid after the nozzle.

Heats of vaporization at normal boiling temperature t_{BP}: $r_{a,BP}$ = 18.8 kJ/mol; $r_{b,BP}$ = 21.3 kJ/mol; $r_{c,BP}$ = 22.4 kJ/mol; $r_{d,BP}$ = 25.8 kJ/mol

Critical temperature: $T_{cr,a}$ = 369.8 K; $T_{cr,b}$ = 408.1 K; $T_{cr,c}$ = 425.2 K; $T_{cr,d}$ = 469.4 K

Solution:

Equation (3.20): $t = t_F - 1/c_{L,m} \cdot \dot{G}/\dot{F} \cdot \Sigma\,(y_i \cdot r_i(t))$.

1. Estimation of end temperature t: t = 0°C. Heat capacity of the liquid $c_{L,m} = \Sigma\, x_{Fi} \cdot c_{Li}$ = 0.13145 kJ/mol · K. Data from Perry.

   Vapor pressures from Antoine equation: p_a^o = 4.73 bar; p_b^o = 1.56 bar; p_c^o = 1.032 bar; p_d^o = 0.245 bar. Boiling point pressure (see Example 3-1): p′ = 1.8153 bar. Dew point pressure: p″ = 0.6232 bar. Relative amount of vapor $\dot{G}/\dot{F}$ (see Example 3-1, Fig. 3-6): $\dot{G}/\dot{F} \approx 0.3872$. Heats of vaporization at 0°C from

$$r_i = r_{i,BP} \cdot \left(\frac{T_{cr,i} - T}{T_{cr,i} - T_{BP}}\right)^{0.38};$$

   r_a = 16.3887 kJ/mol; r_b = 20.6349 kJ/mol; r_c = 22.3749 kJ/mol; r_d = 27.8686 kJ/mol; $r_m \approx \Sigma\, z_{Fi} \cdot r_i$ = 22.1556 kJ/mol.

   Verification of the estimated temperature: t = 20 − 1/0.13145 · 0.3872 · 22.1556 = −45.261°C; the estimation was wrong.

2. Better estimation t = − 10°C. Vapor pressures: p_a^o = 3.445 bar; p_b^o = 1.076 bar; p_c^o = 0.696 bar; p_d^o = 0.152 bar. Boiling point pressure: p′ = 1.2878 bar. Dew point pressure: p″ = 0.3975 bar.

   Relative amount of vapor $\dot{G}/\dot{F}$ (Fig. 3-6): $\dot{G}/\dot{F} \approx 0.0601$. Heats of vaporization at −10°C: r_a = 17.0135 kJ/mol; r_b = 21.2031 kJ/mol; r_c = 22.9230 kJ/mol; r_d = 28.4 kJ/mol; $r_m = \Sigma\, r_i \cdot x_{Fi}$ = 22.7212 kJ/mol. Verification of the estimated temperature: t = 20 − 1/0.13145 · 0.0601 · 22.7212 = 9.611°C; the estimation was wrong.

3. Better estimation by linear interpolation—From Fig. 3-10: t_{est} = −7°C. Vapor pressures: p_a^o = 3.80 bar; p_b^o = 1.207 bar; p_d^o = 0.786 bar; p_c^o = 0.176 bar. Boiling point pressure: p′ = 1.416 bar. Dew point pressure: p″ = 0.4569 bar. Relative amount of vapor $\dot{G}/\dot{F}$ (Fig. 3-6): $\dot{G}/\dot{F} \approx 0.1545$.

   Exact value of $\dot{G}/\dot{F}$ (Equation 3-19)—Equilibrium ratios $K_i = p_i^o/p$: K_a = 3.1662; K_b = 1.0057; K_c = 0.6548; K_d = 0.1466;

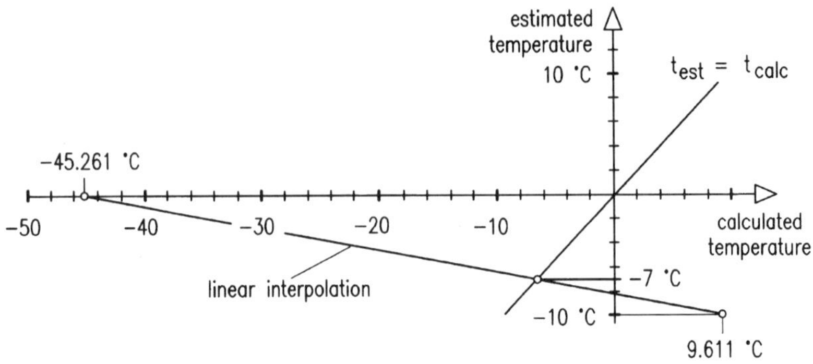

Fig. 3-10. *Auxiliary diagram for linear interpolation of Equations (3.19) and (3.20) for Example 3-3.*

$$f(\dot{G}/\dot{F}) = \sum \frac{x_{Fi} \cdot K_i}{1 + \dot{G}/\dot{F}(K_i - 1)} - 1 = 0;$$

$0.5931 + 0.1507 + 0.2214 + 0.0473 - 1 = 0.0124 \neq 0$. Newton algorithm (See example 3-1): $f'(\dot{G}/\dot{F}) = -[0.9626 + 0.0009 - 0.0807 - 0.0465] = -0.8362$; $\dot{G}/\dot{F} = 0.1545 - \frac{+0.0124}{-0.8362} = 0.1694$; $f(\dot{G}/\dot{F}) = 0.5791 + 0.1507 + 0.2226 + 0.0480 - 1 = 0.0004$ o.k. Vapor compositions are given by the additive terms of the previous equation: $y_a = 0.5791$; $y_b = 0.1507$; $y_c = 0.2226$; $y_d = 0.048$.

Heats of vaporization at -7°C: $r_a = 16.8300$ kJ/mol; $r_b = 21.0352$ kJ/mol; $r_c = 22.7608$ kJ/mol; $r_d = 28.2423$ kJ/mol; $r_m = \sum y_i \cdot r_i = 19.332$ kJ/mol. Verification of the estimated temperature: $t = 20 - 1/0.13145 \cdot 0.1694 \cdot 19.332 = -4.913$°C; estimation was not exact.

The exact values after several iterations are:

$t = -6.6$°C; $\dot{G}/\dot{F} = 0.1811$; $y_a = 0.5722$; $y_b = 0.1523$; $y_c = 0.2262$; $y_d = 0.0493$. Liquid concentration: $x_i = y_i/K_i$; $x_a = 0.1786$; $x_b = 0.1495$; $x_c = 0.3408$; $x_d = 0.3307$.

3.1.3 Application of Flash Equations to a Separation Process

An excellent example for the application of the flash equations is the removal of nitrogen from natural gas. Natural gas contains up to 15% nitrogen that should be removed before feeding the natural gas into a long distance pipeline.

Figure 3-11 displays the flow sheet for such a process. The natural gas is cooled down until most of the organic compounds are condensed and primarily nitrogen is left in vapor state. Then, the vapor-liquid mixture is expanded to cool down further.

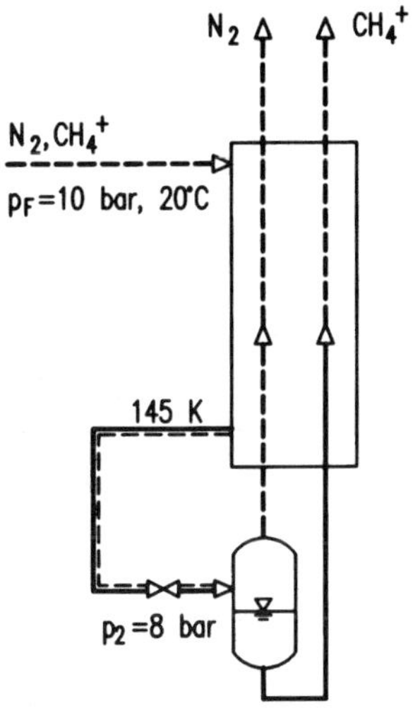

Fig. 3-11. Process for separating nitrogen from natural gas (Example 3-4).

After phase separation, the cold vapor and liquid are warmed up separately against the fresh natural gas. Since no perfect heat exchange can be established, the product streams are a little bit colder than the feed stream (approximately 5°C). This cooling energy is covered by the Joule-Thomson effect that is always connected with expansion of a nonideal gas. The calculation procedure of some essential steps of such a process are explained in detail in Example 3-4.

EXAMPLE 3-4: REMOVAL OF NITROGEN FROM NATURAL GAS BY PARTIAL CONDENSATION

Natural gas ($p_o = 10$ bar, $t_o = 20°C$) is cooled down to $t_1 = 145$ K. Find the relative vapor content $\dot{G}_1/\dot{F}$ and the composition of vapor y_{i1} and liquid x_{i1}.

The vapor-liquid mixture is expanded to $p_2 = 8$ bar. Find the temperature t_2 after the nozzle, the relative amount of vapor $\dot{G}_2/\dot{F}$ and the concentrations of vapor y_{i2} and liquid x_{i2}.

Composition of natural gas: nitrogen (a) $y_a = 0.153$; methane (b) $y_b = 0.732$; ethane (c); $y_c = 0.061$; propane (d) $y_d = 0.032$; butane (e) $y_c = 0.022$.

Solution:

1. State after condensation $T_1 = 145$ K; $p_1 = p_F = 10$ bar. Vapor pressures from Antoine equation: $p_a^o = 65.180$ bar; $p_b^o = 8.1684$ bar; $p_c^o = 0.0619$ bar; $p_d^o = 0.00148$ bar; $p_e^o = 3 \cdot 10^{-5}$ bar. Equilibrium ratios: $K_a = 6.5180$; $K_b = 0.8168$; $K_c = 0.062$; $K_d = 0.000148$; $K_e = 3 \cdot 10^{-6}$. Boiling point pressure: $p' = 15.9556$ bar. Dew point pressure: $p'' = 1.322 \cdot 10^{-3}$ bar.

$\dot{G}/\dot{F}$-ratio—Estimation by Equation (3.15):

$$\dot{G}/\dot{F} \approx \frac{\ln(10/15.9556)}{\ln(1.322 \cdot 10^{-3}/15.9556)} \approx 0.04960.$$

Verification by Equation (3.13): $\sum y_{i1} = 0.7828 + 0.6034 + 3.975 \cdot 10^{-4} + 5 \cdot 10^{-6} + 1 \cdot 10^{-7} = 1.3866 \neq 1$. The estimation was wrong. Newton iteration (see Example 3-1) yields $\dot{G}_1/\dot{F} = 0.3138$: $\sum y_{i1} = 0.3651 + 0.6343 + 5.491 \cdot 10^{-4} + 6.9 \cdot 10^{-6} + 1 \cdot 10^{-7} = 0.999 \approx 1$. Composition of vapor: $y_{a1} = 0.3651$; $y_{b1} = 0.6343$; $y_{c1} = 5.491 \cdot 10^{-4}$; $y_{d1} = 6.9 \cdot 10^{-6}$; $y_{e1} = 1 \cdot 10^{-7}$. Composition of liquid: $x_i = y_i/K_i$; $x_{a1} = 0.056$; $x_{b1} = 0.7766$; $x_{c1} = 0.0886$; $x_{d1} = 0.0466$; $x_{e1} = 0.0321$.

2. State after the nozzle: pressure $p_2 = 8$ bar. Estimated temperature (see Example 3-3): $T_{est} = 140.66$ K. Vapor pressure: $p_a^o = 56.441$ bar; $p_b^o = 6.539$ bar; $p_c^o = 0.0410$ bar; $p_d^o = 8.232 \cdot 10^{-4}$ bar; $p_e^o = 1.333 \cdot 10^{-5}$ bar. Equilibrium ratios: $K_a = 7.055$; $K_b = 0.817$; $K_c = 5.127 \cdot 10^{-3}$; $K_d = 1.029 \cdot 10^{-4}$; $K_e = 1.667 \cdot 10^{-6}$. Boiling point pressure: $p' = 13.425$ bar. Dew point pressure: $p'' = 5.916 \cdot 10^{-4}$ bar.

Relative content of vapor—Estimation by Equation (3.15):

$$\dot{G}/\dot{F} \approx \frac{\ln(8/13.425)}{\ln(5.916 \cdot 10^{-4}/13.425)} \approx 0.0516.$$

Verification by Equation (3.13): $\sum y_{i2} = 0.822 + 0.604 + 3.297 \cdot 10^{-4} + 3.472 \cdot 10^{-6} + 0.387 \cdot 10^{-7} = 1.427 \neq 1$. The estimation was wrong. Newton iteration (see Example 3-1) yields $\dot{G}_2/\dot{F} = 0.3253$: $\sum y_{i2} = 0.364 + 0.636 + 4.623 \cdot 10^{-4} + 4.879 \cdot 10^{-6} + 0.543 \cdot 10^{-7} = 1$ o.k. Composition of vapor: $y_{a2} = 0.364$; $y_{b2} = 0.636$; $y_{c2} = 4.623 \cdot 10^{-4}$; $y_{d2} = 4.879 \cdot 10^{-6}$; $y_{e2} = 0.543 \cdot 10^{-7}$. Composition of liquid: $x_i = y_i/K_i$; $x_{a2} = 0.0516$; $x_{b2} = 0.778$; $x_{c2} = 0.0902$; $x_{d2} = 0.0474$; $x_{e2} = 0.0326$.

Check of the heat balance (Equation 3.20)—Heat capacity of liquid: $c_{L,m} = \sum y_{Fi} \cdot c_{Li} = 0.0283$ kJ/mol · K. With Data from Perry: $c_{L,a} = 0.0297$ kJ/mol · K; $c_{L,b} = 0.0268$ kJ/mol · K; $c_{L,c} = 0.0291$ kJ/mol · K; $c_{L,d} = 0.0358$ kJ/mol · K; $c_{L,e} = 0.0539$ kJ/mol · K.

Heats of vaporization at $T_1 = 145$ K (see Example 3-3): r_a was calculated with the equation of Clausius-Clapeyron. (Heat of vaporization at normal boiling point and critical temperatures from Perry) $r_a = 5.621$ kJ/mol; $r_b = 6.6578$ kJ/mol; $r_c = 16.3671$ kJ/mol; $r_d = 22.5866$ kJ/mol; $r_e = 26.5939$ kJ/mol; $r_m(T_1) = \sum y_{i1} \cdot r_i(T_1) = 6.2841$ kJ/mol.

Heats of vaporization at $T_2 = 140.66$ K (see Example 3-3)—r_a calculated with the equation of Clausius-Clapeyron: $r_a = 5.6282$ kJ/mol; $r_b = 6.8845$ kJ/mol; $r_c = 16.5287$ kJ/mol; $r_d = 22.7461$ kJ/mol; $r_e = 26.7544$ kJ/mol; $r_m(T) = \sum y_{i2} \cdot r_i(T) = 6.4376$ kJ/mol; $t = 145 - 1/0.0283 \cdot (0.3253 \cdot 6.4376 - 0.3138 \cdot 6.2841) = 140.68$ K. The estimation was good enough.

3.2 BATCHWISE OPEN DISTILLATION AND OPEN CONDENSATION

In continuous open distillation and in batchwise distillation, the gas is removed from the system immediately after its generation. Both modes of operation are displayed in Figs. 3-1 and 3-12, respectively. The time dependence of the variables at batch distillation is equal to the locus dependence at continuous open distillation. After elimination of time and locus, respectively, the same equations apply to both modes of operation. For the derivation of the essential equations, the case of batch distillation is considered in this section.

A schematic of batch distillation is shown in Fig. 3-12. The vapor stream $\dot{G}$ produced by heating L moles of liquid is continuously removed and condensed. The distillate D is collected in several receivers.

3.2.1 Binary Mixtures

3.2.1.1 Open Distillation (Batch Distillation) The fundamental laws of batch distillation have been developed as early as at the turn of the century [Ostwald 1900, Schreinemakers 1901, Van der Waals 1902, Rayleigh 1902]. A material balance around the still (boiler) permits calculating the process of batch distillation of a binary mixture of substances *a* and *b*. For a differential element of time, dt, the following equation holds:

$$\dot{G} \cdot dt + dL = 0 \tag{3.21}$$

Here, $\dot{G}$ denotes the molar stream of vapor generated by the heating system. A component balance gives:

$$\dot{G} \cdot dt \cdot y + d(L \cdot x) = 0 \tag{3.22}$$

where the vapor and liquid mole fractions, y and x, refer to the more volatile component *a*. With $\dot{G}$ dt = dG follows from Equation (3.21):

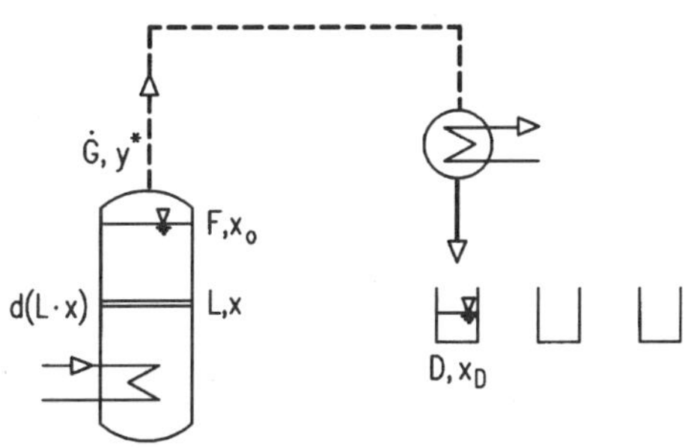

Fig. 3-12. Schematic of batch distillation. Generally, the distillate is collected in several receivers.

$$dG = -dL \tag{3.23}$$

Combining Equations (3.22) and (3.23) yields:

$$-dL \cdot y^* + dL \cdot x + L\,dx = 0 \tag{3.24}$$

After rearrangement:

$$\frac{dL}{L} = \frac{dx}{y^* - x} \tag{3.25}$$

This is the well known Rayleigh equation [Rayleigh 1902]. It formulates the relation between liquid level and liquid composition during open distillation. The vapor arising from the still is supposed to be in equilibrium state with the remaining liquid.

Introducing the equilibrium ratio $K = y^*/x$ yields:

$$\frac{dL}{L} = \frac{dx}{(K - 1) \cdot x} \tag{3.26}$$

Since the equilibrium ratio K is a function of both concentration and temperature, this equation cannot be integrated directly.

For ideal mixtures, the equilibrium concentration, y^*, can be formulated with the relative volatility α, which is often temperature independent:

$$y^* = \frac{\alpha \cdot x}{1 + (\alpha - 1) \cdot x} \tag{3.27}$$

From Equations (3.25) and (3.27) follows:

$$\frac{dL}{L} = \frac{dx}{\alpha\, x/(1 + (\alpha - 1)\, x) - x} \tag{3.28}$$

The solution to the differential equation is:

$$\frac{L}{F} = \left(\frac{x}{x_F}\right)^{1/(\alpha - 1)} \cdot \left(\frac{1 - x_F}{1 - x}\right)^{\alpha/(\alpha - 1)} \tag{3.29}$$

where F represents the liquid in its initial state.

Typical changes in the concentrations of the more volatile component a in the vapor and in the liquid are plotted vs. the relative amount of distillate $D/F = 1 -$

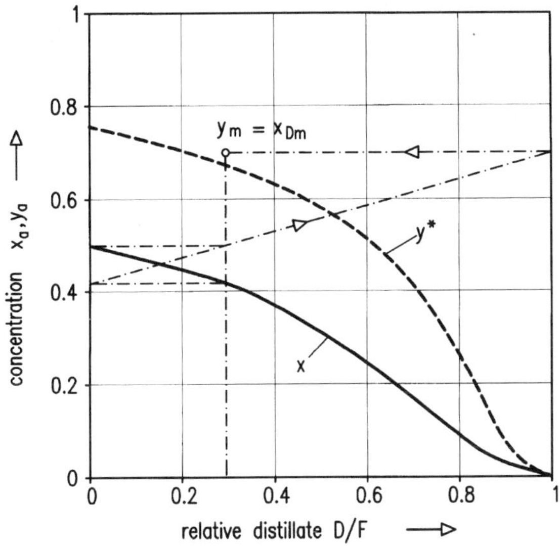

Fig. 3-13. *Batch distillation of an ideal binary mixture calculated with $\alpha = 3$. The system is depleted from light component during operation.*

L/F in Fig. 3-13. During operation, both phases are increasingly depleted of substance a. The mean distillate concentration x_{Dm} ($= y_m$) is given by:

$$x_{Dm} = x + \frac{x_F - x}{D/F} \tag{3.30}$$

This equation can be interpreted graphically with the help of Fig. 3-13. Even for nonideal mixtures, the mean distillate concentration x_{Dm} is easily determined.

EXAMPLE 3-5:

An ideal binary mixture ($z_F = 0.5$, $\alpha = 2$) is distilled in three modes: closed distillation, open distillation and countercurrent distillation. Find the concentrations of gas and liquid and the yield for G/F = 0.5.

1. Closed distillation—Equation (3.5) with $G/F = 0.5$, $x_F = 0.5$, $\alpha = 2$:

$$x = \frac{1}{2 \cdot (1 - 0.5)} \cdot \left[-(1 + (2 - 1) \cdot (0.5 - 0.5)) + \sqrt{(1 + (2 - 1) \cdot (0.5 - 0.5))^2 + 4 \cdot (1 - 0.5) \cdot 0.5} \right] = 0.41$$

$$y = \frac{2 \cdot 0.41}{1 + (2 - 1) \cdot 0.41} = 0.58; \qquad \text{Yield: } Y = \frac{0.5 \cdot 0.58}{1 \cdot 0.5} = 0.58.$$

2. Open distillation—Equation (3.29) with L/F = 0.5, $x_F = 0.5$:

$$x = x_F \cdot \left(\frac{L}{F}\right)^{\alpha - 1} \cdot \left(\frac{1 - x}{1 - x_F}\right)^{\alpha}.$$

Assumption:

$$x = 0.4: x = 0.5 \cdot (0.5)^1 \cdot \left(\frac{1 - 0.4}{1 - 0.5}\right)^2 = 0.36$$

Assumption:

$$x = 0.38: x = 0.5 \cdot 0.5\left(\frac{1 - 0.38}{1 - 0.5}\right)^2 = 0.38 \text{ o.k.}$$

Vapor concentration y from material balance: $F \cdot 0.5 = G \cdot y + L \cdot x$;

$$y = \frac{F \cdot 0.5 - L \cdot 0.38}{G} = 2 \cdot 0.5 - 1 \cdot 0.38 = 0.62;$$

$$\text{Yield: } Y = \frac{0.5 \cdot 0.62}{1 \cdot 0.5} = 0.62.$$

3. Countercurrent distillation—Gas is in equilibrium with the feed:

$$y = \frac{2 \cdot 0.5}{1 + (2 - 1) \cdot 0.5} = 0.667.$$

Liquid concentration x from material balance: $F \cdot 0.5 = G \cdot 0.667 + L \cdot x$;
$x = F/L \cdot 0.5 - G/L \cdot 0.667 = 2 \cdot 0.5 - 1 \cdot 0.667 = 0.334$;

$$\text{Yield: } Y = \frac{0.5 \cdot 0.667}{1 \cdot 0.5} = 0.667.$$

Result: The sharpest separation is reached by countercurrent distillation.

3.2.1.2 Open Condensation In open-partial condensation, the condensate is removed from the system immediately after formation, see Fig. 3-2. The derivation of the fundamental equations is performed in strict analogy to the previous section on open distillation. A material balance yields:

$$dG + dL = 0 \tag{3.31}$$

For the light constituent *a:*

$$d(G\, y) + x^* \, dL = 0 \tag{3.32}$$

Here, x^* denotes the concentration of liquid in equilibrium state with the vapor concentration y. From both equations follows:

$$dG\, y + G\, dy - x^* \, dG = 0 \tag{3.33}$$

After transformation:

$$\frac{dG}{G} = \frac{dy}{x^* - y} \tag{3.34}$$

This equation fully corresponds with the Rayleigh Equation (3.25) derived for open distillation. For ideal mixtures, the vapor-liquid equilibrium is expressed by the relative volatility α (see Equation 2.23):

$$x^* = \frac{\alpha^{-1} \cdot y}{1 + (\alpha^{-1} - 1) \cdot y} \tag{3.35}$$

Combining Equations (3.34) and (3.35) yields:

$$\frac{dG}{G} = \frac{dy}{\alpha^{-1}\, y/(1 + (\alpha^{-1} - 1) \cdot y) - y} \tag{3.36}$$

After integration:

$$\frac{G}{F} = \left(\frac{y}{y_F}\right)^{1/(\alpha^{-1} - 1)} \cdot \left(\frac{(1 - y_F)}{(1 - y)}\right)^{\alpha^{-1}/(\alpha^{-1} - 1)} \tag{3.37}$$

This equation corresponds with Equation (3.29) for open distillation. Just L is replaced by G, x by y and α by α^{-1}. For numerical evaluation, the following version is recommended:

$$\frac{G}{F} = \left(\frac{y}{y_F}\right)^{\alpha/(1 - \alpha)} \cdot \left(\frac{(1 - y_F)}{(1 - y)}\right)^{1/(1 - \alpha)} \tag{3.38}$$

During operation, the vapor G depletes from high boiling constituents and, hence, enriches in low boiling ones. Consequentely, vapor concentration y is always higher than feed concentration y_F.

3.2.2 Ternary Mixtures

3.2.2.1 Open Distillation Open distillation of ternary mixtures has found a wide interest in literature [Hausen 1935, Hausen 1952, Vogelpohl 1963, Vogelpohl 1965]. The Rayleigh equation can be applied directly to ternary mixtures. For a ternary mixture Equation (3.25) becomes:

$$\frac{dL}{L} = \frac{dx_a}{y_a^* - x_a}; \quad \text{and} \quad \frac{dL}{L} = \frac{dx_b}{y_b^* - x_b} \tag{3.39}$$

Combining both equations yields:

$$\frac{dx_a}{dx_b} = \frac{y_a^* - x_a}{y_b^* - x_b} \tag{3.40}$$

This very important differential equation describes the change of liquid concentrations in the still of an open distillation unit. It is called *liquid residuum curve,* or liquid residue [Doherty and Perkins 1979, Petlyuk 1986, Bernot et al. 1990]. Figure 3-14 displays the graphic interpretation of Equation (3.40). The dashed, dotted line is the liquid residuum line, i.e., the course of change of liquid composition during open distillation. The dashed, straight line between liquid concentration x_i and the corresponding equilibrium vapor concentration y_i^* forms a tangent to the liquid residuum line. The liquid is depleted from low boiling constituents during operation. This is indicated by an arrow at the liquid residuum line.

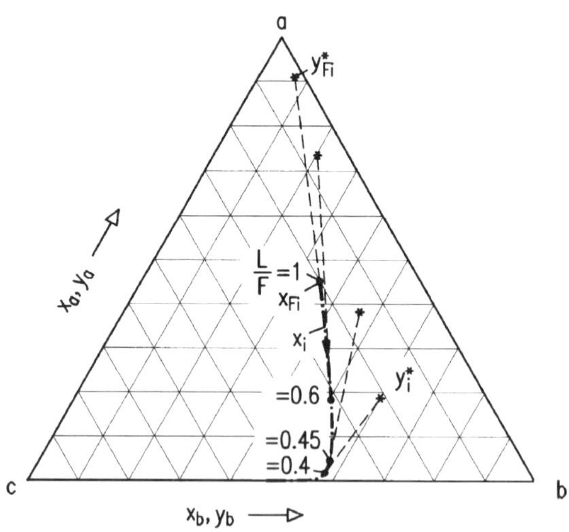

Fig. 3-14. Batch distillation of an ideal ternary mixture (calculated with relative volatilities α_{ac} = 20; α_{bc} = 2). The bold line describes the change of liquid concentrations in the drum. It is called liquid residuum line.

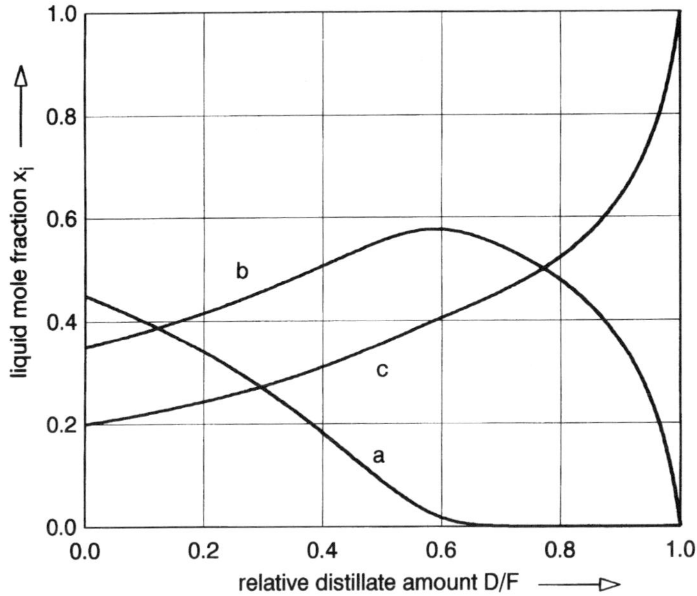

A

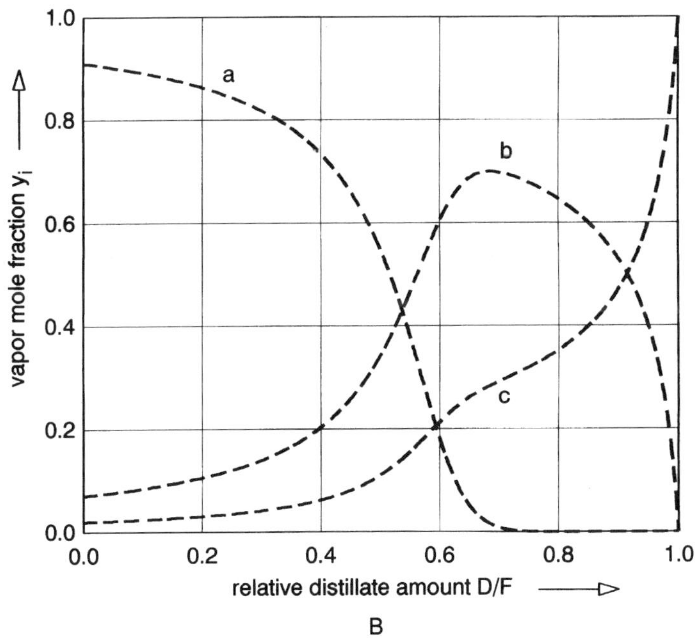

B

Fig. 3-15. *Batch distillation of an ideal ternary mixture (calculated with relative volatilities $\alpha_{ac} = 20$; $\alpha_{bc} = 2$). The concentrations are plotted vs. the relative amount of distillate. The intermediate boiling constituent* b *exhibits a concentration maximum. (A) Liquid concentrations, (B) vapor concentrations.*

For an ideal ternary mixture with constant relative volatilities the vapor concentrations follow from:

$$y_a^* = \frac{\alpha_{ac} x_a}{1 + (\alpha_{ac} - 1)x_a + (\alpha_{bc} - 1)x_b} \quad \text{and}$$

$$y_b^* = \frac{\alpha_{bc} x_b}{1 + (\alpha_{ac} - 1)x_a + (\alpha_{bc} - 1)x_b} \tag{3.41}$$

The solution to the differential Equation (3.40) is [Van der Waals 1902]:

$$\frac{1 - x_a - x_b}{x_a} \cdot \left(\frac{x_a}{x_b}\right)^{(\alpha_{ac}-1)/(\alpha_{ac}-\alpha_{bc})} = \frac{1 - x_{Fa} - x_{Fb}}{x_{Fa}} \cdot \left(\frac{x_{Fa}}{x_{Fb}}\right)^{(\alpha_{ac}-1)/(\alpha_{ac}-\alpha_{bc})} \tag{3.42}$$

The relative amount of liquid in the still, L/F, that corresponds with the actual liquid concentrations, x_a and x_b, is then determined by:

$$\frac{L}{F} = \left(\frac{x_a}{x_{Fa}}\right)^{1/(\alpha_{ac}-1)} \cdot \left(\frac{(1 - x_{Fc})}{(1 - x_c)}\right)^{\alpha_{ac}/(\alpha_{ac}-1)} \tag{3.43}$$

The profiles of concentration over the relative distillation amount D/F = 1 − L/F is displayed in Figs. 3-15A and 3-15B. This plot is analogous to Fig. 3-13 for a binary mixture. In ternary mixtures, the presentation in a triangular diagram is preferred. Figure 3-16 presents such a plot for the non-ideal ternary mixture octane, 2-ethoxyethanol and ethylbenzene. There exist two termini of residuum lines and, consequentely, a border residuum line. The course of liquid residuum lines is very similar to the course of distillation lines, see Fig. 2-15D. However, the arrow that indicates the direction of change is reverse.

Since liquid residuum lines are only a function of vapor liquid equilibrium, they are sometimes used for vapor-liquid equilibrium presentation. In this book, however, distillation lines are preferred for this purpose because they are simpler to calculate. In addition, the length of each equilibrium step can be marked by points on distillation lines (see Chapter 2).

3.2.2.2 Open Condensation In open condensation, the liquid formed by downcooling the system is immediately removed from the vapor. In analogy to Equation (3.39) we get:

$$\frac{dG}{G} = \frac{dy_a}{x_a^* - y_a} \quad \text{and} \quad \frac{dG}{G} = \frac{dy_b}{x_b^* - y_b} \tag{3.44}$$

Eliminating dG/G yields:

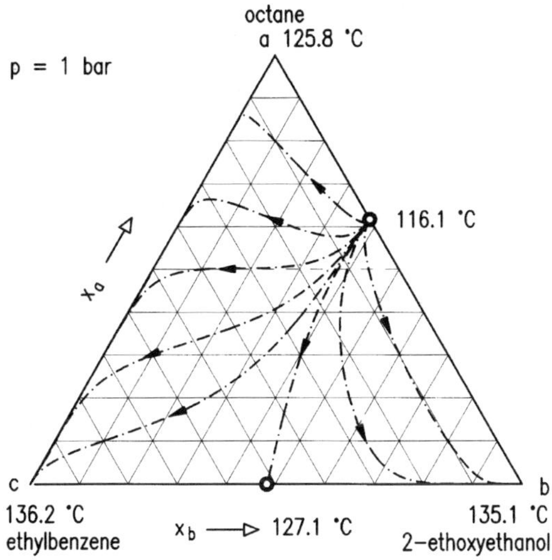

Fig. 3-16. Liquid residuum lines of the mixture octane/2-ethoxyethanol/ethylbenzene. The course of the liquid residuum lines is very similar to that of distillation lines.

$$\frac{dy_a}{dy_b} = \frac{x_a^* - y_a}{x_b^* - y_b} \tag{3.45}$$

This equation describes the course of changes of vapor composition during open condensation. It is called *vapor residuum line,* or vapor residue. It is represented by a dashed, bold line in Fig. 3-17. The straight line between actual vapor composition

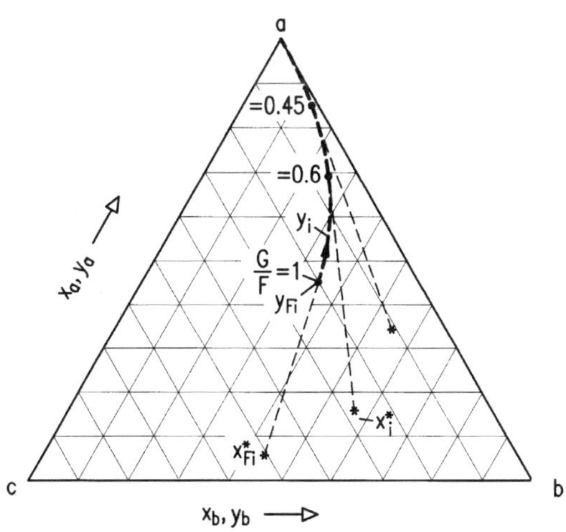

Fig. 3-17. Open condensation of a ternary mixture. The dotted, bold line describes the change of vapor compositions during operation. It is called vapor residuum line.

y_i and the corresponding liquid composition x_i^* forms a tangent to the vapor residuum line.

For ideal systems, the equilibrium is expressed by:

$$x_a^* = \frac{\alpha_{ac}^{-1} y_a}{1 + (\alpha_{ac}^{-1} - 1) \cdot y_a + (\alpha_{bc}^{-1} - 1) \cdot y_b}$$

$$\text{and} \quad x_b^* = \frac{\alpha_{bc}^{-1} y_b}{1 + (\alpha_{ac}^{-1} - 1) \cdot y_a + (\alpha_{bc}^{-1} - 1) \cdot y_b} \tag{3.46}$$

After integration of Equation (3.45):

$$\frac{1 - y_a - y_b}{y_a} \cdot \left(\frac{y_a}{y_b}\right)^{(\alpha_{ac}^{-1} - 1)/(\alpha_{ac}^{-1} - \alpha_{bc}^{-1})} = \frac{1 - y_{Fa} - y_{Fb}}{y_{Fa}} \cdot \left(\frac{y_{Fa}}{y_{Fb}}\right)^{(\alpha_{ac}^{-1} - 1)/(\alpha_{ac}^{-1} - \alpha_{bc}^{-1})} \tag{3.47}$$

The relative amount of remaining vapor is then determined by:

$$\frac{G}{F} = \left(\frac{y_a}{y_{Fa}}\right)^{1/(\alpha_{ac}^{-1} - 1)} \cdot \left(\frac{(1 - y_{Fc})}{(1 - y_c)}\right)^{\alpha_{ac}^{-1}/(\alpha_{ac}^{-1} - 1)} \tag{3.48}$$

These equations fully correspond with Equations (3.42) and (3.43) for open distillation. Only L, x, and α are substituted by G, y and α^{-1}, respectively.

Figure 3-18 displays the course of change of vapor composition during open

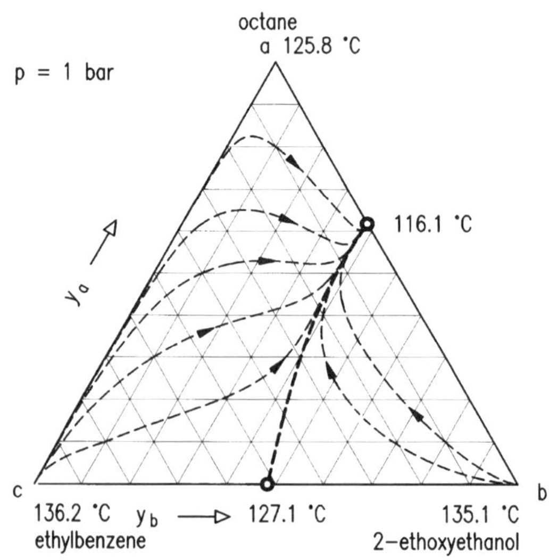

Fig. 3-18. *Vapor residuum lines of the system octane/2-ethoxyethanol/ethylbenzene.*

condensation of the system octane, 2-ethoxyethanol and ethylbenzene. Primarily, the high boiling constituents condense and, hence, the remaining vapor enriches in low boiling constituents. This is indicated by an arrow on the vapor residuum line.

Vapor residuum lines are only a function of vapor-liquid equilibrium alike liquid residuum lines and distillation lines. For close boiling systems the course of these three lines is very similar as can be seen from Figs. 3-19A to 3-19G. Vapor residuum lines and liquid residuum lines show the same characteristics as distillation lines. In an azeotropic system, border residuum lines exist in analogy to border distillation lines. Generally, distillation lines lie between vapor and liquid residuum lines.

3.2.3 Multicomponent Mixtures

3.2.3.1 Open Distillation The Rayleigh Equation (3.25) can be applied to multicomponent systems also. With $y_i^* = K_i x_i$ follows from Equation (3.24):

$$dL \cdot K_i \cdot x_i = d(L \cdot x_i) \tag{3.49}$$

Here, the term $d(L \cdot x_i)$ causes problems because the amount of liquid L, as well as the concentrations x_i, change during operation. More advantageous is the use of component amounts l_i, which are defined by:

$$l_i \equiv L \cdot x_i \tag{3.50}$$

The liquid concentrations x_i are replaced by:

$$x_i = l_i/L \tag{3.51}$$

With (3.50) and (3.51), Equation (3.49) becomes:

$$dL \cdot K_i \cdot l_i/L = dl_i \tag{3.52}$$

Or:

$$\frac{dL}{L} = \frac{1}{K_i} \cdot \frac{dl_i}{l_i} \tag{3.53}$$

This equation is equivalent to Equation (3.26). It also cannot be directly integrated.

For ideal systems with constant relative volatilities α_{ik}, the changes of component i are referred to the heaviest component k:

$$\frac{dL}{L} = \frac{1}{K_i} \cdot \frac{dl_i}{l_i} \quad \text{and} \quad \frac{dL}{L} = \frac{1}{K_k} \cdot \frac{dl_k}{l_k} \tag{3.54}$$

With $K_i/K_k = \alpha_{ik}$:

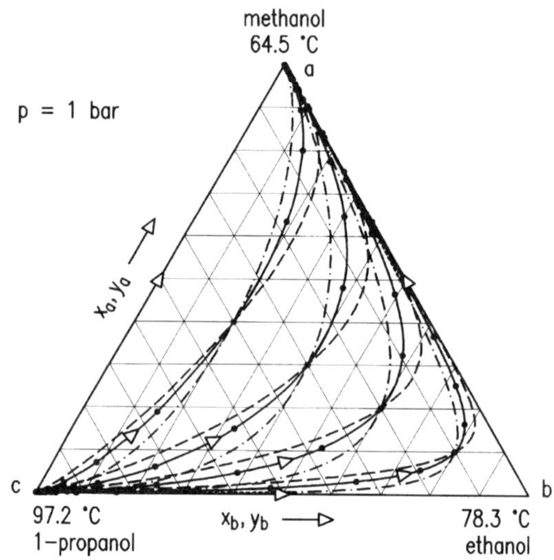

A

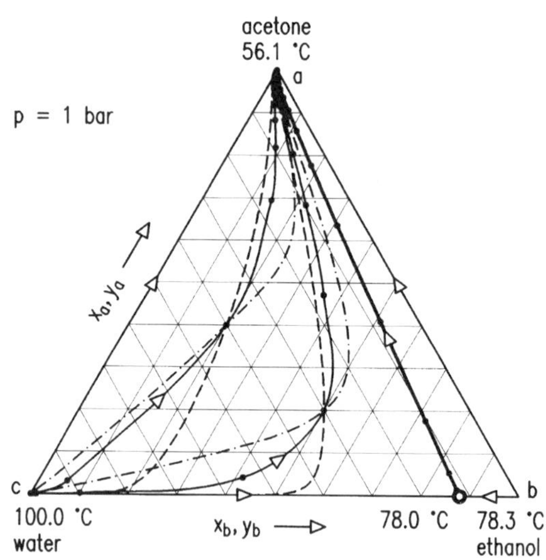

B

Fig. 3-19. *Comparison of the courses of liquid residuum lines (– · –), vapor residuum lines (– – –) and distillation lines (———). Especially in close boiling systems, all three lines lie close together.*

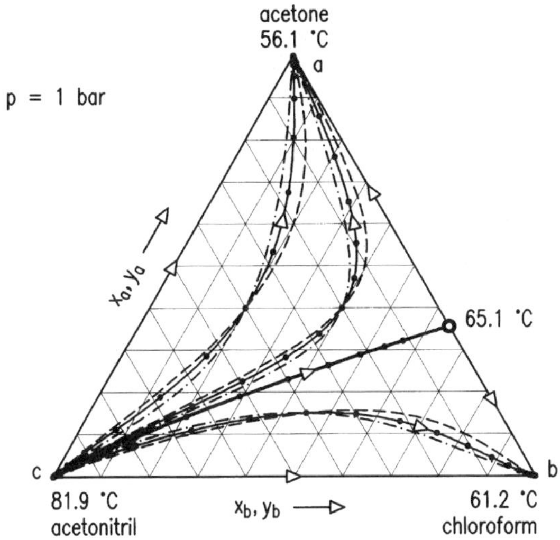

C

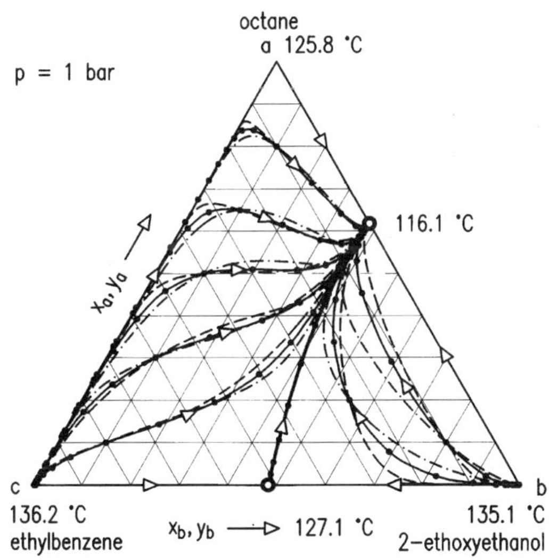

D

Fig. 3-19. *(continued)*

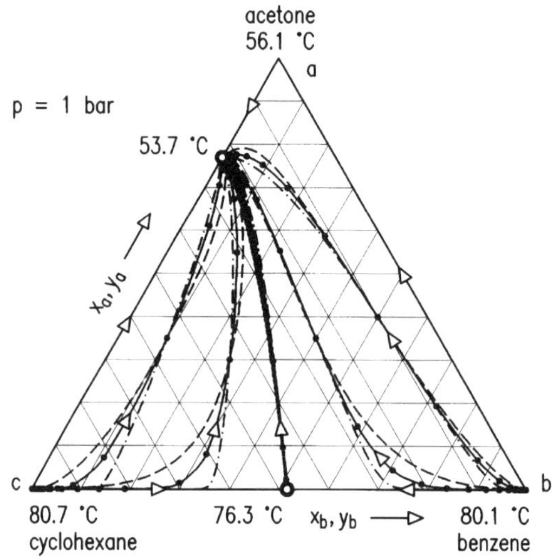

E

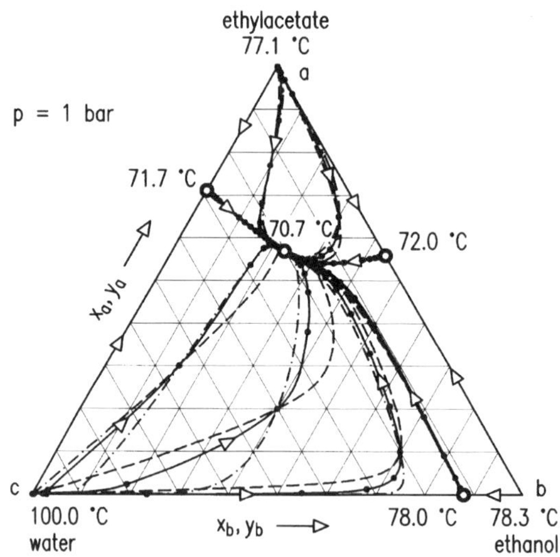

F

Fig. 3-19. *(continued)*

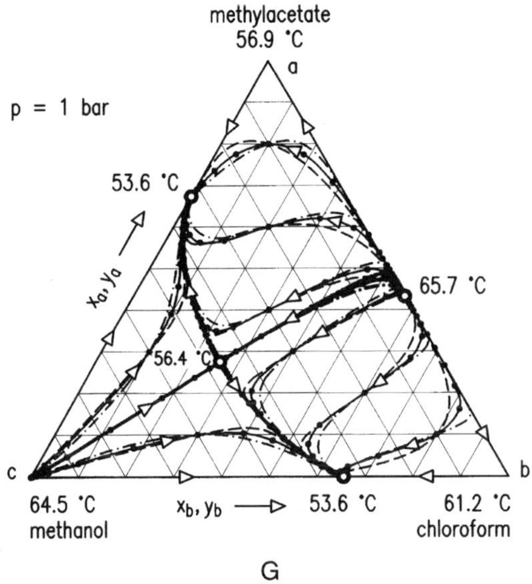

Fig. 3-19. (continued)

$$\frac{dl_i}{l_i} = \alpha_{ik} \cdot \frac{dl_k}{l_k} \tag{3.55}$$

Integration yields:

$$\left(\frac{l_i}{l_{Fi}}\right)^{1/\alpha_{ik}} = \frac{l_k}{l_{Fk}} \tag{3.56}$$

Or:

$$\left(\frac{l_a}{l_{Fa}}\right)^{1/\alpha_{ak}} = \left(\frac{l_b}{l_{Fb}}\right)^{1/\alpha_{bk}} = \left(\frac{l_c}{l_{Fc}}\right)^{1/\alpha_{ck}} = \cdots = \left(\frac{l_k}{l_{Fk}}\right)^{1} \tag{3.57}$$

The amount of liquid L in the still is:

$$L = \sum_{i=1}^{k} l_i \tag{3.58}$$

The concentrations in the still follow from:

$$x_i = l_i/L \tag{3.59}$$

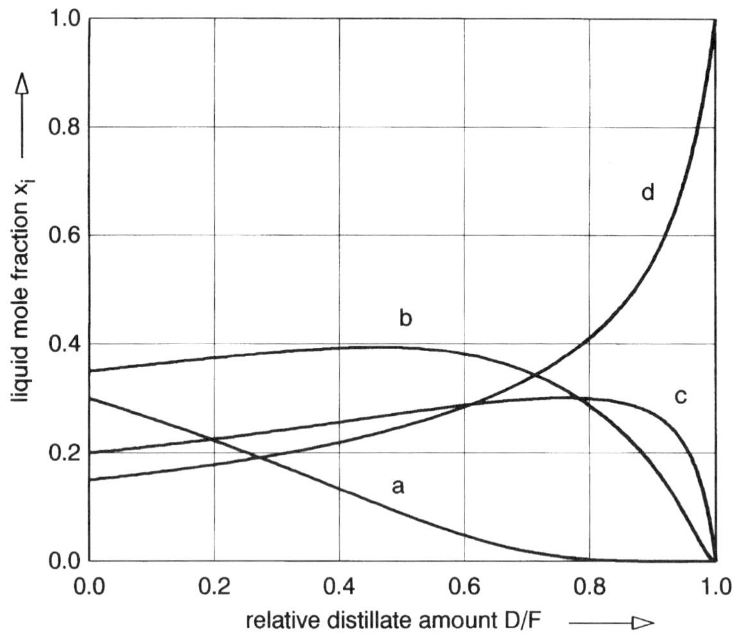

Fig. 3-20. *Open distillation of an ideal quaternary mixture (calculated with relative volatilities $\alpha_{ad} = 10$; $\alpha_{bd} = 3$, $\alpha_{cd} = 2$). The intermediate boiling components exhibit a concentration maximum over the relative amount of distillate.*

Figure 3-20 displays the liquid concentrations in the still for a quaternary ideal mixture with the relative volatilities $\alpha_{ad} = 10$, $\alpha_{bd} = 3$, $\alpha_{cd} = 2$. Note that the intermediate boiling constituents *b, c, d* exhibit a concentration maximum. This is a very essential feature that will be encountered in multistage distillation, too.

3.2.3.2 Open Condensation As has been outlined before the relations of open distillation can be applied to open condensation by simply replacing L, x, α and K by G, y, α^{-1} and K^{-1}, respectively. Hence, Equation (3.49) transfers into:

$$dG \cdot K_i^{-1} \cdot y_i = d(G \cdot y_i) \tag{3.60}$$

Here, the term $d(G \cdot y_i)$ causes problems because both the amount of vapor G and the concentrations y_i change during operation. More advantageous is the use of component amounts g_i:

$$g_i \equiv G \cdot y_i \tag{3.61}$$

From both equations follows:

$$\frac{dG}{G} = K_i \cdot \frac{dg_i}{g_i} \tag{3.62}$$

This equation is equivalent to Equation (3.53). It also cannot be directly integrated.

For ideal systems with constant relative volatilities α_{ik}, the changes of component i are referred to the heaviest component k:

$$\frac{dG}{G} = K_i \cdot \frac{dg_i}{g_i}; \qquad \frac{dG}{G} = K_k \cdot \frac{dg_k}{g_k} \tag{3.63}$$

With $K_i/K_k = \alpha_{ik}$:

$$\frac{dg_i}{g_i} = \alpha_{ik}^{-1} \cdot \frac{dg_k}{g_k} \tag{3.64}$$

Integration yields:

$$\left(\frac{g_i}{g_{Fi}}\right)^{\alpha_{ik}} = \frac{g_k}{g_{Fk}} \tag{3.65}$$

Or:

$$\left(\frac{g_a}{g_{Fa}}\right)^{\alpha_{ak}} = \left(\frac{g_b}{g_{Fb}}\right)^{\alpha_{bk}} = \left(\frac{g_c}{g_{Fc}}\right)^{\alpha_{ck}} = \cdots = \left(\frac{g_k}{g_{Fk}}\right)^{1} \tag{3.66}$$

The amount of remaining vapor is:

$$G = \sum_{i=1}^{k} g_i; \tag{3.67}$$

The concentrations in the still follow from:

$$y_i = g_i/G \tag{3.68}$$

Figure 3-21 displays the liquid concentrations in the still for a quaternary ideal mixture with the relative volatilities: $\alpha_{ad} = 10$, $\alpha_{bd} = 3$, $\alpha_{cd} = 2$. As in Fig. 3-19, the intermediate boiling constituents exhibit a concentration maximum during operation.

3.3 SEMICONTINUOUS SINGLE-STAGE DISTILLATION

Semicontinuous distillation can often be applied to advantage to liquids containing small amounts of impurities with low volatilities. As shown schematically in Fig.

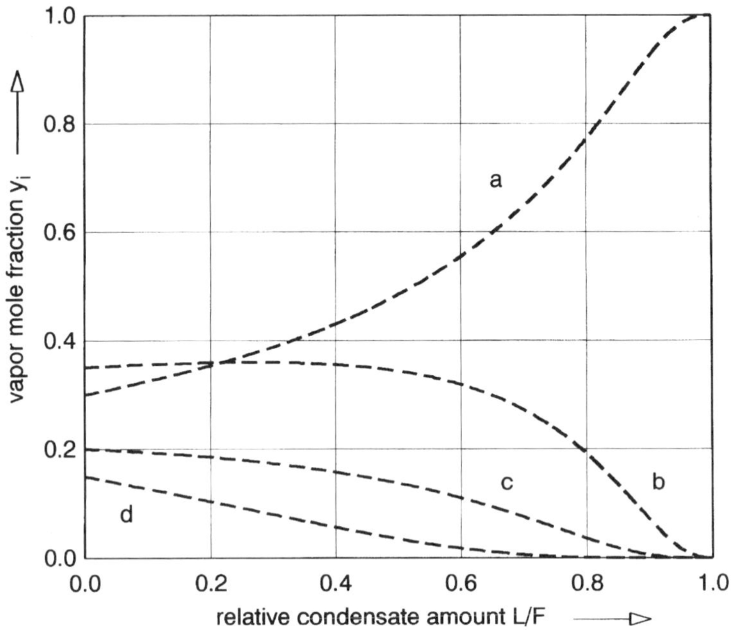

Fig. 3-21. *Open condensation of an ideal quaternary mixture (calculated with relative volatilities $\alpha_{ad} = 10$; $\alpha_{bd} = 3$, $\alpha_{cd} = 2$). The intermediate boiling components exhibit a concentration maximum over the relative amount of condensate.*

3-22, feed $\dot{F}$ is fed continuously into the boiler, and overhead product $\dot{D}$ is withdrawn continuously. During the course of distillation, low-volatility impurities accumulate in the boiler and have to be drained off periodically.

3.3.1 Semicontinuous Single-Stage Distillation of Binary Mixtures

In a manner analogous to that indicated by Equation (3.25), a material balance for a binary mixture yields:

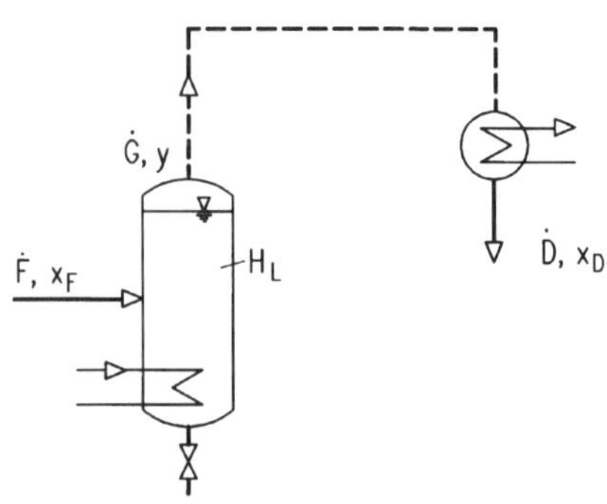

Fig. 3-22. *Schematic of a semicontinuous batch distillation. High boiling constituents accumulate in the boiler and have to be drawn off periodically.*

$$\frac{F}{H_L} = \frac{dx}{x_F - y^*(x)} \tag{3.69}$$

where mole fractions x and y refer to the more volatile component *a*. H_L denotes the moles of liquid holdup in the boiler. For ideal mixtures, integration gives:

$$\frac{F}{H_L} = \frac{\alpha - 1}{x_F(\alpha - 1) - \alpha}(x - x_F) - \frac{\alpha}{(x_F(\alpha - 1) - \alpha)^2} \cdot \ln \frac{x[x_F(\alpha - 1) - \alpha] + x_F}{x_F[x_F(\alpha - 1) - \alpha] + x_F} \tag{3.70}$$

Hence, the concentration changes of impurities in the boiler and in the distillate can be determined. Variation of the concentration with time t is given in the relationship $F = \dot{F} \cdot t$. In the special case of nonvolatile impurites, (i.e., $\alpha \rightarrow \infty$), Equation (3.70) is simplified as follows:

$$\frac{F}{H_L} = \frac{x - x_F}{x_F - 1} \tag{3.71}$$

Rearrangement gives:

$$x = x_F + F/H_L \cdot (x_F - 1) \tag{3.72}$$

Thus, the concentration x of the more volatile substance in the boiler decreases linearly with the amount of feed, while that of the less volatile substance $(1 - x)$ increases linearly.

LITERATURE

Bernot, C.; Doherty, M. F.; and Malone, M. F.: "Patterns of Composition Change in Multicomponent Batch Distillation," *Chem. Eng. Sci.* 45 (1990) No. 5, pp. 1207–1221.

Doherty, M. F. and Perkins, J. D.: "On the Dynamics of Distillation Processes III," *Chem. Eng. Sci.* 34 (1979) No. 6, pp. 1401/1414.

Hausen, H.: "Rektifikation von Dreistoffgemischen," *Forschung auf dem Gebiete des Ingenieurwesen* 6 (1935) pp. 9–22.

Hausen, H.: "Rektifikation von Dreistoffgemischen," *Zeitschrift für angewandte Physik* 4 (1952) pp. 45–51.

Ostwald, W.: Dampfdrucke ternärer Gemische, "Abhandlungen der mathematisch-physischen Classe der Königl," *Sächsischen Gesellschaft der Wissenschaften* 25 (1900) pp. 413–453.

Petlyuk, F. B.: "Rectification Diagrams for Ternary Azeotropic Mixtures," *Theor. Found. Chem. Eng.* 20 (1986) No. 3, pp. 175–185.

Rayleigh, Lord: "On the Distillation of Binary Mixtures," *Philosophical Magazine* (Sixth Series) 4 (1902) No. 23, pp. 521–537.

Schreinemakers, F. A. H.: Dampfdrucke ternärer Gemische, *Zeitschrift für physikalische Chemie* 36 (1901) pp. 257–289 and 413–449.

Van der Waals, J. D.: *Kon. Akad. Wetensch. Amdsterdam,* Proc. 5 (1902) pp. 228.

Vogelpohl, A.: "Der Einfluß der Stoffaustauschwiderstände auf die Rektifikation von Dreistoffgemischen," *Forschung auf dem Gebiete Ingenieurwesen* 29 (1963) pp. 907–915 and 1033–1045.

Vogelpohl, A.: "Offene Verdampfung idealer Mehrstoffgemische," *Chem.-Ing.-Tech.* 37 (1965) pp. 1144–1146.

4

Multistage Continuous Distillation (Rectification)

4.1 Principles
- 4.1.1 Equilibrium-Stage Concept
- 4.1.2 Transfer-Unit Concept
- 4.1.3 Comparison of Equilibrium-Stage and Transfer-Unit Concepts

4.2 Multistage Distillation of Binary Mixtures
- 4.2.1 Calculations Based on Material Balances
- 4.2.2 Calculation Based on Material and Enthalpy Balances
- 4.2.3 Distillation of Binary Mixtures at Total Reflux
- 4.2.4 Distillation of Binary Mixtures at Minimum Reflux
- 4.2.5 Heat Requirements for Distillation of Binary Mixtures

4.3 Multistage Distillation of Ternary Mixtures
- 4.3.1 Calculations Based on Material Balances
  - 4.3.1.1 Problems Arising in Column Calculations
    - 4.3.1.1.1 Choice of Reflux Ratio
    - 4.3.1.1.2 Choice of Bottom State
    - 4.3.1.1.3 Improvement of Estimated Values
    - 4.3.1.1.4 Choice of Feed Point
- 4.3.2 Distillation of Ternary Mixtures at Total Reflux
  - 4.3.2.1 Feasible Bottom and Top Fractions
    - 4.3.2.1.1 Separation Regions at Finite Reflux
  - 4.3.2.2 Feasible Pure Products
- 4.3.3 Distillation of Ternary Mixtures at Minimum Reflux
  - 4.3.3.1 Pinches in Ternary Distillation
    - 4.3.3.1.1 Preferred Separation
    - 4.3.3.1.2 Ideal Systems
    - 4.3.3.1.3 Separation of a Low Boiler—Rich Distillate

4.3.3.1.4 Separation of a High Boiler—Rich Bottom Fraction
4.3.3.2 Localizing the Pinch Point
4.3.3.3 Determination of Minimum Reflux and Minimum Reboil Ratios
4.3.4 Heat Requirement for Ternary Distillation
4.3.4.1 Preferred Separation
4.3.4.2 Low Boiler Separation
4.3.4.3 High Boiler Separation
4.4 Multistage Distillation of Multicomponent Mixtures
4.4.1 Rigorous Column Simulation
4.4.1.1 MESH-Equations
4.4.1.2 Specification of Column and Operating Conditions
4.4.1.3 Procedures for Evaluating the MESH-Equations
4.4.1.3.1 Simultaneous Solution
4.4.1.3.2 Jacobian Matrix
4.4.1.3.3 Partial Derivatives
4.4.1.3.4 Estimation of Starting Values of the Variables
4.4.1.3.5 Iteration History
Literature

4.1 PRINCIPLES

As discussed in Chapter 3, only limited separation of components is achieved in single-stage distillation. The composition difference between the distillation products is often very small; therefore, pure substances are seldom obtained. Figure 4-1A represents a repeated single-stage distillation, i.e., *rectification*. The overhead product of the first distillation is subjected to a second distillation, whose overhead product, in turn, is subjected to a third distillation, and so on. The same is done with the bottom product of the first distillation. As shown in Fig. 4-1B, the vapor and liquid concentrations of each step can be determined easily on the y-x diagram (Fig. 4-1B) by methods developed in Chapter 3.1. This process can yield very pure fractions, but it has two serious disadvantages:

- Only a small part of the starting mixture is recovered in a highly concentrated state.
- At each stage, by-products are obtained which often cannot be processed further.

Both of these disadvantages can be overcome if the by-products from each distillation step are returned to the previous step, as shown in Fig. 4-2A. This process can be further improved, because the vapor need not be condensed before each step, and the liquid does not have to be subsequently revaporized. Vapor and liquid can be brought into direct contact (Fig. 4-2B). In fact, intensive countercurrent contact between the vapor and liquid streams at each stage is important to enable intensive mass transfer.

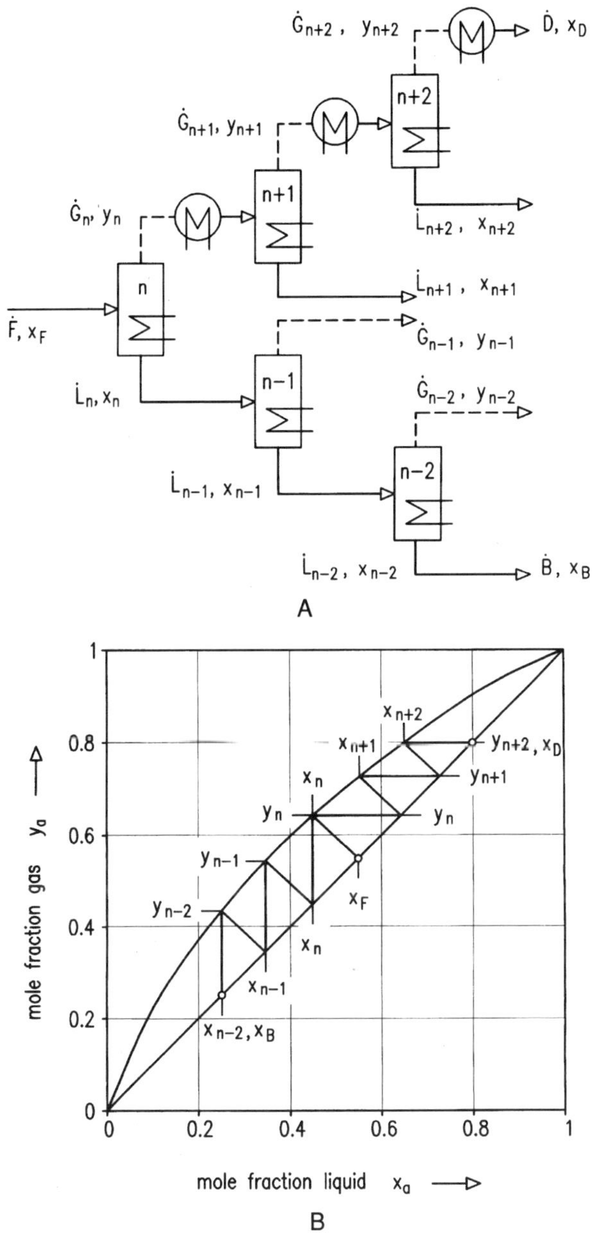

Fig. 4-1. *Multiple distillation of a binary mixture* a-b: *(A) Flow diagram, (B) Determination of vapor and liquid mole fractions of component* a *at each stage on a y-x diagram.*

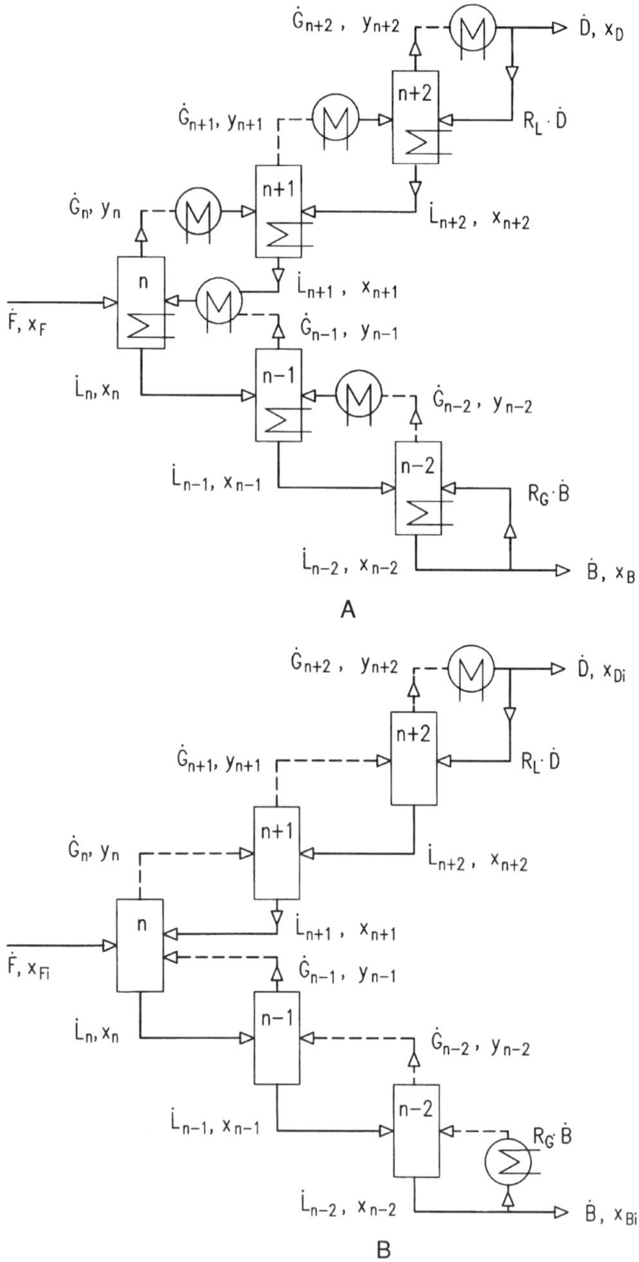

Fig. 4-2. *Improvement of multiple distillation by returning liquid from each distillation step to the previous step: (A) with, or (B) without condensation and vaporization, respectively.*

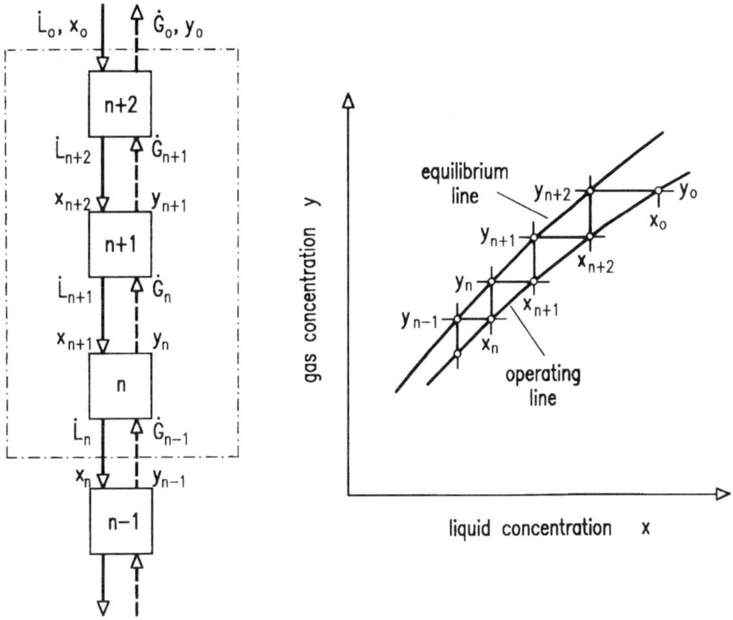

Fig. 4-3. Equilibrium stage concept for determining countercurrent separation efficiency: (A) Schematic showing equilibrium stages and associated parameters—dotted-dashed line indicates the boundary for the material balance described in Equation (4.1). (B) McCabe-Thiele diagram—concentrations at each equilibrium stage and number of stages can be determined by stepping off the stages between the equilibrium line and the operating line.

The concentrations attainable at each step can be determined on the y-x diagram by means of material balances around the top of the column, as shown in Fig. 4-3. The smaller the distance between operating and equilibrium line the more difficult is the separation. Two methods are usually employed to characterize the separation difficulty: The *equilibrium-stage* or the *transfer-unit* concepts.

4.1.1 Equilibrium-Stage Concept

This concept assumes that the vapor and liquid phases are indistinct mass-transfer elements, where gas and liquid are brought into such intimate contact that they reach equilibrium with each other. The phases are subsequently separated; vapor flows upwards to the stage above and liquid runs downwards to the stage below (counter-current phase flow), as shown in Fig. 4-3. A material balance for the more volatile component around the top of the column gives the following relationship for the vapor and liquid concentrations (x, y) and streams ($\dot{G}$, $\dot{L}$) between the individual stages n:

$$y_{n-1} = \frac{\dot{L}_n}{\dot{G}_{n-1}} x_n + \frac{\dot{G}_o}{\dot{G}_{n-1}} y_o - \frac{\dot{L}_o}{\dot{G}_{n-1}} x_o \qquad (4.1)$$

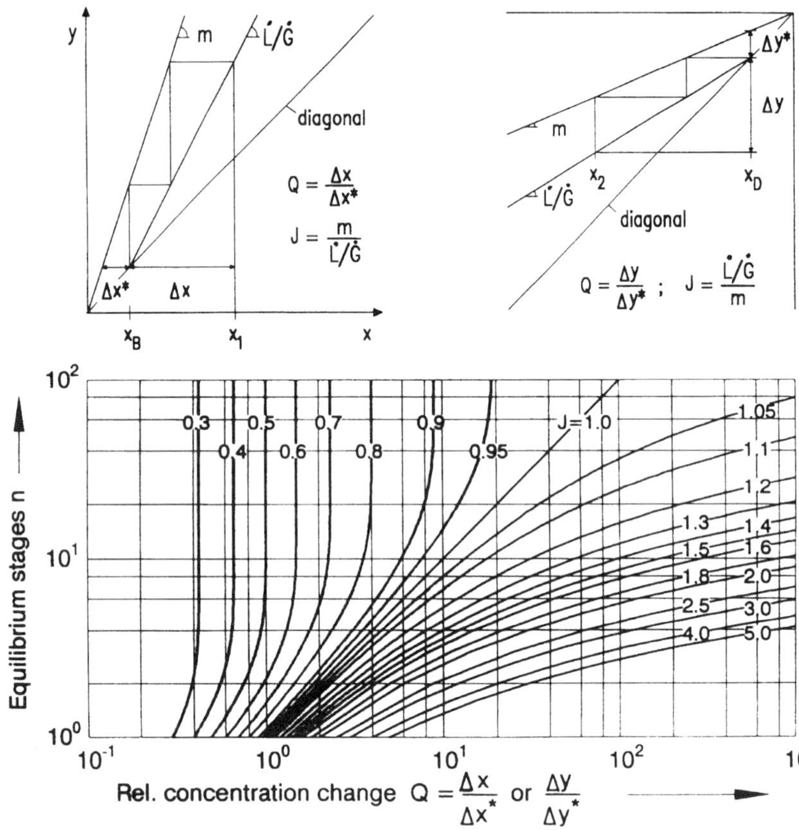

Fig. 4-4. *Graphical determination of the number of equilibrium stages n when both the operating line and the equilibrium line are straight.*

The subscripts o, n, and n − 1 refer to the stages from which the stream originates, o being the overhead. When Equation (4.1) is plotted on the y-x diagram, it furnishes the operating line (Fig. 4-3). The phase equilibrium is represented by a curve on the y-x diagram. The liquid concentration x_n is connected to the vapor concentration y_{n-1} by the material balance; the concentrations x_n and y_{n-1} define a point on the operating line. Equilibrium exists between y_n the vapor above stage n, and x_n the liquid concentration in the same stage. Therefore, a vertical line at the liquid concentration x_n must intersect the equilibrium curve at point y_n. Likewise, a horizontal line at the vapor concentration y_n must pass through the point x_{n+1} on the operating line. The point of intersection of x_{n+1} with the equilibrium curve gives the vapor concentration y_{n+1} of the stage above, and so on. Alternate use of the equilibrium curve and the operating line to "step" from one equilibrium stage to the other is illustrated on the y-x diagram in Fig. 4-3. The operating line gives the loci of states between the individual stages.

The number n of equilibrium stages required to achieve a desired concentration change is a good index for the difficulty of separation. The number of equilibrium stages, represented by steps, can easily be determined graphically on the y-x diagram; this is often called the *McCabe-Thiele* method, after the authors of the original graphical concept [McCabe and Thiele 1925]. Analytical solutions are available for a few special cases. A particularly important case occurs when both the equilibrium curve (slope m) and the operating line (slope $\dot{L}/\dot{G}$) are straight; the number of equilibrium stages n is given by, e.g., [King 1980]:

$$n = \frac{1}{\ln J} \cdot \ln[(1 - 1/J) \cdot Q + 1] \tag{4.2}$$

The definitions of J and Q are given in Fig. 4-4 for the stripping section and the rectifying section. When J = 1 (i.e., operating and equilibrium are parallel), then:

$$n = Q \tag{4.3}$$

Equation (4.2) can often be used to determine the number of stages for the separation of minute impurities. Fig. 4-4 shows a graphical plot of Equation (4.2).

4.1.2 Transfer-Unit Concept

The concept of transfer units considers a differential transfer area dA of the column [Chilton and Colburn 1935]: It is illustrated schematically in Fig. 4-5A. Mass transfer can be formulated as follows (see Chapter 8):

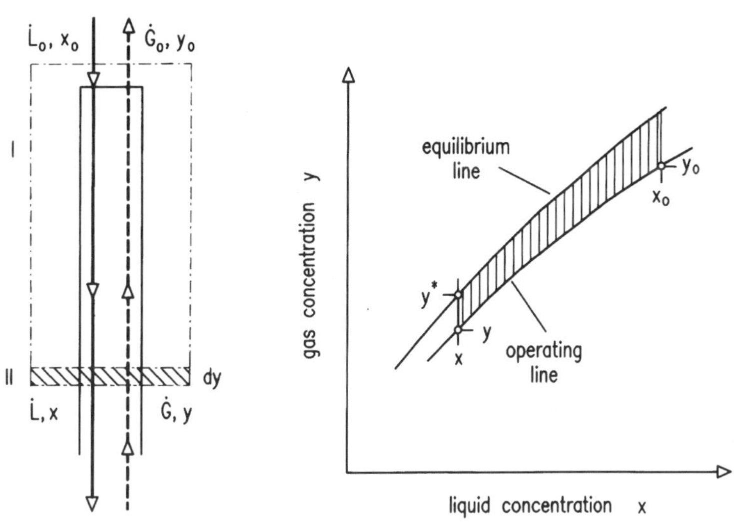

Fig. 4-5. Transfer unit concept for determining countercurrent separation efficiency: (A) Flow diagram showing important parameters and material balance boundaries (I, II), (B) a y-x diagram—the number of transfer units N_{OG} is determined by integrating between the equilibrium line and the operating line (hatched area).

$$d\dot{N} = K_{OG} \cdot (y^* - y) \cdot dA \tag{4.4}$$

where K_{OG} is the overall mass transfer coefficient. The molar mass transfer stream $d\dot{N}$ causes a change dy in vapor composition:

$$\dot{G} \cdot dy = d\dot{N} \tag{4.5}$$

Thus,

$$\dot{G} \cdot dy = K_{OG} \cdot (y^* - y) \cdot dA \tag{4.6}$$

Separating variables gives:

$$\frac{dy}{y^* - y} = K_{OG} \frac{dA}{\dot{G}} \tag{4.7}$$

The term on the right-hand side of Equation (4.7) contains factors pertaining mainly to the apparatus. These factors are of interest primarily for column design and dimensioning (see Chapter 8). The difficulty of separation is characterized sufficiently by the term on the left-hand side, which contains factors pertaining mainly to the system itself. It is defined as the overall number of gas-phase transfer units N_{OG}:

$$dN_{OG} \equiv \frac{dy}{y^* - y} \tag{4.8}$$

The number of transfer units can be interpreted as the ratio of the desired concentration change dy to the driving force $(y^* - y)$ available for mass transfer. At a given liquid state x, gas concentration y lies on the operating line and y* on the equilibrium line. The shaded area in Fig. 4-5 represents the driving force $(y^* - y)$.

The difficulty of the required separation can be characterized in terms of transfer units N_{OG}. If the value of N_{OG} is very large, a very tall column is required. In a sense, transfer units are similar to theoretical stages in that they are indicators of separation difficulty.

The number of transfer units N_{OG} is usually determined by integrating between the operating line and the equilibrium curve, either graphically or numerically. Analytical solutions are available for a few special cases. Again, the solution when both the equilibrium curve and the operating line are straight lines on the y-x diagram is important, e.g., [King 1980]:

$$N_{OG} = \frac{1}{1 - 1/J} \ln[(1 - 1/J) \cdot Q + 1] \tag{4.9}$$

The definitions of J and Q are given in Fig. 4-6 for the stripping and rectifying sections. When $J = 1$, then

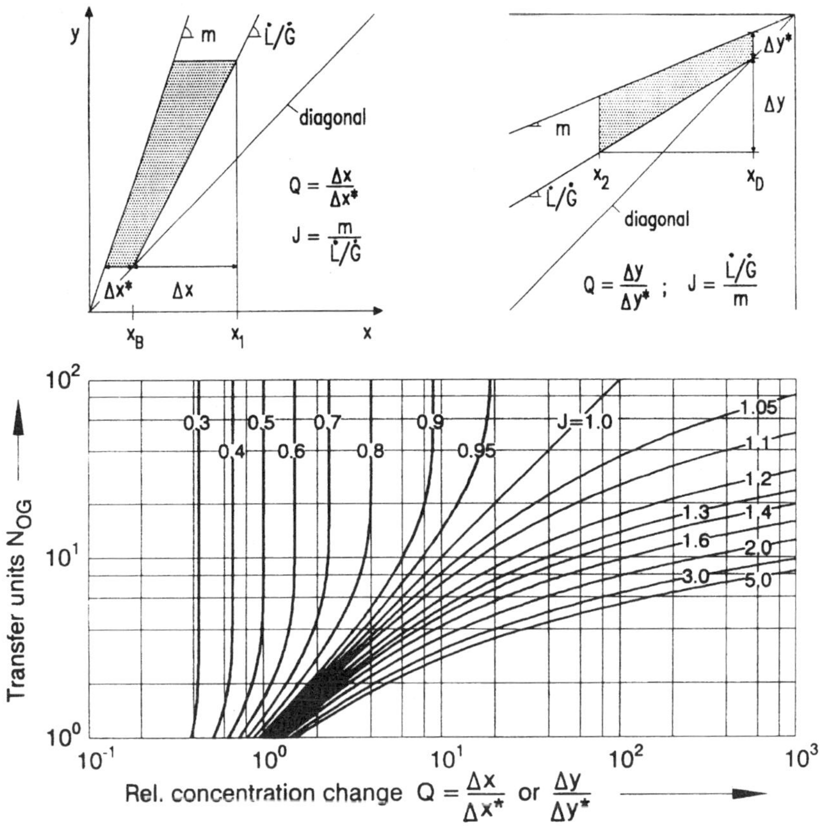

Fig. 4-6. Graphical determination of the number of transfer units N_{OG} when both the operating line and the equilibrium line are straight.

$$N_{OG} = Q \tag{4.10}$$

Equation (4.9) is comparable to Equation (4.2) for the number of equilibrium steps n. Figure 4-6 presents a graphical plot of Equation (4.9).

4.1.3 Comparison of Equilibrium-Stage and Transfer-Unit Concepts

The equilibrium-stage concept for characterization of the difficulty of separation provides an adequate measure of separability for distinct mass-transfer elements in countercurrent columns, e.g., in a highly efficient plate column. On the other hand, the transfer-unit concept is preferred when vapor and liquid are in continuous contact, e.g., in packed columns. However, the equilibrium-stage concept is used almost exclusively in practice because graphical determination of the number of equilibrium steps on the y-x diagram is much easier than integration with respect to the driving force. The equilibrium curve and the operating line can almost always be linearized in narrow concentration ranges. The following relationship is then obtained from Equations (4.2) and (4.9):

$$N_{OG} = n \cdot \frac{\ln(m/\dot{L}/\dot{G})}{m/\dot{L}/\dot{G} - 1} \tag{4.11}$$

In the special case when $m/(\dot{L}/\dot{G}) \rightarrow 1$ (i.e., parallel operating and equilibrium lines), the number of transfer units N_{OG} is equal to the number of equilibrium stages n. This condition is often satisfied to a close approximation in difficult separations that require many equilibrium stages or transfer units.

4.2 MULTISTAGE DISTILLATION OF BINARY MIXTURES

A continuously operated distillation column is shown schematically in Fig. 4-7. The binary feed (molar flow rate $\dot{F}$) is introduced toward the middle of the column. The distillate (molar flow rate $\dot{D}$), which mainly contains the lower boiling component *a*, is withdrawn from the top of the column. The bottom product (molar flow rate $\dot{B}$), which contains the higher boiling component *b* in greater concentration, is removed from the bottom of the column. Some overhead and bottom products are returned into the column to maintain countercurrent flow of vapor and liquid in the column. As discussed in Section 4.1, this is a prerequisite for the separation of feed into substances *a* and *b* with maximum purity. The sections of the column above and below the feed are called the rectifying section and the stripping section, respectively.

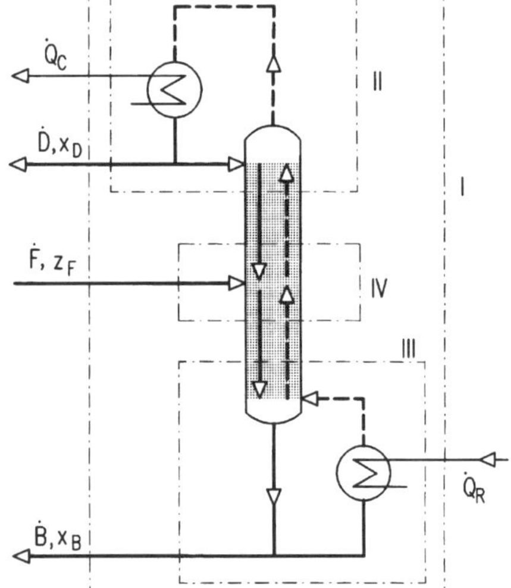

Fig. 4-7. *Schematic of a continuously operated distillation column for separating a binary mixture. Part of the overhead product and part of the bottom product are returned into the column to provide countercurrent flow within the column. Boundaries for material balances discussed in the text are indicated by dotted-dashed lines (I–IV).*

4.2.1 Calculations Based on Material Balances

Distillation can be predicted in the simplest fashion with the help of material balances, by assuming that the vapor and liquid flow rates ($\dot{G}$ and $\dot{L}$, respectively) in the column are constant (see Section 4.2.2). The overall column balance (boundary I in Fig. 4-7) yields:

$$\dot{F} = \dot{D} + \dot{B} \quad \text{and} \quad \dot{F}z_F = \dot{D}x_D + \dot{B}x_B \tag{4.12}$$

where z_F, x_D and x_B are the mole fractions of the more volatile substance *a* in the feed, distillate and bottoms, respectively. Rearrangement gives:

$$\dot{D} = \dot{F}\frac{z_F - x_B}{x_D - x_B} \quad \text{and} \quad \dot{B} = \dot{F}\frac{x_D - z_F}{x_D - x_B} \tag{4.13}$$

These equations can be used to determine the quantities of distillate $\dot{D}$ and bottoms $\dot{B}$ for given product specifications.

A material balance for the top of the column (boundary II in Fig. 4-7) gives a relationship for the states within the rectifying section of the column:

$$\dot{G} = \dot{L} + \dot{D} \qquad \dot{G} \cdot y = \dot{L} \cdot x + \dot{D} \cdot x_D \tag{4.14}$$

where $\dot{L}$ is the molar flow rate of the reflux liquor returned to the column. Thus,

$$y = \frac{\dot{L}}{\dot{G}} x + \left(1 - \frac{\dot{L}}{\dot{G}}\right) \cdot x_D \tag{4.15}$$

The exact position in the column to which Balance II applies is not defined. Hence, Equation (4.15) gives the relationship between the vapor concentration y and the liquid concentration x at any point within the column above the point of feed entry, i.e., in the rectifying section. If $\dot{L}/\dot{G}$ is constant, Equation (4.15) represents a straight line—the *rectifying operating line*—with a slope of $\dot{L}/\dot{G}$ on the y-x diagram, which intersects the $y = x$ diagonal at x_D (Fig. 4-8).

The ratio $\dot{L}/\dot{G}$ is known as the internal reflux ratio. The use of the external reflux ratio R_L is more advantageous. The definition of R_L is:

$$R_L \equiv \dot{L}/\dot{D} \tag{4.16}$$

Thus, Equation (4.15) becomes:

$$y = \frac{R_L}{R_L + 1} x + \frac{1}{R_L + 1} x_D \tag{4.17}$$

Analogously, material Balance III around the bottom of the column yields the following relationship for the stripping section:

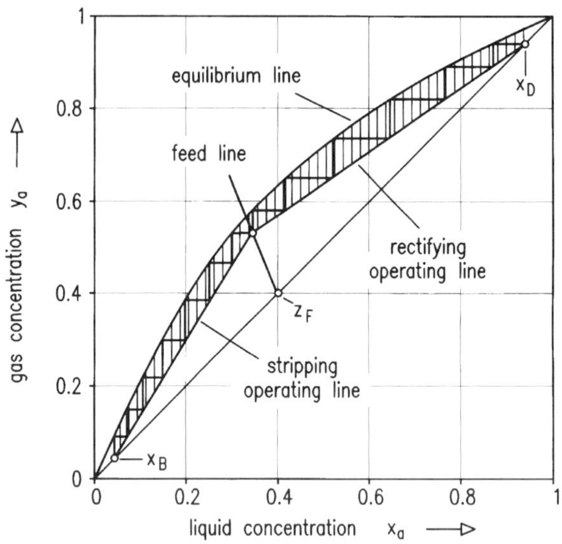

Fig. 4-8. *McCabe-Thiele diagram for a binary mixture* a-b. *Shaded area between the operating line and the equilibrium curve is decisive for operation. Equilibrium stages are also shown.*

$$y = \dot{L}/\dot{G} \cdot x + (1 - \dot{L}/\dot{G}) \cdot x_B \tag{4.18}$$

Use of the external reboil ratio R_G, according to the definition $R_G \equiv \dot{G}/\dot{B}$, gives:

$$y = \frac{R_G + 1}{R_G} x - \frac{1}{R_G} x_B \tag{4.19}$$

Equation (4.19) defines the relationship between the vapor concentration y and the liquid concentration x. If the flow rates $\dot{G}$ and $\dot{L}$ are assumed to be constant, it represents a straight line, called the *stripping operating line,* on the y-x diagram (Fig. 4-8). Feed is introduced into the column at the point of intersection of the rectifying and stripping operating lines defined by Equations (4.17) and (4.19) (see Fig. 4-8).

Figure 4-9 depicts the internal liquid concentration profiles x_a and x_b in the column. The low boiler *a* enriches in the upper section, and the high boiler *b* enriches in the lower section of the column. The feed $\dot{F}$ enters the column at the point where the internal liquid concentration equals the feed concentration. There is also plotted a typical temperature profile in Fig. 4-9.

Material Balance IV around the feed point gives the following equation for the point of intersection, e.g., [King 1980]:

$$y = \frac{q_F}{q_F - 1} x - \frac{1}{q_F - 1} z_F \tag{4.20}$$

The *caloric factor* q_F characterizes the thermal condition of the feed and is defined by:

$$q_F = \frac{\text{enthalpy required to vaporize 1 mol of feed}}{\text{molar latent heat of vaporization of feed}}$$

If the feed is a vapor-liquid mixture the caloric factor q_F describes the fraction of liquid in the feed. If the feed is introduced as a boiling liquid, the value of q_F is 1 and Equation (4.20) gives a vertical line on the y-x diagram; i.e., the operating lines intersect at $x = z_F$. However, if the feed is introduced as a saturated vapor, the value of q_F is 0 and Equation (4.20) yields a horizontal line; the two operating lines intersect at $y = z_F$. If the feed is introduced into the column as a vapor-liquid mixture, the point of intersection of the two operating lines will be between the two extreme cases described above, as shown in Fig. 4-8. The "q-line" can also represent a subcooled liquid ($q_F > 1$) and a superheated vapor feed ($q_F < 0$).

According to Equation (4.20), the q-line or feed line defines the loci of all points of intersection of the two operating lines in the rectifying and stripping sections of the column. Thus, if the feed conditions and the composition of overhead and bottom products are specified, and one operating line is established, the location of the second operating line is also known. This means that a relationship exists between the external reflux ratio R_L and the external reboil ratio R_G.

From the definitions of R_L, R_G and q_F follows (see Example 4-1):

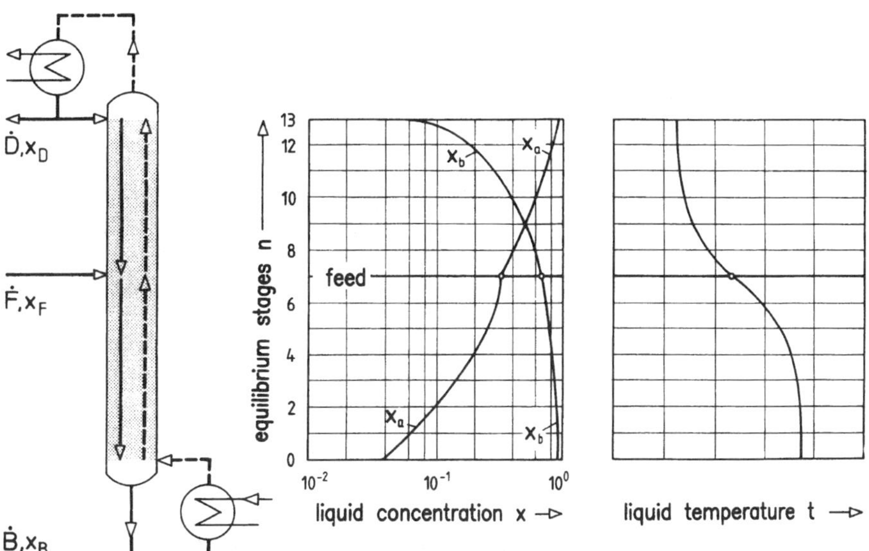

Fig. 4-9. Internal concentration and temperature profiles for a binary mixture.

$$R_G = \frac{\dot{D}/\dot{F} \cdot (R_L + 1) - (1 - q_F)}{1 - \dot{D}/\dot{F}} \tag{4.21}$$

For the two important cases of boiling liquid feed ($q_F = 1$) and saturated vapor feed ($q_F = 0$) this equation simplifies:

$$R_G = \frac{\dot{D}/\dot{F} \cdot (R_L + 1)}{1 - \dot{D}/\dot{F}} \quad \text{for } q_F = 1 \quad \text{or}$$
$$R_G = \frac{\dot{D}/\dot{F} \cdot (R_L + 1) - 1}{1 - \dot{D}/\dot{F}} \quad \text{for } q_F = 0 \tag{4.22}$$

The term $\dot{D}/\dot{F}$ is the relative quantity of overhead product, which can be determined by using Equation (4.13) after the product concentrations have been established.

The calculation of a rectification column with multiple feed entry and side stream withdrawal is conducted in an analogous manner (Fig. 4-10). The operating line breaks at each feed or withdrawal point. Figure 4-10B shows the course of the operating lines when $\dot{F}_1$ is a vapor feed stream ($\dot{F}_1 > 0$); Fig. 4-10C shows the corresponding lines for a vapor side stream ($\dot{F}_1 < 0$).

Sometimes intermediate reboilers and/or intermediate condensers are used in distillation columns. The intermediate heat supply and heat removal causes changes of internal gas and liquid flows. In consequence, the liquid-to-gas ratio $\dot{L}/\dot{G}$ is changed and the operating line breaks. Depending on the details of heat supply or removal

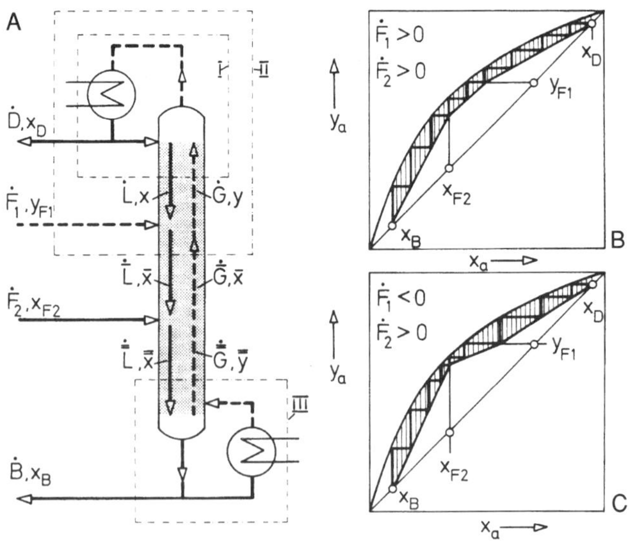

Fig. 4-10. *Distillation column and McCabe-Thiele diagram for multiple feed entries and side-stream withdrawal, respectively: (A) Flow diagram, (B) McCabe-Thiele diagram for two feeds, (C) McCabe-Thiele diagram for one feed and one side stream.*

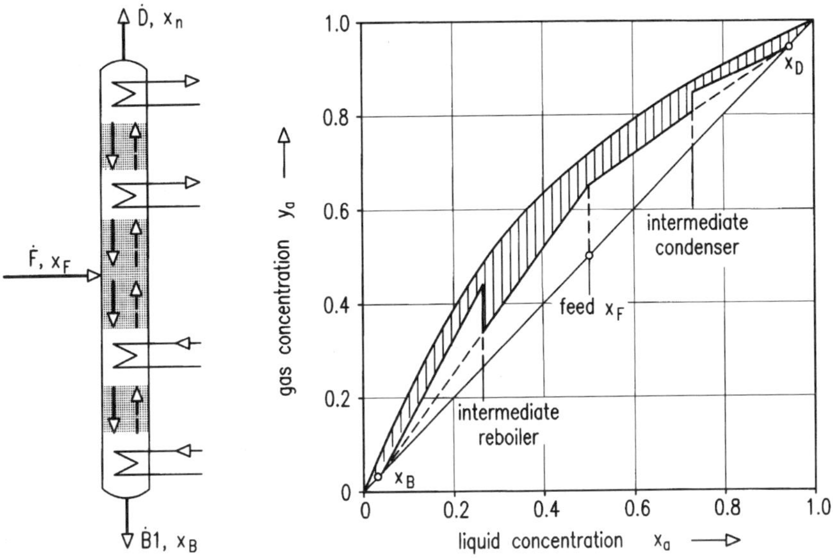

Fig. 4-11. *McCabe-Thiele diagram for a distillation column with intermediate condenser in the rectifying section and intermediate heater in the stripping section.*

(e.g., total evaporation of a partial liquid stream or partial evaporation of the total liquid stream), the position of the operating line might also be shifted. As can be seen in Fig. 4-11 the operating lines are moved closer to the equilibrium line by the use of intermediate condensers or reboilers. The driving force for mass transfer becomes much smaller but, as will be shown in Chapter 7, the exergy losses decrease.

If the positions of the operating lines and the equilibrium curve on the y-x diagram are known, the difficulty of separation is determined by using the number of equilibrium stages n (shown as steps in Fig. 4-8) or the number of transfer units N_{OG} (shaded area in Fig. 4-8). The desired separation is possible only when the operating lines neither touch nor intersect the equilibrium curve. The smaller the space between equilibrium curve and operating lines, the more difficult is the separation.

The operating lines can only lie in the region between the equilibrium line and the diagonal of the y-x diagram (shaded area in Fig. 4-12). The end points of the operating lines are fixed on the diagonal because the reflux $\dot{L}$ and the reboil $\dot{G}$ are generated by condensing or reboiling part of the top and bottom product, respectively. This mode of providing the second phase necessary for countercurrent flow also results in a limitation of the amount of reflux and reboil. Consequently, the slope of the equilibrium line in the rectifying section is always equal to or less than 1, and in the stripping section equal to or greater than 1.

Mass transfer between the phases is possible only when the operating line does not coincide with the equilibrium line. There always has to be a driving force $y^* - y \neq 0$. Consequently, an azeotrope is a barrier for distillation (see Section 5.5).

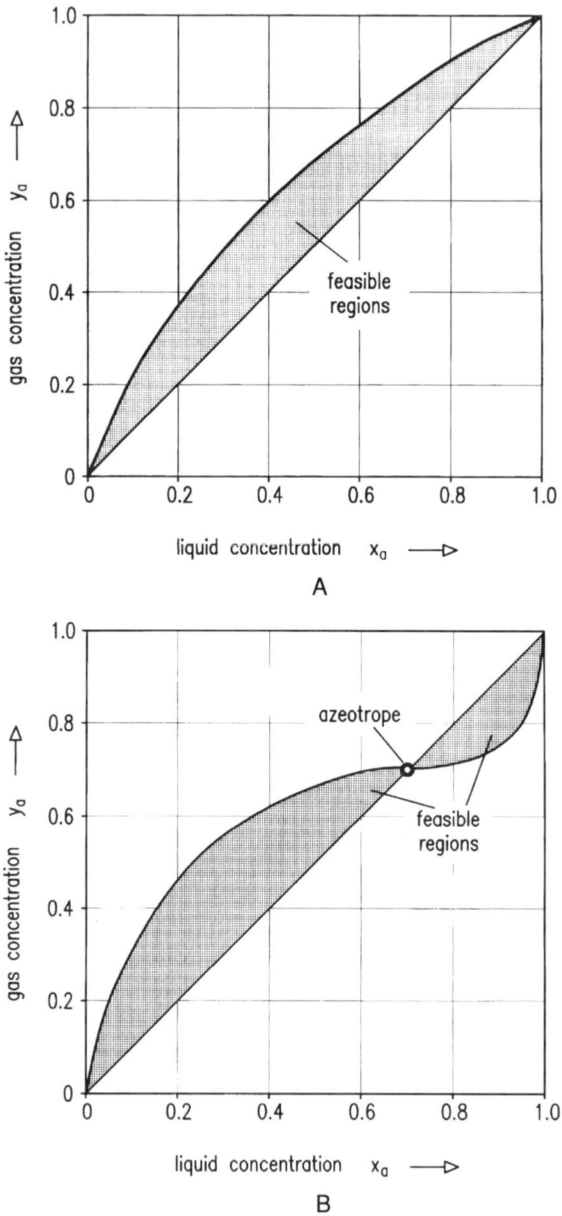

Fig. 4-12. *Feasible regions for operating lines of a binary distillation: (A) zeotropic mixture, (B) azeotropic mixture.*

EXAMPLE 4-1:

Derive a relation between reflux ratio R_L and reboil ratio R_G for constant molar overflow

Solution:

1. Definition of reflux ratio: $R_L \equiv \dot{L}/\dot{D} \rightarrow \dot{L} = \dot{D} \cdot R_L$.
2. Definition of reboil ratio: $R_G \equiv \dot{G}'/\dot{B}$ with $\dot{G}'$ is vapor flow at the bottom.
3. From (1) and (2) follows: $R_G/R_L = \dot{G}'/\dot{B} \cdot \dot{D}/\dot{L} = \dot{D}/\dot{B} \cdot \dot{G}'/\dot{L}$.
4. The vapor flow at the top, $\dot{G}$, is increased by the vapor $\dot{F}_G$ in the feed. $\dot{G} = \dot{G}' + \dot{F}_G \rightarrow \dot{G}' = \dot{G} - \dot{F}_G$ with $\dot{F}_G = f(q_F)$.
5. The caloric factor q_F describes the fraction of liquid $\dot{F}_L$ in the feed. $q_F \equiv \dot{F}_L/\dot{F} = (1 - \dot{F}_G/\dot{F}) \rightarrow \dot{F}_G = (1 - q_F) \cdot \dot{F}$.
6. Combining equations (1) to (5) yields:

$$\frac{R_G}{R_L} = \frac{\dot{D}}{\dot{B}} \cdot \frac{\dot{G} - \dot{F}_G}{\dot{L}} = \frac{\dot{D}}{\dot{B}} \cdot \frac{\dot{G} - (1 - q_F) \cdot \dot{F}}{\dot{D} \cdot R_L}.$$

7. After elimination of the denominator R_L:

$$R_G = \frac{\dot{D}}{\dot{B}} \cdot \left(\frac{\dot{G}}{\dot{D}} - (1 - q_F)\frac{\dot{F}}{\dot{D}}\right).$$

8. With $\dot{G} = \dot{L} + \dot{D} = R_L \cdot \dot{D} + \dot{D} = (R_L + 1)\,\dot{D}$:

$$R_G = \frac{\dot{D}/\dot{F}}{\dot{B}/\dot{F}} \cdot \left(\frac{(R_L + 1) \cdot \dot{D}}{\dot{D}} - (1 - q_F)\frac{1}{\dot{D}/\dot{F}}\right).$$

With $\dot{B}/\dot{F} = 1 - \dot{D}/\dot{F}$ follows the result:

$$R_G = \frac{\dot{D}/\dot{F} \cdot (R_L + 1) - (1 - q_F)}{1 - \dot{D}/\dot{F}}.$$

This relation is very important for developing the control configurations of distillation columns (see Chapter 9).

4.2.2 Calculation Based on Material and Enthalpy Balances

The calculation of multistage distillation based on material balances alone is possible only if the vapor and liquid flow rates ($\dot{G}$ and $\dot{L}$, respectively) are constant in the individual column sections (*constant molar overflow*). If this is not the case, the enthalpy balance must also be taken into consideration. An enthalpy-concentration

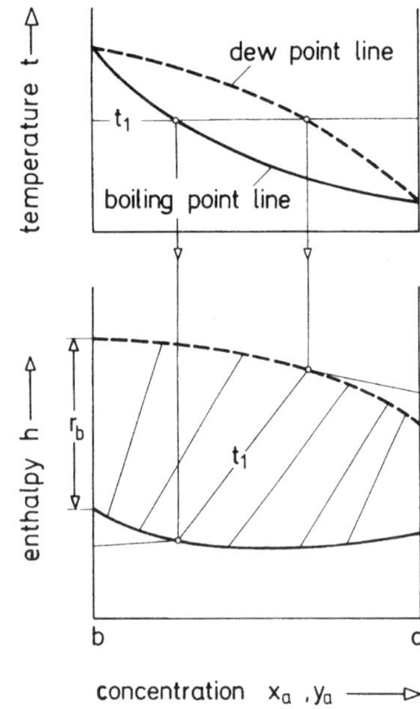

Fig. 4-13. Enthalpy-concentration diagram developed from the dew point and boiling point lines.

(h-x) diagram, shown in Fig. 4-13 for a binary mixture *a-b,* is very convenient for this procedure. [Bosnjakovic and Knoche 1997].

The overall enthalpy balance I for the column shown in Fig. 4-7 gives the following relationship:

$$\dot{F} \cdot h_F + \dot{Q}_R = \dot{D} \cdot h_D + \dot{B} \cdot h_B + \dot{Q}_C \quad (4.23)$$

where h_F, h_D, and h_B are the specific enthalpies of the feed, distillate, and bottoms, respectively; $\dot{Q}_R$ is the heat flow supplied by the reboiler; and $\dot{Q}_C$ is the heat flow removed by the condenser. Thus,

$$\dot{F} \cdot h_F = \dot{D} \cdot \pi_D + \dot{B} \cdot \pi_B$$

$$\text{with } \pi_D = h_D + \dot{Q}_C/\dot{D} \text{ and } \pi_B = h_B - \dot{Q}_R/\dot{B}. \quad (4.24)$$

Equation (4.24) represents a straight line in Fig. 4-14A that extends from point π_D to point π_B and passes through $\dot{F}$. The amounts of heat $\dot{Q}_R$ and $\dot{Q}_C$ can be determined easily, as shown in the diagram, when the compositions of the feed and of the desired overhead or bottom products are known (i.e., $\dot{F}$, $\dot{D}$, and $\dot{B}$). Both the distillate $\dot{D}$ and the bottom product $\dot{B}$ are assumed to leave the column in the boiling state.

Heat Balance II in Fig. 4-7 at the top of the column (rectifying section) gives:

$$\dot{G} \cdot h_G = \dot{L} \cdot h_L + \dot{D} \cdot \pi_D \quad (4.25)$$

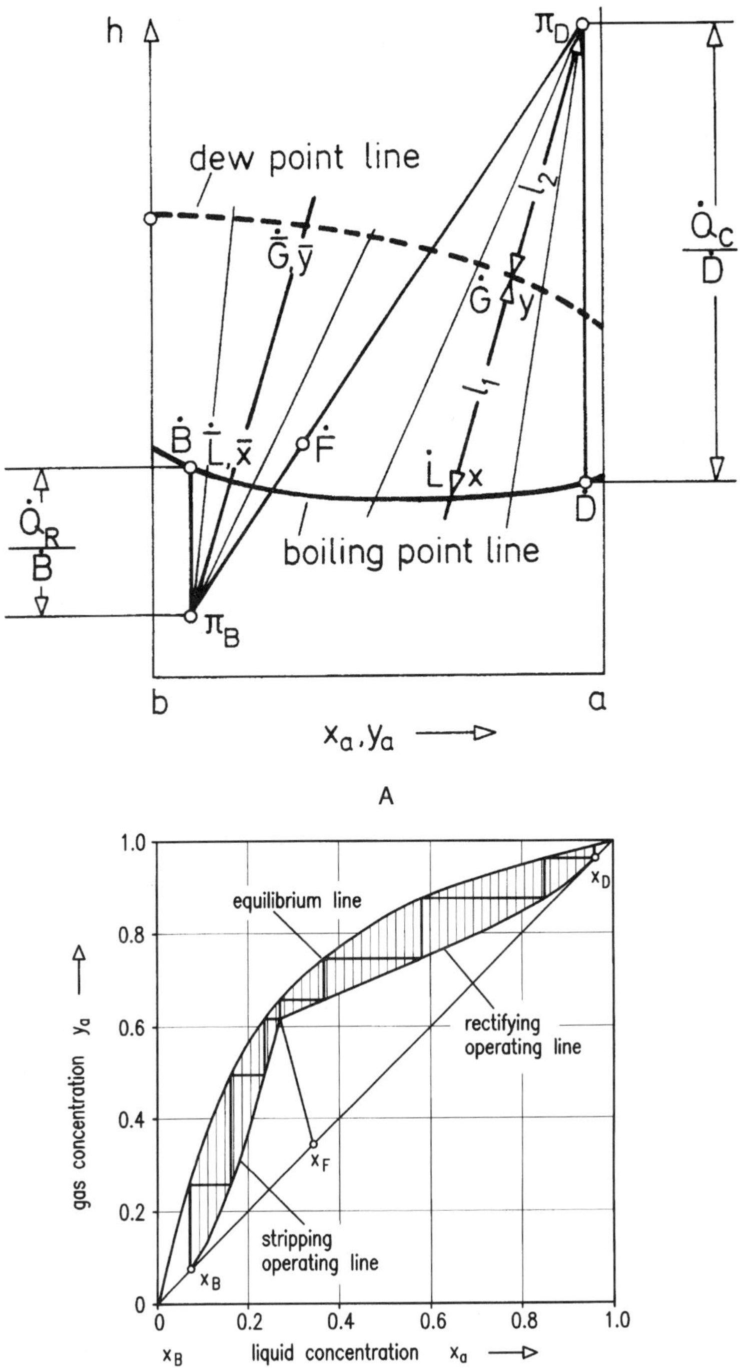

Fig. 4-14. *Representation of distillation in an enthalpy-concentration diagram: (A) The points of intersection of the lines through poles π_D and π_B with the dew point line and the boiling point line give the concentrations x and y, respectively, on the operating line. (B) Operating lines in the McCabe-Thiele diagram developed from enthalpy-concentration diagram.*

Here, h_G and h_L are the specific enthalpies of the vapor and liquid, respectively. This equation also represents a straight line through the point π_D on the enthalpy-concentration diagram. If the vapor in the column is always at the dew point and the liquid always at the boiling point (an assumption that is met closely), the flow rates of the vapor $\dot{G}$ and liquid $\dot{L}$ streams as well as their concentrations y and x can be established as shown in Fig. 4-14A. Concentrations y and x are points on the operating line on the y-x diagram (Fig. 4-14B). Analogously, if the balance lines in the stripping section are allowed to pass through π_B, the vapor and liquid states are obtained from the points of intersection with the dew point and the boiling point lines, respectively, as shown in Fig. 4-14A. The operating lines in the rectifying and stripping sections are obtained by transferring the vapor and liquid concentrations to the y-x diagram. The operating lines are generally slightly curved. The quantities of vapor and liquid can be determined by using the lever arm rule:

$$\dot{G} \cdot l_2 = \dot{L} \cdot (l_1 + l_2) \quad \text{or} \quad \dot{L}/\dot{G} = \frac{l_2}{l_1 + l_2} \tag{4.26}$$

where l_1 is the distance between the boiling point and dew point line and l_2 is the distance between the pole π_D and the dew point line.

For the special case in which the dew point line runs parallel to the boiling point line on the enthalpy-concentration diagram, Equation (4.26) produces a constant value for $\dot{L}/\dot{G}$. Thus, the criteria for constant vapor and liquid overflow can be formulated:

- The heat of vaporization of the two substances *a* and *b* must be equal.
- No heat of mixing should occur.
- The boiling points of the substances should be approximately the same.

The first criterion is the most important one. Equal latent heats of vaporization can often be achieved by using molar enthalpies of vaporization, which, in turn, requires that the compositions be given in mole fractions. Thus, distillation calculations generally employ these molar units.

If the criteria for a constant liquid-vapor ratio $\dot{L}/\dot{G}$ are not satisfied, the following approach is recommended:

- The poles π_D and π_B are established on the enthalpy-concentration diagram.
- The vapor and liquid concentrations y and x are determined, in each case, by extending straight lines that radiate from these poles. The values thus obtained are plotted on the McCabe-Thiele diagram and the curved operating line is obtained, point by point.
- The number of equilibrium stages n or the number of transfer units N_{OG} is determined in the usual manner on the y-x diagram (Fig. 4-14B).

The number of equilibrium stages can also be determined on the h-x diagram, e.g., [Treyball 1968]. However, this method is complex and so inaccurate at low concentrations that it has little practical importance.

4.2.3 Distillation of Binary Mixtures at Total Reflux

If the internal streams $\dot{G}$ and $\dot{L}$ are much larger than those of the feed $\dot{F}$ and the products $\dot{D}$ and $\dot{B}$, the internal reflux ratio $\dot{L}/\dot{G}$ in all parts of the column approaches 1. Thus, according to Equations (4.16) and (4.19), the reflux and reboil ratio, respectively, is infinite:

$$R_L \rightarrow \infty \quad \text{and} \quad R_G \rightarrow \infty \tag{4.27}$$

In this special case, called total reflux, the energy input needed for separation is very large, but a minimum number of equilibrium stages n or transfer units N_{OG} is required. As shown in Fig. 4-15, the operating line coincides with the diagonal and is represented by the simple equation:

$$y = x \tag{4.28}$$

The equilibrium state y^* is determined by starting with the liquid state x_B. The new liquid concentration x_1 is now calculated from the vapor concentration according to Equation (4.28). The vapor state y_1^* in equilibrium with this liquid concentration x_1 is, in turn, determined, and so on. This procedure is identical with the determination of the distillation line described in Section 2.3.3. Equation (2.31) thus applies to ideal binary mixtures with constant relative volatility α:

$$x_D = \frac{\alpha^n \cdot x_B}{1 + (\alpha^n - 1) \cdot x_B} \tag{4.29}$$

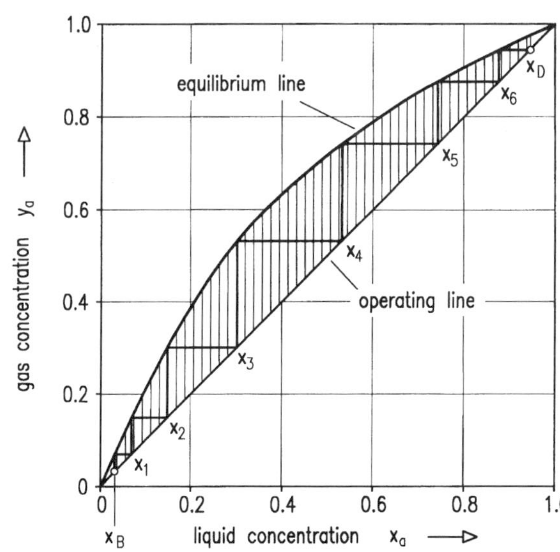

Fig. 4-15. *McCabe-Thiele diagram of a binary mixture at infinite reflux ratio.*

Here, n is the number of equilibrium stages. After rearrangement, the minimum number of equilibrium stages n_{min} is obtained:

$$n_{min} = \frac{1}{\ln\alpha} \ln\left[\frac{x_D}{x_B} \cdot \frac{1 - x_B}{1 - x_D}\right] \tag{4.30}$$

Analogously, the minimum number of transfer units $N_{OG,min}$ is given by:

$$N_{OG,min} = \frac{1}{\alpha - 1} \ln\left[\frac{x_D}{x_B}\left(\frac{1 - x_B}{1 - x_D}\right)\right]^{\alpha} \tag{4.31}$$

4.2.4 Distillation of Binary Mixtures at Minimum Reflux

Column operation with minimum internal gas and liquid flow (i.e., minimum reflux or reboil) has considerable practical importance because it results in separation of a mixture by using the lowest energy input. Such operation is characterized on a McCabe-Thiele plot by the intersection of the rectifying and stripping operating lines at the same point on the equilibrium curve. This point is called a *pinch point,* and an infinite number of equilibrium stages n or transfer units N_{OG} is required to reach this state. Figure 4-16 shows the y-x diagram for a binary system with minimum internal loads. If the vapor and liquid flow rates ($\dot{G}$ and $\dot{L}$, respectively) in the individual sections of the column are constant, the operating lines are straight. The

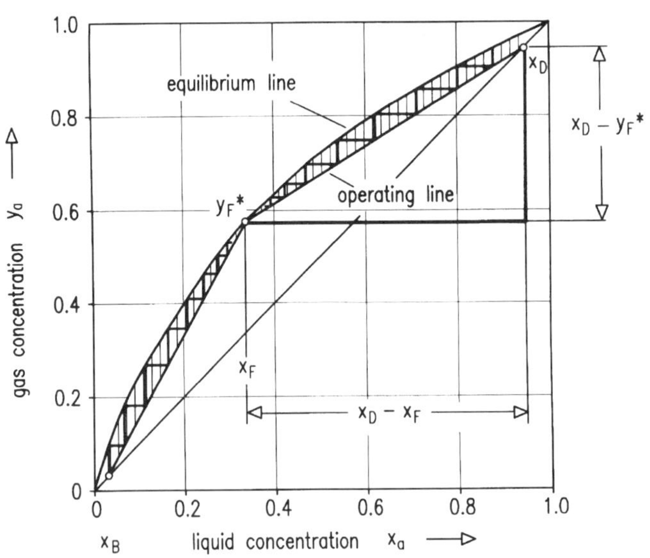

Fig. 4-16. *McCabe-Thiele diagram of a binary mixture at minimum reflux ratio and minimum reboil ratio. The operating lines intersect the equilibrium line at the pinch point.*

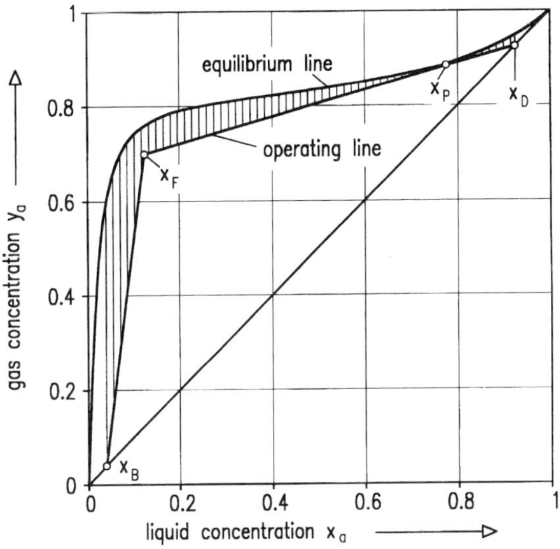

Fig. 4-17. *Minimum reboil ratio with tangential pinch. In such systems the pinch concentration x_{Pi} differs from feed concentration x_{Fi}.*

minimum slope of the rectifying operating line for the more volatile component *a* can be calculated as follows:

$$(\dot{L}/\dot{G})_{min} = \frac{x_D - y_F^*}{x_D - x_F^*} \tag{4.32}$$

By use of the relationship $\dot{L}/\dot{G} = R_L/(R_L + 1)$, the minimum reflux ratio becomes:

$$R_{Lmin} = \frac{x_D - y_F^*}{y_F^* - x_F^*} \tag{4.33}$$

Concentrations y_F^* and x_F^* are related by the phase equilibrium. This equation can be applied to all systems as long as the pinch, i.e., the intersection of operating and equilibrium line, lies at the feed point. Especially in azeotropic systems a tangential pinch may occur (see Fig. 4-17) whose concentration x_P is different from the feed concentration x_F.

By using Equation (2.23), the following relationship is obtained from Equation (4.33) for ideal mixtures:

$$R_{Lmin} = \frac{1}{\alpha - 1}\left(\frac{x_D}{x_F^*} - \alpha\frac{1 - x_D}{1 - x_F^*}\right) \tag{4.34}$$

If the overhead product is required as pure substance *a* (i.e., $x_D = 1$), Equation (4.34) can be simplified:

$$R_{Lmin} = \frac{1}{(\alpha - 1) \cdot x_F^*} \tag{4.35}$$

The point of intersection of the feed line (q-line, Equation 4.20) with the equilibrium line (Equation 2.23) gives the concentration x_F^*. For ideal systems a quadratic equation is obtained which can be solved as follows:

$$x_F^* = -\frac{B}{2} + \sqrt{\frac{B^2}{4} - C}$$

$$\text{where } B = \frac{1}{q_F}\left(\frac{\alpha}{\alpha - 1} - q_F - z_F\right) \text{ and } C = \frac{z_F}{q_F(\alpha - 1)} \tag{4.36}$$

Here, z_F is the mole fraction of substance *a* in a liquid-vapor feed and q_F is the thermal state of the feed (caloric factor) defined in Section 4.2.1.

If the feed is a boiling liquid with composition x_F, then:

$$x_F^* = x_F \tag{4.37}$$

If the feed is a saturated vapor with composition y_F, then:

$$x_F^* = \frac{y_F \cdot \alpha^{-1}}{1 + (\alpha^{-1} - 1) \cdot y_F} \tag{4.38}$$

Analogous equations can be derived for the minimum reboil ratio $R_G \equiv \dot{G}/\dot{B}$. The slope of the stripping operating line can be calculated using Fig. 4-16 as follows:

$$(\dot{L}/\dot{G})_{max} = \frac{y_F^* - x_B}{x_F^* - x_B} \tag{4.39}$$

From the relationship $\dot{L}/\dot{G} = (R_G + 1)/R_G$, the minimum reboil ratio is:

$$R_{Gmin} = \frac{x_F^* - x_B}{y_F^* - x_F^*} \tag{4.40}$$

The relationship $y_F^* = f(x_F^*)$ can be formulated easily for ideal mixtures. Here, Equation (4.40) transfers into:

$$R_{Gmin} = \left[\frac{1}{1 - x_B/x_F^*}\left(\frac{\alpha}{1 + (\alpha - 1) \cdot x_F^*} - 1\right)\right]^{-1} \tag{4.41}$$

An important case exists when pure substance *b* has to be separated as the bottom product (i.e., $x_{Ba} = 0$):

$$R_{Gmin} = \left[\frac{\alpha}{1 + (\alpha - 1) \cdot x_F^*} - 1\right]^{-1} \tag{4.42}$$

The concentration x_F^* can generally be calculated from the composition z_F and the thermal state q_F of the feed (Equation 4.36). Again, the special cases in which the feed is a boiling liquid or a saturated vapor are very important. The relationship between the minimum reflux ratio and the minimum reboil ratio can then be expressed by using Equations (4.21) and (4.22).

Separation at minimum reflux requires an infinite number of equilibrium stages n or transfer units N_{OG}. However, even a slight increase in the reflux of ca. 5 to 10% results in economical operation. Thus, the special case and laws discussed in this section closely approximate actual distillations and have immediate practical importance.

4.2.5 Heat Requirements for Distillation of Binary Mixtures

In distillation columns, a quantity of heat $\dot{Q}_R$ is supplied in the reboiler at the bottom of the column and a quantity $\dot{Q}_C$ is removed by the condenser at the top. The following equations hold:

$$\dot{Q}_R = r \cdot (\dot{G})' \quad \text{or} \quad \dot{Q}_C = -r \cdot \dot{G} \tag{4.43}$$

Here, $(\dot{G})'$ denotes the vapor stream at the bottom and $\dot{G}$ denotes the vapor stream at the top of the column. Approximately equal molar heats r of vaporization of the substances involved are assumed in the following. Important relationships can be established for the quantities of heat $\dot{Q}_R$ and $\dot{Q}_C$. The heat required by the reboiler $\dot{Q}_R$ is formulated by using the reboil ratio $R_G = (\dot{G})'/\dot{B}$. The heat removed by the condenser $\dot{Q}_C$ is determined by using the reflux ratio $R_L = \dot{L}/\dot{D}$. Thus:

$$\dot{Q}_R = r \cdot R_G \cdot \dot{B} \quad \text{or} \quad \dot{Q}_C = -r \cdot (R_L + 1) \cdot \dot{D} \tag{4.44}$$

The term $(R_L + 1)$ arises because both the reflux liquid and the overhead product must be condensed at the top of the column. The product flow rates $\dot{B}$ and $\dot{D}$ in binary mixtures can be expressed in terms of the feed stream $\dot{F}$ and the composition of the products by using Equation (4.13):

$$\frac{\dot{Q}_R}{\dot{F} \cdot r} = R_G \cdot \frac{x_D - z_F}{x_D - x_B} \quad \text{and} \quad \frac{\dot{Q}_C}{\dot{F} \cdot r} = -(R_L + 1) \cdot \frac{z_F - x_B}{x_D - x_B} \tag{4.45}$$

Thus, $\dot{Q}_R$ and $\dot{Q}_C$ are directly proportional to the reboil ratio R_G and the reflux ratio R_L, respectively. These ratios cannot fall below a limiting value (for details, see Section 4.2.4); the minimal quantities of heat are calculated as follows:

$$\frac{\dot{Q}_{R,min}}{\dot{F}r} = R_{Gmin} \frac{x_D - z_F}{x_D - x_B} \quad \text{and} \quad \frac{\dot{Q}_{C,min}}{\dot{F} \cdot r} = -(R_{Lmin} + 1) \frac{z_F - x_B}{x_D - x_B} \tag{4.46}$$

If fractionation into pure products is required during distillation of a binary mixture (i.e., $x_{Da} = 1$ and $x_{Ba} = 0$), use of the appropriate relationships for the reboil and reflux ratios gives:

$$\frac{\dot{Q}_{R,min}}{\dot{F} \cdot r} = \frac{1}{\alpha - 1}\left[1 + \alpha \frac{x_F^*}{1 - x_F^*}\right](1 - z_F) \quad \text{and} \tag{4.47}$$

$$\frac{\dot{Q}_{C,min}}{\dot{F} \cdot r} = -z_F\left[\frac{1}{(\alpha - 1)\, x_F^*} + 1\right]$$

The concentration x_F^* can be determined from the concentration of the feed by using Equation (4.36). In the special case of a boiling-liquid feed having composition x_F, the following equations are obtained:

$$\frac{\dot{Q}_{R,min}}{\dot{F} \cdot r} = \frac{1}{\alpha - 1} + x_F \quad \text{and} \quad \frac{\dot{Q}_{C,min}}{\dot{F} \cdot r} = -\left[\frac{1}{\alpha - 1} + x_F\right] \tag{4.48}$$

The heat requirement $\dot{Q}_{R,min}$ consists of two terms. The first term $1/(\alpha - 1)$ accounts for the heat required to maintain countercurrent flow in the column. This term is independent of feed composition. The second term x_F, however, is directly proportional to the composition of the feed. It represents the heat required for vaporization of the overhead product; this quantity of heat would also be required for single stage distillation. If the relative volatility α is low, the heat requirement is nearly independent of feed concentration. This fact is very important in process design.

For a saturated-vapor feed of composition y_F, Equation (4.47) becomes:

$$\frac{\dot{Q}_{R,min}}{\dot{F} \cdot r} = \frac{1}{\alpha - 1} \quad \text{and} \quad \frac{\dot{Q}_{C,min}}{\dot{F} \cdot r} = \frac{1}{\alpha - 1} + 1 \tag{4.49}$$

Both heat streams are independent of the composition of the feed. If the distillate is withdrawn in the liquid state, more heat must, of course, be removed from the top of the column than is supplied at the bottom.

In general, either the quantity of heat $\dot{Q}_R$ to be added or the quantity of heat $\dot{Q}_C$ to be removed is determined by using the above equations. The remaining quantity of heat is established by using an enthalpy balance around the column. This enthalpy balance can be carried out especially easily on the enthalpy-concentration diagram in Fig. 4-14A.

If the feed is introduced in the boiling state and the distillate and bottom products are removed in the boiling state, then the following equation holds:

$$\dot{Q}_R \cong -\dot{Q}_C \tag{4.50}$$

This relationship is fully valid only when the boiling point line on the enthalpy-concentration diagram, on which points $\dot{F}$, $\dot{B}$, and $\dot{D}$ are located, is a straight line.

The latent heats of vaporization of components *a* and *b*, r_a and r_b, need not be equal in this case.

4.3 MULTISTAGE DISTILLATION OF TERNARY MIXTURES

The binary mixtures discussed in Section 4.2 represent a very special case. In general, multicomponent mixtures have to be separated. Distillation of a ternary mixture of substances *a, b* and *c* can be represented relatively clearly on a triangular diagram. Relationships derived for ternary mixtures are very helpful for a better understanding of distillation. Ternary distillation is more general than the special case of binary distillation.

4.3.1 Calculations Based on Material Balances

Calculations of ternary mixtures follow the same basic principles as have been used for binary mixtures in Section 4.2. However, the graphical approach in the McCabe-Thiele diagram must be replaced by a numerical approach.

Distillation of ternary mixtures is conveniently calculated by first specifying the column and the operating conditions and then determining concentrations at the top and bottom of the column. If the required product specifications are not achieved, either the operating conditions or the column configuration have to be altered and the calculation repeated.

The number of equilibrium stages n and the position of the feed entry point are required for column specification. The operating conditions are characterized sufficiently by the reflux ratio R_L and the relative amount of overhead product $\dot{D}/\dot{F}$. The calculation is best carried out in a manner analogous to the stepwise graphical method used for binary mixtures described in Section 4.1.1 (McCabe-Thiele diagram). By starting with a set liquid composition (e.g., at the bottom), the equilibrium vapor-phase state y_i^* is calculated. For ideal mixtures:

$$y_a^* = \frac{\alpha_{ac} \cdot x_a}{1 + (\alpha_{ac} - 1) \cdot x_a + (\alpha_{bc} - 1) \cdot x_b}$$

$$y_b^* = \frac{\alpha_{bc} \cdot x_b}{1 + (\alpha_{ac} - 1) \cdot x_a + (\alpha_{bc} - 1) \cdot x_b} \qquad (4.51)$$

$$y_c^* = 1 - y_a - y_b$$

For non-ideal mixtures, the equilibrium state can be determined by more complex methods that have been described in Chapter 2.

The next step involves elucidation of the new liquid state from the vapor state by using the operating line. A material balance around the bottom of the column gives the following equation for the operating line (see Equation (4.19)):

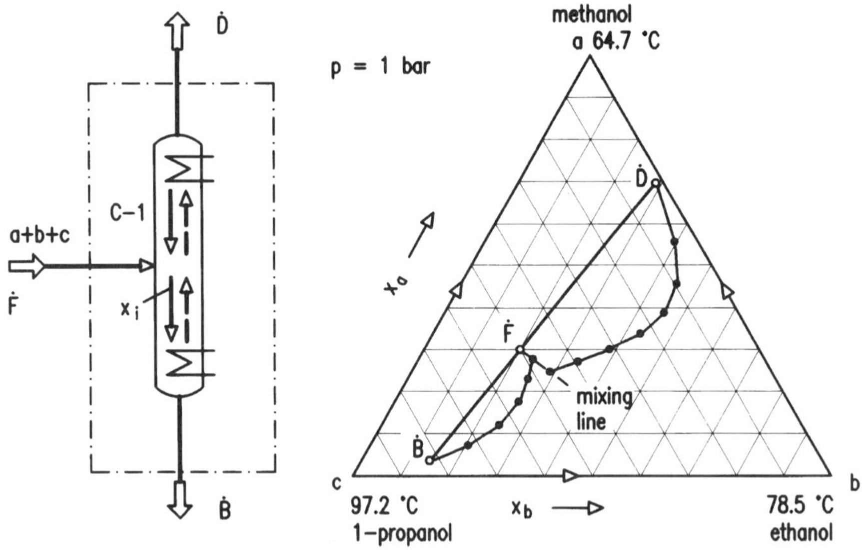

Fig. 4-18. Separation of a ternary mixture at finite reflux ratio. A concentration jump occurs at the feed point due to admixing of the feed: (A) –·– Balance around the column, (B) ––○–– changes in liquid concentration in the column.

$$
\begin{aligned}
x_a &= \frac{R_G}{R_G + 1} y_a^* + \frac{1}{R_G + 1} x_{Ba} \\
x_b &= \frac{R_G}{R_G + 1} y_b^* + \frac{1}{R_G + 1} x_{Bb} \\
x_c &= 1 - x_a - x_b
\end{aligned} \tag{4.52}
$$

The alternate use of Equations (4.51) and (4.52) produces a stepwise mathematical approach similar to the graphical construction for binary mixtures on the y-x diagram (see Section 4.2.1). In Fig. 4-18, the concentration profile observed in the stripping section of the column is represented by dots.

When the feed point has been reached, admixing of the feed must be considered in the calculation process. In most cases the feed concentrations are different from the internal concentrations. The following relationship is obtained from a material balance for the liquid composition $(x_i)'$ directly above the feed point by assuming a boiling-liquid feed and negligible temperature differences at the feed entry point:

$$
\begin{aligned}
(x_a)' &= \frac{(R_G + 1) \cdot \dot{B}/\dot{F}}{(R_G + 1) \cdot \dot{B}/\dot{F} - 1} (x_a - x_{Fa}) + x_{Fa} \\
(x_b)' &= \frac{(R_G + 1) \cdot \dot{B}/\dot{F}}{(R_G + 1) \cdot \dot{B}/\dot{F} - 1} (x_b - x_{Fb}) + x_{Fb}
\end{aligned} \tag{4.53}
$$

Equation (4.53) represents a straight line (mixing line) through $\dot{F}$ in the triangular diagram. The states of the liquid below ($\dot{L}$, x_i) and above ($\dot{L}$, x_i)′ the feed point must lie on the mixing line, as indicated in Fig. 4-18.

The equation of the operating line changes above the feed point because the quantity of liquid is altered. The following holds:

$$
\begin{aligned}
x_a &= \frac{R_G \cdot y_a^* + x_{Ba} - x_{Fa} \cdot \dot{F}/\dot{B}}{R_G + 1 - \dot{F}/\dot{B}} \\
x_b &= \frac{R_G \cdot y_b^* + x_{Bb} - x_{Fb} \cdot \dot{F}/\dot{B}}{R_G + 1 - \dot{F}/\dot{B}}
\end{aligned}
\qquad (4.54)
$$

The calculation is continued by alternating Equations (4.51) and (4.54) and working upward step-by-step to the top of the column.

The concentrations of components *a, b* and *c* calculated in this way are plotted as a function of the number of equilibrium stages n, as shown in Fig. 4-18. The medium-boiling substance *b* can pass through a concentration maximum in both sections of the column. Furthermore, a concentration jump occurs at the feed point.

The result of this stepwise calculation is the internal concentration profile. In addition, the external material balance must be fulfilled:

$$
\begin{aligned}
\dot{F} &= \dot{D} + \dot{B} \\
\dot{F} \cdot z_{Fa} &= \dot{D} \cdot x_{Da} + \dot{B} \cdot x_{Ba} \\
\dot{F} \cdot z_{Fb} &= \dot{D} \cdot x_{Db} + \dot{B} \cdot x_{Bb}
\end{aligned}
\qquad (4.55)
$$

Equation (4.55) is represented by a straight line in Fig. 4-18 on which the Feed $\dot{F}$, the overhead $\dot{D}$ and the bottoms $\dot{B}$ have to lie. Usually, this condition will not be met on the first attempt, and it is necessary to repeat the whole procedure (see Example 4-2).

A very important feature of ternary distillation is the fact that the intermediate boiling component *b* can exhibit a concentration maximum in both the stripping and the rectifying section of the column. Such a concentration maximum never occurs in binary distillation. At the maximum, the effective equilibrium line crosses the operating line. Figure 4-19 presents the McCabe-Thiele plot for each component of a ternary mixture. In each plot the operating lines take different positions; however, their slopes are always the same. The effective equilibrium line approaches the binary *a-b* equilibrium line at the top and the *b-c* equilibrium line at the bottom.

4.3.1.1 Problems Arising in Column Calculations Calculation of a ternary multistage distillation, as discussed in Section 4.3.1, is basically very simple and can be done with a calculator or personal computer. However, a few problems often complicate the calculation.

4.3.1.1.1 Choice of Reflux Ratio The values used in practice for the reflux ratio R_L and the reboil ratio R_G are about 10 to 30% higher than the minimum

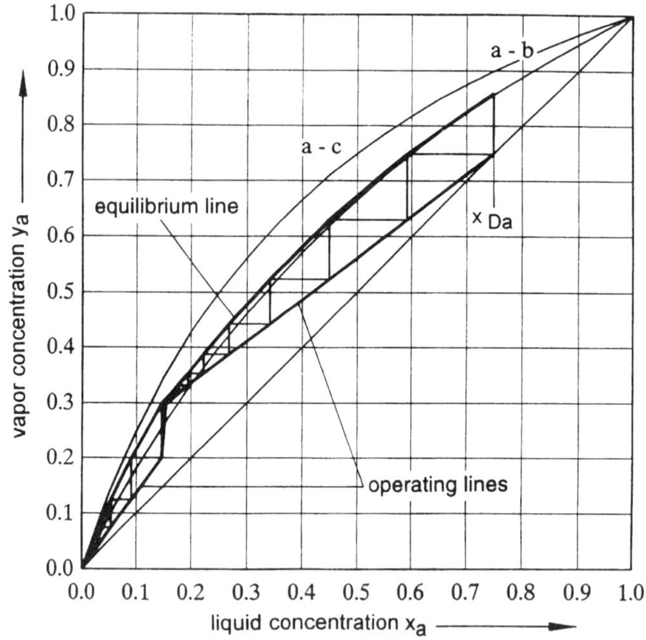

A

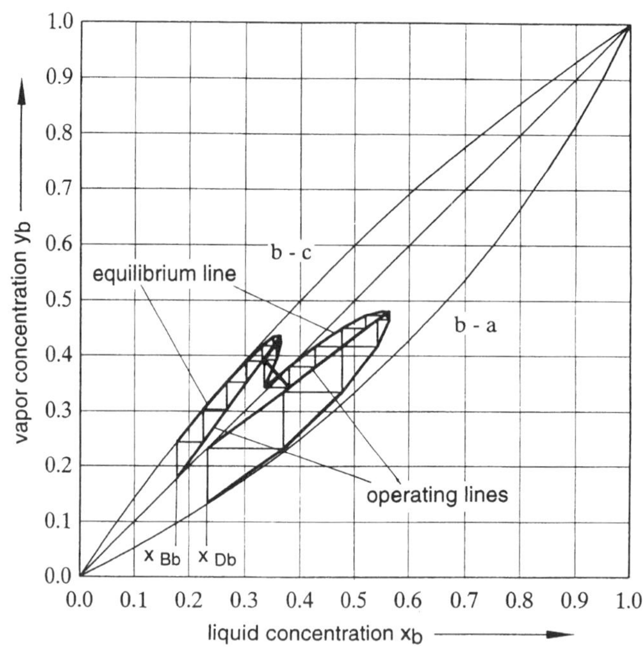

B

Fig. 4-19. *McCabe-Thiele plots for all components of a ternary mixture. The effective equilibrium line lies between the two binary equilibrium lines. For the intermediate boiler* b *the effective equilibrium line circles the operating line.*

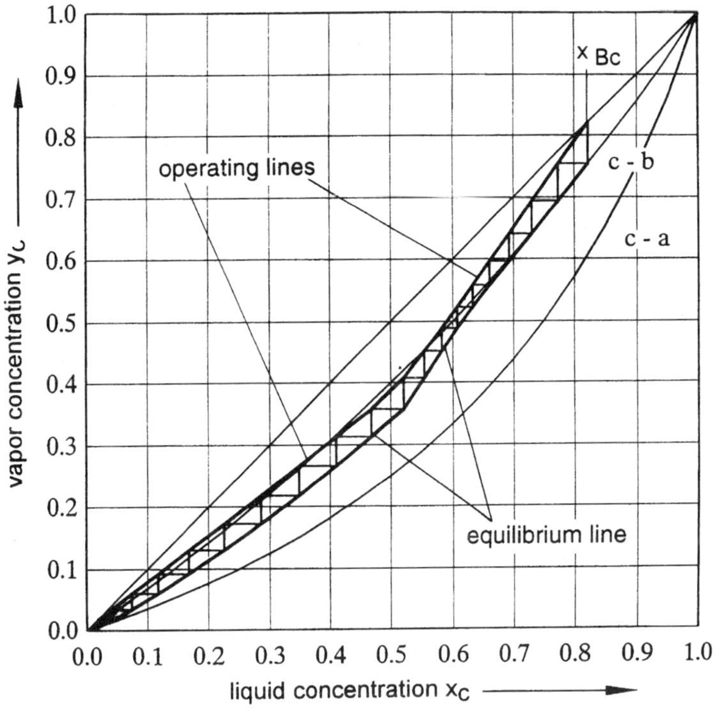

C

Fig. 4-19. (continued)

values. Relationships for calculating R_{Lmin} and R_{Gmin} will be developed in Section 4.3.3; they apply to ideal mixtures and to weakly non-ideal mixtures. However, mixtures that deviate more strongly from ideal behavior can produce a pinch point above or below the feed point (tangential pinch). The reflux ratio can then be determined only by iterative column calculations.

4.3.1.1.2 Choice of Bottom State The bottom composition x_{Bi} is the iteration variable in the calculation process described in Section 4.3.1. A poorly estimated value of x_{Bi} often gives unreasonable concentrations in the first iteration, e.g., concentrations >1 or <0. A useful approach requires knowledge of separation regions (Section 4.3.2.1). Even if the choice of bottom composition is satisfactory, unreasonable concentrations can still be obtained in the rectifying section of the column. This can be avoided by starting with estimated values of the distillate composition x_{Di} for calculation of the rectifying section. The calculations begin at both the top and bottom of the column, and continue step-by-step until the feed point is reached. Estimated values are checked with the mixing calculation at the feed point. For real mixtures, this procedure is effective; however, iterations are required at each step of the calculation in the rectifying section because equilibrium relationships cannot be solved for liquid concentrations.

4.3.1.1.3 Improvement of Estimated Values The material balance around the column is generally not fulfilled after the first calculation. A new bottom composition must be chosen which complies better with the balance around the column. No physically based criteria are available for improving the estimated values; this is the most severe handicap associated with column calculations. Indeed, serious convergence problems often occur at this point, especially in the case of highly nonideal systems.

4.3.1.1.4 Choice of Feed Point For binary mixtures, the feed is introduced at that point of the column at which the internal liquid concentration is identical to the composition of the feed. This is not possible in ternary distillation because the compositions in the column usually differ from that of the feed. As a rule, the concentration of only one substance can be made equal to that of the feed. Other substances undergo irreversible changes of state during mixing, which energetically impair separation. A series of criteria are described by [Vogelpohl 1975] for optimal choice of feed point; most of the criteria aim at decreasing the irreversibilities.

EXAMPLE 4-2:

Find by tray-to-tray calculation the internal concentration profile of a column for the ternary mixture nitrogen(*a*)/argon(*b*)/oxygen(*c*).

Data—Feed concentration: $x_{FN2} = 0.25$; $x_{FAr} = 0.30$; $x_{FO2} = 0.45$. Relative volatilities: $\alpha_{ac} = 4.032$; $\alpha_{bc} = 1.339$; $\alpha_{cc} = 1.0$. Distillate concentration: $x_{DN2} = 0.98$; $x_{DAr} = 0.02$; $x_{DO2} = 0.0$. Bottoms concentration: $x_{BN2} = 0.05$.

Solution:

1. Composition of bottoms—A material balance around the column yields for nitrogen:

$$\dot{D}/\dot{F} = \frac{x_{FN2} - x_{BN2}}{x_{DN2} - x_{BN2}} = \frac{0.25 - 0.05}{0.98 - 0.05} = 0.215.$$

For oxygen:

$$\dot{D}/\dot{F} = \frac{x_{FO2} - x_{BO2}}{x_{DO2} - x_{BO2}} = \frac{0.45 - x_{BO2}}{0.00 - x_{BO2}} = 0.215;$$

$$x_{BO2} = 0.45/(1 - 0.215) = 0.5733;\ x_{BAr} = 1 - 0.05 - 0.5733 = 0.376.$$

2. Reboil ratio R_G—The minimum reboil ratio is $R_{Gmin} = 0.696$ (see Example 4-3). The selected reboil ratio is $R_G = 1.1$.
3. Operating line in the stripping section (Equation 4.52):

$$x_{N2} = \frac{R_G}{R_G + 1} y^*_{N2} + \frac{1}{R_G + 1} x_{BN2} = \frac{1.1}{1.1 + 1} y^*_{N2} + \frac{1}{1.1 + 1} 0.05$$

$$= 0.5238\ y^*_{N2} + 0.0238;$$

$$x_{Ar} = \frac{R_G}{R_G + 1} y^*_{Ar} + \frac{1}{R_G + 1} x_{BAr} = \frac{1.1}{1.1 + 1} y^*_{Ar} + \frac{1}{1.1 + 1} 0.376$$

$$= 0.5238\ y^*_{Ar} + 0.179.$$

4. Tray-to-tray calculation in the stripping section:

Equilibrium Line	Operating Line
$N = 1 + 3.032 \cdot x_{N2} + 0.339 \cdot x_{Ar}$	$x_{N2} = 0.5238 \cdot y^*_{N2} + 0.0238$
$y^*_i = \alpha_{i,O2} \cdot x_i/N$	$x_{Ar} = 0.5238 \cdot y^*_{Ar} + 0.179$
$N = 1 + 3.032 \cdot 0.05 + 0.339 \cdot 0.376 = 1.279$	
$y^*_{N2} = \frac{4.032 \cdot 0.05}{1.279} = 0.1576$	
$y^*_{Ar} = \frac{1.339 \cdot 0.376}{1.279} = 0.394$	
	$x_{N2} = 0.5238 \cdot 0.1576 + 0.0238 = 0.1064$
	$x_{Ar} = 0.5238 \cdot 0.394 + 0.179 = 0.3854$
	$x_{O2} = 1 - 0.1064 - 0.3854 = 0.5082$
$N = 1 + 3.032 \cdot 0.1064 + 0.339 \cdot 0.3854 = 1.453$	
$y^*_{N2} = \frac{4.032 \cdot 0.1064}{1.453} = 0.2952$	
$y^*_{Ar} = \frac{1.339 \cdot 0.3854}{1.453} = 0.355$	
	$x_{N2} = 0.5238 \cdot 0.2952 + 0.0238 = 0.178$
	$x_{Ar} = 0.5238 \cdot 0.355 + 0.179 = 0.365$
	$x_{O2} = 1 - 0.178 - 0.365 = 0.457$
$N = 1 + 3.032 \cdot 0.178 + 0.339 \cdot 0.365 = 1.663$	
$y^*_{N2} = \frac{4.032 \cdot 0.178}{1.663} = 0.432$	
$y^*_{Ar} = \frac{1.339 \cdot 0.365}{1.663} = 0.294$	
	$x_{N2} = 0.5238 \cdot 0.432 + 0.0238 = 0.250$
	$x_{Ar} = 0.5238 \cdot 0.294 + 0.179 = 0.333$
	$x_{O2} = 1 - 0.250 - 0.333 = 0.417$

5. Admixing of the feed (Equation 4.53): $\dot{B}/\dot{F} = 1 - \dot{D}/\dot{F} = 1 - 0.215 = 0.785$;

$$(x_{N2})' = \frac{(R_G + 1)\dot{B}/\dot{F}}{(R_G + 1)\,\dot{B}/\dot{F} - 1}(x_{N2} - x_{FN2}) + x_{FN2}$$
$$= \frac{(1.1 + 1)0.785}{(1.1 + 1)0.785 - 1}(0.25 - 0.25) + 0.25 = 0.250;$$
$$(x_{Ar})' = \frac{(R_G + 1)\dot{B}/\dot{F}}{(R_G + 1)\dot{B}/\dot{F} - 1}(x_{Ar} - x_{FAr}) + x_{FAr}$$
$$= \frac{(1.1 + 1)0.785}{(1.1 + 1)\,0.785 - 1}(0.333 - 0.30) + 0.30 = 0.384.$$

6. Operating line in the rectifying section (Equation 4.54: $\dot{F}/\dot{B} = 1/0.785 = 1.274$;

$$x_{N2} = \frac{R_G \cdot y^*_{N2} + x_{BN2} - x_{FN2} \cdot \dot{F}/\dot{B}}{R_G + 1 - \dot{F}/\dot{B}} = \frac{1.1 \cdot y^*_{N2} + 0.05 - 0.25 \cdot 1.274}{1.1 + 1 - 1.274}$$
$$= 1.332\, y^*_{N2} - 0.325;$$
$$x_{Ar} = \frac{R_G \cdot y^*_{Ar} + x_{BAr} - x_{FAr} \cdot \dot{F}/\dot{B}}{R_G + 1 - \dot{F}/\dot{B}} = \frac{1.1 \cdot y^*_{Ar} + 0.376 - 0.30 \cdot 1.274}{1.1 + 1 - 1.274}$$
$$= 1.332\, y^*_{Ar} - 0.0075.$$

7. Tray-to-tray calculation in the rectifying section:

Equilibrium Line	Operating Line
$N = 1 + 3.032 \cdot x_{N2} + 0.339 \cdot x_{Ar}$	$x_{N2} = 1.332 \cdot y^*_{N2} - 0.325$
$y^*_i = \alpha_{i,O2} \cdot x_i/N$	$x_{Ar} = 1.332 \cdot y^*_{Ar} - 0.0075$
$N = 1 + 3.032 \cdot 0.250 + 0.339 \cdot 0.384$	
$N = 1.888$	
$y^*_{N2} = \frac{4.032 \cdot 0.250}{1.888} = 0.534$	
$y^*_{Ar} = \frac{1.339 \cdot 0.384}{1.888} = 0.272$	
	$x_{N2} = 1.332 \cdot 0.534 - 0.325 = 0.386$
	$x_{Ar} = 1.332 \cdot 0.272 - 0.0075 = 0.355$
	$x_{O2} = 1 - 0.386 - 0.355 = 0.259$
$N = 1 + 3.032 \cdot 0.386 + 0.339 \cdot 0.355$	
$N = 2.291$	
$y^*_{N2} = \frac{4.032 \cdot 0.386}{2.291} = 0.679$	
$y^*_{Ar} = \frac{1.339 \cdot 0.355}{2.291} = 0.207$	
	$x_{N2} = 1.332 \cdot 0.697 - 0.325 = 0.579$
	$x_{Ar} = 1.332 \cdot 0.2075 - 0.0075 = 0.2689$
	$x_{O2} = 1 - 0.579 - 0.2689 = 0.1521$
$N = 1 + 3.032 \cdot 0.579 + 0.339 \cdot 0.2689$	
$N = 2.847$	
$y^*_{N2} = \frac{4.032 \cdot 0.579}{2.847} = 0.820$	
$y^*_{Ar} = \frac{1.339 \cdot 0.2689}{2.847} = 0.126$	
	$x_{N2} = 1.332 \cdot 0.820 - 0.325 = 0.767$
	$x_{Ar} = 1.332 \cdot 0.126 - 0.0075 = 0.160$
	$x_{O2} = 1 - 0.767 - 0.160 = 0.073$

8. Check of external material balance (Equation 4.55):

$$x_{FN2} = \dot{D}/\dot{F}\, x_{DN2} + (1 - \dot{D}/\dot{F}) x_{BN2}$$
$$0.25 = 0.215 \cdot 0.767 + (1 - 0.215) \cdot 0.05$$
$$= 0.202 \text{ false!}$$
$$x_{FAr} = \dot{D}/\dot{F} x_{DAr} + (1 - \dot{D}/\dot{F}) x_{BAr}$$
$$0.30 = 0.215 \cdot 0.160 + (1 - 0.215) \cdot 0.376$$
$$= 0.330 \text{ false!}$$

The calculation has to be repeated.

4.3.2 Distillation of Ternary Mixtures at Total Reflux

Operation at total reflux is of minor industrial importance, but by this mode of operation the feasible separations of ternary mixtures can be determined very easily [Stichlmair 1991, Stichlmair and Herguijuela 1992].

4.3.2.1 Feasible Bottom and Top Fractions As demonstrated in Section 4.3.1, the bottom and top fractions of a distillation column must meet two conditions. The first condition reqires the material balance around the column to be fullfilled. This is:

$$\begin{aligned} \dot{F} &= \dot{D} + \dot{B} \\ \dot{F} \cdot x_{Fa} &= \dot{D} \cdot x_{Da} + \dot{B} \cdot x_{Ba} \\ \dot{F} \cdot x_{Fb} &= \dot{D} \cdot x_{Db} + \dot{B} \cdot x_{Bb} \end{aligned} \tag{4.56}$$

This system of linear equations is represented in Fig. 4-18 by a straight line on which the feed $\dot{F}$, the top product $\dot{D}$, and the bottom product $\dot{B}$ have to lie.

The second condition requires that the concentration x_{Di} of the top fraction and x_{Bi} of the bottom fraction are endpoints of the concentration profile developed within the column. In general, the internal concentration profile is a very complex function:

$$x_{in} = f(x_{Fi}, p_i^o, \gamma_i, T, p, n_1, n_2, R_L, D/F, \ldots\ldots\ldots\ldots) \tag{4.57}$$

The calculation of the internal concentration profile is very tedious, as shown in Section 4.4.1.

If the reflux R_L in a column is very high ($R_L \to \infty$) the internal concentration profile becomes identical with a section of a distillation line [Stichlmair 1988]. Distillation lines represent a sequence of equilibrium states that are easily calculated by the following recursion formula:

$$x_n = y^*; \qquad y_i^* = f(x_{n-1}, \ldots) \tag{4.58}$$

Thus, for ideal mixtures, Equation (2.32) gives the course of concentration profiles within the column:

$$\begin{aligned} x_{an} &= \frac{\alpha_{ac}^{n} \cdot x_{Ba}}{1 + (\alpha_{ac}^{n} - 1) \cdot x_{Ba} + (\alpha_{bc}^{n} - 1) \cdot x_{Bb}} \\ x_{bn} &= \frac{\alpha_{bc}^{n} \cdot x_{Bb}}{1 + (\alpha_{ac}^{n} - 1) \cdot x_{Ba} + (\alpha_{bc}^{n} - 1) \cdot x_{Bb}} \\ x_{cn} &= 1 - x_{an} - x_{bn} \end{aligned} \tag{4.59}$$

where the subscripts a, b and c denote the three components of the mixture; n denotes the number of equilibrium stages; and B the bottom product.

In Fig. 4-20A three possible top and bottom fractions of the distillation of a zeotropic mixture are shown. The top and bottom fractions must lie on a straight line through $\dot{F}$ and—due to the assumption of total reflux—on the same distillation line. The straight line representing Equation (4.56) forms a chord to the distillation line representing Equation (4.58 or 4.59). Obviously, there exists a large number of possible chords through $\dot{F}$. The extreme positions go through the origin and the terminus of the distillation lines, respectively. Top and bottom fractions both can lie on various distillation lines, also. It should be noted that the sides of the triangle are themselves distillation lines. The multiplicity of feasible top and bottom fractions forms the separation regions shaded in Fig. 4-20B. These regions are limited by the distillation line through the feed $\dot{F}$ and by a section of the straight line through $\dot{F}$ and the origin and terminus, respectively, of the distillation lines.

This method of separation region determination is applied to a system with a *b-c* minimum azeotrope in Fig. 4-21. This mixture exhibits two different origins of distillation lines (peaks of boiling point surface). Thus, there must exist a valley between these two peaks. The position of the valley is marked by the course of the border distillation line that runs from the minimum azeotrope to the low boiler *a*. Since the straight line through $\dot{F}$ always is a chord of a distillation line, the top and bottom fractions $\dot{D}$ and $\dot{B}$, respectively, can lie only at the same side of the border distillation line as does the feed $\dot{F}$. Hence, the border distillation line forms a barrier that cannot be crossed by distillation alike in the case of a binary azeotrope. As can be seen in Fig. 4-21B the border distillation line divides the ternary mixtures into two regions (distillation fields). Separation by distillation is possible only within the same field.

As a rule, only these constituents can be recovered in pure form from a ternary mixture where distillation lines either begin or end. In systems with more than one origin or terminus, there always exist borderlines that cannot be crossed by distillation.

There are a few exceptions to this rule. One of these exceptions is encountered in the system shown in Fig. 4-22. Here, the origin and terminus of the distillation lines are both on the same side of the triangle and the outer distillation lines pass through two corners of the triangular diagram. In this case there exists a straight line through $\dot{F}$ that intersects the distillation lines on the sides at a corner (here, the *a*-corner). Under these special conditions, pure substance *a* can be separated

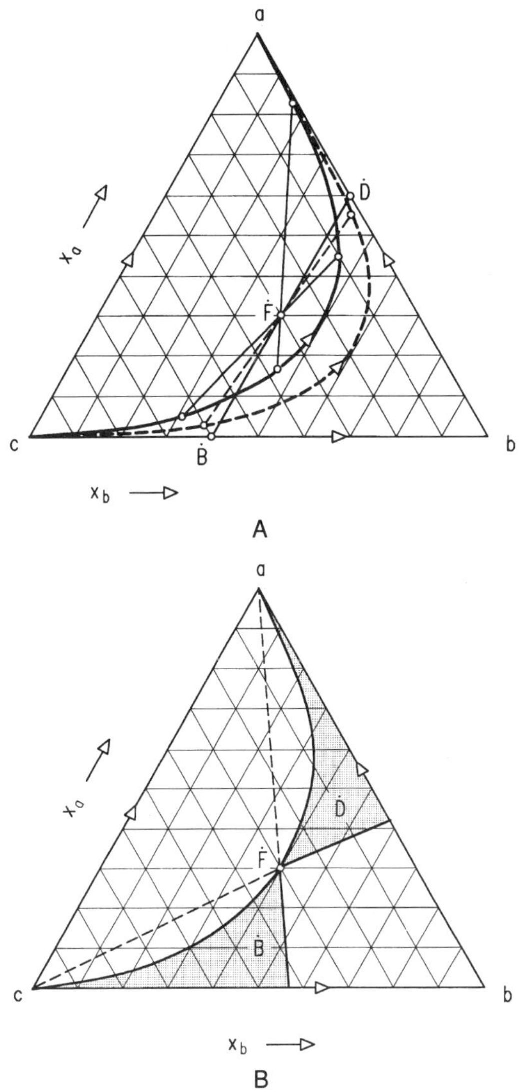

Fig. 4-20. *Separation region of a ternary zeotropic mixture: (A) Determination of possible top and bottom fractions, (B) separation region.*

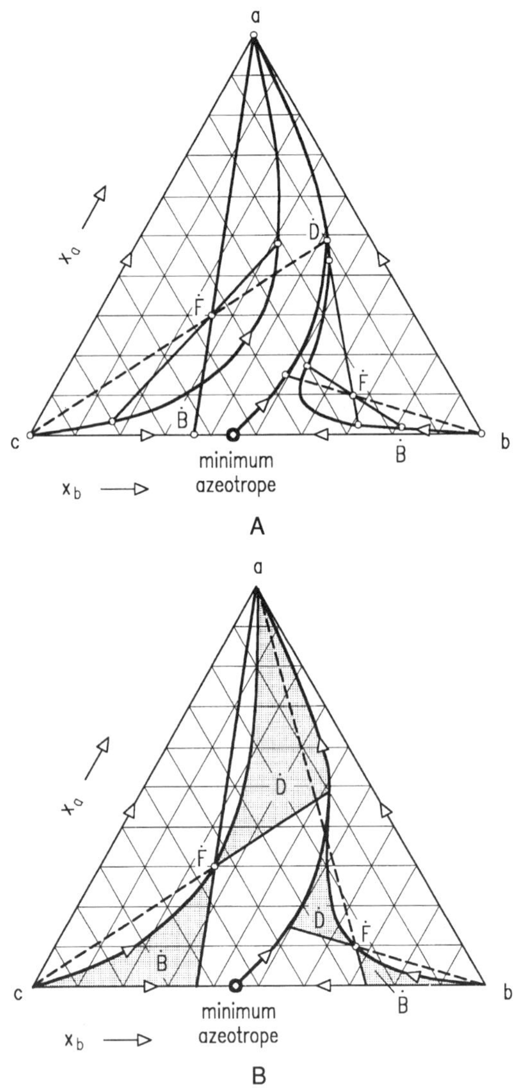

Fig. 4-21. Separation region of a ternary mixture with a b-c minimum azeotrope. The border distillation line forms a barrier that cannot be crossed at infinite reflux: (A) Determination of possible top and bottom fractions, (B) separation region.

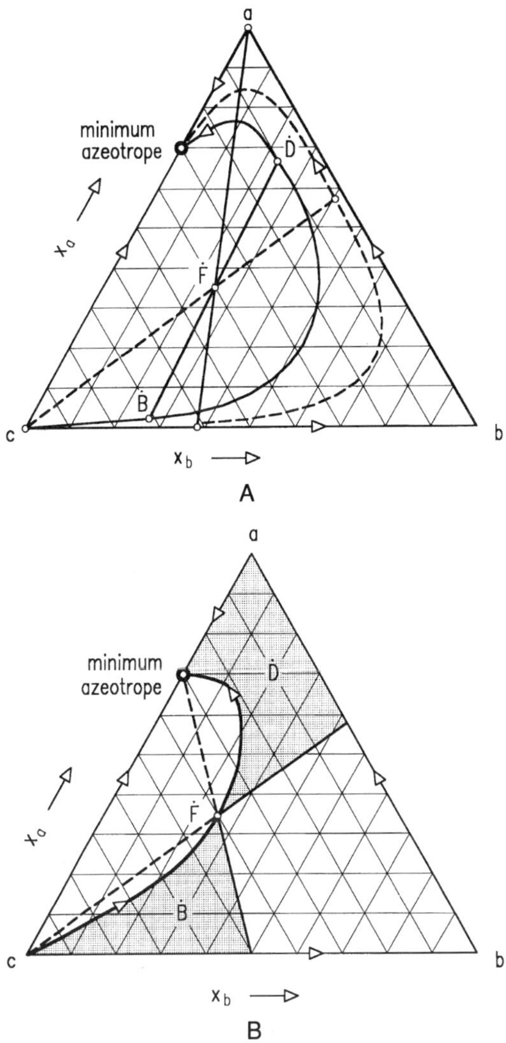

Fig. 4-22. *Separation region of a ternary mixture with an* a-c *minimum azeotrope. Pure component* a *can be recovered even though it is not the endpoint of distillation lines: (A) Determination of possible top and bottom fractions, (B) separation region.*

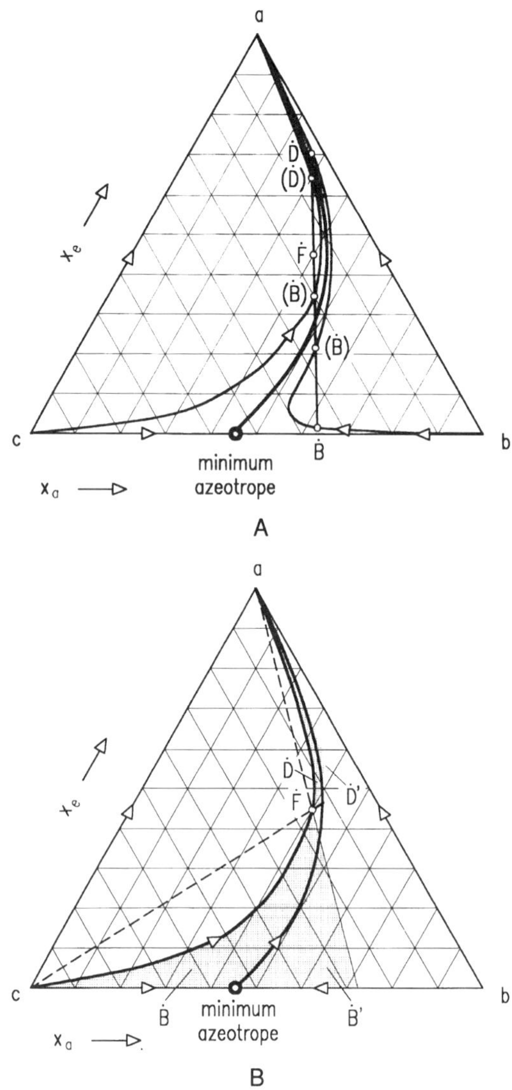

Fig. 4-23. Separation region of a ternary mixture with a b-c *minimum azeotrope when the feed is lying in the concave section of the border distillation line: (A) Determination of possible top and bottom fractions, (B) separation region.*

from a ternary mixture that is not an origin or terminus of distillation lines. Even pure intermediate boiler *b* can be separated when the feed contains only little high boiler *c*.

The second exception is dealt with in Fig. 4-23. If the feed Ḟ lies in the concave region of the border distillation line, the straight line through Ḟ can also form a chord to distillation lines on the other side of the border. Under these special conditions the border distillation line does not form a barrier for distillation. The separation regions expand over both sides of the border distillation line. This exception is very important and is frequently used in distillation processes (see Section 5.6.2).

Border distillation lines are valleys or ridges in the boiling point surface (Fig. 2-8). Both of them have the same consequences for distillation. They act as a barrier that cannot be crossed by distillation at infinite reflux or reboil.

4.3.2.1.1 Separation Regions at Finite Reflux Under finite reflux the top and bottom fractions, $\dot{D}$ and $\dot{B}$, respectively, do not necessarily lie on the same distillation line. Thus, the border distillation line is not a strict barrier for distillation at finite reflux. Several authors report that at low reflux ratio, the border for distillation is not identical with the border distillation line [Hausen 1935, Vogelpohl 1964, Levy et al. 1985, Doherty and Calderola 1985, Bernot et al. 1991, Wahnschafft et al. 1992]. The effective distillation border depends on reflux ratio as well as on concentrations of feed and products. Generally, the border at finite reflux can be determined only by tiresome iterations of rigorous column calculations.

A special case with a non-volatile third component is dealt with in Fig. 4-24. In the system $HCl/H_2O/H_2SO_4$ the vapor does not contain any H_2SO_4. Thus, concentration profiles may be determined in the McCabe-Thiele diagram alike for a binary system. It is assumed that hydrochloric acid with 18 mol-% HCl (31 wt-%) has to be recovered as overheads by a stripping column.

The sulfuric acid concentration, x_{H2SO4}, is approximately constant within the column. However, a sharp change of sulfuric acid concentration is caused by evaporating part of the liquid in the reboiler. The sulfuric acid concentration is higher in the bottoms than within the column. A material balance yields:

$$x_{BH2SO4} = \frac{\dot{L}/\dot{G}}{\dot{L}/\dot{G} - 1} x_{H2SO4} \tag{4.60}$$

Here, $\dot{L}/\dot{G}$ is the slope of the operating line in the McCabe-Thiele diagram. At minimum reboil ratio the slope is determined by extending the tangent to the equilibrium line at $y_{HCl} = 18$ mol-% in Fig. 4-24A. The intersection of the tangent with the diagonal gives the HCl concentration, x_{BHCl}, in the bottoms. The effective border line in the triangular diagram is found by plotting pairs of x_{BHCl} and x_{BH2SO4} for several internal sulfuric acid concentrations.

The border distillation line is represented in this system by points of intersection of the straight line $y_{HCl} = (y_{HCl})_{azeo}$ with the equilibrium lines in the McCabe-Thiele diagram of Fig. 4-24B. The border distillation line and the effective border line lie very close together in the triangular diagram. The maximum distance is far less than 1 mol-% HCl. However, it is very important that the effective border line reaches $x_{HCl} = 0$ at a sulfuric acid concentration of about 15.8 mol.-% (50 wt.-%) whereas the border distillation line approaches $x_{HCl} = 0$ asymptotically.

According to the present state of the art the border distillation line is a good and easy to find approximation of the effective distillation border. Thus, in the first step of process design, only operating regions determined under the assumption of total

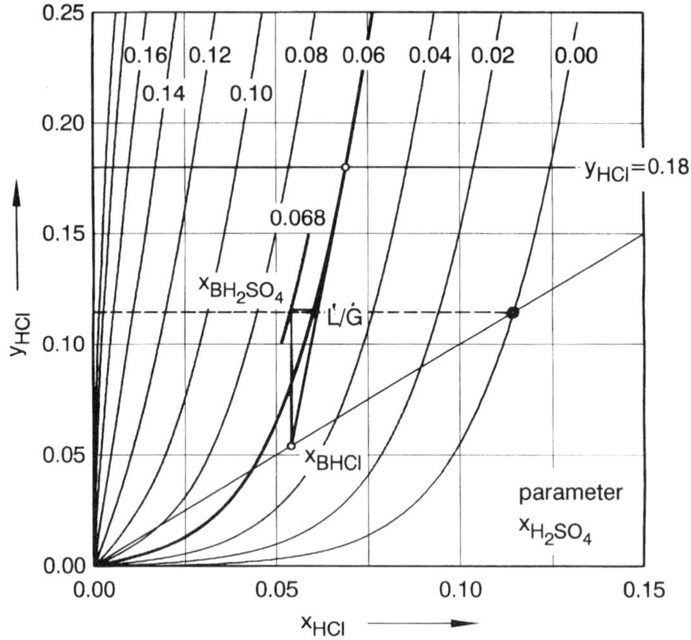

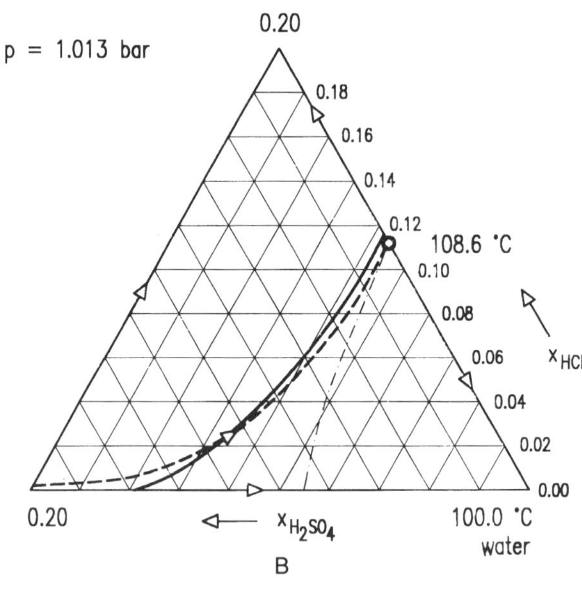

Fig. 4-24. *Determination of the effective distillation border for the system $HCl/H_2O/H_2SO_4$ by recovering hydrochloric acid with 18 mol-% HCl as overhead fraction of a stripping column: (A) McCabe-Thiele diagram, (B) distillation borders: —— border distillation line, --- effective distillation border at minimum reboil ratio, –·–·– line of pseudo azeotropes.*

reflux may be considered. However, one should be aware that the effective separation region might be slightly different. During detail engineering the effective distillation border is determined by rigorous column simulation and an evaluation can be made whether or not different column operation strategies will simplify the process.

4.3.2.2 Feasible Pure Products In most cases pure products have to be separated from a liquid mixture by distillation. The following provides a simple approach to figuring out which components of a mixture can be separated by a single distillation step either as top or as bottom product. Basis for this quick approach is knowledge of the boiling points (low, intermediate or high boiler) of the constituents and the concentration and boiling points of the binary azeotropes existing in the system.

With this information a triangular diagram can be sketched with the low boiler *a* at the top, the intermediate boiler *b* at bottom right and the high boiler *c* at bottom left. The position and type (maximum or minimum) of the azeotropes will be marked in this diagram (see Fig. 4-25). Then, the directions of decreasing boiling points are marked by arrows on the sides of the triangle (arrows pointing to decreasing boiling points). If the arrows on both sides of a corner point away from this corner, then this corner is an origin of distillation lines (peak in boiling point surface), and this substance can be recovered as a bottom product by distillation. If both arrows at the sides of a corner point toward that corner, then this corner is the terminus of distillation lines (hollow in boiling point surface) and that substance can be recovered as a top product by distillation. If one arrow points toward and the other one away from the corner, then this corner is neither an origin (peak) nor terminus (hollow) of distillation lines. That component can, in most cases, not be recovered by a single distillation step from a binary mixture (but see last row of Fig. 4-25).

This approach is demonstrated in Figs. 4-25A to 4-25L. In a zeotropic mixture (A) only the low boiler *a* and the high boiler *c* are end points of distillation lines. The approximate course of distillation lines can be sketched in the diagram. The distillation lines exhibit a sharp turn at the *b*-corner. In system B with an *a*-*b* minimum azeotrope only one substance can be separated in pure form (by a single distillation step). The same holds for systems B, E and H. Two constituents can be separated from systems A, D, F, G and I. All three constituents can be separated from a system with three binary azeotropes of the same type (not shown in Fig. 4-25).

If there exists more than one origin or terminus of distillation lines, then border distillation lines always exist. As they are always running between local boiling point peaks (or hollows) their average course can be sketched easily. In the systems G to I (third row of Fig. 4-25) two border distillation lines occur (one ridge and one valley). At the intersections a ternary saddle azeotrope is formed.

A special situation is met in the systems J to L (lower row of Fig. 4-25). Here, origin and terminus of distillation lines are close together and distillation lines make a sharp turn at two corners. As the straight line through the feed is always a chord of a distillation line (with top and bottom products at the intersection) substances *b* and *a* of system J (*b* and *c* of system K, *b* and *c* of system L) can be recovered.

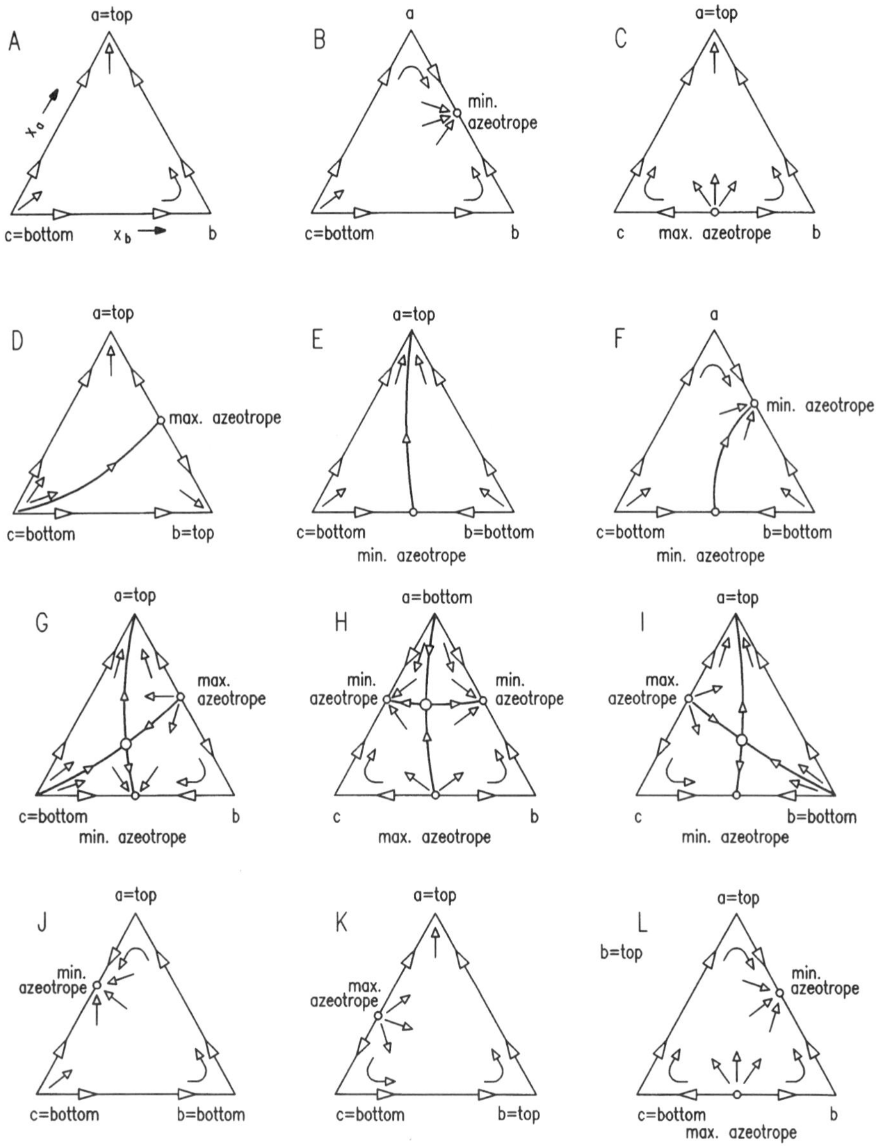

Fig. 4-25. *Determination of feasible pure products of a single distillation column. Arrows point in the direction of decreasing boiling temperatures. If both arrows at a corner point towards this corner, this component can be separated in pure form as an overhead fraction. If both arrows at a corner point away from this corner, this constituent can be recovered in pure form as a bottom fraction. If one arrow points toward the corner and the other one away from that corner, this component cannot be recovered in pure form (but see Fig. 4-22).*

Under such conditions of operation, neither top nor bottom of the column is a boiling point extremum, so column control might be difficult.

This easy evaluation of possible pure top and bottom products is very helpful in process design (see Section 5.3). The results are often quite astonishing. For example, in system H of Fig. 4-25 only the low boiler can be recovered in pure form by a single distillation step and, surprisingly, it emerges as the bottom product. With the technique described here it is very easy to develop separation sequences to separate some constituents from a multicomponent mixture in pure form.

4.3.3 Distillation of Ternary Mixtures at Minimum Reflux

Industrial distillation columns are preferentially operated at a low reflux ratio R_L in order to minimize energy requirement. Hence, the special case of the minimum reflux ratio R_{Lmin} is of considerable interest. A method described by [Underwood 1948] permits determination of the minimum reflux ratio for ideal systems with constant relative volatilities α_{ik}. However, the equations can only be solved iteratively. This section shows that the minimum reflux ratio can be calculated directly and also for nonideal systems.

Operation of a distillation column with minimum reflux or reboil is characterized by the existence of a pinch (i.e., point of intersection of operating and equilibrium line) within the column. At the pinch, the vapor is in equilibrium with the liquid, and consequently, mass transfer between the two phases comes to an end. Approaching the pinch requires an infinite number of equilibrium stages. In binary distillation there exists a double pinch immediately above and below the feed point (except in the cases where a tangential pinch occurs, see Chapter 4.2.4). As will be shown later on, the situation is more complex in ternary and multicomponent mixtures. Two restrictions are made in the following analysis:

- The liquid-to-gas ratio $\dot{L}/\dot{G}$ (slope of operating line) is assumed to be constant in each column section, and
- tangential pinches are not taken into account.

4.3.3.1 Pinches in Ternary Distillation Three classes of separations can be established in ternary and multicomponent distillation, see Fig. 4-26 [Stichlmair et al. 1993]. In the first class of separations, here called preferred separations, the low and high boiler are separated primarily. The intermediate boiler occurs in the overhead fraction as well as in the bottom fraction. In this class of separations, a double pinch exists immediately above and below the feed point. The pinch concentration x_{Pi} is equal to the feed concentration x_{Fi} (in case of a boiling liquid feed). The upper and lower column sections are operated with minimum reflux and reboil, respectively. The situation is nearly identical to the well-understood binary distillation.

In the second class of separations, an overhead fraction enriched in low boiler *a*

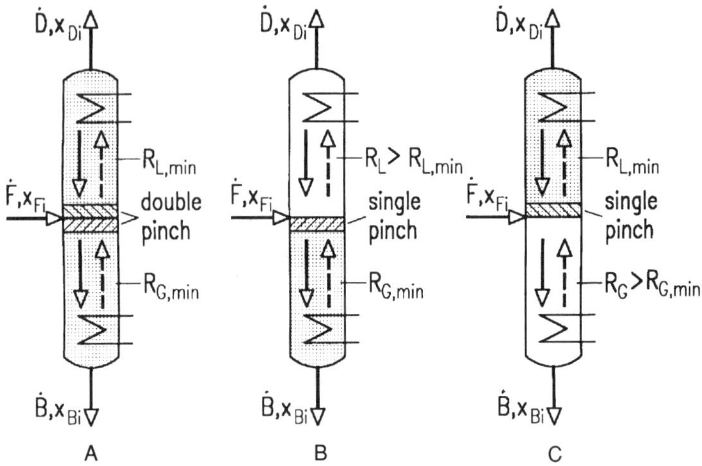

Fig. 4-26. Location of pinches at different modes of column operation with minimum reflux and reboil, respectively. The shaded sections are operated with minimum internal countercurrent flow. (A) Preferred separation. (B) Low boiler—rich distillate. (C) High boiler—rich bottoms.

is recovered. The rectifying section is operated with increased reflux to provide the purer distillate. Only the stripping section is operated with minimum reboil. Hence, there exists a single pinch in the stripping section immediately below the feed point. However, the pinch concentration is different from the feed concentration.

In the third class of separations, a bottom fraction is recovered that is richer in high boiler than in the preferred separation. In order to get the purer bottom product, the stripping section is operated with increased reboil, but the rectifying section is still operated with minimum reflux. In this case, a single pinch exists in the rectifying section immediately above the feed point. The pinch concentration is also different from the feed concentration.

4.3.3.1.1 Preferred Separation The relationships presented in Section 4.2.4 for the minimum reflux ratio of binary mixtures can be applied directly to ternary mixtures. Hence, as in Equation (4.33), the following is valid for each component *i:*

$$R_{Lmin} = \frac{x_{Di} - y^*_{Fi}}{y^*_{Fi} - x^*_{Fi}} \tag{4.61}$$

Equation (4.61) represents a system of three linear equations. By setting the distillate concentration x_{Di} of one component, e.g., the high boiler *c*, the minimum reflux ratio is:

$$R_{Lmin} = \frac{x_{Dc} - y^*_{Fc}}{y^*_{Fc} - x^*_{Fc}} \tag{4.62}$$

The distillate concentration of components *a* and *b* follow from Equation (4.61). A transformation yields:

$$
\begin{aligned}
x_{Da} &= R_{Lmin}\,(y^*_{Fa} - x^*_{Fa}) + y^*_{Fa} \\
x_{Db} &= R_{Lmin}\,(y^*_{Fb} - x^*_{Fb}) + y^*_{Fb}
\end{aligned}
\tag{4.63}
$$

This set of equations is represented by a straight line through x^*_{Fi} and y^*_{Fi} in the triangular diagram on which the feed $\dot{F}$ and the distillate $\dot{D}$ have to lie.

The minimum reboil ratio $R_G \equiv \dot{G}/\dot{B}$ is determined analogously. From Equation (4.40) follows:

$$
R_{Gmin} = \frac{x^*_{Fi} - x_{Bi}}{y^*_{Fi} - x^*_{Fi}} \tag{4.64}
$$

Setting the value x_{Bi} of one component, e.g., the low boiler *a*, the minimum reboil ratio is found by:

$$
R_{Gmin} = \frac{x^*_{Fa} - x_{Ba}}{y^*_{Fa} - x^*_{Fa}} \tag{4.65}
$$

For components *b* and *c*, Equation (4.64) yields:

$$
\begin{aligned}
x_{Bb} &= -R_{Gmin}(y^*_{Fb} - x^*_{Fb}) + x^*_{Fb} \\
x_{Bc} &= -R_{Gmin}(y^*_{Fc} - x^*_{Fc}) + x^*_{Fc}
\end{aligned}
\tag{4.66}
$$

This equation is also represented by a straight line through x^*_{Fi} and y^*_{Fi}. Thus, both products, $\dot{D}$ and $\dot{B}$, have to lie on the same straight line through the feed $\dot{F}$. This agrees with the external material balance represented by Equation (4.56).

If the feed is a boiling liquid, x^*_{Fa} becomes x_{Fa} (see Equation 4.37) and the preferred separation is determined by a straight line through the feed composition x_{Fi} and its equilibrium y^*_{Fi} as shown in Fig. 4-27. If the feed is saturated vapor, x^*_{Fa} is in equilibrium with the vapor concentration y_{Fa}. All these concentrations are subsequent points on the distillation line through the feed.

Under minimum reflux and reboil operation with the pinch point at the feed concentration, the separation region of a ternary system shrinks to a straight line that lies within the separation region at total reflux. This special class of separations is called ''preferred separation'' in the following. These separations represent a certain natural, preferred separation which is achieved at minimum energy input in an infinitely long distillation column. Here, the reflux ratio has its lowest value of all the separations possible in a ternary system. The preferred separation represents a very important class of separations. As described in Chapter 7, it is in many cases the most favorable path for the separation of a ternary system.

The internal liquid concentration profile is plotted in Fig. 4-28 for a sharp separa-

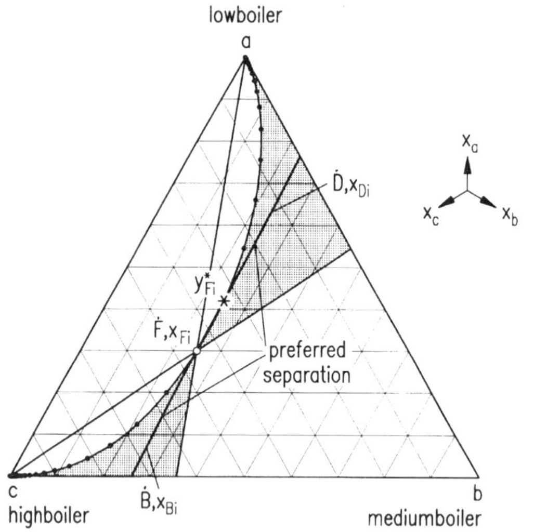

Fig. 4-27. *Preferred separation of a ternary mixture at minimum reflux. Both separation products, $\dot{D}$ and $\dot{B}$, can lie only on the straight line through x_{Fi} and y^*_{Fi} (feed at boiling point). Hence, the separation region of preferred separation shrinks to a straight line.*

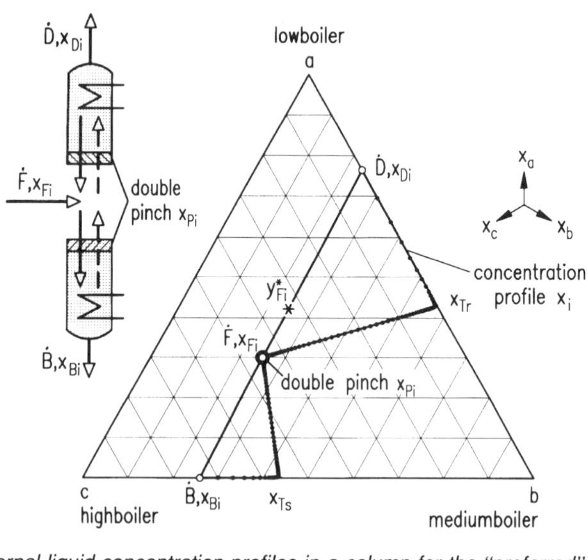

Fig. 4-28. *Internal liquid concentration profiles in a column for the "preferred" separation of a ternary mixture at minimum reflux. The most important characteristics are the transition points x_{Tr} and x_{Ts}: —— balance around the column, —·—·—·— profiles of liquid concentration within the column.*

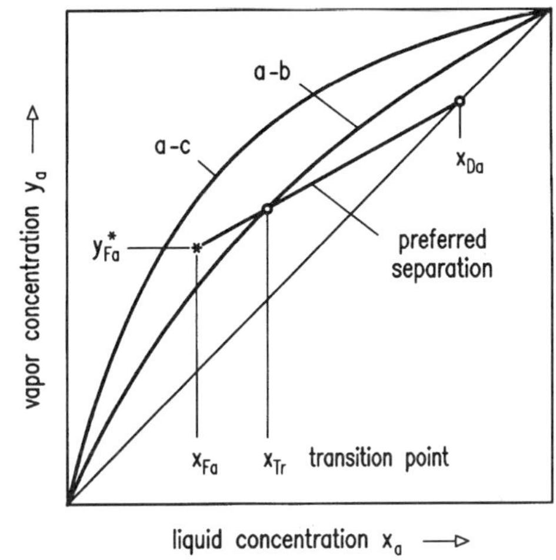

Fig. 4-29. *Determination of the transition point x_{Tr} of the rectifying section in a McCabe-Thiele plot.*

tion (i.e., no high boiler *c* in the overhead and no low boiler *a* in the bottoms) of an ideal mixture. From the feed point upwards, the high boiler *c* is primarily removed from the vapor. The right hand side of the triangular diagram is reached at the point x^*_{Tr}, where the ternary mixture transfers into a binary one. From this point upward, the intermediate boiler *b*, now being the highest boiling constituent, is removed.

From the feed point downward, the low boiler *a* is separated primarily (but not exclusively) from the liquid. The base of the triangular diagram is reached at the transition point x^*_{Ts}. From there on, the intermediate boiler *b* is separated from the liquid until the bottoms' concentration x_{Bi} is established.

The internal liquid concentration profile is characterized by the double pinch point located at the feed and, in case of sharp separations, the transition points. The concentrations of the transition points are easily determined in the McCabe-Thiele diagrams for the low boiler *a* and the high boiler *c*. They represent the intersection of the operating line of preferred separation with the equilibrium line of the binary mixtures *a-b* (see Fig. 4-29) and *b-c* (see Fig. 4-30), respectively.

4.3.3.1.2 Ideal Systems For ideal systems, the relationship between x_{Fi} and y^*_{Fi} can be formulated in general terms and inserted into Equations (4.61) and (4.64). For a sharp separation in the rectifying section, characterized by $x_{Dc} = 0$, this equation holds:

$$R_{Lmin} = \frac{1}{(\alpha_{ac} - 1) \cdot x^*_{Fa} + (\alpha_{bc} - 1) \cdot x^*_{Fb}} \qquad (4.67)$$

If the feed concentration x_{Fb} of the intermediate boiler *b* approaches 0, Equation (4.67) becomes identical to Equation (4.35), valid for binary mixtures.

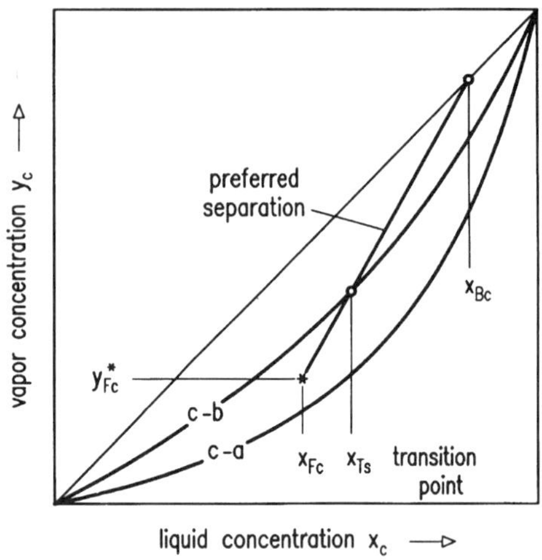

Fig. 4-30. *Determination of the transition point x_{Ts} of the stripping section in a McCabe-Thiele plot.*

The concentration of the distillate follows from Equation (4.63):

$$x_{Da} = \frac{(\alpha_{ac} - 1) \cdot x^*_{Fa}}{(\alpha_{ac} - 1) \cdot x^*_{Fa} + (\alpha_{bc} - 1) \cdot x^*_{Fb}} \quad \text{or} \quad x_{Da} = R_{Lmin}(\alpha_{ac} - 1) \cdot x^*_{Fa}$$

$$x_{Db} = \frac{(\alpha_{bc} - 1) \cdot x^*_{Fb}}{(\alpha_{ac} - 1) \cdot x^*_{Fa} + (\alpha_{bc} - 1) \cdot x^*_{Fb}} \quad \text{or} \quad x_{Db} = R_{Lmin}(\alpha_{bc} - 1) \cdot x^*_{Fb} \tag{4.68}$$

Analogous equations can be derived for the minimum reboil ratio R_{Gmin} and the composition of the bottoms. The reboil ratio for a sharp separation in the stripping section, i.e., $x_{Ba} = 0$, is:

$$R_{Gmin} = \left[\frac{\alpha_{ac}}{1 + (\alpha_{ac} - 1) \cdot x^*_{Fa} + (\alpha_{bc} - 1) \cdot x^*_{Fb}} - 1\right]^{-1} \tag{4.69}$$

The equation for the concentrations of the bottoms' product is:

$$x_{Bb} = \frac{(\alpha_{ac} - 1) \cdot x^*_{Fa}}{(\alpha_{ac} - 1) \cdot x^*_{Fa} + (\alpha_{bc} - 1) \cdot x^*_{Fb}}$$

$$x_{Bc} = \frac{(\alpha_{bc} - 1) \cdot x^*_{Fb}}{(\alpha_{ac} - 1) \cdot x^*_{Fa} + (\alpha_{bc} - 1) \cdot x^*_{Fb}} \tag{4.70}$$

Concentrations x^*_{Fi} can be determined from the point of intersection of the feed line Equation (4.20) with the equilibrium line, as described in Section 4.2.4. In the especially important case of a boiling-liquid feed with composition x_{Fi}, the following holds (see Equation 4.37):

$$x^*_{Fi} = x_{Fi} \tag{4.71}$$

Introduction of a saturated-vapor feed with composition y_{Fi} represents another important case. Here:

$$\begin{aligned} x^*_{Fa} &= \frac{y_{Fa} \cdot \alpha_{ac}^{-1}}{1 + (\alpha_{ac}^{-1} - 1) \cdot y_{Fa} + (\alpha_{bc}^{-1} - 1) \cdot y_{Fb}} \\ x^*_{Fb} &= \frac{y_{Fb} \cdot \alpha_{bc}^{-1}}{1 + (\alpha_{ac}^{-1} - 1) \cdot y_{Fa} + (\alpha_{bc}^{-1} - 1) \cdot y_{Fb}} \end{aligned} \tag{4.72}$$

The Equations (4.67) and (4.69), respectively, allow for the direct determination of minimum reflux ratio and reboil ratio, respectively, for the preferred separations. They are also a rough approximation for other separations possible in ternary systems.

4.3.3.1.3 Separation of a Low Boiler—Rich Distillate The preferred separation dealt with in the previous chapter is a special case of all the separations possible in a ternary system. It cannot provide a pure product. In most cases, pure constituent *a*, or at least an *a*-rich fraction, has to be recovered as overheads. Such separations are feasible (see Section 4.3.2.1) and there has to be a lower limit for the reflux ratio, too.

The internal liquid concentration profile calculated by rigorous column simulation is plotted in Fig. 4-31. The concentration of the low boiler in the distillate is higher than in preferred separation. Again, a sharp separation with no low boiler in bottom and no high boiler in overhead fraction is considered. The rectifying section is operated with a reflux higher than minimum reflux in order to establish the purer overhead product. There are several effects not encountered in binary and preferred separations:

- The internal liquid concentration profile never reaches the concentration of the feed. Hence, a concentration jump is caused by admixing the boiling liquid feed.
- The concentration profile immediately above the feed point is identical with the profile at preferred separation, see Fig. 4-28. The location of the transition point x_{Tr} remains unchanged.
- There exists a single pinch immediately below the feed point. The concentration of the pinch differs from feed concentration. The pinch lies on a line between transition point x_{Tr} and feed point x_{Fi}.
- The stripping section of the column is operated with minimum reboil whereas the rectifying section is operated above the minimum reflux ratio. The feed

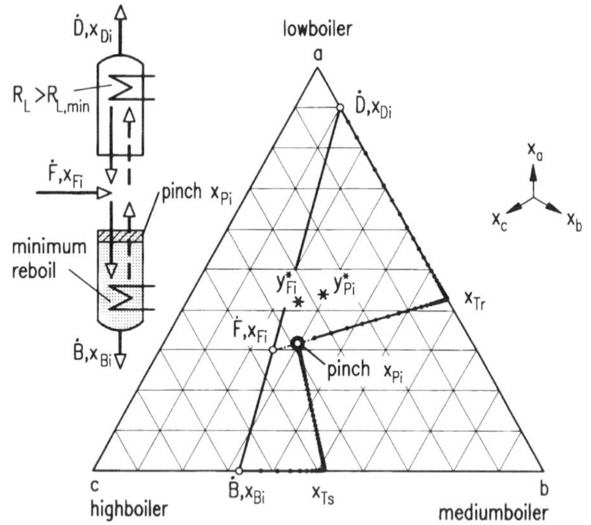

Fig. 4-31. *Separation of a low boiler rich overhead fraction from a ternary mixture at minimum reflux ratio. There exists only a single pinch point located directly below the feed point: — balance around the column, --●-- profile of liquid concentration in the column, –·–·– mixing line at the feed point.*

into this section is a mixture of liquid from rectifying section and column feed. The concentration of this mixture is the pinch concentration.

4.3.3.1.4 Separation of a High Boiler—Rich Bottom Fraction It is well known that a high boiler rich bottom fraction (or even pure high boiler) can be recovered from a ternary mixture. In order to perform such a separation, the reboil ratio has to be increased. The internal profile of liquid concentrations is plotted in the triangular diagram of Fig. 4-32. There exist some noteworthy effects:

- The internal concentration profile is always different from feed concentration. Here, as above, a concentration jump is caused by admixing the feed.
- The concentration profile immediately below the feed is identical to the concentration profile of the preferred separation, see Fig. 4-28. The transition point x_{Ts} has the same location for both separations.
- There exists a single pinch immediately above the feed point. The concentration of the pinch, x_{Pi}, is different from feed concentration. The pinch lies on the line between transition point x_{Ts} and feed x_{Fi}.
- The rectifying section is operated with minimum reflux, whereas the stripping section is operated above the minimum reboil ratio. The feed into that section is a vapor which is in equilibrium with the pinch concentration.

4.3.3.2 Localizing the Pinch Point In preferred separations, the concentration of the pinch is equal to that of the feed, as in binary distillation. In the general case of ternary distillation, however, the pinch concentration differs from the feed concentration. Determination of the pinch concentration makes use of the fact that the pinch lies on a line between transition point x_T and feed x_F [Levy et al. 1985]. This line is determined by the concentration profile immediately above or below the feed point. As is seen from Figs. 4-28, 4-31 and 4-32, this section of the concentration profile is the same in preferred and in low boiler and high boiler separation, respectively. This is rather surprising since reflux and reboil ratios are different.

The internal concentration pofiles have, for sharp separations, three noteworthy points: the pinch point x_{Pi}, the transition point x_{Tr} in the rectifying section and the transition point x_{Ts} in the stripping section. At the pinch point, all three components in liquid and vapor are in the equilibrium state. At the transition point in the rectifying section, x_{Tr}, the vapor and liquid concentrations of substances *a* and *b* are in equilibrium state (but not the high boiler *c* that is always present in small concentrations). By analogy, at the transition point in the stripping section, x_{Ts}, the vapor and liquid concentration of the substances *b* and *c* are in equilibrium (but not the low boiler *a*). The lever arm rule gives for the transition point (see Fig. 4-33):

$$(d_1 \cdot \dot{G})_T = ((d_1 + d_2) \cdot \dot{L})_T \quad \text{or} \quad \left(\frac{\dot{L}}{\dot{G}}\right)_T = \left(\frac{d_1}{d_1 + d_2}\right)_T \tag{4.73}$$

For the pinch point:

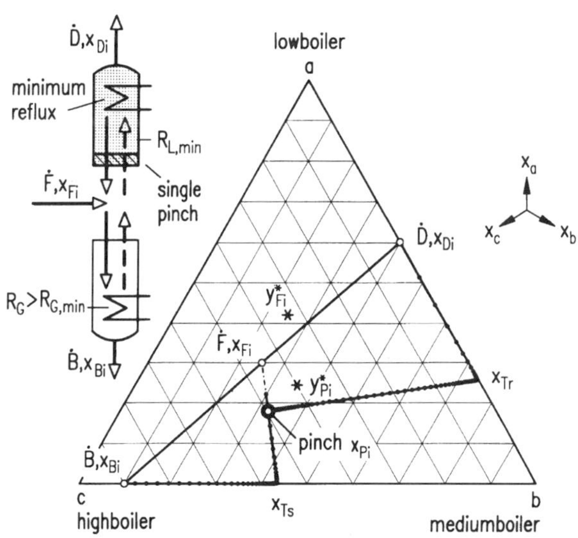

Fig. 4-32. Internal concentration profiles for a high boiler separation. There exists only a single pinch located directly above the feed point.

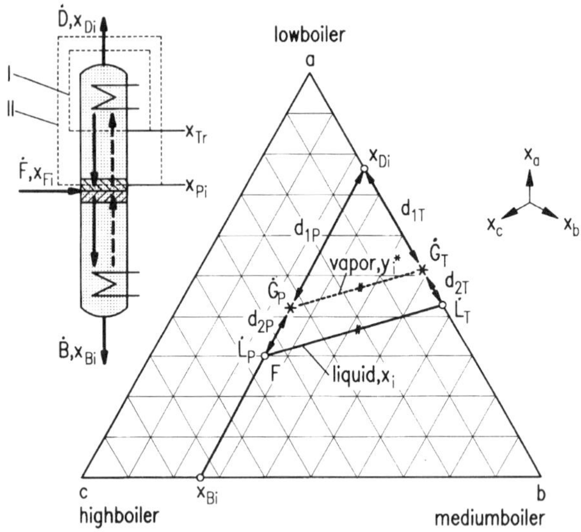

Fig. 4-33. *Geometrical representation of the condition $(\dot{L}/\dot{G})_T = (\dot{L}/\dot{G})_P$. This condition is met when the state of liquid and vapor at both points lie on parallel lines.*

$$(d_1 \cdot \dot{G})_P = ((d_1 + d_2) \cdot \dot{L})_P \quad \text{or} \quad \left(\frac{\dot{L}}{\dot{G}}\right)_P = \left(\frac{d_1}{d_1 + d_2}\right)_P \tag{4.74}$$

The liquid-to-vapor ratio $\dot{L}/\dot{G}$ (= slope of operating line) is the same for both points if constant molar overflow is assumed. Hence:

$$\left(\frac{\dot{L}}{\dot{G}}\right)_T = \left(\frac{\dot{L}}{\dot{G}}\right)_P \quad \text{or} \quad \left(\frac{d_1}{d_1 + d_2}\right)_T = \left(\frac{d_1}{d_1 + d_2}\right)_P \tag{4.75}$$

The ratios of the distances d_1/d_2 have to be the same at the transition point and the pinch point. This condition is met in ideal systems with temperature-independent relative volatilities when the line of equilibrium concentrations y_i^* are parallel to a line of liquid concentrations x_i.

In ideal systems the equilibrium concentrations y_i^* for any straight line of liquid concentrations x_i also form a straight line (see Fig. 4-34A). These lines are, in the general case, not parallel to each other. There exist, however, two directions for each point where the vapor equilibrium concentrations are parallel to liquid concentrations (Fig. 4-34B). The concentration profiles starting from the pinch have to follow this special direction. The condition is:

$$\frac{dx_a}{dx_b} = \frac{dy_a^*}{dy_b^*} \tag{4.76}$$

The total derivatives of vapor-liquid equilibria y_i^* are:

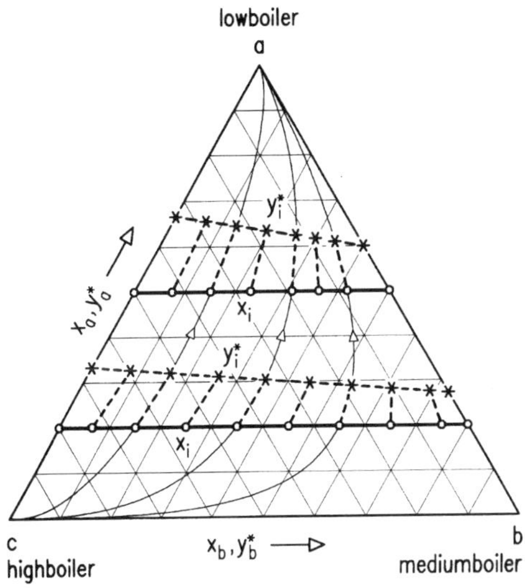

A

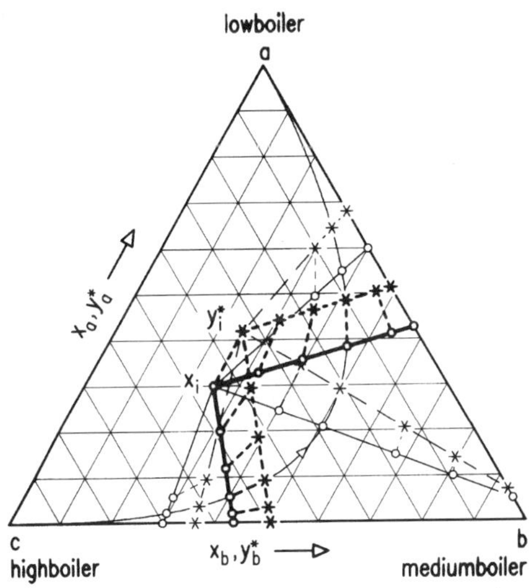

B

Fig. 4-34. *Vapor-liquid equilibrium of an ideal ternary mixture: (A) If the liquid state lies on a straight line, the vapor state must also lie on a straight line. (B) At any point x_i of an ideal mixture, two directions of vapor-liquid equilibrium parallelity (bold lines) exists.*

$$dy_a^* = \frac{\partial y_a^*}{\partial x_a} dx_a + \frac{\partial y_a^*}{\partial x_b} dx_b \quad \text{and} \quad dy_b^* = \frac{\partial y_b^*}{\partial x_a} dx_a + \frac{\partial y_b^*}{\partial x_b} dx_b \tag{4.77}$$

Combining Equations (4.76) and (4.77) yields:

$$\frac{dx_a}{dx_b} = \frac{C_1 \cdot (dx_a/dx_b) + C_2}{C_3 \cdot (dx_a/dx_b) + C_4} \quad \text{:with } C_1 = \frac{\partial y_a^*}{\partial x_a}; \quad C_2 = \frac{\partial y_a^*}{\partial x_b}; \quad C_3 = \frac{\partial y_b^*}{\partial x_a}; C_4 = \frac{\partial y_b^*}{\partial x_b} \tag{4.78}$$

After rearrangement:

$$C_3 \left(\frac{dx_a}{dx_b}\right)^2 + (C_4 - C_1)\left(\frac{dx_a}{dx_b}\right) - C_2 = 0 \tag{4.79}$$

Solution of the quadratic equation:

$$\frac{dx_a}{dx_b} = -\frac{C_4 - C_1}{2 \cdot C_3} \pm \sqrt{\left(\frac{C_4 - C_1}{2 \cdot C_3}\right)^2 + \frac{C_2}{C_3}} \tag{4.80}$$

For ideal systems, the parameters C_1, C_2, C_3 and C_4 are:

$$\begin{aligned} C_1 &= \alpha_{ac} \cdot (1 + (\alpha_{bc} - 1) \cdot x_{Fb})/N^2 \\ C_2 &= (1 - \alpha_{bc}) \cdot x_{Fa} \cdot \alpha_{ac}/N^2 \\ C_3 &= (1 - \alpha_{ac}) \cdot x_{Fb} \cdot \alpha_{bc}/N^2 \\ C_4 &= \alpha_{bc}(1 + (\alpha_{ac} - 1) \cdot x_{Fa})/N^2 \quad \text{with } N = 1 + (\alpha_{ac} - 1) \cdot x_{Fa} + (\alpha_{bc} - 1) \cdot x_{Fb} \end{aligned} \tag{4.81}$$

The value of N, i.e., the denominator of the equilibrium line, need not be evaluated since it can be eliminated from Equation (4.80).

The concentration profile between transition point x_T and feed x_{Fi} is straight, and the same for all low boiler rich products in ideal systems (Fig. 4-31). In non-ideal systems, however, vapor-liquid equilibrium is a non-linear function and, thus, no direction exists where liquid concentrations and equilibrium vapor concentrations form parallel straight lines. In such systems, the curved concentration profile is linearized near the feed point (Fig. 4-35). Since the pinch always lies close to the feed, this is quite a good approximation for the pinch point loci. For non-ideal systems the partial derivatives have to be found numerically.

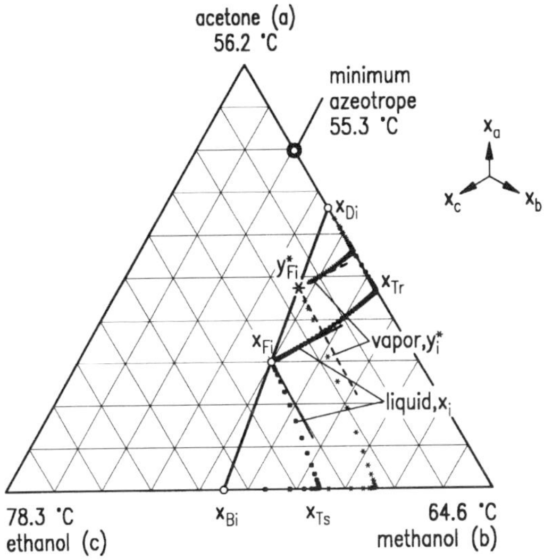

Fig. 4-35. In nonideal mixtures, the concentration profiles between transition and feed point are curved. They have to be linearized near the feed point.

$$C_1 = \left(\frac{\Delta y^*_a}{\Delta x_a}\right)_{x_b} = \text{const.}; \qquad C_2 = \left(\frac{\Delta y^*_a}{\Delta x_b}\right)_{x_a} = \text{const.};$$
$$C_3 = \left(\frac{\Delta y^*_b}{\Delta x_a}\right)_{x_b} = \text{const.}; \qquad C_4 = \left(\frac{\Delta y^*_b}{\Delta x_b}\right)_{x_a} = \text{const.}; \tag{4.82}$$

This requires the determination of vapor-liquid equilibrium at the feed x_{Fi} and two additional concentrations. Recommended are:

$$\Delta x_a = 0.1 \cdot (x_{Fa} - x^*_{Fa}) \qquad \Delta x_b = 0.1 \cdot (x_{Fb} - y^*_{Fb}) \tag{4.83}$$

A more rigorous approach uses analytically developed partial derivatives of the vapor-liquid relationship. This requires the derivatives of the activity coefficient γ_i. These lengthy expressions are available for some excess enthalpy models (e.g., Wilson model) in literature. A good listing is presented in [Taylor and Kooijman 1991].

As a consequence, the pinch point lies either at the feed concentration or at one of the straight lines that meet the condition of parallelity near the feed, see Fig. 4-36. The exact location on the pinch point lines is determined using the fact that one column section is operated with minimum reflux (or reboil). The product composition of this column section has to lie on a straight line through the pinch x_{Pi} and its vapor equilibrium y^*_{Pi} (Figs. 4-37 and 4-38). It is important that the low boiler rich distillate is binary, otherwise no transition point occurs. The bottom fraction, however, might be ternary. Analogously, the high boiler rich bottom fraction must be binary whereas the overhead fraction might be ternary.

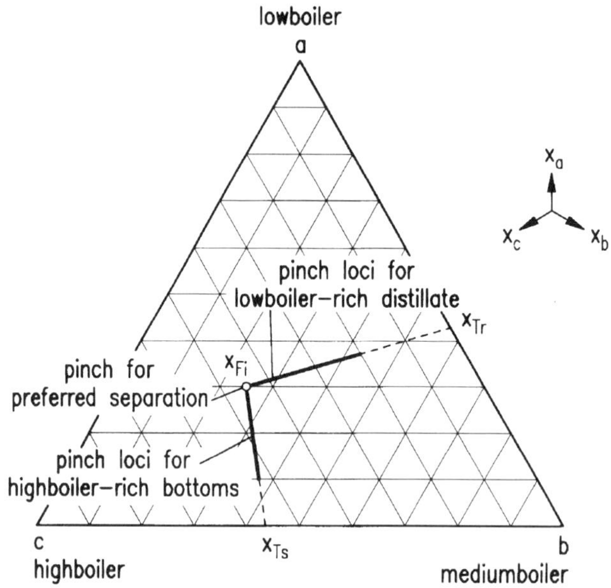

Fig. 4-36. Loci of pinch points in sharp separations. The pinch can only lie on one of these straight lines.

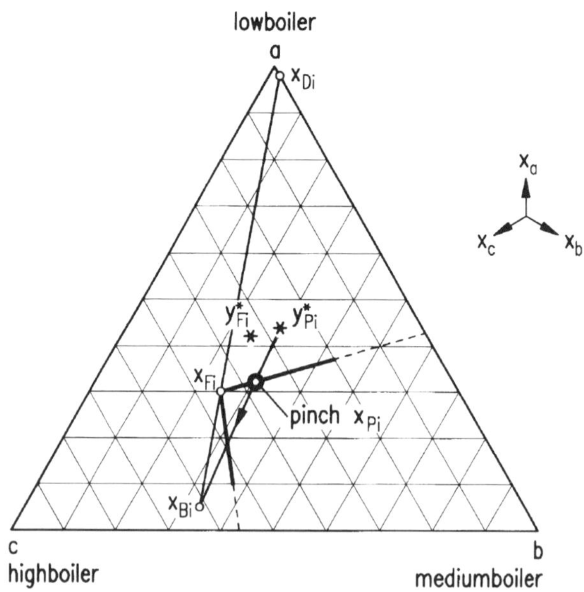

Fig. 4-37. Localizing the pinch point on the pinch point loci line for a low boiler separation. The concentrations x_{Pi}, y^*_{Pi} and x_{Bi} have to lie on the same straight line.

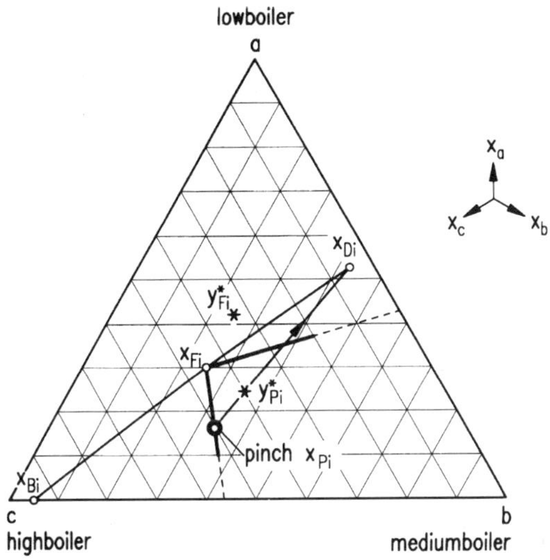

Fig. 4-38. *Localizing the pinch point on the pinch point loci line for a high boiler separation. The concentrations x_{Pi}, y^*_{Pi} and x_{Di} have to lie on the same straight line.*

4.3.3.3 Determination of Minimum Reflux and Minimum Reboil Ratios In principle, minimum reflux and minimum reboil ratios, respectively, can be determined by the equations developed in Section 4.3.3.1 for preferred separations after replacing feed concentrations x_{Fi} by pinch concentrations x_{Pi}.

For ideal system, the relevant equation for separating a low boiler rich distillate is:

$$R_{Gmin} = \left[\frac{\alpha_{ac}}{1 + (\alpha_{ac} - 1) \cdot x_{Pa} + (\alpha_{bc} - 1) \cdot x_{Pb}} - 1 \right]^{-1} \qquad (4.84)$$

The minimum reflux ratio R_{Lmin} follows from Equation (4.21) that formulates the relation between reflux and reboil ratio. For separating a high boiler rich bottom fraction, the relevant equation is:

$$R_{Lmin} = \frac{1}{(\alpha_{ac} - 1) \cdot x_{Pa} + (\alpha_{bc} - 1) \cdot x_{Pb}} \qquad (4.85)$$

To non-ideal mixtures the more fundamental Equations (4.61) and (4.64) have to be applied. Hence, for a low boiler rich distillate:

$$R_{Gmin} = \frac{x^*_{Pi} - x_{Bi}}{y^*_{Pi} - x^*_{Pi}} \qquad (4.86)$$

And for a high boiler rich bottom fraction:

$$R_{Lmin} = \frac{x_{Di} - y^*_{Pi}}{y^*_{Pi} - x^*_{Pi}} \qquad (4.87)$$

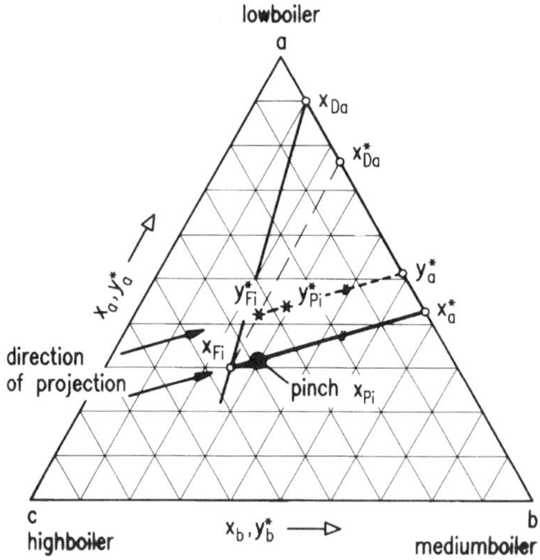

Fig. 4-39. *Projection of the states of feed and pinch in the direction of vapor-liquid equilibrium parallelity.*

However, the application of these equations require determination of the pinch concentration x_{Pi}. This is sometimes tiresome since localization of the pinch point on the pinch loci line (see Figs. 4-37 and 4-38) requires several iterations.

Tiresome iterations are avoided if the essential points x_{Fi}, y^*_{Fi}, x_{Pi} and y^*_{Pi} are projected to the right hand side of the triangular diagram in Fig. 4-39 in such a manner that x_{Fi} and x_{Pi} fall together to form point x_a^*. Thus, the projection has to be made in the direction of vapor-liquid equilibrium parallelity, see Fig. 4-34 and Equation (4.80). From Fig. 4-39 follows:

$$(R_{Lmin})_a = R^*_{Lmin} \frac{x_{Da} - y_a^*}{x^*_{Da} - y_a^*} \quad \text{with } R^*_{Lmin} = \frac{1}{(\alpha_{ac} - 1) \cdot x_{Fa} + (\alpha_{bc} - 1) \cdot x_{Fb}}; \tag{4.88}$$

$$x^*_{Da} = R^*_{Lmin}(\alpha_{ac} - 1) \cdot x_{Fa}$$

$$y_a^* = \frac{\alpha_{ab} + (\alpha_{ab} - 1)(R^*_{Lmin} + x^*_{Da})}{2(\alpha_{ab} - 1) \cdot (R^*_{Lmin} + 1)} - \sqrt{\left(\frac{\alpha_{ab} + (\alpha_{ab} - 1) \cdot (R^*_{Lmin} + x^*_{Da})}{2(\alpha_{ab} - 1) \cdot (R^*_{Lmin} + 1)}\right)^2 - \frac{x^*_{Da}\alpha_{ab}}{(\alpha_{ab} - 1) \cdot (R^*_{Lmin} + 1)}}$$

Equation (4.88) is valid for separating a distillate richer in low boiler than the preferred separation, i.e., $x_{Da} > x^*_{Da}$. Thus, the reflux ratio required for such a separation is always higher than that for the preferred separation.

The analogous equation for separating a high boiler rich or pure bottom fraction is:

$$(R_{Gmin})_c = R^*_{Gmin} \frac{x^*_b - x_{Bb}}{x^*_b - x^*_{Bb}} \tag{4.89}$$

$$\text{with } R^*_{Gmin} = \left(\frac{\alpha_{ac}}{1 + (\alpha_{ac} - 1) \cdot x_{Fa} + (\alpha_{bc} - 1) \cdot x_{Fb}} - 1 \right)^{-1}$$

$$x^*_{Bb} = \frac{(\alpha_{ac} - \alpha_{bc}) \cdot x_{Fb}}{\alpha_{ac} - (1 + (\alpha_{ac} - 1) \cdot x_{Fa} + (\alpha_{bc} - 1) \cdot x_{Fb})}$$

$$x^*_b = -\frac{1 - (\alpha_{bc} - 1) \cdot (R^*_{Gmin} + x^*_{Bb})}{2(\alpha_{bc} - 1) \cdot (R^*_{Gmin} + 1)} + \sqrt{\left(\frac{1 - (\alpha_{bc} - 1) \cdot (R^*_{Gmin} + x^*_{Bb})}{2(\alpha_{bc} - 1) \cdot (R^*_{Gmin} + 1)} \right)^2 + \frac{x^*_{Bb}}{(\alpha_{bc} - 1) \cdot (R^*_{Gmin} + 1)}}$$

To non-ideal systems, the more fundamental Equations (4.61) and (4.64) have to be applied. Here, evaluating the equilibrium vapor concentration and the direction of vapor-liquid equilibrium parallelity are more tiresome. However, the solution strategy is essentially the same.

With the equations developed in this section, the minimum reflux and minimum reboil ratio of all preferred separations can be calculated. In addition, minimum reflux and reboil, respectively, of low boiler and high boiler separations can be calculated if the distillate and bottoms, respectively, are binary or pure. The range of validity of the equations is marked by the heavily shaded area in Fig. 4-40. It

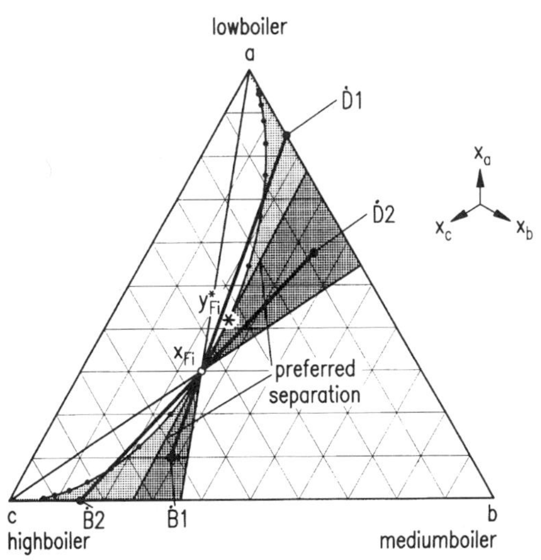

Fig. 4-40. *Range of validity of equations for minimum reflux and minimum reboil ratio. The low boiler-rich distillate or the high boiler-rich bottoms have to be either binary or pure.*

does not cover the full separation range (lightly shaded area). However, the equations give a good estimation of the required minimum reflux or reboil ratio even outside their range of validity.

EXAMPLE 4-3: MINIMUM REBOIL RATIO FOR SEPARATING PURE NITROGEN FROM A NITROGEN/ARGON/OXYGEN MIXTURE

Find the minimum reboil ratio R_{Gmin} for separating pure nitrogen from a nitrogen(a)-argon(b)-oxygen(c) mixture. The relative volatilities are $\alpha_{ac} = 4.032$; $\alpha_{bc} = 1.339$. Feed concentrations: $x_{Fa} = 0.25$; $x_{Fb} = 0.3$. Product concentrations; $x_{Da} = 1.0$; $x_{Ba} = 0.05$.

Solution:

Equation (4.88):

$$R^*_{Lmin} = \frac{1}{(\alpha_{ac}-1)x_{Fa} + (\alpha_{bc}-1)x_{Fb}} = \frac{1}{(4.032-1)0.25 + (1.339-1)0.3} = 1.1632;$$

$$x^*_{Da} = R^*_{Lmin}(\alpha_{ac} - 1)x_{Fa} = 1.1632(4.032-1)0.25 = 0.8817;$$

$$\alpha_{ab} = \alpha_{ac}/\alpha_{bc} = 4.032/1.339 = 3.0112;$$

$$y^*_a = \frac{\alpha_{ab} + (\alpha_{ab} - 1)\cdot(R^*_{Lmin} + x^*_{Da})}{2(\alpha_{ab} - 1)\cdot(R^*_{Lmin} + 1)} - \sqrt{\left(\frac{\alpha_{ab}+(\alpha_{ab}-1)\cdot(R^*_{Lmin}+x^*_{Da})}{2(\alpha_{ab}-1)\cdot(R^*_{Lmin}+1)}\right)^2 - \frac{x^*_{Da}\alpha_{ab}}{(\alpha_{ab}-1)\cdot(R^*_{Lmin}+1)}}$$

$$= \frac{3.0112 + (3.0112 - 1)\cdot(1.1632 + 0.8817)}{2(3.0112 - 1)\cdot(1.1632 + 1)} - \sqrt{\left(\frac{3.0112 + (3.0112 - 1)\cdot(1.1632 + 0.8817)}{2(3.0112 - 1)\cdot(1.1632 + 1)}\right)^2 - \frac{0.8817\cdot 3.0112}{(3.0112 - 1)\cdot(1.1632 + 1)}} = 0.5736;$$

$$(R_{Lmin})_a = R^*_{Lmin}\frac{x_{Da} - y^*_a}{x^*_{Da} - y^*_a} = 1.1632\frac{1.0 - 0.5736}{0.8817 - 0.5736} = 1.61;$$

$$\dot{D}/\dot{F} = \frac{x_{Fa} - x_{Ba}}{x_{Da} - x_{Ba}} = \frac{0.25 - 0.05}{1.0 - 0.05} = 0.2105;$$

$$R_{Gmin} = \frac{\dot{D}/\dot{F}}{1 - \dot{D}/\dot{F}}(R_{Lmin} + 1) = \frac{0.2105}{1 - 0.2105}(1.61 + 1) = 0.696.$$

The recommended value is approximately 30 to 50% higher (see Example 4-2).

4.3.4 Heat Requirement for Ternary Distillation

The minimum heat required for distillation can be evaluated from knowledge of minimum reflux ratio R_G and minimum reboil ratio R_G, respectively. The basic Equation (4.44) derived for binary mixtures also holds for ternary mixtures:

$$\dot{Q}_R = r \cdot R_G \cdot \dot{B} \quad \text{or} \quad \dot{Q}_C = -r \cdot (R_L + 1)\dot{D} \tag{4.90}$$

If the feed enters the column at its boiling point and the products are removed at their boiling points, the heat required in the reboiler is equal to the heat to be removed in the condenser ($/\dot{Q}_c/ = /\dot{Q}_R/ = \dot{Q}$). The heat $\dot{Q}$ required is proportional to the heat of vaporization r (which is assumed to be the same for all components), the product streams $\dot{B}$ and $\dot{D}$, respectively, and the reboil ratio R_G and the reflux ratio R_L, respectively. The heat requirement is lowest when the column is operated with minimum reboil R_{Gmin} and minimum reflux R_{Lmin}, respectively. The number of equilibrium stages and transfer units are supposed to be infinite.

4.3.4.1 Preferred Separation For a sharp separation (no high boiler in the top fraction, $x_{Dc} = 0$, and no low boiler in the bottom fraction, $x_{Ba} = 0$) a material balance around the column yields:

$$\dot{F} \cdot x_{Fa} = \dot{D} \cdot x_{Da} \tag{4.91}$$

With x_{Da} according to Equation (4.63):

$$x_{Da} = R_{Lmin} \cdot (y^*_{Fa} - x_{Fa}) + y^*_{Fa} \tag{4.92}$$

From Equations (4.90) to (4.92) it follows:

$$\frac{\dot{Q}_{min}}{\dot{F} \cdot r} = \frac{(R_{Lmin} 1) \cdot x_{Fa}}{R_{Lmin} \cdot (y^*_{Fa} - x_{Fa}) + y^*_{Fa}} \quad \text{with } R_{Lmin} = \frac{-y^*_{Fc}}{y^*_{Fc} - x_{Fc}} \tag{4.93}$$

For ideal systems with Equations (4.67) and (4.68):

$$\frac{\dot{Q}_{min}}{\dot{F} \cdot r} = \frac{1 + (\alpha_{ac} - 1) \cdot x_{Fa} + (\alpha_{bc} - 1) \cdot x_{Fb}}{\alpha_{ac} - 1} \tag{4.94}$$

For $x_{Fb} = 0$, i.e., binary mixtures, this equation becomes identical with the well known Equation (4.48):

$$\frac{\dot{Q}_{min}}{\dot{F} \cdot r} = \frac{1}{\alpha_{ac} - 1} + x_{Fa} \tag{4.95}$$

Thus, the energy requirement for an ideal ternary system with constant relative volatilities is a linear function of feed composition x_{Fi} (see Chapter 7).

4.3.4.2 Low Boiler Separation As has been shown before, the low boiler *a* can be separated in enriched or even pure form from an ideal ternary mixture. Generally, distillation is performed to gain pure substances. Therefore, this mode of operation of a distillation column is especially important. For a sharp separation (no high boiler in distillate, $x_{Dc} = 0$, and no low boiler in bottoms, $x_{Ba} = 0$), the following holds:

$$\dot{D} = \dot{F} \cdot x_{Fa} \tag{4.96}$$

Inserting in Equation (4.90) yields:

$$\frac{\dot{Q}_{min}}{\dot{F} \cdot r} = ((R_{Lmin})_a + 1) \cdot x_{Fa} \tag{4.97}$$

The value of $(R_{Lmin})_a$ is determined from Equation (4.88).

As the minimum reflux ratio $(R_{Lmin})_a$ for the low boiler rich separation is always higher than for the preferred separation, the heat requirement for low boiler separation is higher.

4.3.4.3 High Boiler Separation Another important mode of column operation is the removal of a high boiler rich bottom fraction from the ternary mixture. For a sharp separation ($x_{Ba} = 0$, $x_{Dc} = 0$), the material balance around the column yields:

$$\dot{B} = \dot{F}x_{Fc} \tag{4.98}$$

With Equation (4.90) we get:

$$\frac{\dot{Q}_{min}}{\dot{F} \cdot r} = (R_{Gmin})_c \cdot x_{Fc} \tag{4.99}$$

Here, the minimum reboil ratio $(R_{Gmin})_c$ is determined from Equation (4.89). In this case also, the heat requirement is always higher than in the case of the preferred separation.

In contrast to preferred separations, the energy requirement of low boiler and high boiler separations is a non-linear function of feed composition. This is outlined in more detail in Chapter 7.

4.4 MULTISTAGE DISTILLATION OF MULTICOMPONENT MIXTURES

Highly sophisticated models for rigorous column simulation have been developed since the broad availability of high capacity computers. Most models are based on the concept of equilibrium stages [Thiele and Geddes 1933]. However, the concept of transfer units has become more important in recent years, e.g., [Krishnamurthy and Taylor 1985, Taylor et al. 1993, Gorak 1990].

Rigorous column simulation refers to the evaluation of the thermodynamic parameters of the column. The results are the internal concentration profiles of each constituent, the temperature profile and the vapor and liquid flow rate profiles as functions of equilibrium stages. Typically, rigorous column simulation does not account for the problems of two-phase flow and mass transfer rate of the real column. However, these problems arise in column dimensioning (see Chapter 8).

4.4.1 Rigorous Column Simulation

Important contributions to rigorous column simulation have been made by [Wang and Henke 1966, Holland 1963, Goldstein and Stanfield 1970, Naphtali and Sandholm 1971, Block and Hegner 1976 and 1977]. The models developed until now differ primarily in the iteration procedure, i.e., in the iteration variables, e.g., flow rate profiles or concentration profiles and the mathematically iteration routines. Some models account for liquid phase splitting [Block and Hegner 1976] and for liquid phase chemical reaction [Block and Hegner 1977].

Rigorous column simulation is mathematically very complex. Therefore, it is generally not advisable to develop one's own program. There are several companies that offer such programs either on a sales or leasing basis (e.g., Aspen, Science Simulation, Process, Design 2000).

4.4.1.1 MESH-Equations The concept of equilibrium stages leads to the well known MESH-equations for each stage j, e.g., [Henley and Seader 1981]. With the symbols from Fig. 4-41, we get the
Material balances for i = 1 . . . k components:

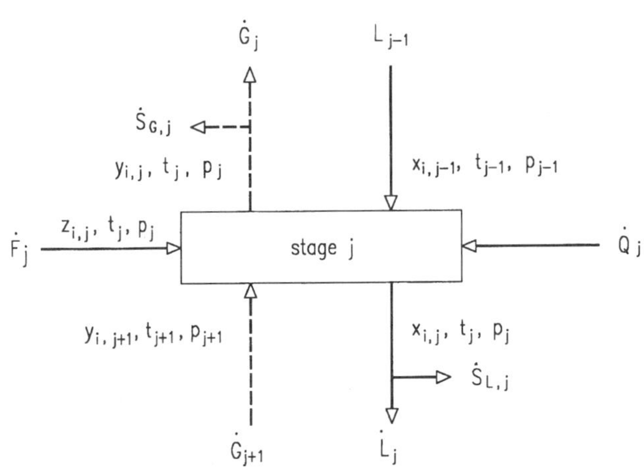

Fig. 4-41. Schematic of a single equilibrium stage for the derivation of the MESH-equations.

$$\dot{L}_{j-1} \cdot x_{i,j-1} + \dot{G}_{j+1} \cdot y_{i,j+1} + \dot{F}_j \cdot z_{i,j} - (\dot{L}_j + \dot{S}_{L,j}) \cdot x_{i,j} - (\dot{G}_j + \dot{S}_{G,j}) \cdot y_{i,j} = 0 \quad (4.100)$$

Equilibria for i = 1 .... k components:

$$y_{i,j} - K_{i,j} \cdot x_{i,j} = 0 \quad \text{with } K_{i,j} = f(t_j, p_j, x_{i,j}, y_{i,j}) \quad (4.101)$$

Summation of mole fractions in liquid and in vapor phase for i = 1 .... k components

$$\sum_{i=1}^{k} x_{i,j} - 1.0 = 0 \quad (4.102)$$

$$\sum_{i=1}^{k} y_{i,j} - 1.0 = 0 \quad (4.103)$$

Heat balances:

$$\dot{L}_{j-1} \cdot h_{L,j-1} + \dot{G}_{j+1} \cdot h_{G,j+1} + \dot{F}_j \cdot h_{F,j} - (\dot{L}_j + \dot{S}_{L,j}) \cdot h_{L,j} - (\dot{G}_j + \dot{S}_{G,j}) \cdot h_{G,j} + \dot{Q}_j = 0 \quad (4.104)$$

$$\text{with } h_{L,j} = f(t_j, p_j, x_{i,j}) \text{ and } h_{G,j} = f(t_j, p_j, y_{i,j})$$

In a column, the index j runs from j = 1 to j = n. The MESH-equations represent a set of $n \cdot (2k + 3)$ equations. Herein, k denotes the number of components and n the number of stages. The separation of a five component mixture in a column with 50 equilibrium stages gives, for instance, a system of 650 equations. In addition, complex functions for vapor-liquid equilibrium ratios K_i and for vapor and liquid enthalpies, h_L and h_G, are required.

As written in Equations (4.100) to (4.104) the MESH-equations look like an uncoupled linear system. However, even in ideal systems the equilibrium ratio K_i is a non-linear function of temperature. The higher the variation of temperature, the higher is the non-linearity of the MESH-equations. In addition, the equations are coupled since the temperature change caused by the concentration change of one component influences the concentrations of the other components via the equilibrium ratio (which strongly depends on temperature). In non-ideal systems the degree of non-linearity is even higher, since the equilibrium ratio K_i of each component is a complex function of concentration. Because the equilibrium ratio of component *i* depends on the concentration of all components, the equations are even more coupled in non-ideal systems. If the degree of non-linearity of the MESH-equations is very high, more than one solution may exist as, for instance, in any ordinary quadratic system [Bekiaris, Meski et al. 1993].

The number of equations can be reduced by the introduction of component flow rates for all streams:

$$l_{i,j} \equiv \dot{L}_j \cdot x_{i,j} \qquad g_{i,j} \equiv \dot{G}_j \cdot y_{i,j} \qquad f_{i,j} \equiv \dot{F}_{i,j} \cdot z_{i,j} \tag{4.105}$$

The use of component rates l_i and g_i is especially advantageous when the flow rates $\dot{G}$ and $\dot{L}$ are not constant within the column because concentration changes are caused by flow rate changes. Consequently, the material balance of component *i* depends on the concentration of all components. This is avoided by component rates.

With Equation (4.105) the MESH-equations get the following form:

$$\begin{aligned}
M_{i,j} &= l_{i,j-1} + g_{i,j+1} + f_{i,j} - l_{i,j}\cdot(1 + s_{L,j}) - g_{i,j}\cdot(1 + s_{G,j}) = 0 \\
E_{i,j} &= g_{i,j}\cdot\sum_i^k l_{i,j} - K_{i,j}\cdot l_{i,j}\cdot\sum_i^k g_{i,j} = 0 \\
&\quad \text{with } K_{i,j} = f(l_{1,j}, \ldots l_{k,j}, g_{1,j} \ldots g_{k,j}, t_j) \\
H_j &= h_{L,j-1}\cdot\sum_i^k l_{i,j-1} + h_{G,j+1}\cdot\sum_i^k g_{i,j+1} + h_{F,j}\cdot\sum_i^k f_{i,j} - h_{L,j}\cdot\sum_i^k l_{i,j}\cdot(1 + s_{L,j}) \\
&\quad - h_{G,j}\cdot\sum_i^{\kappa} g_{i,j}\cdot(1 + s_{G,j}) + Q_j = 0
\end{aligned} \tag{4.106}$$

with $h_{L,j} = f(l_{1,j}, \ldots l_{k,j}, t_j)$ and $h_{G,j} = f(g_{1,j} \ldots g_{k,j}, t_j)$

The number of equations in a system with k components and n equilibrium stages is now $n \cdot (2k + 1)$.

The functions H_j, M_j and E_j are discrepancy functions, that is, they are a quantitative measure of the failure of the values of $l_{i,j}$, $g_{i,j}$ and t_j to satisfy the physical relationship. A solution to the problem has been obtained when one has values of the variables which make H_j, M_j and E_j zero [Naphtali and Sandholm 1971]. It is important to be aware of the dependencies of the functions on the variables involved.

$$\begin{aligned}
M_{i,j} &= f(l_{i,j-1}, l_{i,j}, g_{i,j}, g_{i,j+1}) \\
E_{i,j} &= f(l_{1,j}, \ldots l_{k,j}, g_{1,j}, \ldots g_{k,j}, t_j) \\
H_j &= f(l_{1,j-1}; \ldots l_{k,j-1}, l_{1,j}, \ldots l_{k,j}, g_{1,j}, \ldots g_{k,j}, g_{1,j+1}, \ldots g_{k,j+1}, t_{j-1}, t_j, t_{j+1})
\end{aligned} \tag{4.107}$$

The functions depend only on the variables of stage j and the neighbor stages j − 1 and j + 1, except the equilibrium functions, which contain variables of stage j only. The material balances of component *i* depends on the flow rates of this component only, whereas the equilibrium and enthalpy functions are influenced by all constituents.

4.4.1.2 *Specification of Column and Operating Conditions* Rigorous column simulation requires specifying the column and the operating conditions. Several possibilities exist.

Typically, the column is fully specified. This includes setting the number n of equilibrium stages (those different from the number of actual trays), the definition of the feed stage n_F (or in the case of multiple feed, the definition of the feed stages) and, if necessary, the stages of the withdrawal of side products. In the case of intermediate reboilers or condensers their locations and duties have to be set also.

The operating conditions are defined in most cases by setting the reflux or reboil ratio, or equivalent; the duties of condenser and reboiler; the relative amount of overhead and bottom products $\dot{D}/\dot{F}$ and $\dot{B}/\dot{F}$, respectively; and the sidestreams $\dot{S}_L$ and $\dot{S}_G$.

In most cases it is not advisable to set both the condenser and the reboiler duties since they are very interpendent. When specifying the reflux ratio R_L instead of condenser duty, the enthalpy functions of the top stage (j = 1) and the bottom stage (j = n) have to be replaced by:

$$H_1 = \sum_i^k l_{i,1} - R_L \cdot \sum_i^k g_{i,1} = 0 \qquad H_n = \sum_i^k l_{i,n} - \dot{B} = 0 \qquad (4.108)$$

When the reboil ratio R_G is specified, the enthalpy functions of stages 1 and n are modified to:

$$H_1 = \sum_i^k g_{i,1} - \dot{D} = 0 \qquad H_n = \sum_i^k g_{i,n} - R_G \cdot \sum_i^k l_{i,n} = 0 \qquad (4.109)$$

Solution of the MESH-equations yields the internal profiles of temperature and component flow rates. Automatically included are the compositions and amounts of the overhead and bottom products. If these concentrations do not meet the product purities required, the simulation has to be repeated with altered specifications of the column and/or the operating conditions. Because of its high complexity, it is not advisable to specify the product purities and to search the number of equilibrium stages by evaluating the MESH-equations.

4.4.1.3 Procedures for Evaluating the MESH-Equations There exist numerous strategies for evaluating the MESH-equations. They are classified in three groups; the sequential methods (e.g., tray-to-tray calculation) [Lewis and Matheson 1932], the semi-sequential methods, e.g., [Wang and Henke 1966] and the simultaneous methods, which are today's most important strategies.

The tray-to-tray calculation has been demonstrated in Example 4-2. It is very easy to perform, but instabilities are often encountered. A serious problem arises in systems containing very high or very low boiling components because their concentrations are very low at one end of the column.

In the semisequential methods the equations are divided into groups. One set is solved as the overall problem, with the other sets being solved as inner subproblems.

4.4.1.3.1 Simultaneous Solution A modified Newton-Raphson-iteration procedure is used to solve the non-linear set of equations simultaneously [Naphtali and Sandholm 1971]. The advantage is that this iteration procedure accelerates as the solution is approached.

The Newton-Raphson algorithm is a very effective iteration procedure for solving systems of non-linear coupled equations. Its application to a system of two quadratic equations, f_1 and f_2, is demonstrated in detail in Example 4-4.

First, an estimation of the variables $x_1 = x_{1m}$ and $x_2 = x_{2m}$ has to be made. Then, the non-linear equations are linearized in the vicinity of the estimated values x_{1m}, x_{2m} by Taylor series:

$$\begin{aligned} f_1 + \left(\frac{\partial f_1}{\partial x_1}\right) \cdot \Delta x_1 + \left(\frac{\partial f_1}{\partial x_2}\right) \cdot \Delta x_2 &= 0 \\ f_2 + \left(\frac{\partial f_2}{\partial x_1}\right) \cdot \Delta x_1 + \left(\frac{\partial f_2}{\partial x_2}\right) \cdot \Delta x_2 &= 0 \end{aligned} \tag{4.110}$$

After the partial derivatives of the functions with respect to all variables have been performed, Equations (4.110) form a system of linear equations for Δx_1 and Δx_2 that can be solved rigorously. The solution of the linear system is:

$$\Delta x_1 = \Delta x_{1,m}; \qquad \Delta x_2 = \Delta x_{2,m} \tag{4.111}$$

The solution of the non-linear system is:

$$x_{1,m+1} = x_{1,m} + \Delta x_{1,m} \qquad x_{2,m+1} = x_{2,m} + \Delta x_{2,m} \tag{4.112}$$

Since the linear system is only an approximation of the non-linear system, the values $x_{1,m+1}$ and $x_{2,m+1}$ do not represent the exact solution. Thus, the procedure has to be repeated. Figure 4-42 displays the values of the functions f_1 and f_2 during the iteration. The values are driven to zero with increasing speed.

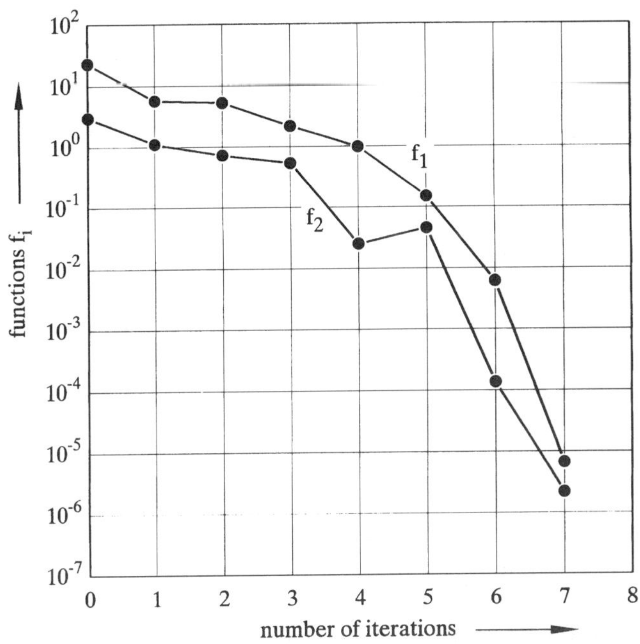

Fig. 4-42. *Iteration history for solving the two nonlinear equations of Example 4-3 by the Newton-Raphson algorithm.*

EXAMPLE 4-4: SOLUTION OF A SYSTEM OF TWO QUADRATIC EQUATIONS BY THE NEWTON-RAPHSON ALGORITHM

System of non-linear equations: $f_1 = 3 \cdot (x_1^2 + x_2^2) - 10 \cdot x_1 - 14 \cdot x_2 + 23 = 0$; $f_2 = x_1^2 - 2 \cdot x_1 - x_2 + 3 = 0$.

Solution:

Estimation of starting values: $x_1 = x_{1,m}$; $x_2 = x_{2,m}$. Transformation into a linear system by Taylor series around $x_{1,m}$ and $x_{2,m}$:

$$f_1 + \left(\frac{\partial f_1}{\partial x_1}\right)_{x_{1,m}} \cdot \Delta x_1 + \left(\frac{\partial f_1}{\partial x_2}\right)_{x_{2,m}} \cdot \Delta x_2 = 0$$

$$\text{with } \left(\frac{\partial f_1}{\partial x_1}\right)_{x_{1,m}} = 6 \cdot x_{1,m} - 10 \quad \left(\frac{\partial f_1}{\partial x_2}\right)_{x_{2,m}} = 6 \cdot x_{2,m} - 14;$$

$$f_2 + \left(\frac{\partial f_2}{\partial x_1}\right)_{x_{1,m}} \cdot \Delta x_1 + \left(\frac{\partial f_2}{\partial x_2}\right)_{x_{2,m}} \cdot \Delta x_2 = 0$$

$$\text{with } \left(\frac{\partial f_2}{\partial x_1}\right)_{x_{1,m}} = 2 \cdot x_{1,m} - 2 \quad \left(\frac{\partial f_2}{\partial x_2}\right)_{x_{2,m}} = -1.$$

Solution of linear system: $\Delta x_1 = \Delta x_{1,m}$; $\Delta x_2 = \Delta x_{2,m}$. Improved values of non-linear system: $x_{1,m+1} = x_{1,m} + \Delta x_{1,m}$; $x_{2,m+1} = x_{2,m} + \Delta x_{2,m}$.

Evaluation:

Estimated values: $x_{1,0} = 0$; $x_{2,0} = 0$.

1. Iteration—Linear system (Taylor series): $23 - 10 \cdot \Delta x_1 - 14 \cdot \Delta x_2 = 0$; $3 - 2 \cdot \Delta x_1 - 1 \cdot \Delta x_2 = 0$. Solution of linear system: $\Delta x_1 = 1.0555$; $\Delta x_2 = 0.88888$. Improved values of non-linear system: $x_1 = 0 + 1.0555 = 1.0555$; $x_2 = 0 + 0.8888 = 0.8888$.
2. Iteration—Linear system (Taylor series): $5.7129 - 3.6666 \cdot \Delta x_1 - 8.6666 \cdot \Delta x_2 = 0$; $1.1141 + 0.1111 \cdot \Delta x_1 - 1.0 \cdot \Delta x_2 = 0$. Solution of linear system: $\Delta x_1 = -0.85177$; $\Delta x_2 = 1.0195$. Improved value of non-linear system: $x_1 = 1.0555 - 0.85177 = 0.20377$; $x_2 = 0.8888 + 1.0195 = 1.9084$.
3. Iteration—Linear system (Taylor series): $5.2950 - 8.7773 \cdot \Delta x_1 - 2.5493 \cdot \Delta x_2 = 0$; $0.72552 - 1.5924 \cdot \Delta x_1 - 1.0 \cdot \Delta x_2 = 0$. Solution of linear system: $\Delta x_1 = 0.73033$; $\Delta x_2 = -0.43748$. Improved values of non-linear system: $x_1 = 0.93410$; $x_2 = 1.4709$.

4. Iteration—Linear system (Taylor series): $2.1743 - 4.3953 \cdot \Delta x_1 - 5.1742 \cdot \Delta x_2 = 0$; $0.53338 - 0.13178 \cdot \Delta x_1 - 1.0 \cdot \Delta x_2 = 0$. Solution of linear system: $\Delta x_1 = -0.15767$; $\Delta x_2 = 0.55416$. Improved values of non-linear system: $x_1 = 0.77643$; $x_2 = 2.0251$.
5. Iteration—Linear system (Taylor series): $0.99586 - 5.3414 \cdot \Delta x_1 - 1.8492 \cdot \Delta x_2 = 0$; $0.02486157 - 0.44713 \cdot \Delta x_1 - 1.0 \cdot \Delta x_2 = 0$. Solution of linear system: $\Delta x_1 = 0.21040$; $\Delta x_2 = -0.069219$. Improved values of non-linear system: $x_1 = 0.98684$; $x_2 = 1.9559$.
6. Iteration—Linear system (Taylor series): $0.14718 - 4.0789 \cdot \Delta x_1 - 2.2645 \cdot \Delta x_2 = 0$; $0.04427 - 0.02631 \cdot \Delta x_1 - 1.0 \cdot \Delta x_2 = 0$. Solution of linear system: $\Delta x_1 = 0.01167$; $\Delta x_2 = 0.043964$. Improved values of non-linear system: $x_1 = 0.99851$; $x_2 = 1.9998$.
7. Iteration—Linear system (Taylor series): $0.0062075 - 4.0088 \cdot \Delta x_1 - 2.0008 \cdot \Delta x_2 = 0$; $0.00013634 - 0.0029663 \cdot \Delta x_1 - 1.0 \cdot \Delta x_2 = 0$. Solution of linear system: $\Delta x_1 = 0.00148260$; $\Delta x_2 = 0.00013194$. Improved values of non-linear system: $x_1 = 0.99999$; $x_2 = 1.9999$.
8. Iteration—Linear system (Taylor series): $0.00000665 - 4.00000338 \cdot \Delta x_1 - 2.00001319 \cdot \Delta x_2 = 0$; $0.00000220 - 0.00000113 \cdot \Delta x_1 - 1.0 \cdot \Delta x_2 = 0$. Solution of linear system: $\Delta x_1 = 0.00000056$; $\Delta x_2 = 0.00000220$. Improved values of non-linear system: $x_1 = 1.0$; $x_2 = 2.0$.

Because of the large number of variables and functions in the modified MESH-equations (4.106) the Newton-Raphson algorithm has to be written with matrices. If the equations and variables are grouped according to stages, the matrix of partial derivatives needed in the Newton-Raphson method assumes a form that is easy to solve. The amount of storage and the number of calculations are drastically reduced.

The vector $\mathbf{x}$ of variables is:

$$\mathbf{x} = \begin{pmatrix} \mathbf{x}_1 \\ \mathbf{x}_2 \\ \cdots \\ \cdots \\ \mathbf{x}_j \\ \cdots \\ \cdots \\ \mathbf{x}_n \end{pmatrix} \text{with} \begin{array}{l} \mathbf{x}_1 = [l_{1,1} \ldots\ldots l_{k,1}, g_{1,1} \ldots\ldots gk_{,1}, t_1]^T \\ \mathbf{x}_2 = [l_{1,2} \ldots\ldots l_{k,2}, g_{1,2} \ldots\ldots gk_{,2}, t_2]^T \\ \cdot\cdot \\ \cdot\cdot \\ \mathbf{x}_j = [l_{1,j} \ldots\ldots l_{k,j}, g_{1,j} \ldots\ldots g_{k,j}, t_j]^T \\ \cdot\cdot \\ \cdot\cdot \\ \mathbf{x}_n = [l_{1,n} \ldots\ldots l_{k,n}, g_{1,n} \ldots\ldots g_{k,n}, t_n]^T \end{array} \tag{4.113}$$

The vector $\mathbf{F}$ of functions is:

$$\mathbf{F} = \begin{pmatrix} \mathbf{F}_1 \\ \mathbf{F}_2 \\ \cdots \\ \cdots \\ \mathbf{F}_j \\ \cdots \\ \cdots \\ \mathbf{F}_n \end{pmatrix} \text{with} \begin{array}{l} \mathbf{F}_1 = [M_{1,1} \ldots\ldots M_{k,1}, E_{1,1} \ldots\ldots E_{k,1}, H_1]^T \\ \mathbf{F}_2 = [M_{1,2} \ldots\ldots M_{k,2}, E_{1,2} \ldots\ldots E_{k,2}, H_2]^T \\ \cdot\cdot \\ \cdot\cdot \\ \mathbf{F}_j = [M_{1,j} \ldots\ldots M_{k,j}, E_{1,j} \ldots\ldots E_{k,j}, H_j]_T \\ \cdot\cdot \\ \cdot\cdot \\ \mathbf{F}_n = [M_{1,n} \ldots\ldots M_{k,n}, E_{1,n} \ldots\ldots E_{k,n}, H_n]^T \end{array} \tag{4.114}$$

With this notation, the Taylor series developed around the estimated values $\mathbf{x}_m$ is:

$$\mathbf{F}(\mathbf{x}_m) + \mathbf{J}(\mathbf{x}_m) \cdot \Delta\mathbf{x} = 0 \tag{4.115}$$

Herein is **J** the Jacobian matrix of the partial derivatives $\partial\mathbf{F}/\partial\mathbf{x}$ of all functions with respect to all variables.

Figure 4-43 presents the Jacobian matrix in general form. It contains n^2 terms $\partial\mathbf{F}/\partial\mathbf{x}$ (n is the number of stages). Each term $\partial\mathbf{F}/\partial\mathbf{x}$ represents a submatrix of the partial derivatives of all functions of a single stage with respect to all variables on any stage. These submatrices have $(2k + 1)^2$ elements (k is the number of components). Thus, the Jacobian matrix has $n^2 \cdot (2k + 1)^2$ elements. For example, the separation of a five-component system in a column with 50 stages yields an Jacobian matrix with 302 500 elements. However, as will be outlined later, most elements are zero.

Equation (4.115) represents a system of coupled equations that are linear with respect to Δx. In order to solve this system, the invers matrix $\mathbf{J}^{-1}$ has to be formed. This is a very complex problem of matrix calculation. With the invers matrix we get the solution of the Taylor series.

$$\Delta\mathbf{x} = -\mathbf{F}(\mathbf{x}_m) \cdot \mathbf{J}^{-1}(\mathbf{x}_m) \tag{4.116}$$

Improved values of the non-linear system (4.114) are easily evaluated.

$$\mathbf{x}_{m+1} = \mathbf{x}_m + \Delta\mathbf{x} \tag{4.117}$$

Here, $\Delta\mathbf{x}$ is the calculated correction. If the functions **F** were linear, this correction

$$\mathbf{J} = \begin{matrix}
\frac{\partial F_1}{\partial x_1} & \frac{\partial F_1}{\partial x_2} & \frac{\partial F_1}{\partial x_3} & \cdots & \frac{\partial F_1}{\partial x_{n-2}} & \frac{\partial F_1}{\partial x_{n-1}} & \frac{\partial F_1}{\partial x_n} \\
\frac{\partial F_2}{\partial x_1} & \frac{\partial F_2}{\partial x_2} & \frac{\partial F_2}{\partial x_3} & \cdots & \frac{\partial F_2}{\partial x_{n-2}} & \frac{\partial F_2}{\partial x_{n-1}} & \frac{\partial F_2}{\partial x_n} \\
\frac{\partial F_3}{\partial x_1} & \frac{F_3}{\partial x_2} & \frac{\partial F_3}{\partial x_3} & \cdots & \frac{\partial F_3}{\partial x_{n-2}} & \frac{\partial F_3}{\partial x_{n-1}} & \frac{\partial F_3}{\partial x_n} \\
\cdots & \cdots & \cdots & \cdots & \cdots & \cdots & \cdots \\
\cdots & \cdots & \cdots & \cdots & \cdots & \cdots & \cdots \\
\cdots & \cdots & \cdots & \cdots & \cdots & \cdots & \cdots \\
\cdots & \cdots & \cdots & \cdots & \cdots & \cdots & \cdots \\
\cdots & \cdots & \cdots & \cdots & \cdots & \cdots & \cdots \\
\frac{\partial F_{n-1}}{\partial x_1} & \frac{\partial F_{n-1}}{\partial x_2} & \frac{\partial F_{n-1}}{\partial x_3} & \cdots & \frac{\partial F_{n-1}}{\partial x_{n-2}} & \frac{\partial F_{n-1}}{\partial x_{n-1}} & \frac{\partial F_{n-1}}{\partial x_n} \\
\frac{\partial F_n}{\partial x_1} & \frac{\partial F_n}{\partial x_2} & \frac{\partial F_n}{\partial x_3} & \cdots & \frac{\partial F_n}{\partial x_{n-2}} & \frac{\partial F_n}{\partial x_{n-1}} & \frac{\partial F_n}{\partial x_n}
\end{matrix}$$

Fig. 4-43. *Jacobian matrix in the general form. The matrix contains n^2 submatrices (n is the number of stages).*

would make the value of each function zero. For non-linear equations, the values $\mathbf{x}_{m+1}$ are just an approximation, and the procedure has to be repeated.

Because of the large number of variables and functions, the procedure of solving the MESH-equations is very involved and time consuming. Some details are presented in the following.

4.4.1.3.2 Jacobian Matrix The Jacobian matrix is very large in the general case, see Fig. 4-43. However, in the modified MESH-equations of a distillation column, the functions $\mathbf{F}_j$ depend only on the variables of three neighboring stages $j - 1$, j, and $j + 1$. Thus, only the terms in the main diagonal and the diagonal below and above are relevant. The non-zero elements are shaded in the matrix Fig. 4-44. All unshaded elements in the Jacobian matrix are zero. Hence, the Jacobian matrix is of the special type of a tridiagonal matrix. Since the terms $\partial\mathbf{F}/\partial\mathbf{x}$ are matrices or blocks, the Jacobian matrix is a block tridiagonal matrix.

The tridiagonal matrix contains $n + 2\,(n - 1)$ non-zero submatrices each with $(2k + 1)^2$ elements. However, many of these elements are zero. The submatrix for the term $\partial\mathbf{F}_j/\partial\mathbf{x}_{j-1}$, i.e., the derivatives of the functions of stage j with respect to the variables of stage $j - 1$, is presented in Fig. 4-45. In this submatrix most elements are zero because the functions of stage j are only influenced by the state of the liquid of stage $j - 1$, i.e., liquid flow rates $l_{i,j-1}$ and temperature t_{j-1}. As can be seen from Equation (4.107) the material balance of component *i* depends only on the liquid rates of the same component of stage $j - 1$. The equilibrium function does not depend on any variable on stage $j - 1$, but the enthalpy function depends on

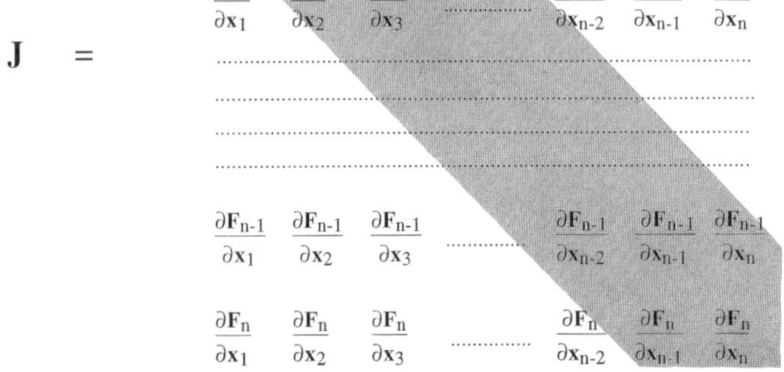

Fig. 4-44. Jacobian matrix of a distillation column. The matrix has a tridiagonal structure with $n + 2(n - 1)$ non-zero submatrices (shaded elements).

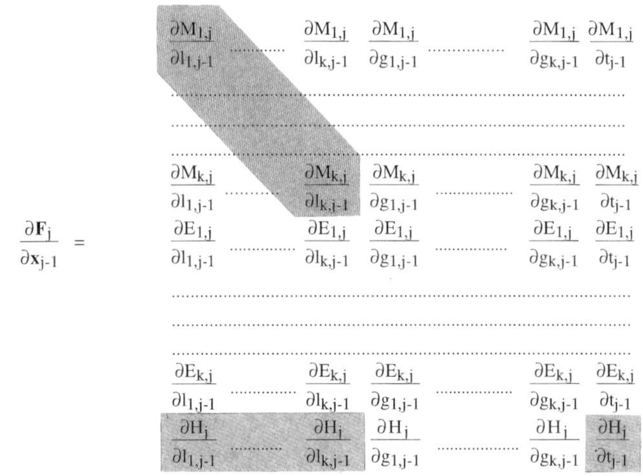

Fig. 4-45. Structure of the submatrices in the upper side diagonal of the Jacobian matrix. Only 2k + 1 elements are non-zero (k is the number of components).

all liquid rates on stage j − 1 and on temperature t_{j-1}. Thus, all but the shaded elements are zero in the submatrix of Fig. 4-45.

The submatrix for the term $\partial \mathbf{F}_j/\partial \mathbf{x}_j$, i.e., the derivatives of the functions of stage j with respect to the variables of the same stage j, is depicted in Fig. 4-46. According to Equation (4.107) the material balance of component *i* depends on the liquid gas rates of the same component only. The equilibrium function $E_{i,j}$ depends on all liquid

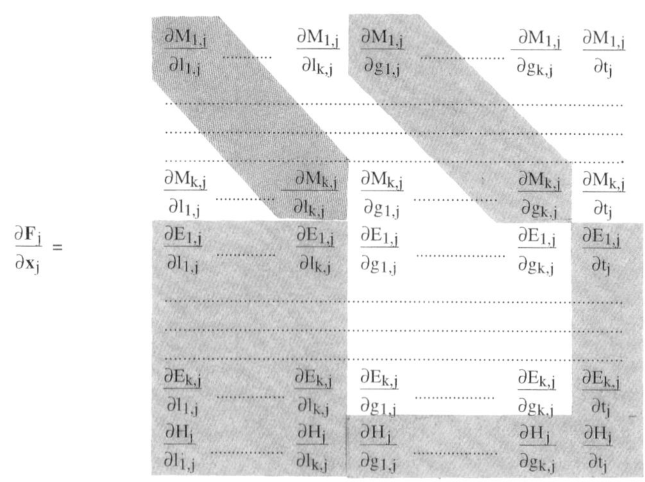

Fig. 4-46. Structure of the submatrices in the main diagonal. The number of non-zero elements is k^2 + 5k + 1.

$$\frac{\partial \mathbf{F}_j}{\partial \mathbf{x}_{j+1}} = \begin{bmatrix} \frac{\partial M_{1,j}}{\partial l_{1,j+1}} & \cdots & \frac{\partial M_{1,j}}{\partial l_{k,j+1}} & \frac{\partial M_{1,j}}{\partial g_{1,j+1}} & \cdots & \frac{\partial M_{1,j}}{\partial g_{k,j+1}} & \frac{\partial M_{1,j}}{\partial t_{j+1}} \\ \cdots & \cdots & \cdots & \cdots & \cdots & \cdots & \cdots \\ \cdots & \cdots & \cdots & \cdots & \cdots & \cdots & \cdots \\ \cdots & \cdots & \cdots & \cdots & \cdots & \cdots & \cdots \\ \frac{\partial M_{k,j}}{\partial l_{1,j+1}} & & \frac{\partial M_{k,j}}{\partial l_{k,j+1}} & \frac{\partial M_{k,j}}{\partial g_{1,j+1}} & \cdots & \frac{\partial M_{k,j}}{\partial g_{k,j+1}} & \frac{\partial M_{k,j}}{\partial t_{j+1}} \\ \frac{\partial E_{1,j}}{\partial l_{1,j+1}} & & \frac{\partial E_{1,j}}{\partial l_{k,j+1}} & \frac{\partial E_{1,j}}{\partial g_{1,j+1}} & \cdots & \frac{\partial E_{1,j}}{\partial g_{k,j+1}} & \frac{\partial E_{1,j}}{\partial t_{j+1}} \\ \cdots & \cdots & \cdots & \cdots & \cdots & \cdots & \cdots \\ \cdots & \cdots & \cdots & \cdots & \cdots & \cdots & \cdots \\ \cdots & \cdots & \cdots & \cdots & \cdots & \cdots & \cdots \\ \frac{\partial E_{k,j}}{\partial l_{1,j+1}} & \cdots & \frac{\partial E_{k,j}}{\partial l_{k,j+1}} & \frac{\partial E_{k,j}}{\partial g_{1,j+1}} & \cdots & \frac{\partial E_{k,j}}{\partial g_{k,j+1}} & \frac{\partial E_{k,j}}{\partial t_{j+1}} \\ \frac{\partial H_j}{\partial l_{1,j+1}} & \cdots & \frac{\partial H_j}{\partial l_{k,j+1}} & \frac{\partial H_j}{\partial g_{1,j+1}} & \cdots & \frac{\partial H_j}{\partial g_{k,j+1}} & \frac{\partial H_j}{\partial t_{j+1}} \end{bmatrix}$$

Fig. 4-47. *Structure of the submatrices in the lower side diagonal of the Jacobian matrix. Only 2k + 1 elements are non-zero.*

and gas rates on stage j and on the temperature on stage j. The same holds for the enthalpy function. Here again, the non-zero elements are shaded in Fig. 4-46.

The submatrix for the term $\partial \mathbf{F}_j/\partial \mathbf{x}_{j+1}$, i.e., the derivatives of the functions of stage j with respect to the variables of stage j + 1, is presented in Fig. 4-47. The functions of stage j are only influenced by the state of the gas of stage j + 1, i.e., $g_{i,j+1}$ and t_{j+1}. The M_j equations of components *i* depend only on the gas flow rates $g_{i,j+1}$ of the same component. However, the enthalpy function H_j is influenced by all gas flow rates of stage j + 1 and the temperature of stage j + 1. These relevant elements are shaded in the matrix in Fig. 4-47. All the other elements are zero.

Each submatrix in the side diagonals has 2k + 1 non-zero elements, whereas each submatrix in the main diagonal matrix contains $k^2 + 5k + 1$ non-zero elements. In total, the Jacobian matrix contains $2 \cdot (n - 1) \cdot (2k + 1) + n \cdot (k^2 + 5k + 1)$ non-zero elements. In a system with 5 components and 50 stages, only 3628 out of 302 500 elements are non-zero. Hence, approximately 99% of all elements of the Jacobian matrix are zero. Consequentely, spars techniques are advantageously applied to matrix transformations [Naphtali and Sandholm 1971].

4.4.1.3.3 Partial Derivatives For the set up of the Jacobian matrix the non-zero elements of the partial derivatives $\partial \mathbf{F}/\partial \mathbf{x}$ have to be generated for all functions. The results are listed in Table 4.1.

The partial derivatives of the material balances $M_{i,j}$ are easily performed. Their values are +1, $(1 + s_{L,j})$ or $(1 + s_{G,j})$ according to the rules of differentiation with a constant factor.

The partial derivatives of the equilibrium functions are more complex because differentiation of products has to be performed. Just a few of the derivatives can be performed analytically. The terms listed in Table 4.1 contain the partial derivatives

Table 4.1. Partial Derivatives of Material, Equilibrium and Enthalpy Functions

M_i-Functions:

$$\left(\frac{\partial M_{i,j}}{\partial l_{i,j-1}}\right) = 1; \quad \left(\frac{\partial M_{i,j}}{\partial l_{i,j}}\right) = -(1 + s_{L,j}); \quad \left(\frac{\partial M_{i,j}}{\partial g_{i,j}}\right) = -(1 + s_{G,j}); \quad \left(\frac{\partial M_{i,j}}{\partial g_{i,j+1}}\right) = 1$$

E_i-Functions:

$$\left(\frac{\partial E_{i,j}}{\partial l_{m,j}}\right) = g_{i,j} - \left(\frac{\partial K_{i,j}}{\partial l_{m,j}}\right) \cdot l_{i,j} \cdot \sum_{z=1}^{k} g_{z,j} - K_{i,j} \cdot \delta_{i,m} \cdot \sum_{z=1}^{k} g_{z,j}$$

with $\delta_{i,m} = 1$ for $i = m$ and 0 for $i \neq m$

$$\left(\frac{\partial E_{i,j}}{\partial g_{m,j}}\right) = \delta_{i,m} \cdot \sum_{z=1}^{k} l_{z,j} - \left(\frac{\partial K_{i,j}}{\partial g_{m,j}}\right) \cdot l_{i,j} \cdot \sum_{z=1}^{k} g_{z,j} - K_{i,j} \cdot l_{i,j}$$

$$\left(\frac{\partial E_{i,j}}{\partial t_j}\right) = \left(\frac{\partial K_{i,j}}{\partial t_j}\right) \cdot l_{i,j} \cdot \sum_{z=1}^{k} g_{z,j}$$

H-Functions:

$$\left(\frac{\partial H_j}{\partial l_{m,j-1}}\right) = \left(\frac{\partial h_{L,j-1}}{\partial l_{m,j-1}}\right) \cdot \sum_{z=1}^{k} l_{z,j-1} + h_{L,j-1}; \quad \left(\frac{\partial H_j}{\partial t_{j-1}}\right) = \left(\frac{\partial h_{L,j-1}}{\partial t_{j-1}}\right) \cdot \sum_{z=1}^{k} l_{z,j-1}$$

$$\left(\frac{\partial H_j}{\partial l_{m,j}}\right) = -\left(\left(\frac{\partial h_{L,j}}{\partial l_{m,j}}\right) \cdot \sum_{z=1}^{k} l_{z,j} + h_{L,j}\right) \cdot (1 + s_{L,j})$$

$$\left(\frac{\partial H_j}{\partial g_{m,j}}\right) = -\left(\left(\frac{\partial h_{G,j}}{\partial g_{m,j}}\right) \cdot \sum_{z=1}^{k} g_{z,j} + h_{G,j}\right) \cdot (1 + s_{G,j})$$

$$\left(\frac{\partial H_j}{\partial t_j}\right) = -\left(\frac{\partial h_{L,j}}{\partial t_j}\right) \cdot (1 + s_{L,j}) \cdot \sum_{z=1}^{k} l_{z,j} - \left(\frac{\partial h_{G,j}}{\partial t_j}\right) \cdot (1 + s_{G,j}) \cdot \sum_{z=1}^{k} g_{z,j}$$

$$\left(\frac{\partial H_j}{\partial g_{m,j+1}}\right) = \left(\frac{\partial h_{G,j+1}}{\partial g_{m,j+1}}\right) \cdot \sum_{z=1}^{k} g_{z,j+1} + h_{G,j+1}; \quad \left(\frac{\partial H_j}{\partial t_{j+1}}\right) = \left(\frac{\partial h_{G,j+1}}{\partial t_{j+1}}\right) \cdot \sum_{z=1}^{k} g_{z,j+1}$$

of the equilibrium ratio $K_{i,j}$ with respect to liquid concentration, gas concentration and temperature. The expressions for these derivatives depend on the equations used to describe the gas-liquid equilibria either by exess enthalpies (e.g., Wilson, Uniquac) or by equations of state (e.g., Redlich Kwong). In most programs for column simulation, these partial derivatives are evaluated numerically by:

$$\frac{\partial \mathbf{F}}{\partial \mathbf{x}} = \frac{f(\mathbf{x} + \Delta\mathbf{x}) - f(\mathbf{x} - \Delta\mathbf{x})}{2 \cdot \Delta\mathbf{x}} \tag{4.118}$$

The same is true for the partial derivatives of the enthalpies h_L and h_G, especially with respect to concentrations.

A good approximation for the partial derivative of the equilibrium ratio $K_i = \gamma_i \cdot p_i^o/p_t$ with respect to temperature t can be made because the temperature dependency of the vapor pressure. p_i^o is much stronger than that of the activity coefficient γ_i. With the Antoine equation $p_i^o = \exp(A_i - B_i/(t + C_i))$ follows:

$$\left(\frac{\partial K_i}{\partial t}\right) = \frac{\gamma_i}{p_t} \cdot \exp\left(A_i - \frac{B_i}{(t + C_i)}\right) \cdot \frac{B_i}{(t + C_i)^2} \tag{4.119}$$

An approximation for partial derivatives of enthalpies h_L and h_G with respect to temperature is:

$$\left(\frac{\partial h_L}{\partial t}\right) = c_L \quad \text{and} \quad \frac{\partial h_G}{\partial t} = c_p \tag{4.120}$$

Here, c_L and c_p are the heat capacities of the liquid and gas mixture, respectively. They can often be considered as independent on temperature.

4.4.1.3.4 Estimation of Starting Values of the Variables The most essential prerequisite for good convergence of the Newthon-Raphson algorithm is the setting of good starting values of all the variables. One possibility is the use of a distillation line through the feed. The higher the number of equilibrium stages, the more difficult convergence is. A good strategy is to start with only a few equilibrium stages and to search the solution. Then, the number of equilibrium stages is increased and the profiles of temperature and component flow rates are proportionally transferred to the higher number of stages. With these starting profiles the calculation is performed again. This strategy can be applied several times until the correct number of equilibrium stages is reached.

4.4.1.3.5 Iteration History A computer program has been developed by the authors for rigorous column simulation with simultanous solution of the modified MESH-equations as described in the previous chapters. The program can be operated on a MS-DOS personal computer with 640 K storage.

The iteration history for the simulations of the system methanol/ethanol/1-propanol is presented in Fig. 4-48. Column specifications are: number of trays, location of feed tray, relative amount of distillate D/F, thermal state and concentration of the feed and reflux ratio.

Starting values for the internal liquid and vapor states were generated by calculating the distillation line through the feed. Liquid and vapor concentrations, x_i and y_i, were transformed into component flow rates under the assumption of constant liquid and vapor flow rates within the column.

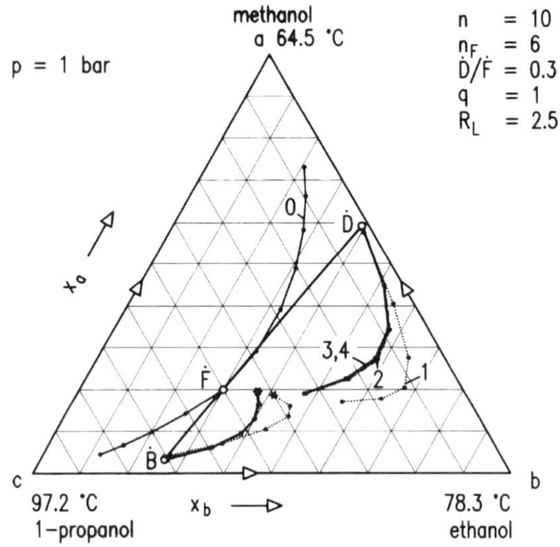

A

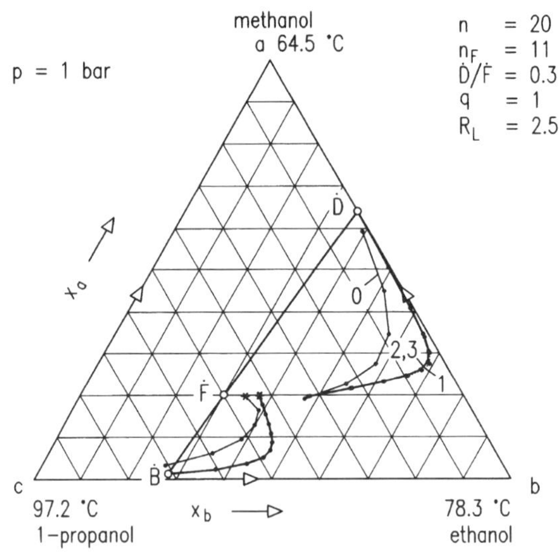

B

Fig. 4-48. *Internal concentration profiles for the system methanol/ethanol/1-propanol: (A) Iteration history for 10 stages with the feed distillation line as starting profile, (B) iteration history for 20 stages with the 10 stage solution as starting profile.*

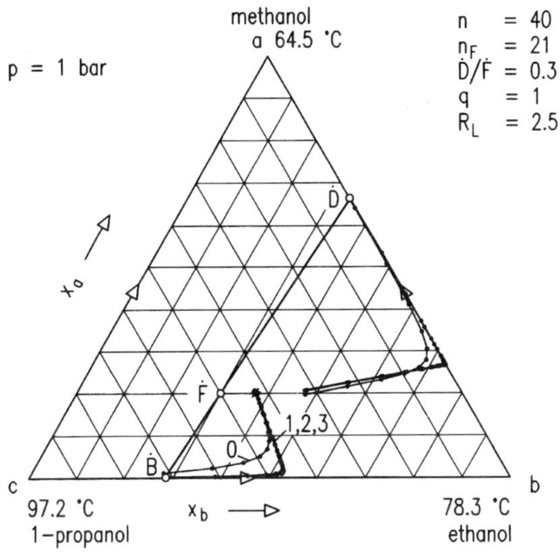

C

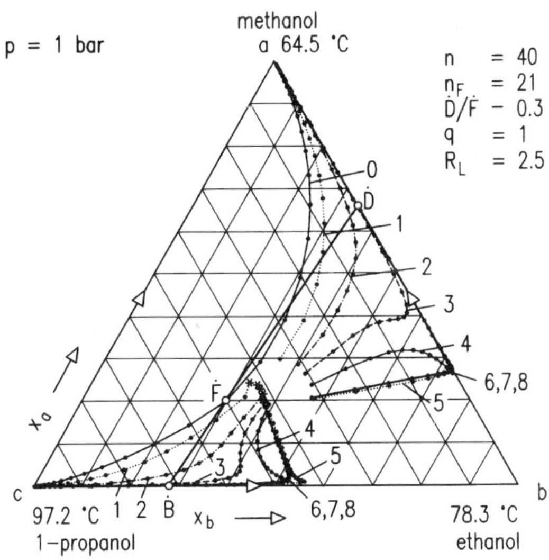

D

Fig. 4-48. *(C) Iteration history for 40 stages with the 20 stage solution as starting profile, (D) iteration history for 40 stages with the feed distillation line as starting profile.*

The starting profile is indicated in the triangular diagram of Fig. 4-48A by 0. The first iteration yields the dotted liquid concentration profile indicated by 1. Number 2 indicates the result of the second iteration. This profile is nearly identical with the final one reached in four iterations.

The iteration history for a column with 20 stages is presented in Fig. 4-48B. Starting values were the previous results with 10 stages. In only three iterations convergence was reached. The transition from 20 to 40 stages also requires 3 iterations only as is seen in Fig. 4-48C.

The iteration history for a column with 40 stages with a distillation line through the feed as starting profile is presented in Fig. 4-48D. Due to the large discrepancy between starting profile and end profile, eight iterations are required to reach convergence. Large changes occur from iteration to iteration and it is a little bit astonishing that convergence is finally reached.

Similar results for the system acetone/chloroform/benzene are presented in Figs. 4-49A to 4-49D. This system exhibits a binary maximum azeotrope and a border distillation line that runs from the high boiler to the maximum azeotrope. Here again, starting values were generated by calculation of the distillation line through the feed (indicated by 0 in Fig. 4-49A). This profile lies far left from the final result. The first iteration yields the dotted profile on the right hand side of the final one. The profiles of the third, fourth and fifth iteration lie very close together. Starting with this profile the solution for a column with 20 stages is found in 4 iterations, see Fig. 4-49B. The same number of iterations is required for the transition from 20 to 40 stages that is presented in Fig. 4-49C. However, iteration history is much more complex for a 40 stage column when the starting profile is the feed distillation line, see Fig. 4-49D.

The system methanol/ethanol/water, presented in Figs. 4-50A to 4-50C, forms a binary minimum azeotrope with a border distillation line between azeotrope and low boiler. When starting with the profile of a distillation line through the feed, 6 iterations are required to reach convergence for a column with 10 stages (Fig. 4-50A). In particular, the liquid concentration profile after the first iteration is not a very good one. But in spite of that, convergence is reached. With the solution of the 10 stage column as starting profile, simulation of a 20 stage column converges in 3 iterations (Fig. 4-50B). The transition from 20 to 40 stages is displayed in Fig. 4-50C. No convergence has been reached for direct simulation of the 40 stage column with the feed distillation line as starting profile.

Computer programs for column simulation are a valuable means for the design of distillation processes, especially for nonideal multicomponent mixtures. However, they cannot replace the distillation expert. One cannot hope that a complex problem can be solved without intensive work of an expert even if a good program is available. The program cannot substitute the engineer. The program is just an efficient tool for the engineer.

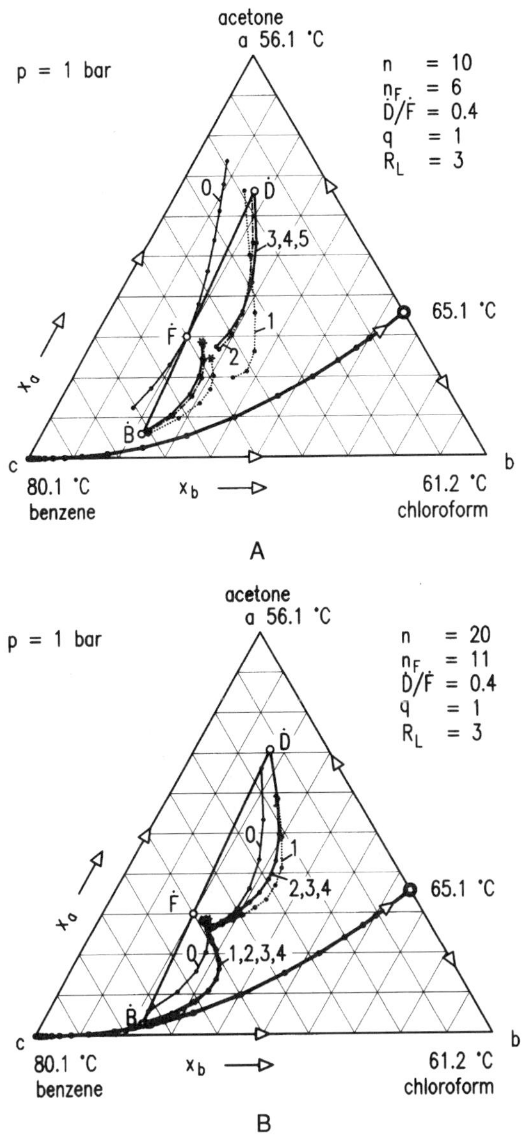

Fig. 4-49. *Internal concentration profiles for the system acetone/chloroform/benzene: (A) Iteration history for 10 stages with the feed distillation line as starting profile, (B) iteration history for 20 stages with the 10 stage solution as starting profile.*

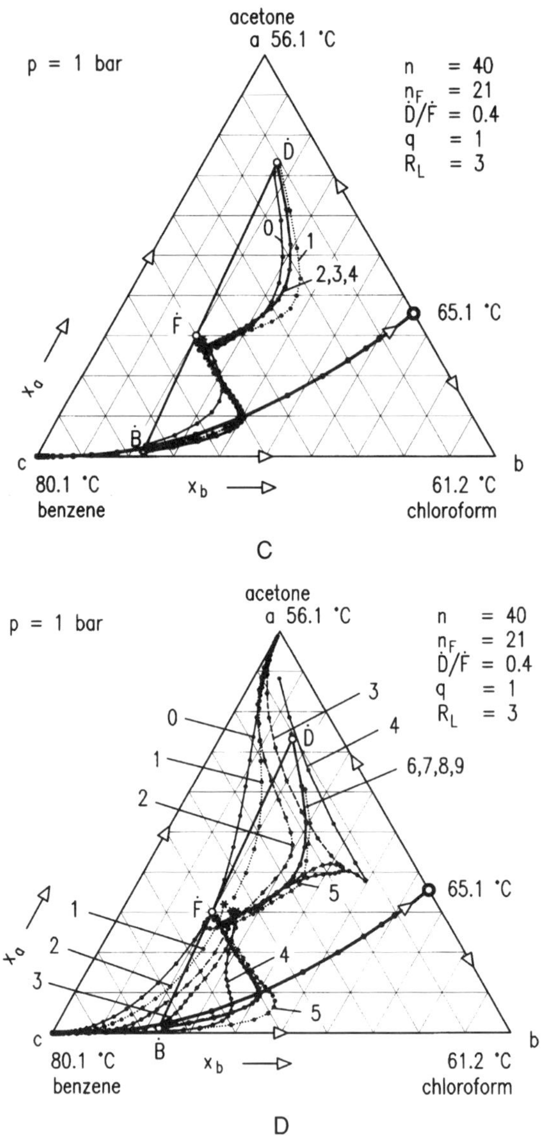

Fig. 4-49. *(C) iteration history for 40 stages with the 20 stage solution as starting profile, (D) iteration history for 40 stages with the feed distillation line as starting profile.*

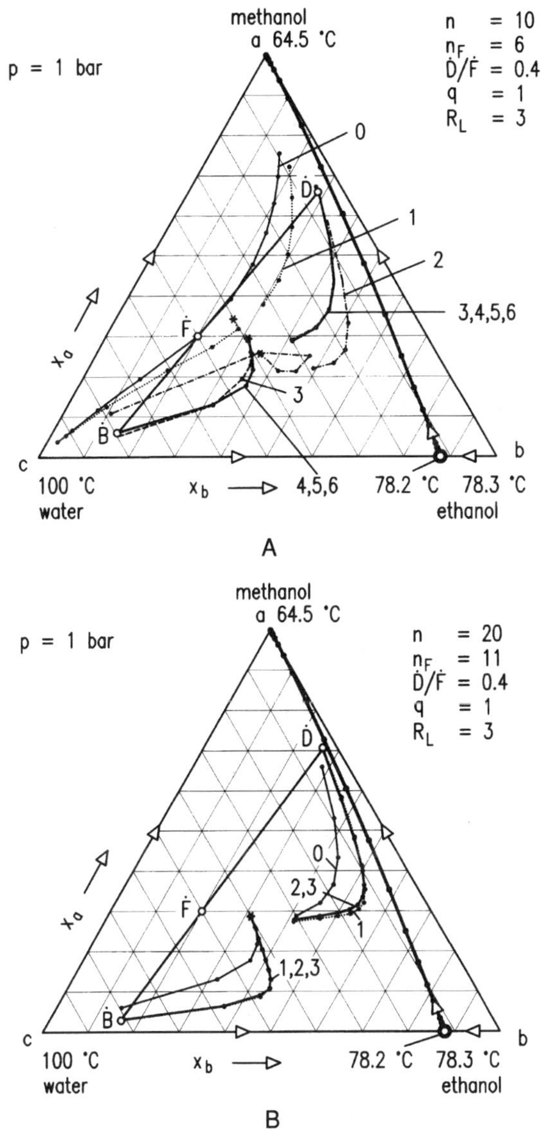

Fig. 4-50. *Internal concentration profiles for the system methanol/ethanol/water: (A) Iteration history for 10 stages with the feed distillation line as starting profile, (B) iteration history for 20 stages with the 10 stage solution as starting profile.*

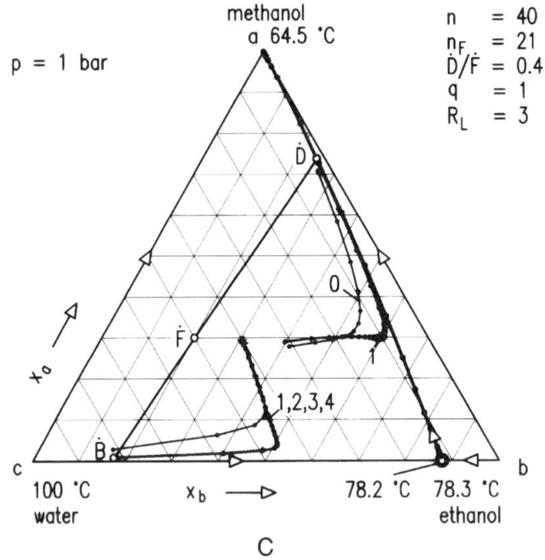

Fig. 4-50. *(C) iteration history for 40 stages with the 20 stage solution as starting profile.*

LITERATURE

Bekiaris, N.; Meski, G. A.; Radu, C., M. and Morari, M.: "Multiple Steady States in Homogenous Azeotropic Distillation," *Ind. Eng. Chem. Res.* 32 (1993) No. 9, pp. 2023/2038.

Bernot, Chr.; Doherty, M. F.; and Malone, M. F.: "Feasibility and Separation Sequencing in Multicomponent Batch Distillation," *Chemical Engineering Science* 46 (1991) No. 5/6, pp. 1311/1326.

Block, U.; and Hegner, B.: "Development and Application of a Simulation Model for Three-Phase Distillation," *AIChE Journal* 22 (1976) No. 3, pp. 582/589.

Block, U.; and Hegner, B.: "Simulationsmethode für homogene Flüssigkeitsphasenreaktionen mit überlagerter Destillation," *Vt Verfahrenstechnik* 11 (1977) No. 2, pp. 101/104.

Bosnjakovic, F.; and Knoche, K. F.: Technische Thermodynamik, Sternkopf Verlag, Darmstadt 1997.

Chilton, T. H.; and Colburn, A. P.: "Distillation and Absorption in Packed Columns," *Industrial and Engineering Chemistry* 27 (1935) No. 3, pp. 255/260.

Doherty, M. F.; and Calderola, G. A.: "Design and Synthesis of Homogenous Azeotropic Distillation, 3. The Sequencing of Columns for Azeotropic and Extraktive Distillations," *Ind. Eng. Chem. Fundam.* 24 (1985) No. 4, pp. 474/485.

Goldstein, R. P.; and Stanfield, R. B.: "Flexible Method for the Solution of Distillation Design Problems Using Newton-Raphson Technique," *Ind. Eng. Chem. Process Des. Develop.* 9 (1970) No. 1, pp. 78/84.

Gorak, A.: Berechnungsmethoden der Mehrstoffrektifikation—Theorie und Anwendungen, Thesis RWTH-Aachen 1990.

Hausen, H. "Rektifikation von Dreistoffgemischen," *Forschung auf dem Gebiet des Ingenieurwesens.* 6 (1935) No. 1, pp. 9/22.

Henley, E. J.; and Seader, J. D.: "*Equilibrium Stage Separation Operations in Chemical Engineering.* John Wiley & Sons, New York, 1981.

Holland, C. D.: *Multicomponent Distillation.* Interscience, New York, 1963.

King, C. J.: *Separation Processes.* McGraw-Hill, New York, 1980.

Krishnamurthy, R.; and Taylor, R.: "Simulation of Packed Distillation and Absorption Columns," *Ind. Eng. Chem. Process Des. Dev.* 24 (1985) No. 3, pp. 513/524.

Levy, S. G.; van Dongen, D. B.; and Doherty, M. F.: "Design and Synthesis of Homogenous Azeotropic Distillations 2." "Minimum Reflux Calculations for Non-ideal and Azeotropic Columns," *Ind. Eng. Chem. Res.* 24 (1985) No. 4, pp. 463/474.

Lewis, W. K.; and Matheson, G. L.: "Studies in Distillation, Design of Rectifying Columns for Natural and Refinery Gasoline," *Industrial and Engineering Chemistry* 24 (1932) No. 5, pp. 494/498.

McCabe, W.L.; and Thiele, E. W.: "Graphical Design of Fractionating Columns," *Industrial and Engineering Chemistry* 17 (1925) No. 6, pp. 606/611.

Naphtali, L. M.; and Sandholm, D. P.: "Multicomponent Separation Calculation by Linearization," *AIChE Journal* 17 (1971) No. 1, pp. 148/153.

Stichlmair, J.: "Zerlegung von Dreistoffgemischen durch Rektifikation," *Chem.-Ing.-Tech.* 60 (1988) No. 10, pp. 747/754.

Stichlmair, J.: "Separation of Ternary Mixtures by Rectification," *International Chemical Engineering* 31 (1991) No. 3, pp. 423/433.

Stichlmair, J. G.; and Herguijuela, J. R.: "Separation Regions and Processes of Zeotropic and Azeotropic Ternary Distillation," *AIChE Journal* 38 (1992) No. 10, pp. 1523/1535.

Stichlmair, J.; Offers, H.; and Potthoff, R. W.: "Minimum Reflux and Minimum Reboil in Ternary Distillation," *Ind. Eng. Chem. Res* 32 (1993) No. 10, pp. 2438/2445.

Taylor, R.; and Kooijman, H. A.: "Composition Derivatives of Activity Coefficient Models," *Chem. Eng. Comm.* 102 (1991) pp. 87/106.

Taylor, R.; Kooijman, H. A.; and Woodman, M. R.: "Industrial Applications of a Non-Equilibrium Model of Distillation and Absorption Operations," *ICHEME SYMPOSIUM SERIES,* (1993) No. 128, pp. A415/A427.

Thiele, E. W.; and Geddes, R. L.: "Computation of Distillation Apparatus for Hydrocarbon Mixtures," *Industrial and Engineering Chemistry* 25 (1933) No. 3, pp. 289/295.

Treyball, R. G.: *Mass-Transfer Operations.* McGraw-Hill, New York, 1968.

Underwood, A. J. V.: "Fractional Distillation of Multicomponent Mixtures," *Chem. Eng. Prog.* 44 (1948) No. 8, pp. 603/614.

Vogelpohl, A. "Rektifikation von Dreistoffgemischen," "Teil 1: Rektifikation als Stoffaustauschvorgang und Rekftifikationslinien idealer Gemische," *Chemie-Ing.-Tech.* 36 (1964) No. 9, pp. 907/915.

Vogelpohl, A.: "Rektifikation von Dreistoffgemischen," "Teil 2: Rektifikationslinien realer Gemische und Berechnung der Dreistoffrektifikation," *Chemie-Ing.-Techn.* 36 (1964) No. 10, pp. 1033/1045.

Vogelpohl, A.: "Die optimale Lage des Zulaufbodens bei der Vielstoffrektifikation," *Chem. Ing. Tech.* 47 (1975) No. 21, pp. 895.

Wahnschafft, O. M.; Köhler, J. W.; Blaβ, E.; and Westerberg, A. W.: "The Product Composition Regions of Single-Feed Azeotropic Distillation Columns," *Ind. Eng. Chem. Res.* 31 (1992) No. 10, pp. 2345/2362.

Wang, J. C.; and Henke, G. E.: "Tridiagonal Matrix for Distillation," *Hydrocarbon Processing* 45 (1966) No. 8, pp. 155/163.

5

Industrial Distillation Processes

5.1 Constraints for Industrial Distillation Processes
5.1.1 Feasible Temperatures
5.1.2 Feasible Pressures
5.1.3 Feasible Dimensions of Columns
5.2 Separation of Binary Mixtures
5.2.1 Concentration of Sulfuric Acid
5.2.2 Ammonia Recovery from Waste Water
5.2.3 Hydrogenchloride Recovery from Inert Gases
5.2.4 Linde Process for Air Separation
5.2.5 Process Water Purification
5.2.6 Steam Distillation
5.3 Separation of Multicomponent Zeotropic Mixtures
5.3.1 Separation Paths
5.3.2 Side Columns
5.3.2.1 Argon Recovery from Air
5.3.2.2 Crude Oil Distillation
5.4 Separation of Heterogeneous Azeotropic Mixtures
5.5 Separation of Binary Azeotropic Mixtures by Changing System Pressure
5.6 Separation of Azeotropic Mixtures by an Entrainer
5.6.1 Processes in One Distillation Field
5.6.1.1 Criteria for Entrainer Selection
5.6.1.2 Processes
5.6.2 Processes in Two Distillation Fields
5.6.2.1 Criteria for Entrainer Selection
5.6.2.2 Generalized Process
5.6.2.3 Simplifications of the Generalized Process

5.6.3 Hybrid Processes
5.6.3.1 Distillation Combined with Decantation (Azeotropic Distillation)
5.6.3.2 Distillation Combined with Absorption (Extractive Distillation)
5.6.3.3 Distillation Combined with Desorption
5.6.3.4 Distillation Combined with Extraction
5.6.3.5 Distillation Combined with Adsorption
5.6.3.6 Distillation Combined with Membrane Permeation
5.7 Reactive Distillation
5.7.1 Fundamentals
5.7.1.1 Chemical Equilibrium
5.7.1.2 Stoichiometric Lines
5.7.1.3 Non-Reactive and Reactive Distillation Lines
5.7.1.4 Reactive Azeotropes
5.7.2 Topology of Reactive Distillation Lines
5.7.2.1 Reactions in Ternary Systems
5.7.2.2 Reactions in Ternary Systems with Inerts
5.7.2.3 Reactions with Side Products
5.7.2.4 Reactions in Quaternary Systems
5.7.3 Topology of Reactive Distillation Processes
5.7.3.1 Single Product Reactions
5.7.3.2 Decomposition Reactions
5.7.3.3 Side Reactions
5.7.4 Industrial Processes of Reactive Distillation
5.7.4.1 Synthesis of MTBE
5.7.4.2 Synthesis of Mono-Ethylene Glycol
5.7.4.3 Synthesis of TAME
Literature

In this chapter some typical examples of industrial distillation processes are presented. However, a few comments on additional non-thermodynamic constraints for industrial processes have to be made first.

5.1 CONSTRAINTS FOR INDUSTRIAL DISTILLATION PROCESSES

Industrial distillation processes underlie some restrictions due to operability of the units, to economic conditions and to environmental constraints. Some aspects are dealt with in this chapter.

5.1.1 Feasible Temperatures

In industrial distillation processes there exists an upper limit for feasible operating temperatures. One reason is the thermal stability of the species in the mixtures to be separated. Many substances decompose gradually at higher temperatures. Some species are not even stable at their normal boiling points. Maximum allowable tem-

Table 5-1. Maximum Allowable Operating Temperatures of Distillation Columns to Avoid Thermal Degradation or Polymerization.

Substance	Maximum Operation Temperature	Operating Pressure	Comments
Sulfuric acid, H_2SO_4	170°C	1 bar	formation of SO_3
Hydrogen peroxide	60 to 80°C	1 bar	explosive vapor
Crude oil	350°C	2.1 bar	degradation
Reduced crude oil	420°C	0.013 bar	degradation
Styrene	100°C		polymerization
Tall oil	260°C	$5\ 10^{-3}$ bar	degradation
Nitric acid	below normal boiling point		formation of NO_x that can be removed by air stripping
Acrylic acid	130°C		polymerisation

peratures for selected substances are shown in Table 5-1. Thermal stability depends on time of exposure as well as on temperature level. The data in Table 5-1 are approximate values for typical distillation conditions.

A second reason for a maximum temperature limitation is the means of heat supply in industrial processes. Except for a few very special cases, the heat required in distillation columns is supplied by condensing steam. The pressure of the steam available places an upper limit on temperature levels that can be achieved on the process side of the reboilers and heaters. A practical temperature difference for heat transfer is in the range of about 10 to 30°C. In special cases, e.g., in crude oil distillation or sulfuric acid concentration, higher temperatures may be realized by heating with hot oil or with flue gases. Maximum feasible temperatures are as high as 300 and 400°C, respectively. Electric heating of distillation columns is normally not practical. Table 5-2 lists the maximum product side temperatures and the relative costs of some important heating media for distillation columns. The values of relative costs are very vague because they are not correctly evaluated in most companies

Table 5-2. Maximum Product Side Temperatures and Relative Costs of Some Heating Media for Distillation Columns.

Medium	Product Side Temperatures	Relative Costs
Steam		
Exhaust	90°C	50%
Low pressure (7 bar)	140°C	100%
Medium pressure (15 bar)	170°C	130%
High pressure (30 bar)	200°C	150%
High pressure (80 bar)	260°C	
Hot oil	300°C	
Hot oil (synthetic)	400°C	
Flue gas	450°C	

Table 5-3. Minimum Product Side Temperatures and Relative Costs of Some Cooling Media for Distillation Columns.

Medium	Product Side Temperatures	Relative Costs
Air	50–70°C	60%
Cooling tower water	40–50°C	100%
Chilled water	20–30°C	150%
Ammonia	10°C	300%
	−5°C	400%
	−30°C	500%
Sole	0°C	300%
Propane/propylene	−30°C	
Ethane/ethylene	−75°C	
Methane	−150°C	

[Cooper 1989]. In particular, there is a tendency to underevaluate high pressure steam.

A lower limit of feasible temperatures in distillation columns is dictated by the temperature of the coolant for the overhead condenser. In most cases water is used as coolant and practical temperature differences in the condenser lead to a minimum temperature in the column as low as 40 to 50°C. With increasing tendency, air is used as coolant, and this leads to minimum temperatures of about 60 to 70°C. Seldom used coolants are chilled water or evaporating refrigerants such as ammonia, propane or nitrogen. Even when available, such coolants are relatively expensive, as may be noted from Table 5-3.

5.1.2 Feasible Pressures

The temperature constraints set by the heating and cooling media can be met by the selection of a proper operating pressure or the addition of low boiling constituents. To decrease operating temperatures the separation may require use of a vacuum. It is technically feasible to operate distillate columns at a pressure as low as 0.06 to 0.08 bar with a single-stage rotating vacuum pump and as low as about 0.013 bar with a double-stage vacuum pump. Vacua higher than $(2 \text{ to } 5) \cdot 10^{-3}$ bar are very seldom used in distillation columns [Strigle 1987] because of the high operating and capital costs of the vacuum-producing equipment.

Instead of operating at a vacuum, a low-boiling substance can be added to decrease the boiling temperature of the material to be distilled. Often used additives are inert gases (air and nitrogen) and—especially in organic systems—steam, since it can be easily removed after distillate condensation (see Section 5.2.6, Steam Distillation).

To increase the boiling point temperatures of low boiling mixtures the column can be operated at higher pressures. However, the higher the pressure the lower, in most cases, is the concentration difference between gas and liquid. An upper limit for the operating pressure lies in the range of the critical pressures of the constituents. Very low boiling hydrocarbons such as methane are usually distilled at pressures of about 70% of their critical one. Ethylene and ethane are typically distilled at 40 to 55% of their critical pressures while for propylene and propane, 35% to 50% of the critical pressures are used [Strigle 1987].

The increase of operating pressure is a very fundamental problem in industrial processes since gas and liquid streams have to be compressed and pumped, respectively, to the higher pressure. Whenever possible, a pressure increase should be established in the liquid phase instead of in the vapor phase since power consumption is much lower then. A comparison of power consumption per mass flow is presented in Fig. 5.1 for air and water. The diagram clearly shows that power consumption for compression of air is by two to three orders of magnitude higher than for pumping the same mass stream of water. This large difference results, in essence, from the density difference of air and water, and exists, therefore, in all systems.

5.1.3 Feasible Dimensions of Columns

The feasible number of theoretical stages of industrial columns is also limited. As can be seen in Table 5-4, more than 100 stages are very seldom utilized in commercial distillation columns.

A very important problem in column design is proper selection of the construction material. For very corrosive systems, e.g., acids, pyrex glass has to be used. In this case, the diameter of the column is restricted to 1 m or less and the operating pressure is 1 bar or less. The largest diameters of industrial distillation columns are as high as 13 to 15 m.

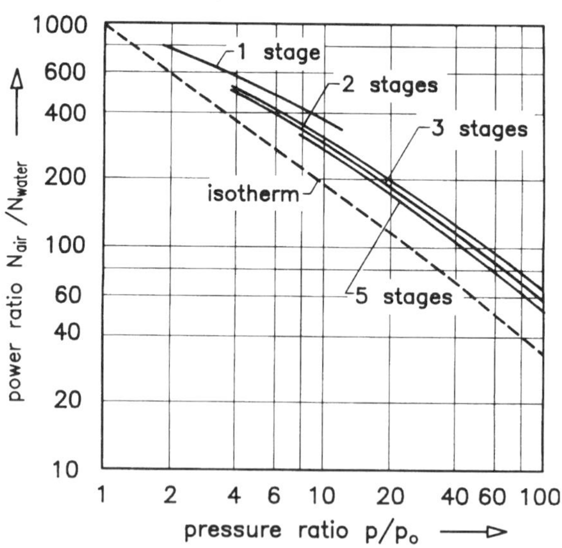

Fig. 5-1. Comparison of power consumption for compressing of air and pumping of water. Power consumption (kW/kg) for compressing a gas is much higher than for pumping the same mass stream of liquid due to the large density difference of gases and liquids.

Table 5-4. Key Components for Distillation Processes of Industrial Importance and Typical Number of Trays (modified after [Perry 1984]).

Key Components	Typical Number of Trays
Nitrogen/oxygen	100
Argon/oxygen	150
Hydrocarbon Systems	
crude oil (atmospheric tower inclusive side columns)	51
ethylene/ethane	73
propylene/propane	138
propyne/1–3-butadiene	40
1–3 butadiene/vinyl acetylene	130
benzene/toluene	34, 53
benzene/ethyl benzene	20
benzene/diethyl benzene	50
toluene/ethyl benzene	28
toluene/xylenes	45
ethyl benzene/styrene	70
o-xylene/m-xylene	130
Organic Systems	
methanol/formaldehyde	23
dichloroethane/trichloroethane	30
acetic acid—acetic anhydride	50
acetic anhydride/ethylene diacetate	32
vinyl acetate/ethyl acetate	90
ethylene glycol/diethylene glycol	16
cumene/phenol	38
phenol/acetophenone	39, 54
Aqueous Systems	
HCN/water	15
acetic acid/water	40
methanol/water	60
ethanol/water	60
isopropanol/water	12
vinyl acetate/water	35
ethylene oxide/water	50
ethylene glycol/water	16

5.2 SEPARATION OF BINARY MIXTURES

In the following paragraphs, a few examples of industrial binary distillation processes are given. The examples have been selected to demonstrate the importance of the constraints discussed in the previous section.

5.2.1 Concentration of Sulfuric Acid

Sulfuric acid is one of the most important basic substances of the chemical industry. It is used in many processes to enhance a chemical reaction and, in most cases,

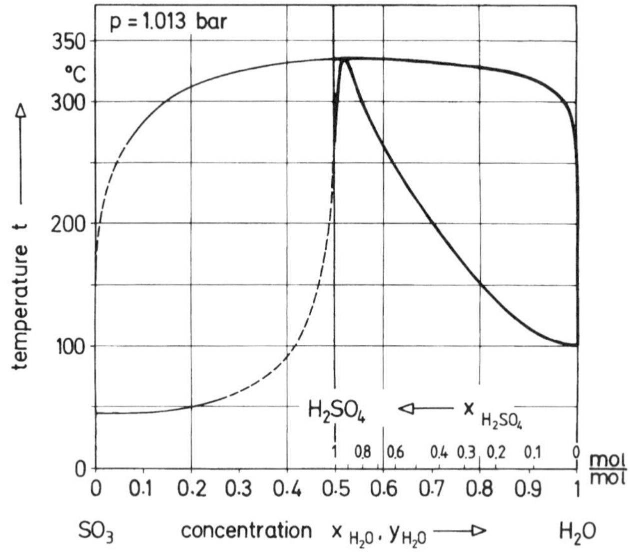

Fig. 5-2. Vapor-liquid equilibrium and dew and boiling point of H_2O/H_2SO_4 mixture at p = 1 bar [Perry et al. 1984]. There exists a maximum azeotrope at high H_2SO_4 content.

leaves the process as a diluted impure acid. The boiling and dew point temperatures of H_2O/H_2SO_4 mixtures at a pressure of 1.0 bar are shown in Fig. 5-2. A maximum azeotrope exists at a weight fraction of 98.4%. Up to a liquid concentration of about 75 wt.-% H_2SO_4 the vapor contains only water. The vapor pressures of H_2O/H_2SO_4 mixtures at various temperatures are shown in Fig. 5-3 [Perry et al. 1984]. This plot is very helpful for defining conditions of technical concentration processes.

If cooling water is used to condense the overhead product, the minimum overhead temperature is 40 to 50°C (see Table 5-3). As nearly pure water is the overhead product, the minimum operating pressure is about 0.12 bar. The maximum bottom concentration of the acid at this pressure depends on the heating medium. With medium pressure steam (15 bar, 200°C) the maximum process temperature is as high as 170°C, which results in an acid concentration of about 86 wt-%. Because of these temperature constraints, the feasible sulfuric acid concentrations in an industrial plant are by far lower than the thermodynamic limit set by the azeotrope.

A two-pressure process for sulfuric acid concentration is shown in Fig. 5-4. Initially, the diluted acid is concentrated to 72 wt-% at normal pressure by single-stage distillation. The vapor does not contain any H_2SO_4 (or SO_3) under these conditions. In a second step, the acid is further concentrated up to 86% at a vacuum of 0.12 bar. As the vapor contains some H_2SO_4 (in form of SO_3) the separation is performed in a multistage distillation column. The SO_3 in the vapor is absorbed by the sulfuric acid entering at the top of the column.

The problem of recycling the H_2SO_4 is not the mere concentration of H_2SO_4 but the removal of impurities. Organic substances are decomposed during concentration

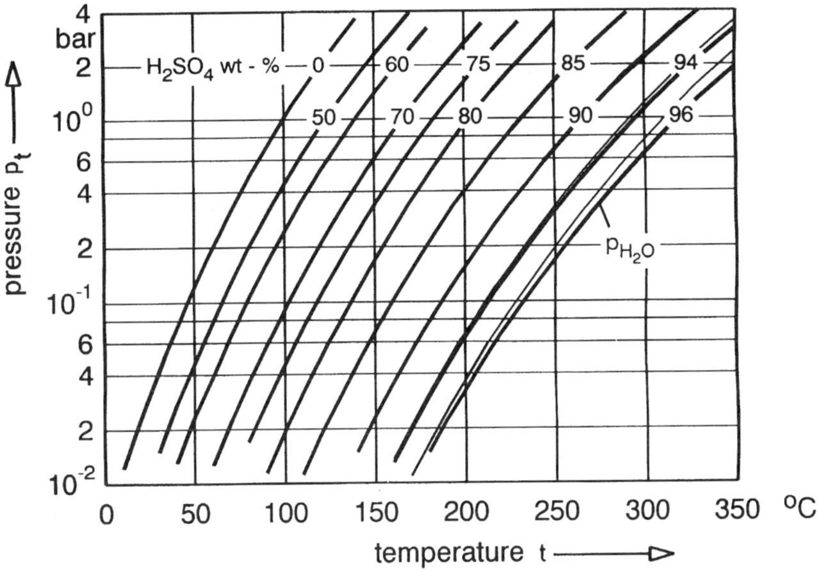

Fig. 5-3. *Vapor pressure over a H_2O/H_2SO_4 mixture [Perry et al. 1984]. The parameter is the H_2SO_4 concentration in weight percent.*

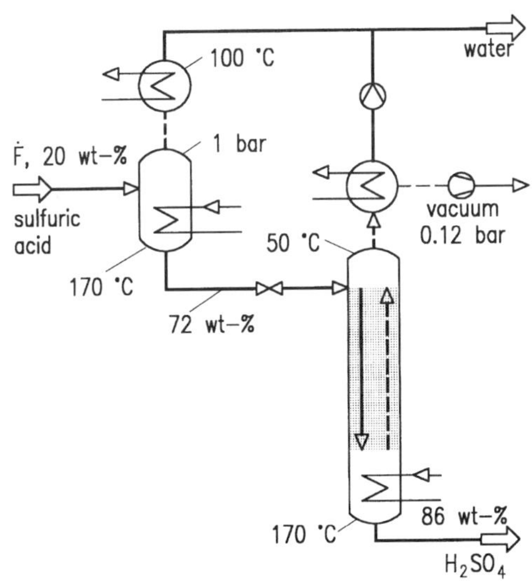

Fig. 5-4. *Process for concentrating diluted sulfuric acid.*

and stripped from the acid by the vapor. Inorganic substances often precipitate during the concentration step, what happens preferably in the concentration range of about 60 wt-% H_2SO_4.

5.2.2 Ammonia Recovery from Waste Water

Ammonia is a toxic chemical that is often contained in waste water and must be separated and recovered as a pure liquid. The content of NH_3 in waste water is typically as low as one weight percent. As may be seen from the diagram in Fig. 5-5. [Rizvi and Heidemann 1987], the dew point of ammonia is very low (−33.4°C) at a pressure of 1 bar. To increase the dew point to 50°C, the minimum temperature of a water-cooled condenser, the pressure must be increased to about 20 bar. At this pressure the boiling point of water is as high as 212°C, what is too high for a reboiler heated by steam. Therefore, the separation requires two steps (Fig. 5-6) [Wunder 1990]. In the first step, most of the water is removed as bottoms from column C-1, which is operated at normal pressure. The distillate must have a boiling point not lower than 45°C. From Fig. 5-5 the maximum NH_3-content of the distillate can

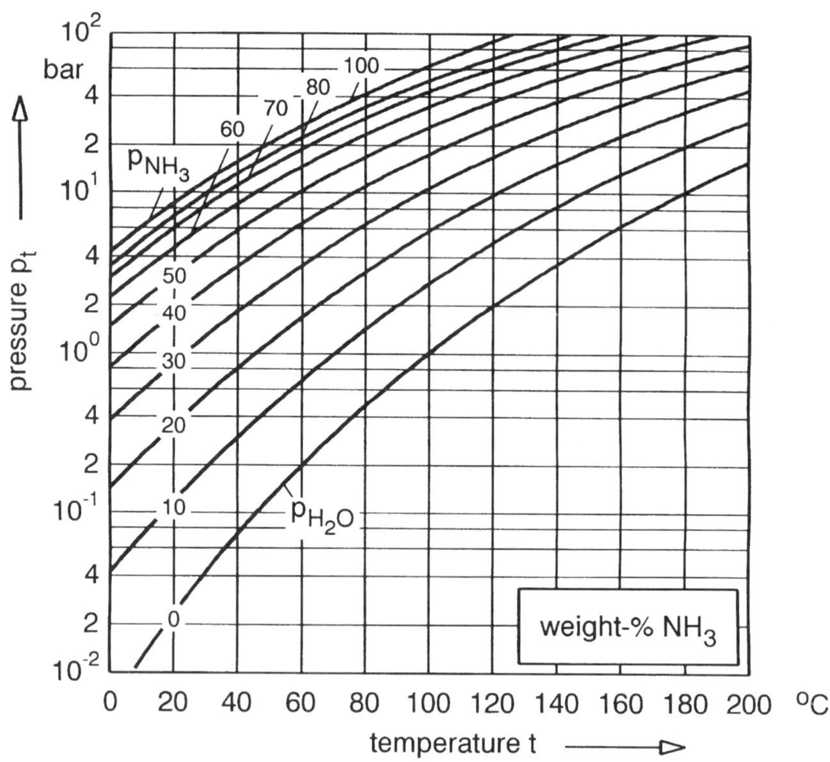

Fig. 5-5. Vapor pressure of ammonia/water solutions versus temperature [Rizvi and Heidemann 1987]. The parameter is the NH_3 concentration in the liquid in weight percent.

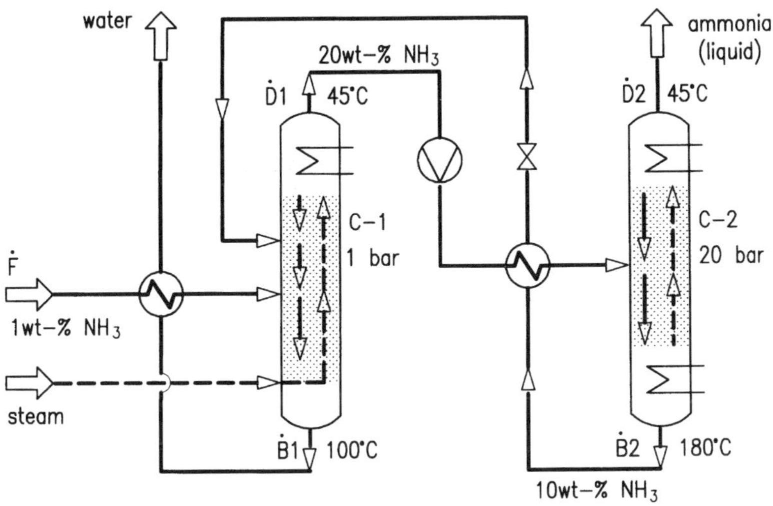

Fig. 5-6. Process for recovering pure ammonia from waste water [Wunder 1990].

be determined to be 20 wt-%. The distillate is pumped to 20 bar and fed into column C-2. The overhead $\dot{D}2$ is pure liquid ammonia. Considering the temperature limitation of 180°C in the reboiler, the minimum NH_3 content of the bottoms is as low as 10 wt-%. Therefore, the bottoms have to be recycled into column C-1. As water is the bottom fraction of column C-1 the reboiler can be discarded and heating can be provided by exhaust steam.

EXAMPLE 5-1:

Find the flow rates of the streams $\dot{D}1$ and $\dot{B}2$ of the process of Fig. 5-6 for an ammonia concentration of 1 wt-% in the feed.

Solution:

A material balance of the unit gives: $$\dot{D}2 = \dot{F}\, x_F = \dot{F}\, 0.01. \quad (1)$$

The material balance over column C-2 yields: $$\dot{D}1 = \dot{D}2 + \dot{B}2 \quad (2)$$

A water balance over column C-2 gives: $$\dot{D}1\, 0.8 = \dot{B}2\, 0.9 \quad (3)$$

From Equations (5.1) to (5.3) we get: $$\dot{B}2 = 0.08\,\dot{F} \quad \text{and} \quad \dot{D}1 = 0.09\,\dot{F} \quad (4)$$

By recycling $\dot{B}2$, the feed to column C-1 is increased by 8%. The feed to column C-2 is as low as 9% of the feed $\dot{F}$ to the process.

5.2.3 Hydrogenchloride Recovery from Inert Gases

Figure 5-7 presents a very interesting process for the recovery of hydrogenchloride, HCl, from inert gases, e.g., [Stichlmair 1979]. This process can be applied to HCl rich gases with no or low water content. The normal boiling point of HCl is −85°C. To avoid a cryogenic process the HCl is first absorbed in water and then recovered from the liquid solution by distillation. The solubility of HCl in water is very high and a hydrochloric acid concentration up to 31 wt-% can be achieved in the absorber. A large heat of absorption is set free in the absorber and temperature rises. At higher temperature the water becomes volatile and mass transfer of water from liquid to gas phase takes place, too. At the top of the absorber the water is condensed and recycled into the column.

The hydrochloric acid formed in the absorber is fed via a heat exchanger into distillation column C-2 to be separated into pure HCl gas as overhead and azeotropic acid (~21 wt-%) as bottom product, which is recycled into the lower section of the absorber. Because water is continuously fed into the process, water or a water-rich fraction has to be withdrawn. In most cases the bottoms product from the absorber is partly withdrawn as side products, because it meets the standards of commercial hydrochloric acid. The recovery of pure HCl gas is then as low as 45%.

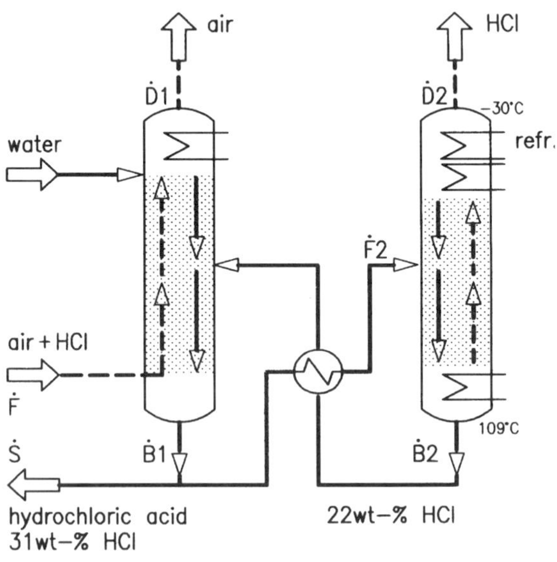

Fig. 5-7. Process for recovery of HCl from waste inert gases [Stichlmair 1979].

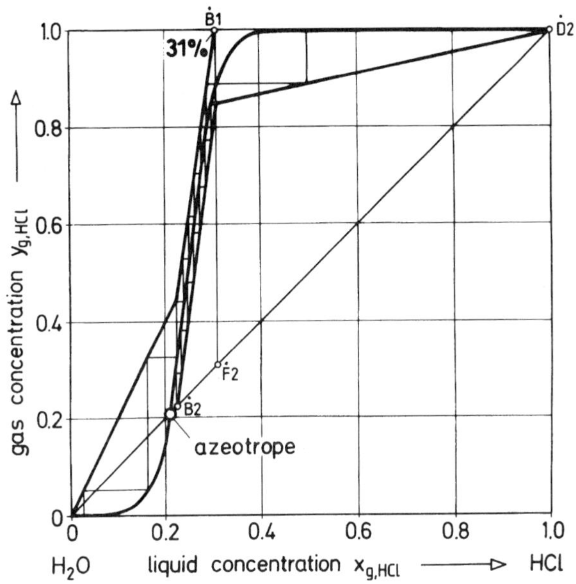

Fig. 5-8. *McCabe-Thiele diagram for the process in Fig. 5-7.*

The y/x-diagram of the process is shown in Fig. 5-8. The only effect of the inert gas in the absorber is the reduction of the partial pressures of water and HCl. Consequently, the inert gas does not need to be accounted for in the calculations. Only HCl and water have to be considered at a pressure $p_t - p_{inert}$. The boiling point of the liquid has to be determined for this reduced pressure, like in a vacuum column. As the heat of HCl absorption is nearly equal to the heat of water vaporization, the flow rate of the gas (inert gas neglected) is nearly constant, and the equilibrium line for HCl water can be determined for constant pressure $p_t - p_{inert}$. As long as the partial pressure of HCl in the feed is higher than 0.1 bar the equilibrium line is nearly the same in the absorption and in the distillation step.

The process can be modified by withdrawal of the bottoms of the distillation C-2 instead of the absorber. Since $\dot{B}2$ has a lower content of HCl and a higher content of water, the yield of pure HCl gas increases up to 70%. But the bottoms $\dot{B}2$ are less worthy than the 31 wt-% side stream.

EXAMPLE 5-2:

Find the yield of pure HCl gas and the flow rates of the streams of the process in Fig. 5-7.

Solution:

The slope of the operating line in the upper section of the absorber can be determined in Fig. 5-8 to be $\dot{L}/\dot{G}_{HCl} = 2$ (inert gas neglected). In the lower section of the absorber we find $\dot{L}/\dot{G}_{HCl} = 4.7$. The flow rate of water into the process is:

$$\dot{L}_{H2O} = \dot{G}_{HCl} \quad \text{with } \dot{G}_{HCl} = \dot{F}\, y_{HCl} \tag{1}$$

Consequently, half of the liquid flow in the absorber is provided by the condenser. The rest is tap water.

In the side stream $\dot{S}$ of hydrochloric acid all the water has to be withdrawn.

A water balance over the process yields: $\dot{S}\,(1 - 0.31) = \dot{L}_{H2O}$ (2)

With (1) we get: $\dot{S} = 1.45\, \dot{G}_{HCl}$ (3)

A HCl balance over the process yields: $\dot{G}_{HCl} = \dot{D}2.1 + \dot{S}\, 0.31$ (4)

With Eqs. (3) and (4) we find: $\dot{D}2 = 0.45\, \dot{G}_{HCl}$ (5)

The yield of pure HCl is as low as 45%. The rest of the HCl in the feed gas is in the side stream $\dot{S}$ of hydrochloric acid.

The feed into column C-2 is: $\dot{F} = 4.7\, \dot{G}_{HCl} - 1.45\, \dot{G}_{HCl} = 3.25\, \dot{G}_{HCl}$ (6)

For the bottoms $\dot{B}2$ we get: $\dot{B}2 = 3.25\, \dot{G}_{HCl} - 1.45\, \dot{G}_{HCl} = 1.8\, \dot{G}_{HCl}$ (7)

5.2.4 Linde Process for Air Separation

The separation of air into nitrogen and oxygen (and often argon) is performed on a large scale in cryogenic processes. In 1980 the production of oxygen was as high as 112.2 10^6 t/a making air separation the most important separation process second only to crude oil refinery [Baldus et al. 1983].

Figure 5-9 shows a simplified flow sheet of the Linde process [Baldus at al. 1983]. First, air is compressed to about 6 bar and cooled down against nitrogen and oxygen. Water, carbon dioxide and hydrocarbons are removed from the air by automatically reversing molecular sieve adsorbers. The air is cooled down until about 20% of the feed is liquified. The air enters the so-called Linde double column to be separated into nitrogen and oxygen. Both products are warmed up in the heat exchangers E-1, E-2 and E-3 against the incoming air. The low temperatures required for the process are generated by expanding some nitrogen (approximately 12% of the feed) in an expansion machine from 5.7 bar to 1.3 bar.

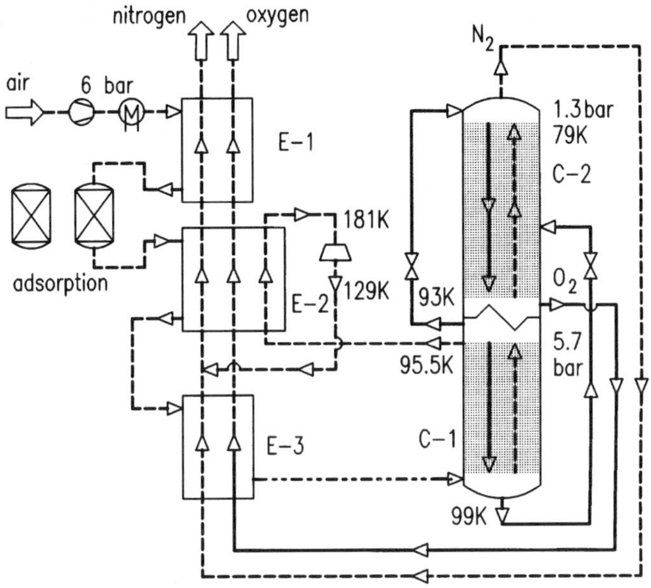

Fig. 5-9. *Linde process for air separation.*

The peculiarity of the process is the Linde double column. It is a distillation column without a reboiler at the bottom and a condenser at the top. The function of the column is explained best in Fig. 5-10. The separation of the binary mixture is performed in two steps. In the first step some pure nitrogen is separated from the air. As most of the feed is gaseous, only a rectifying section with a top condenser is required. The distillate of the column C-1 is used as reflux in column C-2. The bottoms of C-1 are fed into column C-2 to be separated into nitrogen and oxygen. Since the reflux is provided by the first column C-1, no top condenser is required. Column C-1 has only a condenser and column C-2 only a reboiler.

Condenser and reboiler may be thermally coupled (see Section 7.3.2). The heat removed in the condenser of C-1 is used for heating the reboiler of column C-2. As the temperature at the top of column C-1 is lower than at the bottom of column C-1, the thermal coupling is only possible if column C-1 is operated at a higher pressure than column C-2. The minimum pressure of column C-2 is 1.3 bar to cover the pressure drop in the heat exchangers E-1, E-2 and E-3. At a temperature difference of 2.5 K for heat transfer, the pressure of column C-1 has to be 5.7 bar (see Fig. 5-11).

In the Linde double column the column C-2 is installed at the top of column C-1 so that condenser and reboiler can be combined in a single heat exchanger.

Air separation by cryogenic distillation is a highly sophisticated process. The energy required for separation is provided by the compressor of the feed. The low temperatures are generated by the Joule-Thomson effect which utilizes the fact that the enthalpy of low pressure gas is higher than of high pressure gas. In modern air

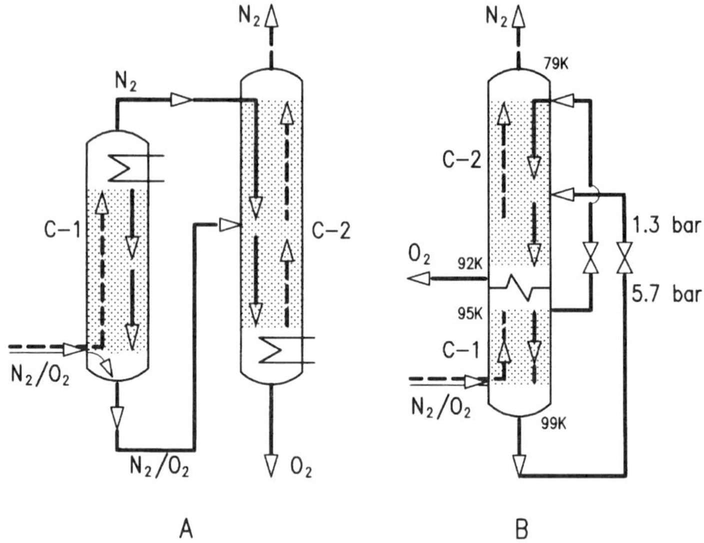

Fig. 5-10. *Recovery of oxygen and nitrogen from air by distillation: A) With two separate columns C-1 and C-2, B) with Linde double column.*

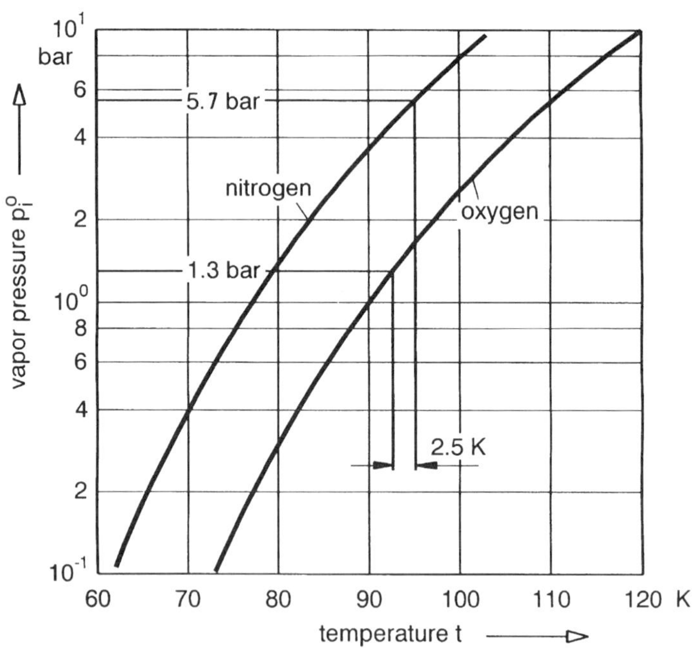

Fig. 5-11. *Determination of the operating pressure for thermally coupling of the columns C-1 and C-2 of the process in Fig. 5-10.*

separation units, especially when parts of the products are withdrawn as cryogenic liquids, the expansion valve is replaced by an expansion machine. The specific energy demand of modern cryogenic air separation is as low as 1.9 $kWh/kmol_{air}$ with expansion machine (but gaseous products) [Streich 1975]. By replacing the sieve trays in the column (typically 40 trays in the high pressure section and 60 trays in the low pressure section) by structured packings (see Chapter 8.3), the energy requirement of the process is reduced by about 8% [Agrawal et al. 1992].

5.2.5 Process Water Purification

From many processes there results an aqueous side stream polluted by organic components with limited solubility in water. Such mixtures often form a heteroazeotrope, which is always a low boiling one. Consequently, these organics can be removed as overheads of a distillation column, even when they are higher boiling than water.

As an example, the removal of toluene from water is considered. It is assumed that the water is saturated with toluene at a temperature of 80°C. The toluene concentration in water is as high as 1.6 mol-%. The boiling point of a two-liquid phase toluene/water mixture is 84.1°C. The vapor pressure of water at 84.1°C is 0.55 bar, the vapor pressure of toluene is 0.45 bar (adding up to 1 bar, see Fig. 5-17). In the region of immiscibility the equilibrium is independent on liquid composition, as can be seen in Fig. 5-12. The azeotrope has a water content of 0.55 mol-%. Important for the separation is that part of the equilibrium line that lies close by pure water, especially the region of toluene solubility in water. This region is depicted separately in Fig. 5-12.

The equilibrium line runs below the diagonal of the McCabe-Thiele-Diagram in this concentration range. Thus, the concentration of water (low boiler) is lower in the vapor than in the liquid phase. Toluene enriches up to 45% in the vapor phase. After condensation two-liquid phases are formed. The water rich phase is recycled

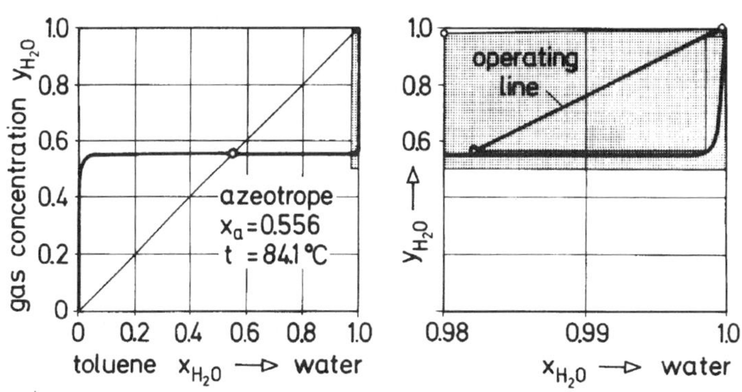

Fig. 5-12. McCabe-Thiele diagram for the separation of toluene from water.

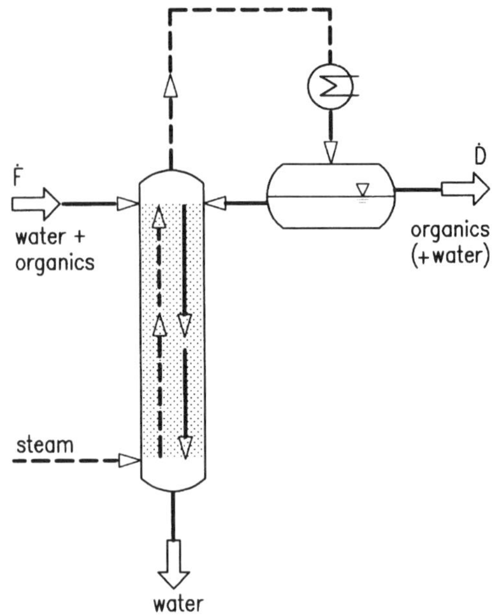

Fig. 5-13. Process for separating high boiling organics from water. In the range of immiscibility even high boiling organics go overhead.

as reflux into the column, Fig. 5-13. The toluene rich phase contains about 0.02 mol-% water [Sorensen and Arlt 1979].

It is advantageous to heat the column by life steam because cheap exhaust steam can be utilized and the reboiler can be omitted. It must be recognized, however, that the end point of the operating line does not lie on the diagonal in this case, but rather at $y_{H2O} = 1$ and $x = x_B$, as shown in Fig. 5-12.

The process of Fig. 5-13 is a standard process very often used for the separation of organic constituents from water. The process can be applied when the main organic constituent is at least partly immiscible with water. The amount of steam required is typically as low as 5 to 10% of the liquid feed.

If the organic phase contains too much water it can be purified by a similar separation step. An example of such a process is dealt with in Section 5.4.

In many cases, the organic phase contains light components that are fully soluble in water (e.g., methanol or acetone). These constituents can be separated from the water as top fraction of a previous distillation step, see Fig. 5-14. An example of such a system is the mixture acetone/water/toluene. It is assumed that the mixture is saturated with toluene. Acetone is separated in column C-1 as top fraction $\dot{D}1$. The bottom fraction $\dot{B}1$ lies in the region of immiscibility. This fraction is split in the decanter S-1 into an organic phase and a water phase. The water is purified by steam stripping in analogy to the process in Fig. 5-13.

In the process of Fig. 5-14 vapor is condensed at the top of column C-2 and vapor of the same composition is generated in the reboiler of column C-1. The

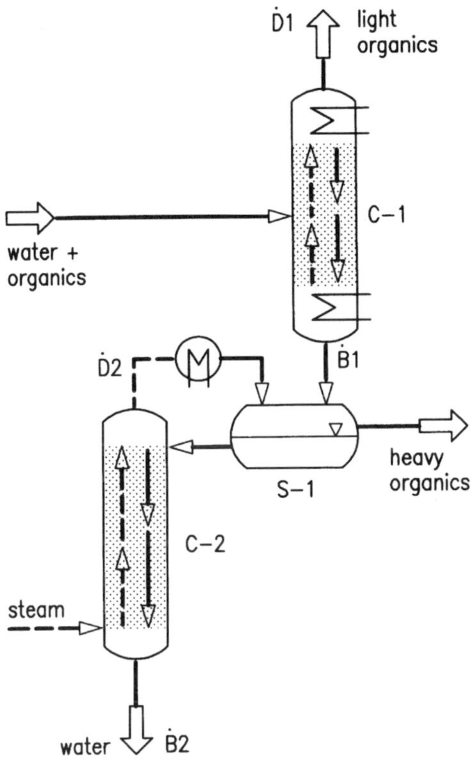

Fig. 5-14. *Process for separating low boiling and high boiling organics from water (see also Fig. 5-30).*

process can be improved by direct coupling of columns C-1 and C-2 (see Section 7.3.1). The vapor at the head of column C-2 is fed directly into column C-1, whereby the top condenser of C-2 and the reboiler of C-1 can be discarded (Fig. 5-15). In the combined column the top product is acetone and the bottom product is pure water. The heavy organics (e.g., toluene) concentrate in the middle of the column and two-liquid phases are formed there. The liquid is drawn off a few stages below the feed point and separated in a decanter. The water phase is recycled into the column. The organic phase is the side product.

Depending on the solubility of water in the organic phase, the water content might be too high in many cases. The water can easily be removed by a second column (Fig. 5-16). The overhead is recycled into the main column. The bottoms is pure organic product.

5.2.6 Steam Distillation

Steam distillation is a special case of a distillation process which is primarily applied to high boiling organic systems that are immiscible with water. As has been shown

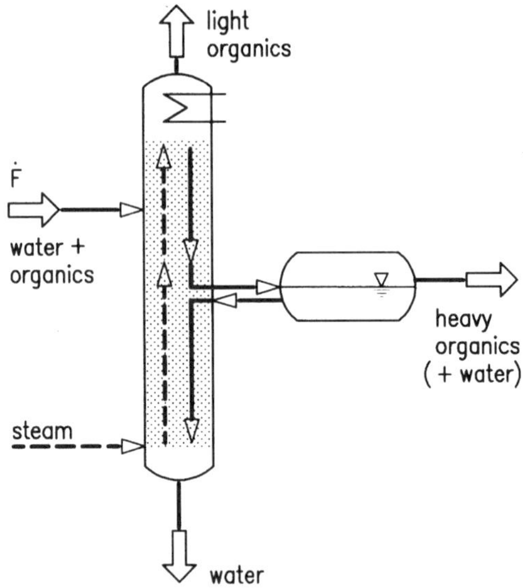

Fig. 5-15. *Direct coupling of columns of the process in Fig. 5-14 (See Section 7.3.1).*

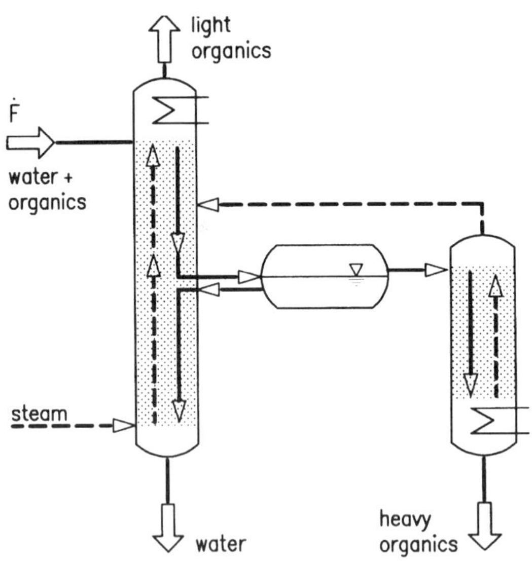

Fig. 5-16. *Process for separating low and high boiling organics from water with direct coupling of columns.*

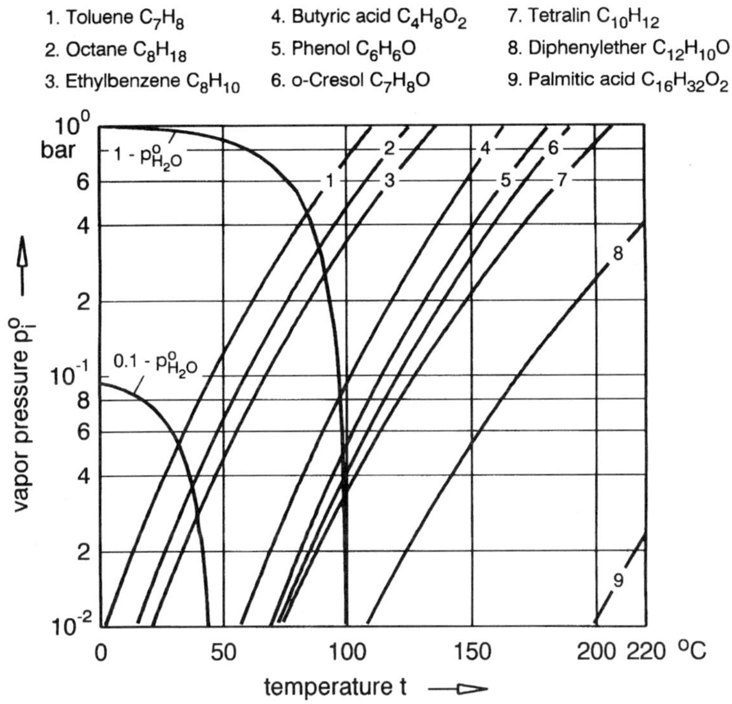

Fig. 5-17. Vapor pressure curves for some high boiling organic compounds that are immiscible with water. The vapor pressure of water p^o_{H2O} is presented as difference to the total pressure p_t p^o_{H2O}. The intersection of the $p_t - p^o_{H2O}$ line with the vapor pressure curves of the organic compounds gives the boiling point of the mixture.

in Chapter 2 the existence of a region of immiscibility always results in a decrease of boiling temperature. The boiling point of such a mixture is always lower than the boiling points of the constituents. Even very high boiling organic systems have, in a mixture with water, a boiling point lower than pure water. Thus, the problem of thermal degradation of organic compounds can often be avoided by steam distillation.

The boiling points of substances immiscible with water can be easily determined by the diagram in Fig. 5-17. The difference of system pressure p_t and vapor pressure p^o_{H2O} of water is plotted versus the temperature (here $p_t = 1$ bar and $p_t = 0.1$ bar). The intersection of the $(p_t - p^o_{H2O})$-line with the vapor pressure line of the organic compound gives the boiling point of the mixture. At the point of intersection the vapor pressure p^o_i of the organic substance allows for the calculation of the vapor composition by $y^*_i = p^o_i/p_t$. As can be seen from Fig. 5-17 even very high boiling organic substances boil below the boiling point of water if liquid water is present.

The schematic of a batch steam distillation process is shown in Fig. 5-18 [Sattler 1988]. The steam is directly blown into the liquid in the still and partly condensed. As soon as the boiling point of the mixture is reached the volatile organic constituents

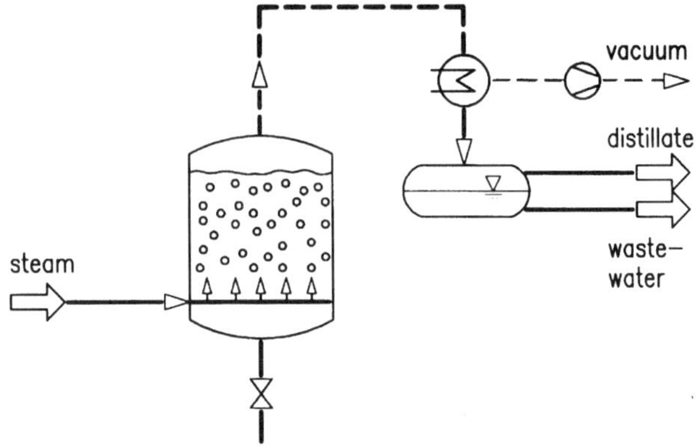

Fig. 5-18. Schematic of batch steam distillation.

are stripped off. The vapor is condensed and two-liquid phases are formed that are split in a decanter. Water, the lower phase in the decanter, is drawn off. It contains at least small amounts of the organic constituents. This formation of polluted waste water is the main disadvantage of steam distillation.

The schematic of a continuous process of steam distillation is presented in Fig. 5-19. In order to minimize the amount of live steam that will be polluted, a reboiler with indirect heating is very often used. In addition, countercurrent flow of liquid and vapor in a column increases mass transfer efficiency. Superheated steam is also very often used.

Steam distillation is a very simple but effective means for purification of high

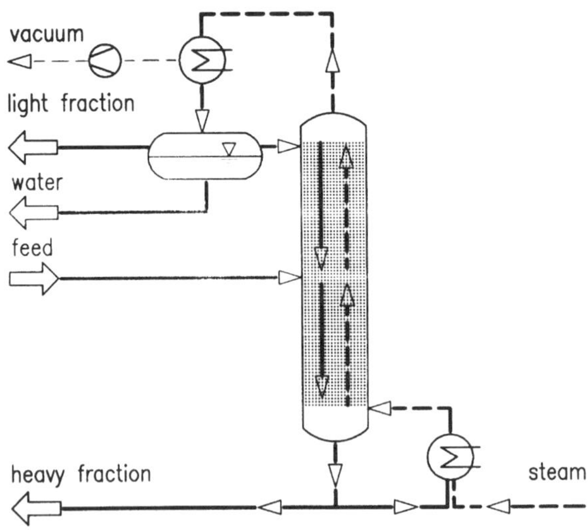

Fig. 5-19. Schematic of continuous multistage steam distillation.

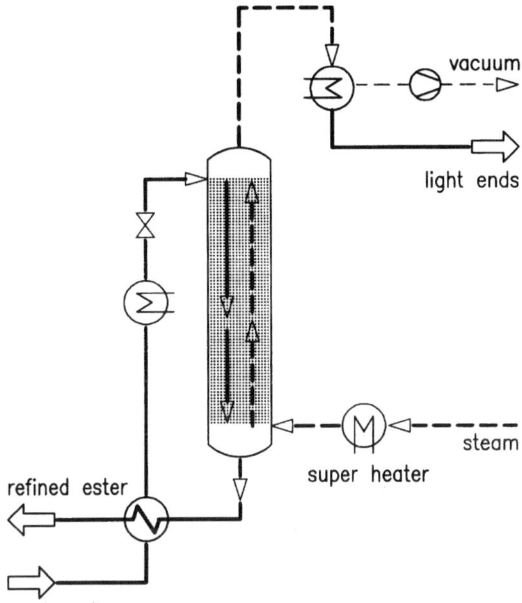

Fig. 5-20. *Process for deodorizing of crude ester [Johannisbauer and Jeromin 1992].*

boiling organic substances, e.g., essential oils for perfumery or long-chained fatty acids, and for deodorization or decolorization of fats and oil [Ellerbe 1979].

Figure 5-20 shows a process for deodorization of esters by steam distillation. The volatile impurities in the crude ester have a concentration of typically less than 1%. The amount of steam required for stripping is about 5 to 10% of the crude feed [Johannisbauer and Jeromin 1990].

An important field of application of steam distillation is the raffination of palm oil [Drew and Probst 1981]. A very modern version of a process for deodorization and deacidification of palm oil is presented in Fig. 5-21 [Stage 1988]. Typical operating conditions are a pressure of 2 to 3 Torr and a temperature of about 260°C. The system is heated by hot oil. The reboiler is a bundle type heat exchanger. The liquid is running as a thin film countercurrently to the steam inside the tubes. The time of residence of the product is as low as about 10 minutes, thus avoiding the risk of thermal degradation. The countercurrent flow of liquid and steam reduces the amount of steam required for effective stripping. It is typically less than 1 wt% of the crude feed. The condensation of the vapor at the column head is performed by quenching with subcooled top product.

Steam stripping is also widely used in crude oil distillation (see Section 5.3.2.2). The advantage of steam distillation over conventional distillation is the low operation temperature which is essential to avoid thermal degradation of the organic compounds, the use of low pressure steam (exhaust steam) and, in many cases, a simple device without a reboiler. The disadvantage of steam distillation is the high steam

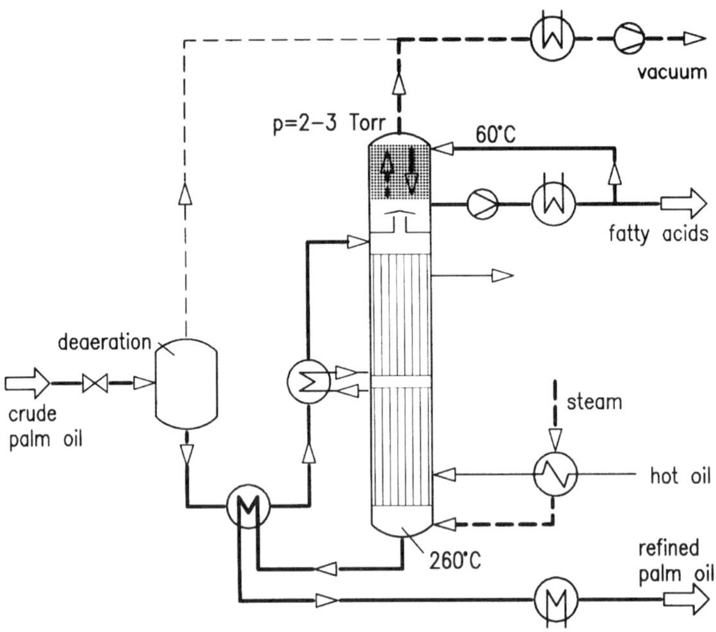

Fig. 5-21. Process for refining palm oil [Stage 1988].

requirement, the generation of waste water and the risk of foam or emulsion formation.

5.3 SEPARATION OF MULTICOMPONENT ZEOTROPIC MIXTURES

Complete separation of a multicomponent liquid mixture into individual components requires several distillation steps. As discussed in Chapter 4, a ternary mixture can be separated to only a limited extent into an overhead product $\dot{D}$ and a bottom product $\dot{B}$. Figure 4-20 shows the possible range of compositions of the two fractions on a triangular diagram. In special cases, pure overhead product *a* or pure bottom product *c* can be separated. The remaining binary fraction must subsequently be separated into the pure substances by a second column.

5.3.1 Separation Paths

Figure 5-22 shows the low boiler path (*a*-path) and the high boiler path (*c*-path), respectively, for separating a zeotropic ternary mixture. In the *a*-path the light constituent *a* is separated first. The bottoms is fed into a second column, where the *b*-c mixture is split into pure substances. In the c-path the high boiler is separated first and the overhead fraction, mixture *a-b,* is vaporously fed into column C-2.

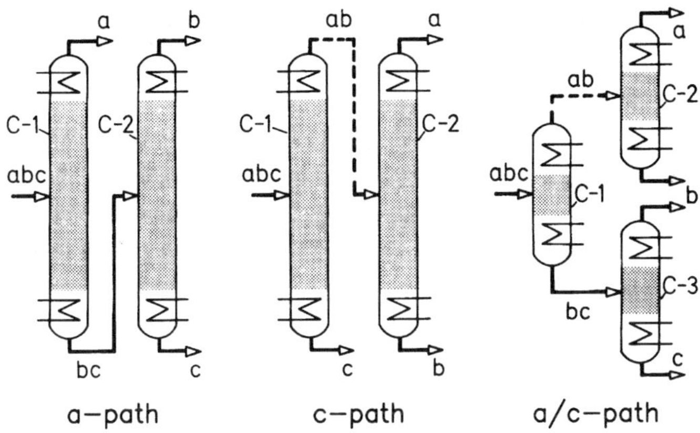

Fig. 5-22. Paths for complete separation of a ternary mixture.

In an alternative approach, the intermediate boiler path (*a*/*c*-path), the first column produces an overhead product without the high boiler *c* and a bottom product without the low boiler *a*. Two additional columns are then required to separate the remaining binary fractions. The medium boiler *b* appears, in one case, as the bottom product and, in the second case, as the overhead product. As discussed in Chapter 7.2, this alternative is energetically the most favorable one at higher concentrations of medium boiler *b*.

For the complete separation of 4-component mixture, 22 different separation paths, shown in Fig. 5-23, exist. Analogously, 719 different separation paths are feasible for the separation of a 5-component mixture. Each of these separation paths has its own energy requirement that is primarily influenced by the composition of the feed. Extensive studies are necessary to find the path with the lowest energy requirement.

In practice, the separation sequence of multicomponent mixtures is often determined not by energy requirement but by other limiting conditions, e.g., stability or corrosiveness of individual substances, danger of explosion, toxicity, tendency to stick to column internals, etc. In all these cases, the most difficult substance should, if possible, be separated first. Thus, the energy requirement is only one out of several aspects determining the choice of a separating sequence.

5.3.2 Side Columns

The separation of multicomponent mixtures often involves side columns. A side column does not represent another separating path but merely a different equipment design for low boiler or high boiler path. The lower section of the column C-2 is combined with column C-1, as is depicted in Fig. 5-24 for the low boiler path. Thereby, the reboiler of C-2 can be eliminated. This type of design offers no energetic advantages. On the contrary, heat $\dot{Q}_{R2}$ must be supplied at a higher temperature,

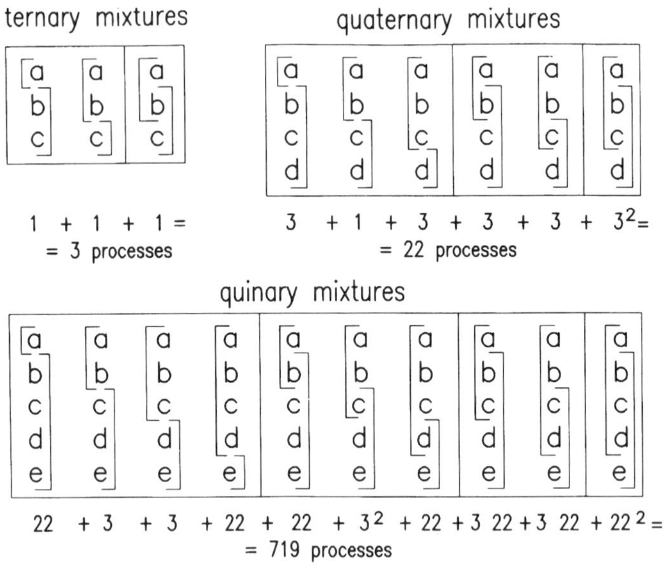

Fig. 5-23. *Determination of number of separation paths of ternary, quaternary and quinary mixtures. The brackets sign the two fractions of the first separation step. Several constituents can exist in both fractions.*

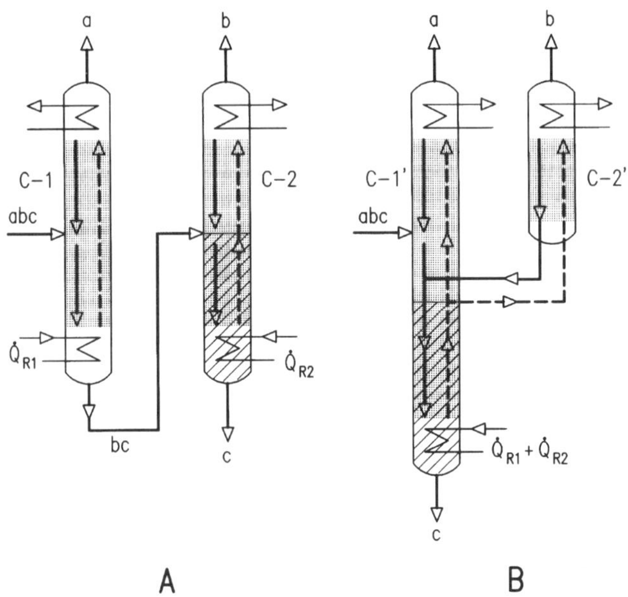

Fig. 5-24. *Use of a side column in the separation of a ternary mixture: (A) Conventional low boiler path. (B) The shaded section of column C-2 is combined with column C-1. The reminder C-2' is a rectifying side column.*

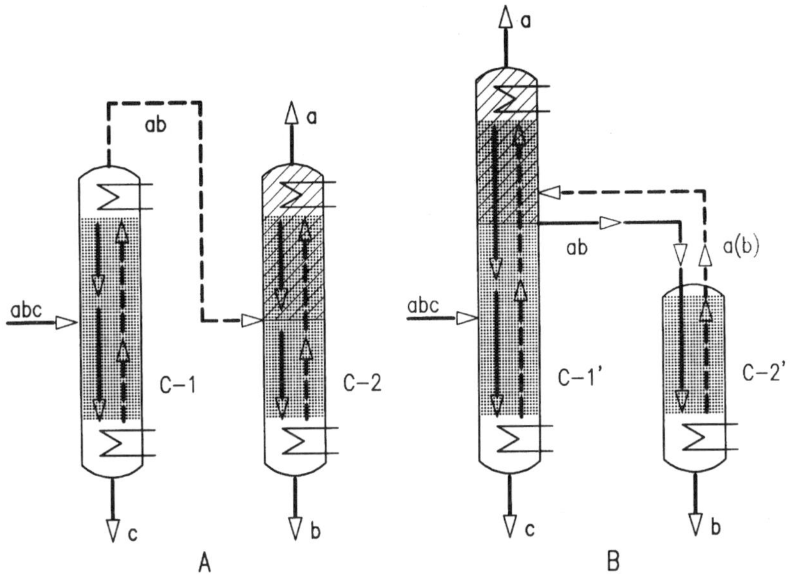

Fig. 5-25. *Use of a side column in the separation of a ternary mixture: (A) Conventional high boiler path. (B) The shaded section of column C-2 is combined with column C-1. The reminder C-2' is a stripping side column.*

which is an exergetic disadvantage. In addition, internal loads in the shaded part of the column are increased. This design is advantageous only when the amount of the medium-boiling substance *b* is relatively low.

Figure 5-25 shows the use of a side column for the high boiler path of a ternary mixture. Here the upper part of column C-2 is combined with column C-1 by discarding the condenser. Gas and liquid load in the shaded section are increased. The liquid is partially withdrawn and fed into the side column, which is the residuum of column C-2.

Figure 5-26 depicts the use of side columns for the complete separation of a quaternary mixture by the high boiler path. In each column the actual heaviest component is separated as bottoms, first *d*, then *c* and then *b*. The upper section of column C-2 is combined with column C-1, the upper section of column C-3 is combined with the modified column C-1. The gas and liquid load in column C-1 is highest at the upper section and—due to the side withdrawals of liquid and side feeds of vapor—lowest in the bottom section.

The advantage of side columns is the possibility of discarding condensers (heavy path) or reboilers (light path). However, the total heat charge is not affected significantly.

5.3.2.1 Argon Recovery from Air An example for the use of a side column in an industrial distillation process is the argon recovery in an air separation process

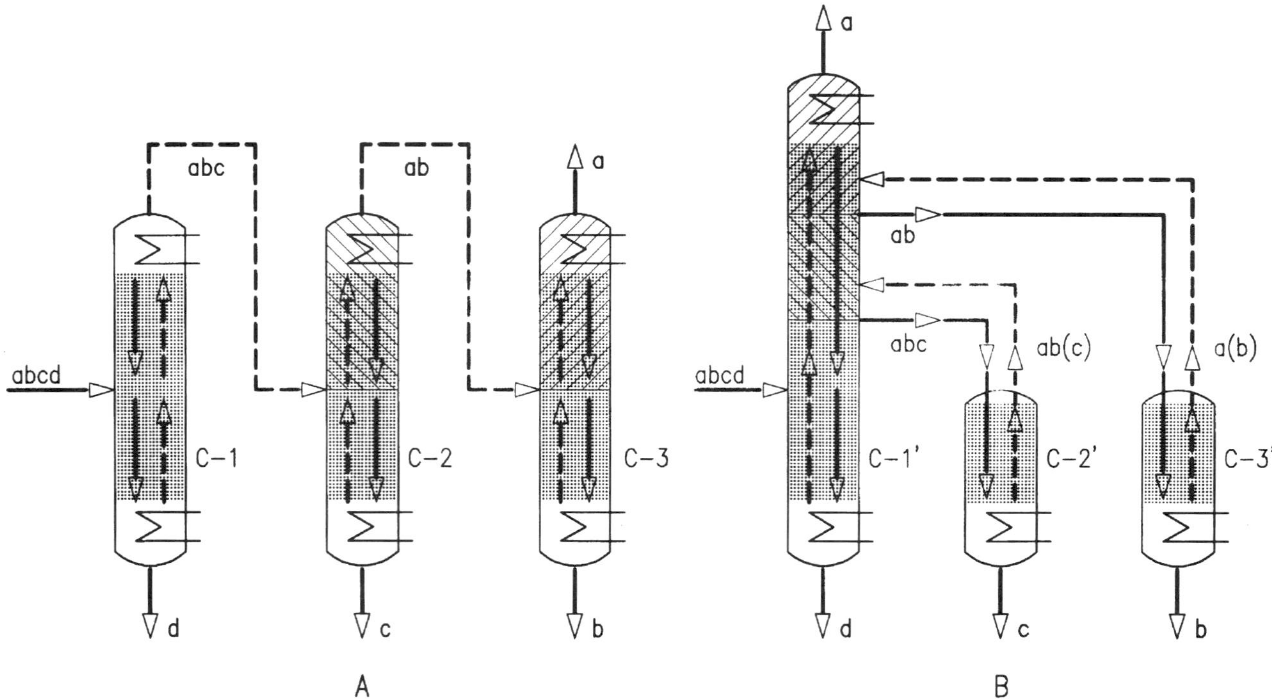

Fig. 5-26. *Use of side columns in the separation of a quaternary mixture: (A) Conventional high boiler path. (B) The shaded section of columns C-2 and C-3 are combined with column C-1. The reminders are stripping side columns.*

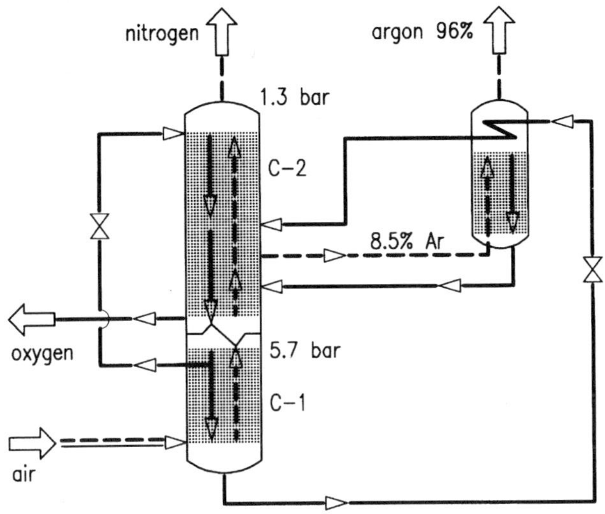

Fig. 5-27. Use of a rectifying side column for argon recovery in the Linde air separation process (see Fig. 5-9).

(see Section 5.2.4). Air contains about 1 mol-% argon. Argon is the intermediate boiler in the nitrogen/argon/oxygen mixture as can be seen in Fig. 2-7.

In the low pressure section of the Linde double column, Fig. 5-9, oxygen is the bottom and nitrogen the top fraction. Argon concentrates in the middle section typically up to 8.5 mol-% in the vapor. As shown in Fig. 5-27 vaporous side stream is withdrawn and fed to a rectifying side column. The top fraction is highly concentrated argon with a purity of about 96 mole-%, the bottoms is recycled to the main column. The cooling medium of the side column's condenser is the bottoms of the high pressure column, after expansion to 1.3 bar. The liquid is rich in nitrogen and, in consequence, boils lower than argon.

If high-purity argon is required, the oxygen contained in the crude argon can be removed by catalytic combustion with hydrogen and condensation of the water [Baldus et al. 1983]. Application of low pressure structured packings to the argon side column enables direct removal of high purity argon as overhead fraction, thus avoiding the subsequent purification of crude argon. Due to the vapor-liquid equilibrium of argon and oxygen, up to 180 equilibrium stages are required in the side column.

5.3.2.2 Crude Oil Distillation The first step in any petroleum refinery is the separation of crude oil into various fractions by distillation. These fractions may be products in their own right or may be feedstocks for other refining or processing units [Watkins 1979]. Crude oil distillation is by far the most important distillation process.

Crude oil consists primarily of a large number of different types of hydrocarbons.

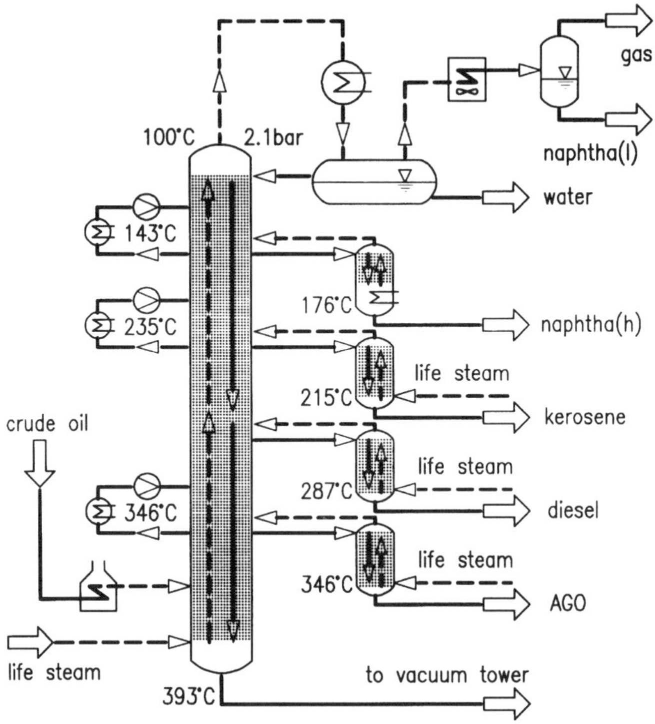

Fig. 5-28. Use of stripping side columns in the atmospheric tower of crude oil distillation.

The normal boiling points of the constituents range from −150°C to more than 500°C. This complex mixture has to be split in up to 10 fractions in a refinery. The fractions are not pure constituents but complex mixtures with about the same boiling point or boiling point range. The separation is performed in two distillation towers, the so-called atmospheric tower and the vacuum tower. The oil is first heated to the maximum temperature allowable to avoid thermal degradation, and then fed to a fractionating tower which operates slightly above atmospheric pressure. This tower is called the atmospheric tower. It yields several distillate products and a bottoms product which is the residual liquid material which could not be vaporized under the conditions of temperature and pressure existing in the atmospheric tower. This bottoms liquid is then reheated to its maximum allowable temperature (about 420°C) and fed to a second fractionating tower which operates at subatmospheric pressure (minimum top tower pressure about 0.013 bar). This tower is usually called the vacuum tower.

Figure 5-28 shows the atmospheric tower of a petroleum distillation unit. The crude oil is considered as a mixture with six pseudo components to be separated into six fractions by the high boiler path. The actual highest boiling fraction is separated first (see Figs. 5-25 and 5-26). The rectifying sections of all columns are combined in the main tower and the stripping sections are formed by side columns. Typically, the main tower has about 35 actual trays and each side column has about 4 actual trays. The products are light naphtha, heavy naphtha, light distillate (kerosene), heavy distillate (diesel), atmospheric gas oil (AGO) and reduced crude which is processed further in the vacuum tower. The amount of reduced crude is about 50 to 60% of the crude oil feed.

The crude oil enters the column after being heated in a furnace as a vapor-liquid mixture. The liquid fraction containing primarily the high boilers but also some lower boiling constituents is stripped by superheated life steam (steam distillation, see Section 5.2.6). The products of the side columns are either reboiled or stripped by life steam. Due to the compilation of several rectifying sections in the main column, the amount of vapor increases in the upper part of the column. This is disadvantageous because the diameter of the tower has to be increased and all the heat has to be withdrawn at the lowest temperature. To overcome these disadvantages, heat is removed from the main column by several *pump arounds*. Some liquid is removed from the tower, cooled down against the cold crude oil and fed three trays above the withdrawal point into the tower again.

5.4 SEPARATION OF HETEROGENEOUS AZEOTROPIC MIXTURES

Azeotropic points constitute a barrier to separation by distillation. A heterogeneous azeotrope can be overcome by use of a decanter. For example, in Fig. 5-29, separation

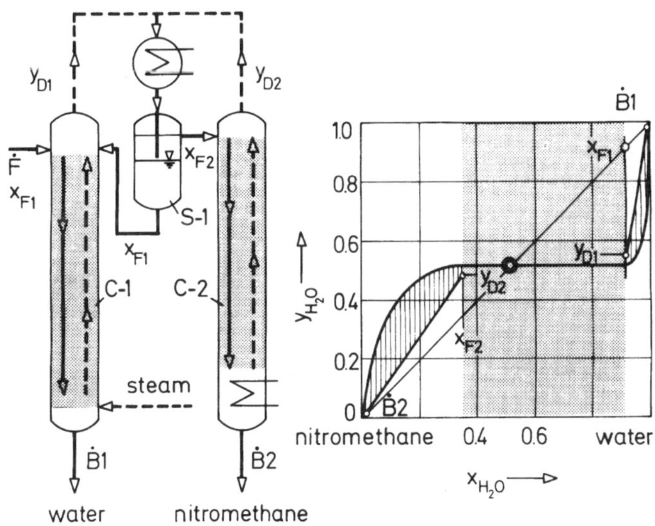

Fig. 5-29. *Flowsheet and McCabe-Thiele diagram for separation of the binary mixture water/nitromethane. The water/nitromethane mixture forms an immiscibility region (shaded area) with a heterogeneous azeotrope. Complete separation is possible by using two columns, C-1 and C-2, and a decanter S-1. The vertically hatched region on the y-x diagram represents the area between the operating line and the equilibrium line in which the number of equilibrium stages n or transfer units* N_{OG} *is determined.*

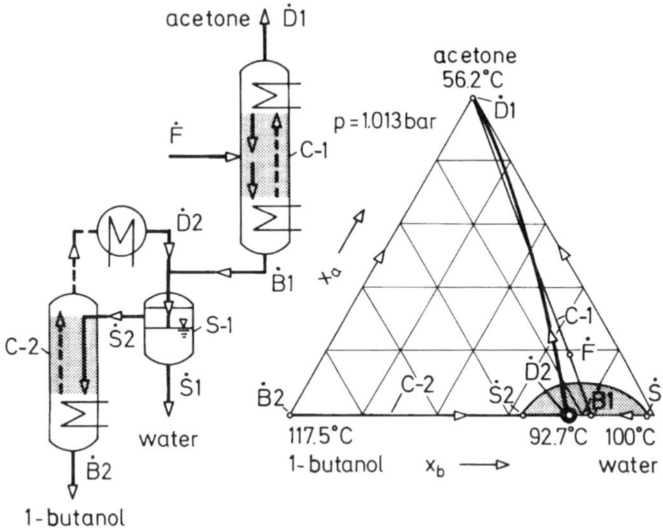

Fig. 5-30. *Flowsheet and phase diagram for separation of the ternary mixture acetone/water/1-butanol (p = 1.013 bar). The existence of an immiscibility region (shaded area) permits complete separation of the mixture in two columns, C-1 and C-2, with an intermediate decanter S-1.*

of the binary nitromethane/water mixture requires two columns, C-1 and C-2, but both pure substances are removed as bottom products. The vapor composition at the top of each column is approximately that of the azeotrope. Hence, both vapor streams can be combined and condensed in the same condenser. The resulting two-phase mixture goes through a decanter S-1.

The nitromethane-rich phase with composition x_{F2} is fed into column C-2, and the water-rich phase is returned to column C-1. Because water is the bottom product in column C-1, steam used for heating can be fed directly into the column and no reboiler is required. This process can be applied to many systems. Examples are the systems heptane/water, i-butanol/water, n-butanol/water and ethylacetate/water. According to [Drew 1979] the organic constituent should be withdrawn a few plates above the column bottom as a vapor side stream when dirt or any high boilers are present in the feed.

In ternary mixtures, several distillation fields can occur which are divided by distillation boundaries (see Sections 4.3.2.1 and 5.6). As in binary azeotropes, these distillation boundaries cannot be crossed by distillation. However, if phase splitting occurs, the distillation boundary can be crossed by separation in a decanter.

Separation of an acetone/water/butanol mixture illustrates this procedure (Fig. 5-30) [Pucci et al. 1986]. The water/butanol mixture forms a heterogenous azeotrope; the distillation boundary runs from the azeotrope to the low boiler acetone. Separation is carried out in two columns. Acetone is withdrawn as overhead fraction $\dot{D}1$ from the first column C-1. The bottom fraction $\dot{B}1$ lies in the shaded two-phase range and is, consequently, separated into a water-rich phase $\dot{S}1$ and a butanol-rich phase $\dot{S}2$ in the decanter $\dot{S}1$. Because $\dot{S}2$ is located in a different distillation field, butanol can be withdrawn from column C-2 as bottom product $\dot{B}2$. The composition of the overhead fraction from C-2 is approximately that of the heteroazeotrope and can also be fed into the decanter. The water-rich fraction $\dot{S}2$ of the process in Fig. 5-30 is, generally, polluted with 1-butanol. To meet quality standards of waste water, separation of butanol from the mixture might be necessary.

5.5 SEPARATION OF BINARY AZEOTROPIC MIXTURES BY CHANGING SYSTEM PRESSURE

Many mixtures contain an azeotrope whose position can be shifted substantially by changing system pressure. Especially in systems with low boiling azeotropes, the azeotropic concentrations depend significantly on temperature, which can be changed by changing the operating pressure. An important example is the mixture ethanol/water [Gmehling and Kolbe 1988], whose azeotrope even vanishes at low temperatures, i.e., low operating pressures, see Fig. 5-31.

Another example is the mixture tetrahydrofuran/water [Abu-Eishuh and Luyben 1985]. As shown in Fig. 5-32, this mixture forms a binary minimum azeotrope, which shifts towards the low-boiling substance (tetrahydrofuran) at lower system pressure. The binary azeotrope can be crossed by first separating the high-boiling substance (water) at low pressure $p_1 = 1\text{bar}$ in column C-1. The composition of the overhead product $\dot{D}1$ should be as close as possible to that of the azeotrope at this pressure. After the pressure has been increased to $p_2 = 8$ bar (0.8 MPa), $\dot{D}1$ is fed into a second column C-2. At the higher pressure in this column, the azeotrope forms at a lower concentration of low boiler (tetrahydrofuran), which can then be

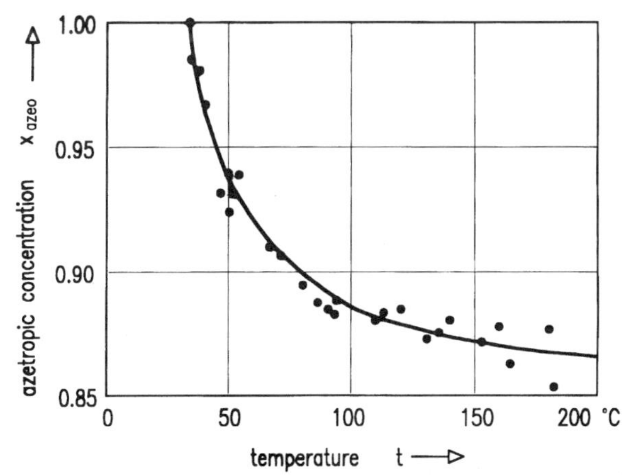

Fig. 5-31. *Concentration of the azeotrope in the mixture ethanol/water as function of temperature [Gmehling and Kolbe 1988].*

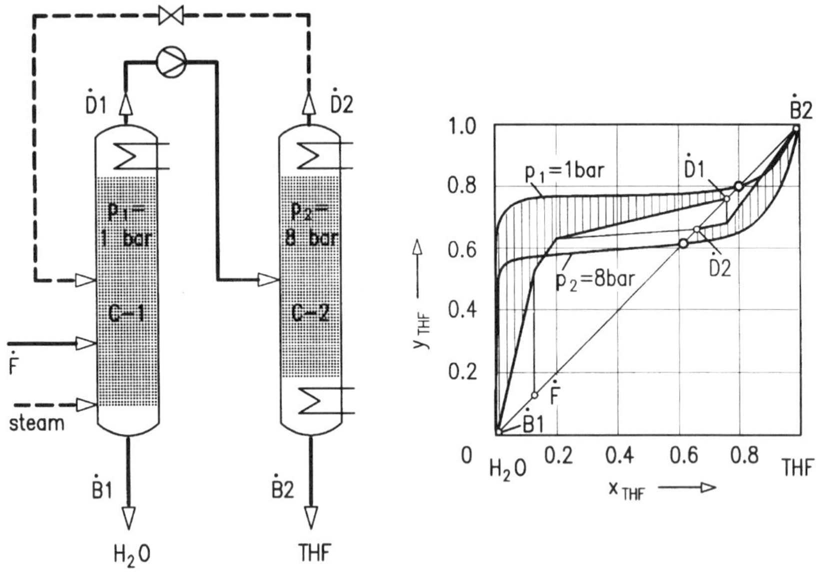

Fig. 5-32. *Flow sheet and McCabe-Thiele diagram for separation of the binary tetrahydrofuran/water mixture in two columns, C-1 and C-2, operating at different pressures. This mixture contains a minimum azeotrope, whose position can be shifted by changing system pressure. The hatched region represents the area between the equilibrium curves and the operating lines in which the number of equilibrium stages n or transfer units N_{OG} is determined.*

removed as bottom product Ḃ2. The overhead product of column C-2 (Ḋ2) is, preferably in the vapor state, returned to column C-1 after pressure reduction. Hence, column C-1 has a liquid feed Ḟ and a vapor feed Ḋ2; its operating line shows two breaks. The overhead temperature of column C-2 is higher than the boiling temperature of column C-1. Therefore, the requirements for indirect (thermal) column coupling are fulfilled (see Section 7.3).

This process can be applied to many systems. Some examples are listed in Table 5-5. In addition, typical operation pressures of the two columns are given.

5.6 SEPARATION OF AZEOTROPIC MIXTURES BY AN ENTRAINER

Azeotropic mixtures can be separated into their pure constituents by the addition of an auxiliary substance called an entrainer [Hoffmann 1964, Tassios 1972]. The choice of the entrainer must permit easy separation of the mixture's constituents and the recovery of the entrainer that will be recycled within the process.

The principles of separating an azeotropic mixture by use of an entrainer are demonstrated for the case of a binary azeotropic mixture in the following. By adding the entrainer a ternary system is formed. The process can be easily understood with the help of the rules of ternary distillation dealt with in Chapter 4.3.

Table 5-5. Examples of Systems to be Separated by the Process in Fig. 5-31.

Systems	Pressure p_1	Pressure p_2	Reference
methylethylketone/water	1 bar	7 bar	[Schweitzer 1979]
methanol/acetone	1 bar	0.25 bar	[Schweitzer 1979]
methanol/methylethylketone	1	7.6	[Schweitzer 1979]
tetrahydrofuran/water	1	7.6	[Schweitzer 1979]
ethanol/water			
water/2-butanone			
water/isobutylalcohol			
ethanol/2-pentanone			
methanol/acetone			
methanol/2-butanone			

There are two different approaches to separate a ternary mixture with azeotropes. In the first approach the entrainer is selected so that no distillation border exists in the ternary system. In the second approach the entrainer is selected so that the constituents *a* and *b* of the starting mixture are both origins or termini of distillation lines. In this case, the azeotrope is the beginning or end of a border distillation line.

5.6.1 Processes in One Distillation Field

5.6.1.1 Criteria for Entrainer Selection Processes without distillation borders have been studied in detail [Doherty and Caldarola 1986]. The entrainer has to be selected so that the constituents *a* and *b* are in the same distillation field. The *a*-*b* azeotrope must be either an origin or a terminus of distillation lines. Table 5-6 gives the criteria for entrainer selection resulting from these requirements. These criteria are minimum requirements. Additional azeotropes may exist in the system. A further possibility for processes without distillation border is the use of a high boiling entrainer for a minimum azeotrope to be separated (see Section 5.6.3.2 Extractive Distillation).

Table 5-6. Criteria for Entrainer Selection for Processes Without Distillation Borders, See Process in Fig. 5-33.

Entrainer for the separation of a mixture with a minimum azeotrope:

- medium boiler.
- low boiler, which forms a medium boiling maximum azeotrope with constituent *a*.

Entrainer for the separation of a mixture with a maximum azeotrope:

- medium boiler.
- high boiler, which forms a medium boiling minimum azeotrope with constituent *b*.

5.6.1.2 Processes Figure 5-33 presents the flowsheet and the triangular composition diagram for the separation of a binary mixture with an *a*-*b* minimum (or maximum) azeotrope with an intermediate boiling entrainer *e*. All distillation lines originate at the high boiler *b* and terminate at the azeotrope because it has the lowest boiling point of the mixture. The feed $\dot{F}$ is mixed with the entrainer *e* to form $\dot{M}$ from which the high boiler *b* is separated in column C-1 as bottom fraction $\dot{B}1$. The top fraction $\dot{D}1$ is (vaporosly) fed into column C-2 that yields the low boiler *a* as top and the entrainer *e* as bottom fraction. The entrainer is recycled to column C-1. This separation sequence is similar to the well-known separation of a zeotropic ternary mixture (see Section 5.3.1).

According to Section 4.3.2.1 the low boiler *a* can also be separated first from the ternary mixture. In such a process the column C-2 splits the bottom $\dot{B}1$ into top fraction $\dot{D}2$ with the entrainer and bottom fraction $\dot{B}2$, which consists of pure high boiler *b*. However, this variant of the process is, in most cases, less advantageous.

A third modification of the process is presented in Fig. 5-34. The feed $\dot{F}$ and the entrainer *e* are fed into column C-1 and split in a top fraction free of *b* and a bottom fraction free of *a*. Both fractions are separately fed into column C-2 whose top fraction consists of pure substance *a* and the bottom fraction of pure substance *b*. The entrainer *e* is removed from the middle of the column to be recycled to column C-1. According to Section 7.3.1 this process is in most cases more favorable due to its lower energy requirement.

A very elegant modification of the process is to establish the separation in a column that is divided in the middle section by a vertical wall, see Fig. 5-34C. In such a column, the starting mixture *ab* is fed into the left compartment and the entrainer is regained in the right compartment of the column. From there the entrainer is internally recycled to the feed. The function of a column with a dividing wall is explained in detail in Section 7.3.1, see Fig. 7-10. Such a modification allows complete separation of an azeotropic mixture in a single column. The separating agent, the entrainer, remains in the column and establishes the separation without any external action.

The triangular diagram for the separation of an *a*-*b* minimum azeotrope with a low boiling entrainer that forms an intermediate boiling maximum azeotrope with the constituent *a* is shown in Fig. 5-35. The process is the same as in Fig. 5-33, just the *e*-*a* azeotrope acts as the intermediate boiling entrainer.

The processes without distillation border consist of two columns and one recycle. The problem of all these processes lies in the existence of an appropriate entrainer. As is well known, the probability of azeotrope formation increases with decreasing boiling point differences of the constituents *a* and *b*. Consequently, most azeotropic mixtures to be separated by distillation are close boiling ones. In many cases it will be difficult or even impossible to find a substance that boils between and that does not form new azeotropes. But even if such a substance exists, the process might be uneconomical because of the small boiling point differences in the system. The entire process has to be performed in the boiling point range of the *a*-*b* mixture. The system presented in Fig. 5-33 is one of the few examples in which this process works economically.

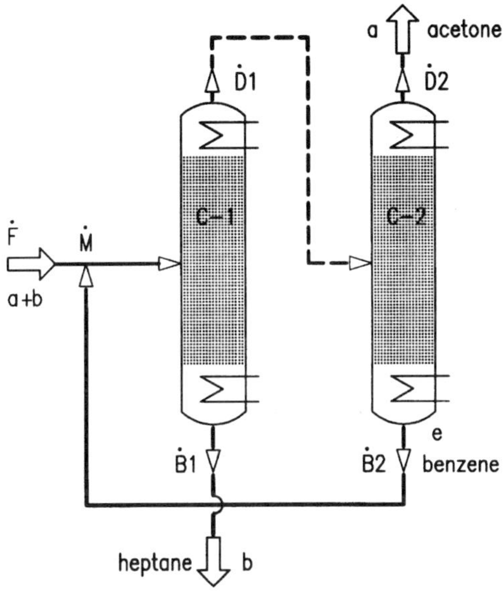

A

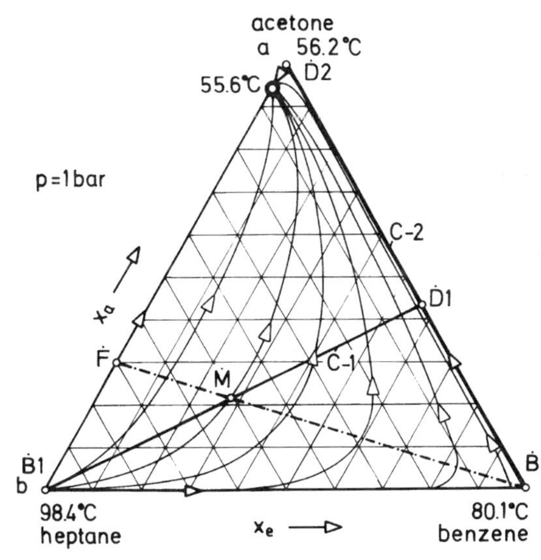

B

Fig. 5-33. *Separation of the azeotropic mixture acetone/heptane using benzene as entrainer. The entrainer boils intermediate and, therefore, no distillation border exists in the ternary mixture: (A) Flowsheet, (B) Triangular diagram. [Doherty and Caldavola 1986].*

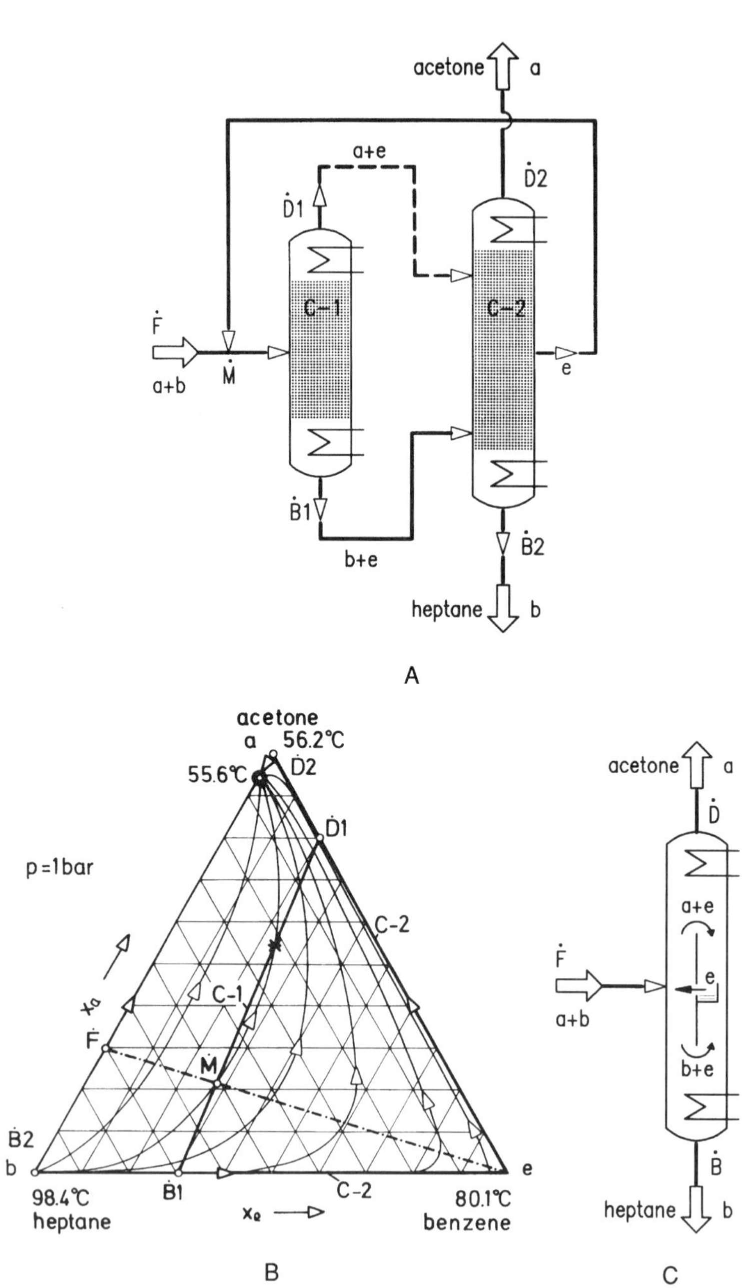

Fig. 5-34. Separation of the azeotropic mixture acetone/heptane using benzene as entrainer. Here, the column C-1 splits the feed in an overhead fraction without high boiler and a bottom fraction without low boiler. This modification of the process shown in Fig. 5-33 is more economical: (A) Flowsheet, (B) triangular diagram, (C) column with internal wall (see Fig. 7-10).

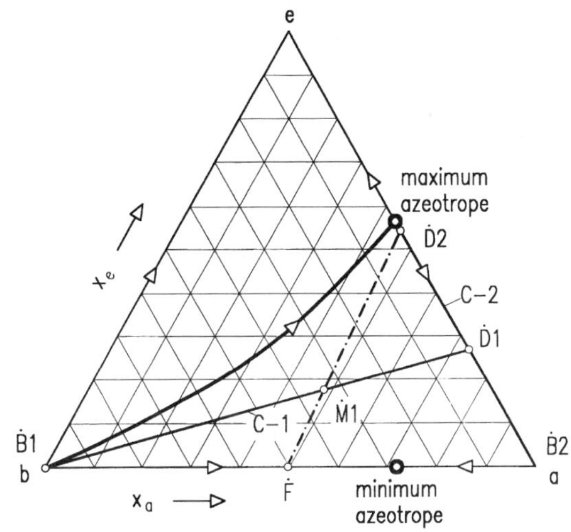

Fig. 5-35. Triangular diagram for a process for separating azeotropic mixtures with a low boiling entrainer that forms an intermediate boiling maximum azeotrope. The flow sheet is the same as in Fig. 5-33A.

5.6.2 Processes in Two Distillation Fields

5.6.2.1 Criteria for Entrainer Selection This group of processes has been studied by [Stichlmair 1988, Stichlmair and Herguijuela 1992]. Recently, a good review has been published by [Widago and Seider 1996]. The entrainer has to be selected so that the constituents *a* and *b* of the azeotropic mixture to be separated are both origins or termini of distillation lines. Such systems always have distillation borders. The entrainer must permit the recovery of products *a* and *b* at the top or bottom of distillation columns. This is possible only if the distillation lines in the triangular diagram either start or end at *a* and *b* (see Section 4.3). Figure 5-36 shows the minimum qualitative requirements that must be met by the entrainer. In the case of a maximum-boiling binary azeotrope (Figs. 5-36A to C), components *a* and *b* in the starting mixture are low-boiling substances; i.e., the distillation lines terminate at points *a* and *b*. The entrainer used must ensure that the distillation lines of both binary mixtures *a-e* and *b-e* also terminate at *a* and *b*, respectively. This condition is fulfilled when entrainer *e* constitutes the high boiler of the system (Fig. 5-36A). When the entrainer is a medium-boiling substance, component *a* can be separated, but component *b* can be recovered only if the mixture *b-e* forms a maximum azeotrope (Fig. 5-36B). With a low-boiling entrainer (Fig. 5-36C), mixtures *b-e* and *a-e* must both form a maximum azeotrope. Products *a* and *b* leave the rectification columns overhead, since they are termini of distillation line, i.e., sinks in the boiling point surface (see Section 2).

If a minimum-boiling binary azeotrope occurs (Figs. 5-36D to F), *a* and *b* are high-boiling substances; i.e., the distillation lines start at *a* and *b*. This must also be

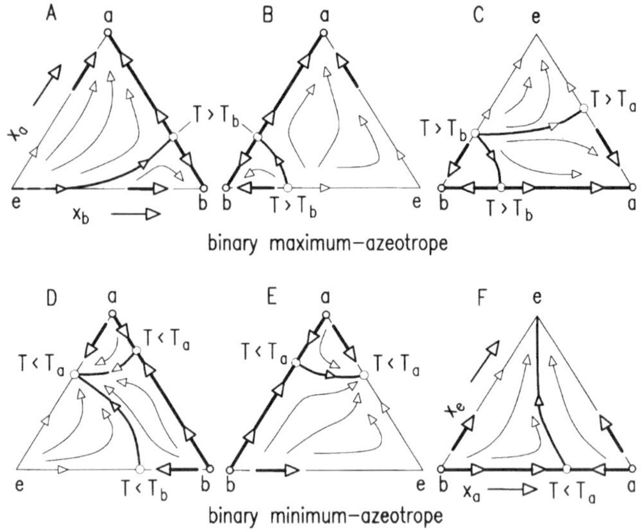

Fig. 5-36. Minimum requirements for the selection for the entrainer e *for the processes with distillation borders. Separation of the azeotropic mixture into pure* a *and* b *is possible only when the distillation lines both either start or terminate at* a *and* b.

the case in a ternary system. This condition is always fulfilled when the entrainer is a low-boiling substance (Fig. 5-36F). If the entrainer has a boiling temperature between that of *a* and *b*, pure *b* can be separated; however, substance *a* can be recovered only if the mixture *a-e* forms a minimum azeotrope (Fig. 5-36E). If the entrainer is a high-boiling substance (Fig. 5-36D), *a* and *b* can be separated only if the entrainer forms a minimum azeotrope with both substances. Substances *a* and *b* are obtained as bottom products, since they are origins of distillation lines.

Thus, a binary mixture with a maximum azeotrope can generally be separated into its two components only if the entrainer is higher boiling than the azeotrope, or if it forms high boiling azeotropes (i.e., maximum azeotropes). On the other hand, a mixture with a minimum azeotrope requires a low-boiling entrainer or the formation of low-boiling azeotropes. In other words, only azeotropes of the same type occur in the process. The criteria listed in Table 5-7 are minimum requirements. As will be shown in the following examples, additional azeotropes and even ternary azeotropes may exist, but they have to be azeotropes of the same type.

5.6.2.2 Generalized Process Figure 5-37 presents the generalized process for completely separating an *a-b* minimum azeotrope. According to Table 5-7 an entrainer that boils lower than the minimum azeotrope has to be used. Distillation lines start at pure substance *a* as well as at pure substance *b* and end at the entrainer *e*. Therefore, the constituents *a* and *b* of the given system can be recovered as bottom fractions of different distillation steps.

First, it is assumed that the feed $\dot{F}$ lies in that distillation field where the low boiler *a* can be separated as bottoms $\dot{B}1$ of column C-1. The top fraction $\dot{D}1$ is close to the azeotrope. By admixing the entrainer *e*, the distillation border is crossed and the mixture $\dot{M}2$ lies on the other side of the border where substance *b* can be separated as bottoms $\dot{B}2$ of column C-2. The top fraction $\dot{D}2$ lies close to the distillation border. Assuming that both the distillation border is curved and $\dot{D}2$ is lying on the concave side, the fraction $\dot{D}2$ can be separated in column C-3 into a top fraction $\dot{D}3$ and a bottom fraction $\dot{B}3$ both lying on the other side of the border line (see Fig. 4-23). The top fraction $\dot{D}3$ is recycled as entrainer. The bottom fraction is recycled to column C-1. Depending on composition, the fraction $\dot{B}3$ can be fed into column C-1 either separated or mixed with the feed $\dot{F}$.

The generalized process of Fig. 5-37 consists of three separation steps and two recycles. A material balance around column C-2 and C-3 requires that fraction $\dot{D}1$, pure substance *b* and fraction $\dot{B}3$ all lie on a straight line (thin dashed line in Fig. 5-37).

It is essential that the amount of entrainer to be recycled is as small as possible. A more detailed study reveals that the amount of fraction $\dot{D}2$ has to be as small as possible because both fractions of C-3 are recycled. Decisive for the amount of $\dot{D}2$ is the split of $\dot{M}2$ in $\dot{D}2$ and $\dot{B}2$. Since the bottom fraction $\dot{B}2$ contains all of substance *b* in the feed the following holds:

$$\dot{B}2 = \dot{F}\, x_{Fb} \tag{5.1}$$

The lever arm rule gives:

$$\dot{D}2 = \dot{F}\, x_{Fb}\, \overline{B2M2}/\overline{D2M2} \tag{5.2}$$

Table 5-7. Criteria for Entrainer Selection for Processes With Distillation Borders, See Process in Fig. 5-37.

Entrainer for the separation of a mixture with a minimum azeotrope:

- low boiler (lower than the minimum azeotrope).
- medium boiler, which forms a new minimum azeotrope with the low boiling constituent of the given mixture.
- high boiler, which forms new minimum azeotropes with both constituents of the given mixture. At least one of them has to boil lower than the azeotrope of the given mixture.

Entrainer for the separation of a mixture with a maximum azeotrope:

- high boiler (higher than the maximum azeotrope).
- medium boiler, which forms a new maximum azeotrope with the high boiling constituent of the given mixture.
- low boiler, which forms new maximum azeotropes with both constituents of the given mixture. At least one of them has to boil higher than the azeotrope of the given mixture.

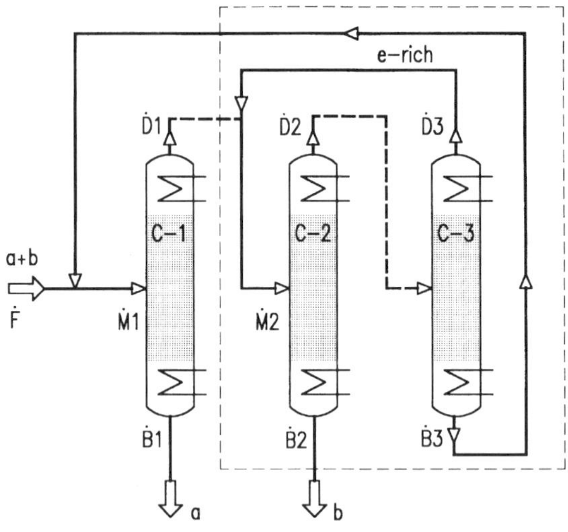

A

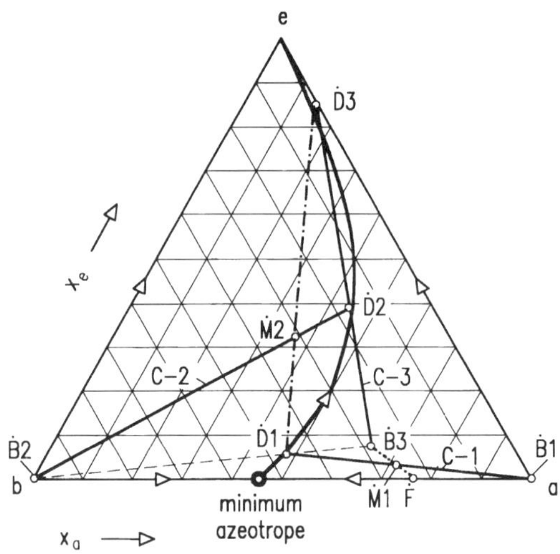

B

Fig. 5-37. Process for the separation of a binary mixture with a minimum azeotrope by using a low boiling entrainer. The feed is rich in a.: *(A) Flowsheet. Components* a *and* b *are obtained as pure bottoms products from columns C-1 and C-2, respectively. (B) Triangular diagram. The ternary mixture is divided into two distillation fields by a distillation border running from the minimum azeotrope to the entrainer. The distillation border is crossed by the admixing of the entrainer* e. *— Distillation border; — External balance of distillation column; –·–·–· Mixing of two streams; –––– Material balance around columns C-2 and C-3.*

The recycles $\dot{D}3$ and $\dot{B}3$ are small when the length $\overline{D2M2}$ is large with respect to the length $\overline{B2M2}$. This requires a heavily curved border distillation line and a position of $\dot{M}2$ far away from the distillation border.

Figure 5-38 presents the generalized process when the feed lies in that distillation field, where the high boiler *b* can be separated. In this case, just the sequence of the separations is different from that in Fig. 5-37. Both modifications of the generalized process are characterized by two separations performed on the convex and one separation performed on the concave side of the border distillation line. The generalized process also works for the separation of a mixture with a maximum azeotrope if an upside down version of the flow sheet is used (Fig. 5-39).

According to the criteria for entrainer selection (Table 5-7) a minimum azeotrope can be separated by an intermediate boiling or a high boiling entrainer instead of a low boiler, if additional azeotropes are formed. Figures 5-40 and 5-41 present the triangular diagrams for these cases. A detailed study reveals that one of the azeotropes formed by the substance *e* acts as entrainer. The process is the same as in Fig. 5-37.

The generalized process presented here allows the separation of two pure constituents from a ternary mixture if the border line is curved. This requirement is more or less met by real systems. However, the more curved the distillation border, the more economical the process.

5.6.2.3 *Simplifications of the Generalized Process* The generalized process with three distillation columns and two recycles is expensive with respect to investment and operating costs. However, there exists a lot of possibilities for process simplification. Some examples are presented in the following sections.

Under special conditions one of the three separation steps of the generalized process can be discarded. This is possible when the distillation border is heavily curved and the stream $\dot{M}2$ is lying close by the entrainer. In this case, the separation performed in column C-2 (Fig. 5-38) yields a bottom fraction $\dot{B}2$ that contains the substance *a* in high purity. Thus, the column C-3 and the recycle $\dot{B}3$ can be omitted.

This advantageous modification of the generalized process requires a very heavily curved distillation border that lies in part on a side of the triangular diagram. Additionally, a high amount of entrainer has to be recycled within the process. This process modification is faciliated by the existence of one or two azeotropes formed by the entrainer with the constituents *a* and *b*. However, no ternary azeotropes should exist.

Figure 5-42 presents the flow sheet and the triangular diagram for concentrating diluted hydrochloric acid [Grewer 1971]. The system HCl/H_2O forms a maximum azeotrope (21 wt-% HCl, 108.6°C). According to Table 5-7 the entrainer has to be the high boiler in the system. Sulfuric acid is used in the process. The distillation border line, a ridge in the boiling point surface, lies in part on the basis of the triangular diagram. The feed $\dot{F}$ with a high water content is mixed with concentrated sulfuric acid (approximately 75 wt-%). The mixture $\dot{M}1$ lies in that distillation field, where HCl (or an HCl-rich mixture) can be separated as top fraction $\dot{D}1$. If $\dot{M}1$ is sufficiently close to the entrainer, then the bottom fraction $\dot{B}1$ is nearly free of HCl.

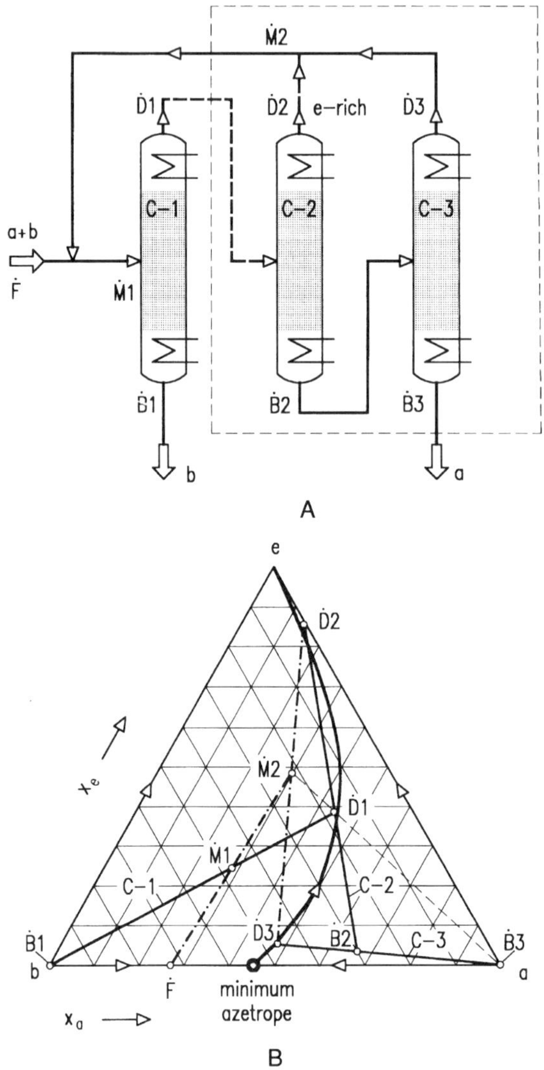

Fig. 5-38. *Process for the separation of a binary mixture with a minimum azeotrope by using a low-boiling entrainer. The feed is rich in* b. *The process is nearly identical with the process in Fig. 5-37. Just the sequence of the columns is different: (A) Flowsheet, (B) Triangular diagram.*

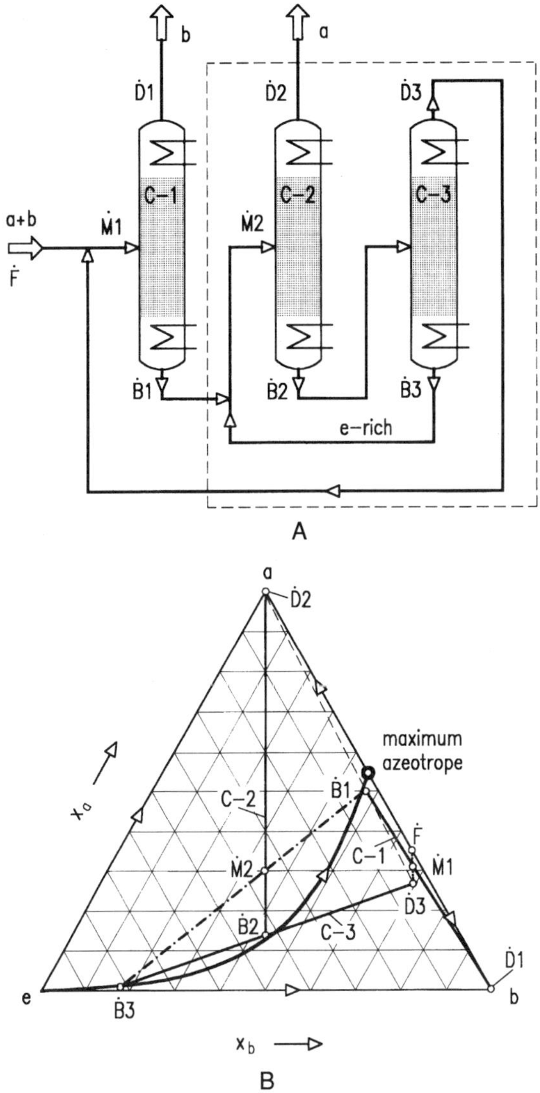

Fig. 5-39. *Process for the separation of a binary mixture with a maximum azeotrope by using a high-boiling entrainer. The feed is rich in* b. *The process is an upside down version of the process in Fig. 5-37A: (A) Flowsheet, (B) triangular diagram.*

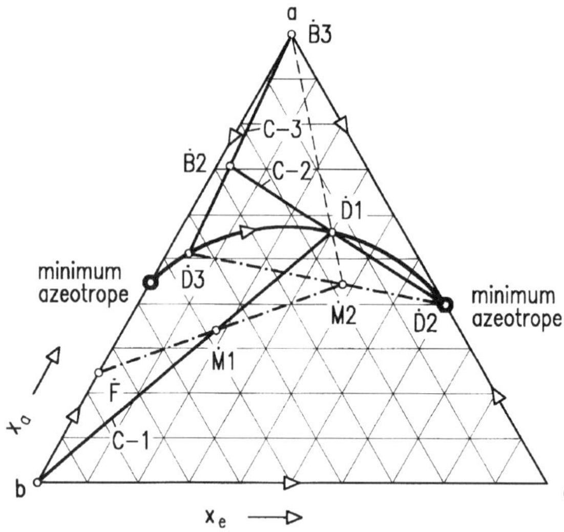

Fig. 5-40. *Process for the separation of a binary mixture with a minimum azeotrope by using an intermediate boiling entrainer* e *that forms a minimum azeotrope with* a. *The flowsheet is the same as in Fig. 5-37A.*

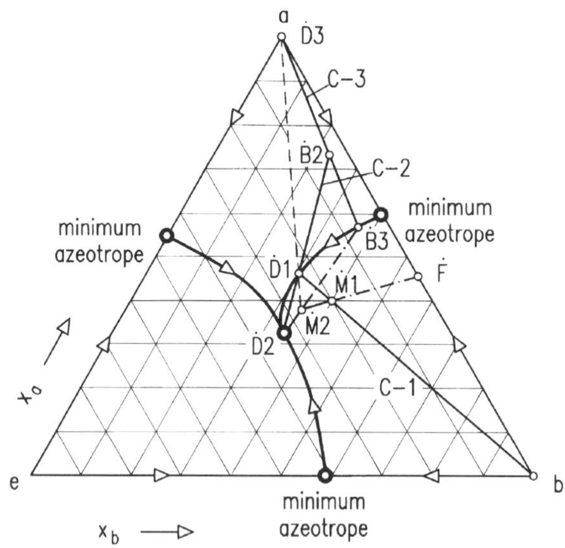

Fig. 5-41. *Process for the separation of a binary mixture with a minimum azeotrope by using a high boiling entrainer* e *that forms binary minimum azeotropes with* a *and with* b. *The flowsheet is the same as in Fig. 5-37A.*

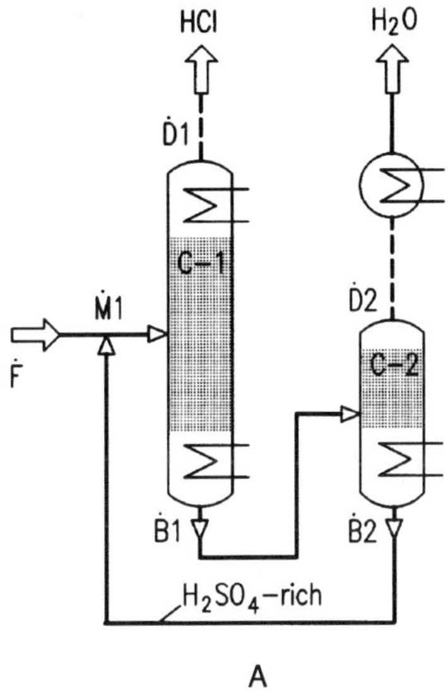

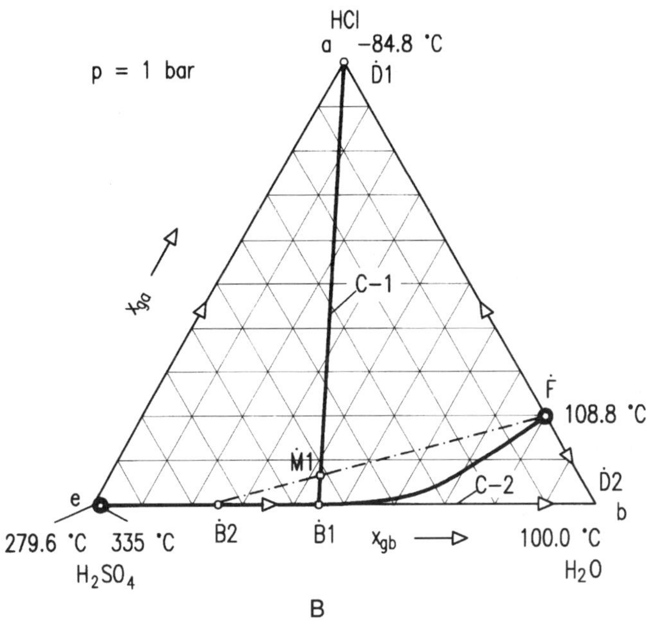

Fig. 5-42. *Two-column process for the separation of hydrogenchloride HCl and water by using hydrogensulfat H_2SO_4 as entrainer. In this sytem the distillation border lies in part on the base of the triangular diagram. Thus, pure water is recoveed in column C-2 and the third column of the generalized process can be discarded: (A) Flowsheet, (B) triangular diagram.*

The subsequent separation step (single-stage distillation) yields an HCl-free water as top fraction. The process can be modified by feeding the entrainer directly on the top of column C-1 instead of mixing it with the feed (see Section 5.6.3.2). A process similar to this one is used for concentrating a diluted nitric acid (Fig. 5-47).

Figure 5-43 presents a two column process for separating a mixture with a minimum azeotrope. The low boiling entrainer *e* forms a minimum azeotrope with substance *b* that is not required by the entrainer selection rules in Table 5-7. The *e*-*b* azeotrope that boils lower than the azeotrope of the starting mixture acts in the process as entrainer. If the distillation border between the two azeotropes is heavily curved and the top fraction $\dot{D}1$ lies on the concave side then substance *a* can be recovered in pure form as bottoms of column C-2.

A similar process with a high boiling entrainer *e* is depicted in Fig. 5-44. A ternary azeotrope not required by the entrainer selection rules is formed in this system. The distillation border that runs from the *a*-*b* azeotrope to the ternary azeotrope is so heavily curved that the separation in column C-2 yields the pure substance *a* as bottoms.

While the generalized process with three separation steps can be performed even in systems with a slightly curved distillation border, the two-column process requires a heavily curved border. Thus, the two-column process is a special case of the generalized process.

5.6.3 Hybrid Processes

The generalized process might be simplified by performing one (or two) of the three separation steps not by distillation but by any other separation technique, e.g., decantation, absorption, stripping, extraction, adsorption or membrane permeation. Some examples will be shown in the following.

5.6.3.1 Distillation Combined with Decantation (Azeotropic Distillation) Figure 5-45 presents a process for ethanol/water separation by toluene. Ethanol and water form a low boiling azeotrope, and toluene is a suitable entrainer because, even though it is the high boiler in the system, it forms minimum azeotropes with water as well as with ethanol. Thus, the criteria for entrainer selection are fulfilled. The water/toluene azeotrope is heterogeneous and forms two liquid phases. In addition, there exists a ternary azeotrope, that has the lowest boiling point of the system. All distillation lines end at the ternary azeotrope, and each corner of the triangular diagram is a starting point for them. Distillation borders are running from each binary azeotrope to the ternary azeotrope dividing the mixture in three distillation fields that differ in the origins of the distillation lines.

The feed $\dot{F}$ is, generally, rich in water. It is lying in that distillation field where water can be separated as bottoms product $\dot{B}1$. This separation is performed in column C-1. The top fraction $\dot{D}1$ lies very close to the distillation border. The admixing of the entrainer produces a stream $\dot{M}2$ located in the concave region of

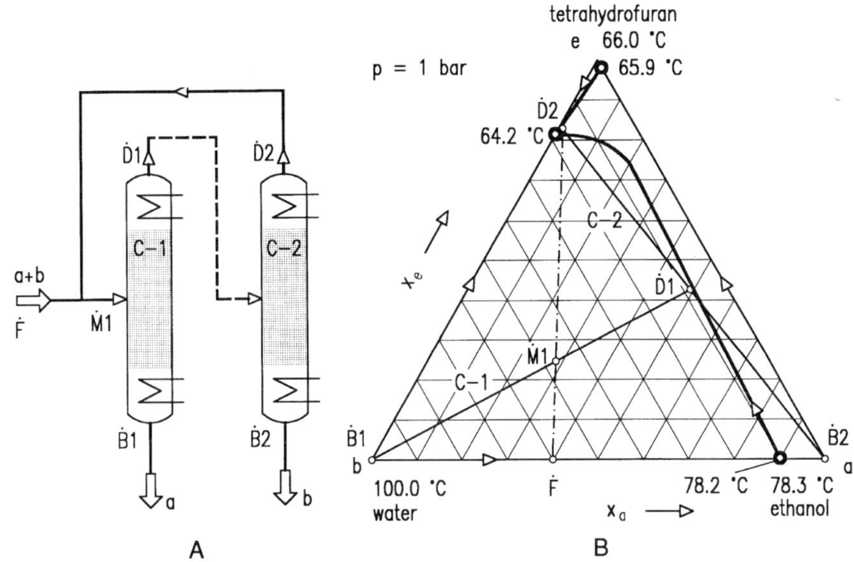

Fig. 5-43. *Two-column process for the separation of a binary mixture (ethanol/water) with a minimum azeotrope by using a low boiling entrainer (tetrahydrofuran) that forms a minimum azeotrope with component* b. *The borderline is so heavily curved that pure* a *is recovered as bottoms of column C-2. The third column of the generalized process can be discarded: (A) Flowsheet, (B) triangular diagram.*

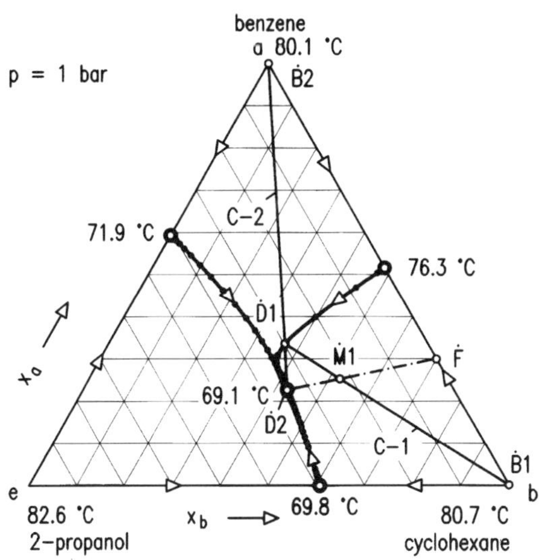

Fig. 5-44. *Two-column process for the separation of a binary mixture with a minimum azeotrope by using a low-boiling entrainer that forms minimum azeotropes with components* a *and* b. *The borderlines are so heavily curved that pure* a *is recovered as bottoms of column C-2. The third column of the generalized process can be discarded.*

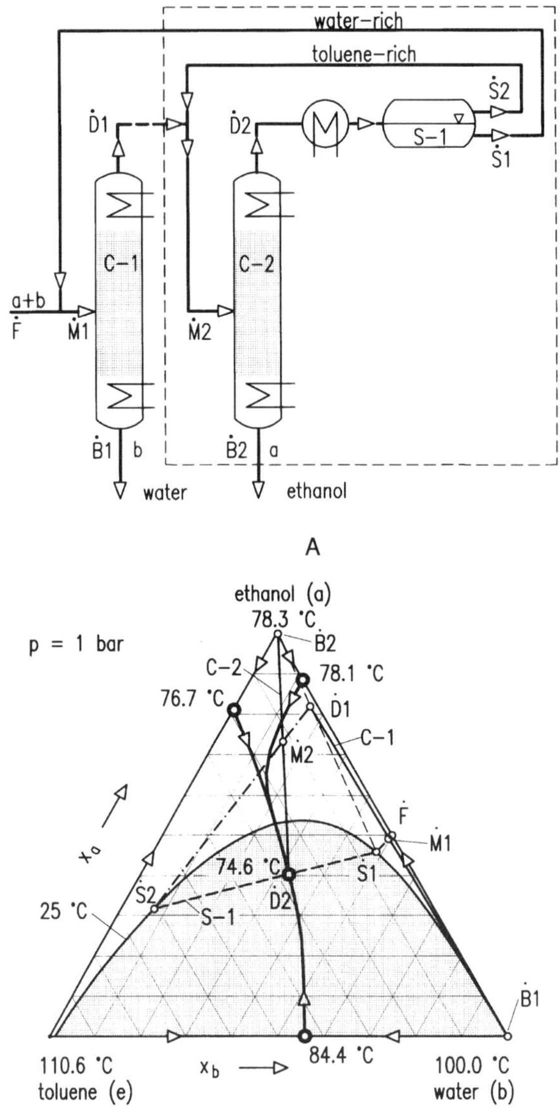

Fig. 5-45. *Separation of the ethanol/water mixture using toluene as entrainer. This process is referred to as* azeotropic distillation: *(A) Flowsheet showing distillation columns C-1 and C-2 and decanter S-1. (B) Triangular diagram. Because the mixture contains three binary minimum azeotropes and one ternary azeotrope, it is divided into three distillation fields. —— Distillation boundary; — External material balance of distillation column; –·–·– Mixing of two streams; –––– Material balance around column C-2 and decanter S-1. ––– Separation in the decanter. Shaded area represents the immiscibility range at 25°C.*

Table 5-8. Systems to be Separated by Combination of Distillation and Decantation, See Process in Fig. 5-45.

Mixture to be Separated	Entrainer	Reference
water/ethanol	benzene	[Keyes 1929]
	toluene	[Pilhofer 1983]
	pentane	[Black 1980]
	trichlorethylene	[Fritzweiler 1938]
	cyclohexane	[Herfurth et al. 1987]
	ethylacetate	[Cairns and Furzer 1981]
	ethylether	[Othmer 1974]
acetic acid/formic acid	chloroform	[Hunsmann and Simmrock 1966]
water/pyridine	benzene	[Berg et al. 1945]
	toluene	[Berg 1969]
water/acetic acid	butylacetate	[Hegner et al. 1973]
	propylacetate	[Hegner et al. 1973]
water/propanol	benzene	[Bril et al. 1977]

the distillation border where pure ethanol can be obtained as a bottom product $\dot{B}2$ in the next column C-2. The top product $\dot{D}2$ is very close to the ternary azeotrope and, when condensed and subcooled, it forms two liquid phases that are separated in the decanter S-1. The toluene-rich fraction $\dot{S}2$ serves as the entrainer, and the water-rich fraction $\dot{S}1$ is recirculated to column C-1. This process is equivalent to the generalized process (Fig. 5-37) with the exception that the third separation step is performed by decantation instead of by distillation.

A material balance around column C-2 and decanter S-1 (thin dashed line) leads to the condition that the states of the streams $\dot{B}2$ (ethanol), $\dot{S}1$ and $\dot{D}1$ are collinear in the triangular diagram of Fig. 5-45B. This condition fixes the state of the overhead fraction $\dot{D}1$ and, in turn, the state of the stream $\dot{M}1$. The only free parameter in the process is the circulation rate of the entrainer-rich fraction $\dot{S}2$. It must be chosen so that the stream $\dot{M}2$ lies on a straight line between the ternary azeotrope and the pure ethanol.

Using a decanter instead of a distillation column results in a very significant simplification of the process. This trick for process simplification works for minimum azeotropic systems only since—as has been shown in Section 5.6.2.1—in a process, only azeotropes of the same type can occur, and azeotropes in a region of immiscibility (heterogenous azeotropes) are always minimum azeotropes.

The process presented in Fig. 5-45 is very often used. Some examples are listed in Table 5-8. It is possible to feed the recycles $\dot{S}1$ and $\dot{S}2$ directly into the top of columns C-1 and C-2, respectively, instead of first mixing them with $\dot{F}$ and $\dot{D}1$. A critical point is sometimes the decanter. Some mixtures form a very stable emmulsion which cannot be separated into the two-liquid phases by decantation. Experiments with the original feedstock should be made.

5.6.3.2 Distillation Combined with Absorption (Extractive Distillation) The use of an absorption step instead of distillation is demonstrated in Fig. 5-46. This process can be applied, for instance, to ethanol/water separation. In col-

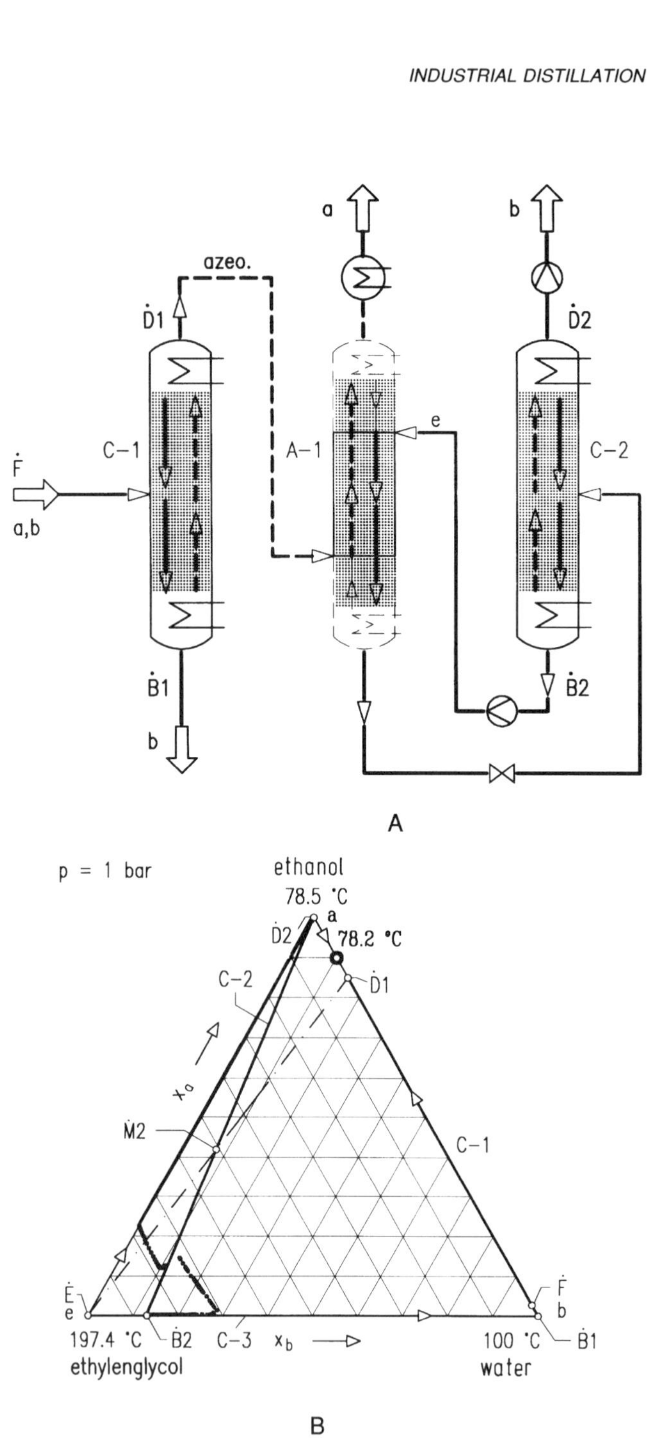

Fig. 5-46. *Separation of a binary mixture* a-b *with a minimum azeotrope by using an absorption step. This process is generally known as extractive distillation although it actually involves absorption and not extraction: (A) Flowsheet, (B) triangular diagram.*

umn C-1 most water is removed as bottoms $\dot{B}1$. The top fraction $\dot{D}1$ with azeotropic composition (approximately 90 mol.-% ethanol) is vaporously fed into the absorber A-1 where the water is removed by an appropriate absorbens (e.g., ethylene glycol, which is hygroscopic). The absorbens is regenerated in column C-2 by vacuum distillation.

The vapor at the top of the absorption column contains small amounts of the absorbens *e*. For separation, part of the top fraction is condensed and recycled as reflux at some stages above the absorbens feed point. Analogously, some of substance *a* is dissolved in the liquid in the bottom, which can be separated by reboiling. Because of the existence of a condenser and a reboiler the middle column of the process depicted in Fig. 5-46 is often misinterpreted as a distillation column. However, since the low boiling fraction $\dot{D}1$ is fed into the bottom, the high boiling liquid $\dot{B}2$ is (subcooledly) fed into the top of the column and both flow rates are much higher than that of reflux and reboil, there is no doubt that the process is absorption. The section above the entrainer feed is a rectifying section to remove the entrainer *e* from the top product. It is the middle section where the azeotrope is crossed. The lower section is a stripping section where the lightest consituent is removed from the liquid. The three sections can be seen very clearly in the concentration profile in Fig. 5-46B.

The azeotrope is crossed by absorption in this process. Therefore, the criteria for entrainer selection (Table 5-7) cannot be applied. Here, the well-known rules for absorbens selection have to be used. They require a high boiling liquid with high capacity and selectivity for substance *a* or *b*. In addition, some other requirements have to be met [Blaß 1989].

The process depicted in Fig. 5-46 is widely used not only for separation of azeotropic mixtures but also for separation of close boiling mixtures. Table 5-9 lists

Table 5-9. Systems to be Separated by Extractive Distillation, See Process in Fig. 5-46.

Mixture to be Separated	Entrainer	Reference
butane/butene/butadiene	furfural	[Peters and Rogwers 1983] [Happel et al. 1946]
	acetonitrile	[Bittrich 1987]
	dimethylacetamid	[Coogler 1967]
	n-methylpyrrolidon	[Volkamer et al. 1981]
	dimethylformamid	[Hebel and Kaspareit 1971]
benzene/cyclohexane	anilin	[Blaß 1989]
	n-methylpyrrolidon	[Mueller 1980]
	n-formylmorpholine	[Lackner 1981]
butene/isoprene	dimethylformamid	[Ushio 1972]
acetone/methanol	water	[Drew 1979]
tetrahydrofuran/water	dimethylformamid	[Drew 1979]
ethanol/water	ethylenglycol	[Black 1980]
	salt (potassium acetate)	[Schweitzer 1979]
propylene/propane	acrylonitrile	[Hafslund 1969]

some examples of industrial important processes. Depending on solubility, either the light constituent *a* or the heavy constituent *b* is absorbed. The process cannot be applied to systems with a maximum azeotrope. This process is generally known as ''Extractive Distillation,'' even though in reality it involves an absorption step and not an extraction step.

5.6.3.3 Distillation Combined with Desorption An example for a hybryd process with the combination of distillation and desorption is the separation of the mixture nitric acid and water. This system forms a maximum aceotrope with a nitric acid concentration of 37 mole-% that boils at 121°C. According to Table 5-7, a high-boiling entrainer has to be selected. Sulfuric acid is used in the process presented in Fig. 5-47 [Geriche 1973].

In this mixture a distillation border runs from the maximum azeotrope to the H_2O/H_2SO_4-corner dividing the ternary mixture in two distillation fields. In one field, pure water can be recovered as overhead fraction. In the other field, HNO_3 can be gained as top fraction. Normally, the feed is rich in water, it lies in the right-hand side distillation field in Fig. 5-47. Most of the water is removed as bottoms in column C-1. By admixing of concentrated sulfuric acid (46 mole-%) to the fraction $\dot{B}1$, the mixture $\dot{M}1$ lies in the left distillation field where pure HNO_3 is recovered as top product of column C-2. The bottom fraction lies at the distillation border at best. This border is crossed by stripping with life steam. The fraction $\dot{B}2$, which is free of HNO_3, is fed to the single-stage distillation unit C-3 to be split into pure water as overheads and concentrated sulfuric acid as bottoms that are recycled to column C-2. The recovery of concentrated sulfuric acid has to be performed under vacuum (see Section 5.2.1).

The process consists of three separation steps, two of them are performed by distillation and one by desorption. Nitric acid, when boiled, always forms some NO_x that colors the liquid. To get clear liquid, the NO_x is removed by stripping with air.

5.6.3.4 Distillation Combined with Extraction The separation of an azeotropic mixture by using a combination of distillation and extraction is also often used in industrial processes. Figure 5-48 shows a process for the separation of tetrahydrofuran (THF) and water [Schoenmakers 1984]. This mixture exhibits a minimum azeotrope at about 81.7 mol-% THF (see Section 5.5).

From the water-rich feed $\dot{F}$, water is removed as bottoms in column C-1. The distillate with azeotropic concentration is fed into an extraction column E-1. Concentrated sodium hydroxide (NaOH) is used as solvent because it is hygroscopic. The loaded solvent is regenerated by single-stage vacuum distillation. The concentrated sodium hydroxide is recycled to the extractor E-1, the distillate containing water and some THF is fed into column C-1 either together with the feed or separate from it.

Solvent extraction has the disadvantage that no pure products can be obtained as the raffinate always contains some solvent. According to Schoenmakers, tetrahydrofuran with 99 wt-% can be gained at best. If a higher purity is required, an additional

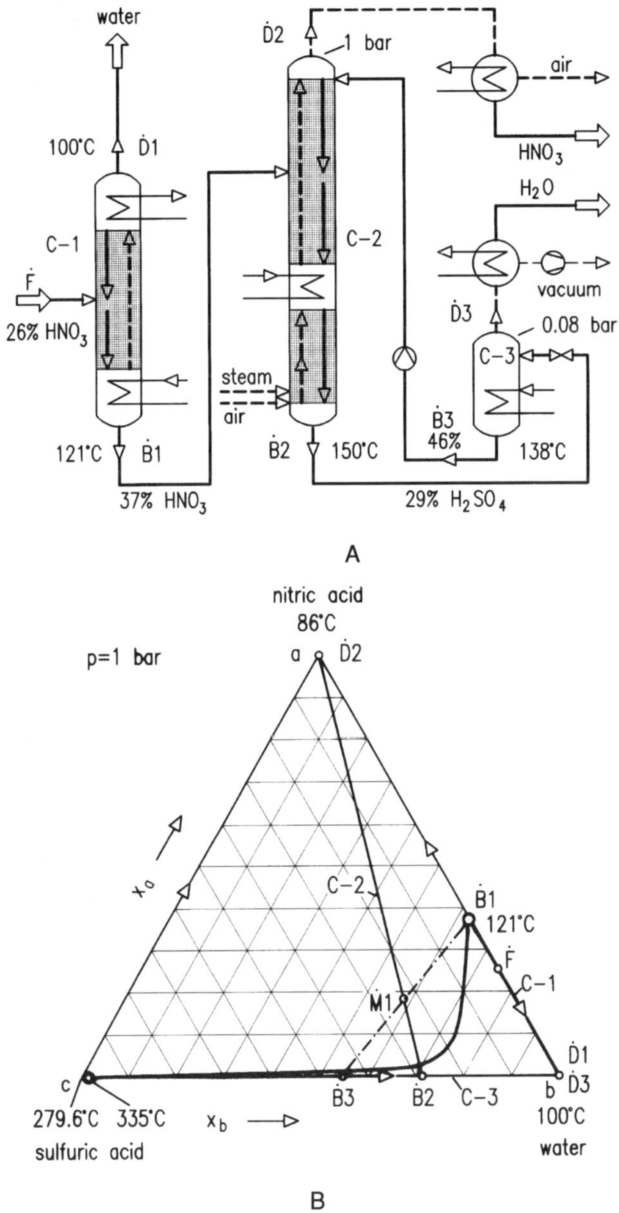

Fig. 5-47. *Concentration of nitric acid and water by using sulfuric acid as entrainer: (A) Flowsheet, (B) triangular diagram. —— Distillation boundary; — External material balance of distillation column; –·–·– Mixing of two streams.*

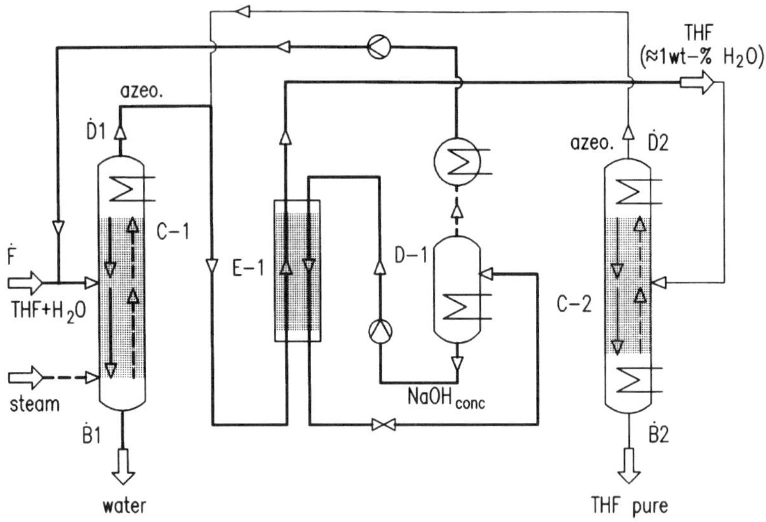

Fig. 5-48. Flowsheet of the process for the separation of the mixture tetrahydrofuran/water by distillation and extraction. The entrainer is concentrated natrium hydroxide NaOH.

distillation step, performed in column C2, has to be used. The bottoms of C-2 is pure tetrahydrofuran, the distillate is recycled to the extractor.

Another example for the combination of distillation and extraction is the separation of the mixture dichloromethane/methanol [Drew 1979]. Here, the given mixture which is usually rich in methanol is fed into the extractor first, see Fig. 5-49. The solvent used in the process is water that extracts methanol. The raffinate is split in column C-1 into dichloromethane (bottoms $\dot{B}1$) and a top fraction $\dot{D}1$ with nearly azeotropic concentration. The extract which is rich in methanol is fed into column C-2 to be separated into water (bottoms $\dot{B}2$) and methanol distillate $\dot{D}2$. This process produces highly pure dichloromethane but the methanol fraction may still contain some dichloromethane.

A very interesting hybrid process is presented in Fig. 5-50 for the separation of a benzene cyclohexane mixture [Seader 1984, pp. 13–58]. This mixture has a minimum azeotrope at about 53 mol-% benzene. Acetone is added that forms a minimum azeotrope with cyclohexane. As can be seen in Fig. 5-50B, the border distillation line is running from the benzene/cyclohexane azeotrope to the acetone/cyclohexane azeotrope. Benzene is the origin of distillation lines and, consequentely, can be removed from the ternary mixture $\dot{M}1$ as bottoms. The overhead fraction is the acetone/cyclohexane azeotrope. This fraction is fed into the extractor E-1 where acetone is removed by extraction with water. As water is only slightly soluble in cyclohexane ($x_{H_2O} < 0.1$ mol-% [Sorensen and Arlt 1979]) nearly pure cyclohexane is gained.

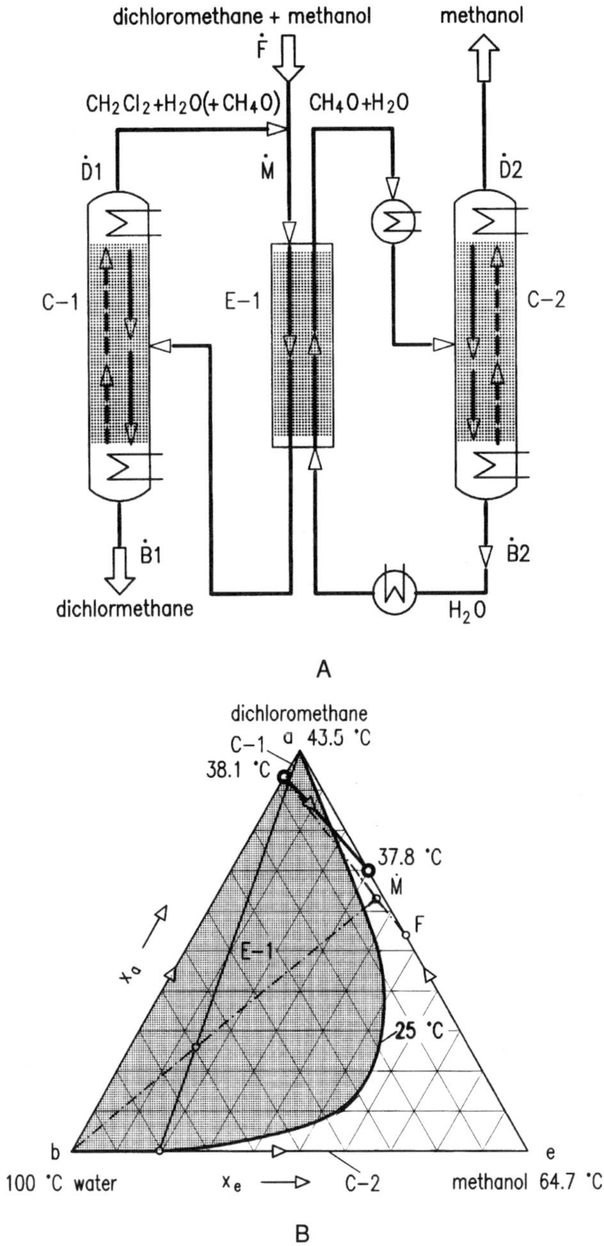

Fig. 5-49. *Process for the separation of the mixture dichloromethane/methanol by distillation and extraction. The entrainer is water: (A) Flowsheet, (B) triangular diagram.*

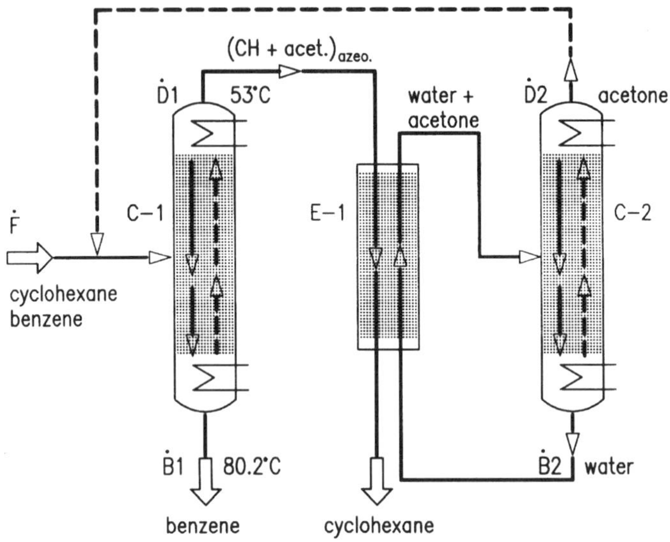

A

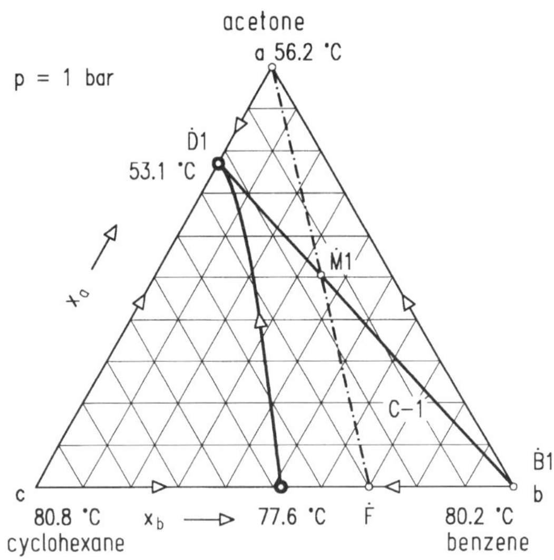

B

Fig. 5-50. *Process for the separation of the mixture benzene/cyclohexane mixture by distillation and extraction. Two entrainers, acetone and water, are used in this process: (A) Flowsheet, (B) triangular diagram for distillation.*

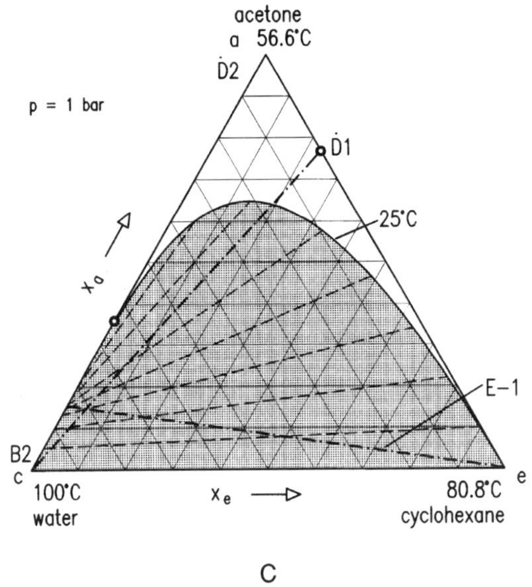

Fig. 5-50. (C) Triangular diagram for extraction.

The acetone/water mixture is fed into the distillation column C-2 to be split into acetone as overhead Ḋ2 and water as bottoms Ḃ2. Water is recycled to serve as solvent in the extractor. Acetone is recycled to column C-1 to serve there as entrainer.

In this process two entrainers are used. By the first entrainer acetone, the benzene/cyclohexane azeotrope is replaced by the acetone/cyclohexane aceotrope. This azeotropic mixture can easily be separated by extraction with water, the second entrainer.

5.6.3.5 Distillation Combined with Adsorption Adsorption is a very effective means for breaking the azeotrope in a separation process. Several authors report on the separation of azeotropic alcohol/water mixtures. Adsorption in liquid phase has been tested by [Bindal and Misra 1986] using commercially available molecular sieves as adsorbers. The separation factors are very high (85 to 100). They appear to be better than results reported for pervaporation (see Section 5.6.3.6). Dehydration of ethanol in vapor phase using a 3A molecular sieve has been studied by [Carton et al. 1987]. According to their results, vapor-phase adsorption is superior to liquid-phase adsorption in both kinetics (i.e., form of the breakthrough curve) and capacity. These findings have been confirmed by [Sowerby and Crittenden 1991]. Very low water concentrations (<0.05 wt.-%) are achieved in the product provided that the temperature was initially not too high.

Figure 5-51 depicts a process for the separation of an ethanol/water mixture by a hybrid distillation/adsorption process [Westphal 1987]. The water-rich feed is separated in distillation column C-1 into pure water as bottoms fraction Ḃ1 and the ethanol-water azeotrope as overhead fraction Ḋ1. The vaporous overhead fraction

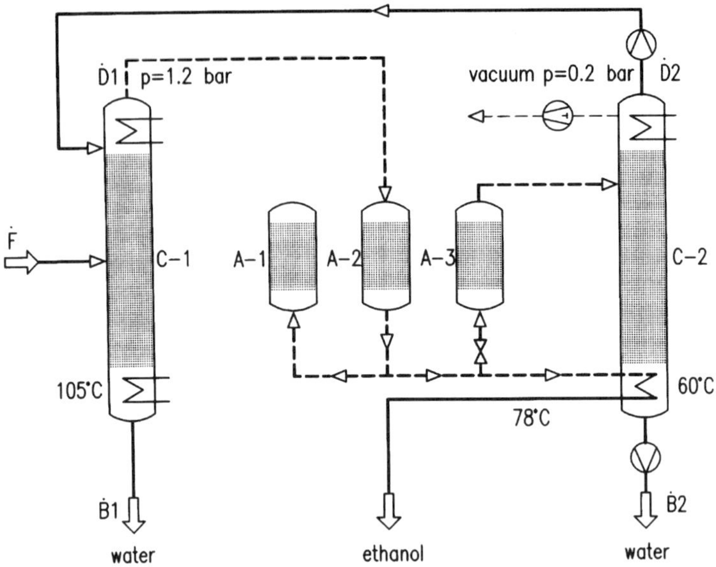

Fig. 5-51. *Process for the separation of the mixture ethanol/water by distillation and adsorption. In order to ensure a pseudo continous product flow, three adsorbers have to be installed. Each adsorber is subsequentially operated in the adsorption, regeneration and pressure build-up modes.*

Ḋ1 with approximately 10 mol-% water is fed into the adsober A-2 where the water is primarily adsorbed in a molecular sieve bed. The water-free ethanol vapor leaving the adsorber is used for heating the distillation column C-2 that is operated at a lower pressure. The liquid ethanol product is practically free of water.

Adsorption in a fixed-bed adsorber is a batch process. In order to provide a pseudo continous operation, three adsorber are required: one for adsorption, one for regeneration and one for pressure build-up. In a full cycle each adsorber is subsequently operated in all three modes.

The regeneration of the loaded adsorber is best performed by desorption at low pressure, approximately 0.2 bar, with pure ethanol vapor as stripping agent. The ethanol/water mixture formed during regeneration is fed into column C-2 to be separated into pure water Ḃ2 and the ethanol/water azeotrope Ḋ2. Column C-2 is operated at a pressure of approximately 0.2 bar. The overhead fraction is recyled as reflux of column C-1.

The adsorbers have to be heated in order to avoid vapor condensation. An elegant possibility is to produce the vacuum in column C-2 by a steam-driven ejector and to utilize the waste steam for adsorber heating. The condensate is recycled into column C-2 because it contains some ethanol [Westphal 1987].

In this process the azeotrope is crossed by adsorption. Adsorption of water by a molecular sieve is so effective that water-free ethanol is produced. The task of column C-2 is only the separation of the ethanol/water mixture formed during desorption. The energy demand of column C-2 is covered by condensating the vaporous overhead fraction of column C-1. Hence, no additional heat is required for producing the water-free ethanol from the azeotropic mixture $\dot{D}1$.

The advantage of the process is that the ethanol product will never be polluted by any entrainer, even in case of operational disturbances. Thus, the process is well suited for the production of absolute ethanol for medical and pharmaceutical purposes. However, the feed must not contain any light organic components like methanol because they may pass the adsorber and collect in the ethanol product.

5.6.3.6 Distillation Combined with Membrane Permeation The combination of distillation and membrane permeation can be a promising hybrid process for separating azeotropic mixtures, e.g., [Rautenbach and Albrecht 1989, Udriot et al. 1994, Rautenbach et al. 1996]. Figure 5-52 shows a process in which the azeotrope is crossed by membrane permeation [Baker 1990]. This process can be applied, for instance, to the separation of ethanol and water. The water-rich feed is split into water and azeotrope by distillation in column C-1. After compression to about 3 bar, the azeotropic mixture is fed into the membrane stack with a vacuum at the permeate site. A water-rich fraction permeates the membrane. Due to the vacuum, the permeate flow is evaporated and first condensed and then pumped into column C-1. The ethanol-rich retentate is fed into column C-2 where pure ethanol is re-

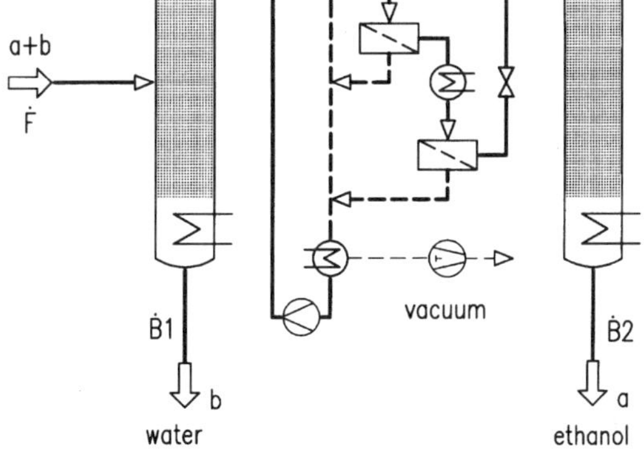

Fig. 5-52. Process for the separation of the mixture ethanol/water by distillation and membrane permeation. Since the permeate side of the membrane is at high vacuum, the permeate evaporates.

covered as bottoms. The top fraction is recycled to the membrane stack. Up to 5 to 7 membrane stacks are required. After each of them the retentate has to be heated to cover the heat of permeate vaporization. In this process the membrane is used only for crossing the azeotrope. The pure products are recovered by conventional distillation.

Decisive for the effectiveness of the process is the availability of membranes with both high capacity and high selectivity. Membranes of cellophane, cellulose acetate, polytetrafluoroethylene, acrylic acid and acyrilamide have been tested for separating alcolhol/water mixtures [Ruckenstein and Park 1990]. Ionomeric materials of perfluorosulfonic acid proofed well for the separation of polar organic compounds from their azeotropic mixtures [Dutta and Sikdar 1991].

5.7 REACTIVE DISTILLATION

A major task of chemical engineering is the down stream processing of the effluents of chemical reactors. Since most chemical reactions are reversible, that is, the educts (i.e., feed components) are not completely converted into the desired products, the effluent of a reactor is a mixture of educts and products. The non-converted educts have to be separated from the products to enable an effective recycling. The products must be purified. The separation of the reactor effluents is in most cases established by distillation. Often, down stream processing is more complex and more expensive than the reaction itself, especially in mixtures with azeotropes.

A significant simplification of such processes is often feasible by simultaneous realization of reaction and distillation in a countercurrently operated column. The objective of this chapter is to demonstrate the potentialities of this so called reactive distillation.

Some essential simplifications are made in the following:

- The reaction takes place in the liquid phase only.
- The reaction is catalyzed either by a homogeneous or a heterogeneous catalyst.
- With catalyst the reaction is instantaneous.
- No reaction takes place without catalyst.
- The reaction is reversible, i.e., both educts and products exist in the effluent.

These assumptions simplify the real conditions. However, these simplifications allow an easy description and, in turn, a good understanding of the fundamental mechanism effective in reactive distillation.

5.7.1 Fundamentals

Knowledge of the equilibrium is the most fundamental prerequisite for a conceptual synthesis of non-reactive as well as reactive distillation processes. However, the equilibrium in reactive distillation systems is more complex since the chemical equilibrium is superimposed on the vapor liquid equilibrium. Firstly, the laws of reactive and non-reactive equilibrium must be presented.

5.7.1.1 Chemical Equilibrium The following reaction mechanism is considered:

$$\nu_a \cdot a + \nu_b \cdot b + \ldots \leftrightarrow \nu_p \cdot p + \nu_q \cdot q + \ldots \tag{5.3}$$

The letter ν denotes the stoichiometric factor of the substance i. The reaction is reversible but instantaneous. In ideal mixtures the following holds for the chemical equilibrium:

$$K_R = \frac{x_p^{\nu_p} \cdot x_q^{\nu_q} \cdot \ldots}{x_a^{\nu_a} \cdot x_b^{\nu_b} \cdot \ldots} = \prod_{i=1}^{k} x_i^{\nu_i} \tag{5.4}$$

By convention, the stoichiometric coefficients ν_i are negative for educts and positive for products. The equilibrium constant of the reaction, K_R, is essentially a function of temperature. Hence, its value can be influenced by the pressure of the reactive distillation column since the mixture is at the boiling point. In nonideal mixtures the concentrations need to be corrected by the activity coefficients γ_i:

$$K_R = \frac{x_p^{\nu_p} \cdot x_q^{\nu_q} \cdot \ldots}{x_a^{\nu_a} \cdot x_b^{\nu_b} \cdot \ldots} \cdot \frac{\gamma_p^{\nu_p} \cdot \gamma_q^{\nu_q} \cdot \ldots}{\gamma_a^{\nu_a} \cdot \gamma_b^{\nu_b} \cdot \ldots} = \prod_{i=1}^{k} (x_i \cdot \gamma_i)^{\nu_i} \tag{5.5}$$

Here, the value of the equilibrium constant K_R is additionally influenced by the concentrations x_i especially at low concentrations.

Figure 5-53 presents the equilibrium for the following reaction mechanism:

$$a + b \leftrightarrow c \quad \text{with } K_R = \frac{x_c}{x_a \cdot x_b} = 8 \tag{5.6}$$

The letters a, b, and c denote the substances in the order of increasing boiling points. Smaller values of the equilibrium constant K_R lead to a smaller conversion rate of the educts into products.

5.7.1.2 Stoichiometric Lines In the determination of the reaction products the stoichiometry of the reaction has to be considered:

$$x_i = \frac{x_{io} \cdot (\nu_k - \nu_t \cdot x_k) + \nu_i \cdot (x_k - x_{ko})}{\nu_k - \nu_t \cdot x_{ko}} \quad \text{with} \quad \nu_t = \sum \nu_i \tag{5.7}$$

Here, k denotes a reference component. Equation (5.7) is represented by dotted lines

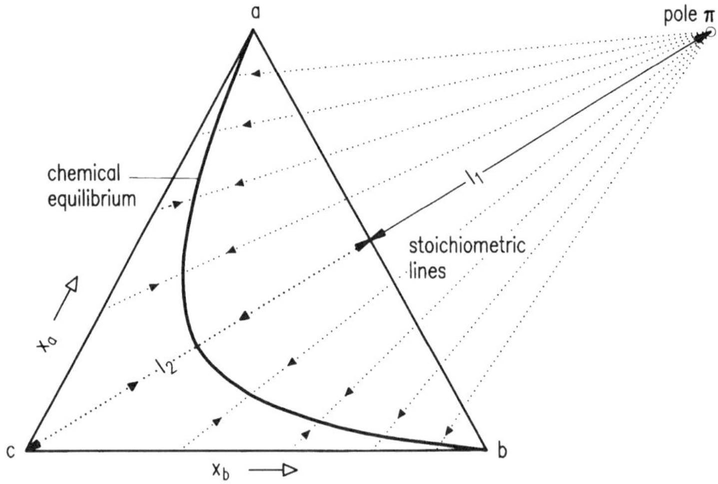

Fig. 5-53. *Chemical equilibrium of the reaction* a + b ↔ c *with* $K_R = 8$. *The stoichiometric lines emerge from a pole π that is located on the main stoichiometric line at a distance* l_1 *from the stoichiometric educt.*

in the triangular diagram of Fig. 5-53. They describe the direction of conversion of the educts by the chemical reaction. All stoichiometric lines emerge from a pole π that can easily be located by:

$$l_1 = l_2 \frac{n_p}{n_e - n_p} \text{ or } x_{i\pi} = \frac{\nu_i}{\nu_t} \tag{5.8}$$

The length l_2 is the distance between the stoichiometric educt (here $x_a = 0.5$ and $x_b = 0.5$) and the product of an irreversible reaction (here pure c). It is taken from the triangular diagram. The terms n_e and n_p denote the number of moles of the educt and the product, respectively. In equimolar reactions (e.g., $a + b \leftrightarrow 2\,c$) the length l_1 becomes infinite, i.e. the stoichiometric lines are parallel. The reaction runs from the educt to the equilibrium as indicated by the arrows on the stoichiometric lines.

5.7.1.3 Non-Reactive and Reactive Distillation Lines The adequate form for the presentation of the vapor liquid equilibrium are distillation lines. They constitute a sequence of equilibrium stages that run from a local boiling point maximum to a local boiling point minimum. Their general definition is:

$$x_{i1} \rightarrow y_{i1}^* = x_{i2} \rightarrow y_{i2}^* = x_{i3} \rightarrow y_{i3}^* \rightarrow \cdots \tag{5.9}$$

Starting at any liquid concentration x_{i1} the equilibrium concentration of the vapor

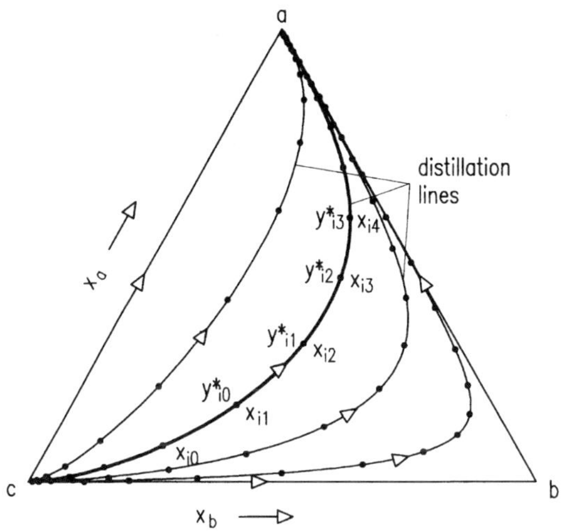

Fig. 5-54. *Distillation lines of an ideal ternary mixture with $\alpha_{ac} = 3$ and $\alpha_{bc} = 2$.*

y^*_{i1} is determined, see Fig. 5-54. Condensation of this vapor generates a new liquid with the concentration $x_{i2} = y^*_{i1}$. A sequence of such steps generates the distillation line that describes the maximal enrichment of low boilers in the mixture. The arrows on the distillation lines point to falling boiling point temperatures. The length of an equilibrium step is marked by points. Distillation lines are very helpful for process synthesis since they represent the concentration profile within a distillation column at high reflux ratio (see Section 4.3.2). The distillation lines of an ideal ternary mixtures follow from equation (4.59):

$$x_{an} = \alpha_{ac}^{n} \cdot x_{ao}/N; \qquad x_{bn} = \alpha_{bc}^{n} \cdot x_{bo}/N$$

$$\text{with } N = 1 + (\alpha_{ac}^{n} - 1) \cdot x_{ao} + (\alpha_{bc}^{n} - 1) \cdot x_{bo} \tag{5.10}$$

The letter n denotes the number of equilibrium stages by integer or real variables.

Figure 5-54 presents the course of distillation lines for an ideal ternary mixture with the relative volatilities $\alpha_{ac} = 3$ and $\alpha_{bc} = 2$. All distillation lines originate at the high boiler *c* and terminate at the low boiler *a*. These substances can be easily separated from a ternary mixture either as bottom fraction (*c*) or as overhead fraction (*a*) of a distillation column. The larger the distance between the points on the distillation lines the easier is the separation.

In reactive systems the effect of the chemical reaction is superimposed on the vapor liquid equilibrium. The states in a column can be described by reactive distillation lines whose definition is:

$$x_{i1} \rightarrow y^*_{i1} = x^*_{i1} \rightarrow x_{i2} \rightarrow y^*_{i2} = x^*_{i2} \rightarrow x_{i3} \rightarrow \cdots \tag{5.11}$$

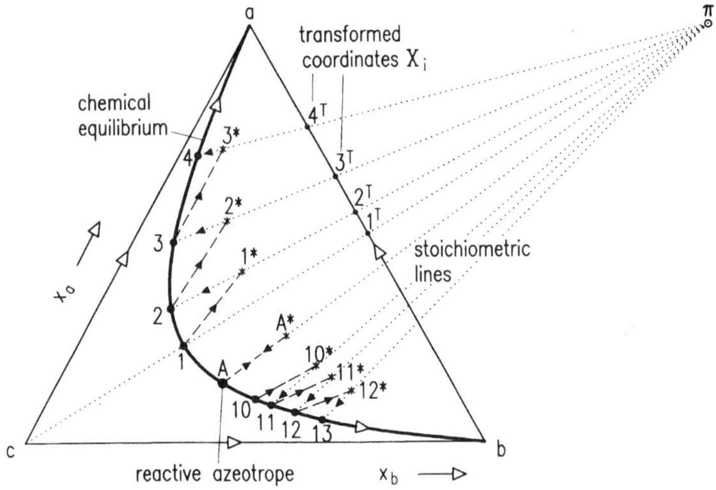

Fig. 5-55. Design of a reactive distillation line of an ideal ternary mixture. ---- vapor-liquid equilibrium, ... stoichiometric lines.

The procedure is graphically explained in Fig. 5-55. The starting point 1 lies on the chemical equilibrium line. The concentration of the vapor that is in equilibrium with the liquid x_{i1} is y_{i1}^*. It is marked in the diagram by point 1*. Condensation of the vapor generates a liquid x_{i1}^* that is converted by the chemical reaction into the state x_{i2} (point 2). The direction of conversion is given by the stoichiometric line through point 1*. Multiple application of this procedure leads to a sequence of liquid states on the chemical equilibrium line that are marked by points. This sequence of liquid states constitutes a reactive distillation line that runs from the starting point 1 to the low boiler *a*. In ternary mixtures the reactive distillation line coincides with the chemical equilibrium curve.

Applying the same procedure to a different starting point, e.g., point 10, leads to a reactive distillation line that runs to the intermediate boiler *b*. The reason for the different direction of movement is obviously the different gradient of the vapor liquid equilibrium.

A very special situation exists when, at any point A on the chemical equilibrium line, the vapor liquid equilibrium is co-linear with the stoichiometric line. In this case the liquid state remains unchanged since the vapor with different concentration (point A*) is, after condensation, converted by the chemical reaction into the original liquid state (Point A). This special liquid state on the reaction line is called ''reactive azeotrope'' (see next section) [Okasinski and Doherty 1997].

Several authors [Ung and Doherty 1995, Espinosa et al. 1995] propose a transformation of the liquid concentrations:

$$X_i = \frac{\nu_k \cdot x_i - \nu_i \cdot x_k}{\nu_k - \nu_t \cdot x_k} \quad \text{with } \nu_t = \sum \nu_i \tag{5.12}$$

The transformed concentrations meet the condition $\Sigma X_i = 1$. In ternary mixtures the values of X_i lie between 0 and 1, in quaternary mixtures one component has a value between -1 and $+1$. This transformation is, as indicated by dotted lines in Fig. 5-55, a projection of the states on the reactive distillation line along the stoichiometric lines onto the right-hand side of the triangle. The reference component k (here c) is not visible in the transformed coordinates. Such a transformation is advantageous since it effects a reduction of the dimensions of the concentration space. A triangle is transformed into a line, a tetrahedron into a plane, etc. In the transformed concentrations the lever arm rules are valid since the effect of the reaction is not visible. The disadvantage, however, of this transformation is the loss of some important informations.

5.7.1.4 Reactive Azeotropes The condition for the formation of reactive azeotropes is that the vapor which emerges from the liquid is, after condensation, converted by the reaction into the same liquid concentration. The effect of the distillation is compensated by the reaction. In mathematical terms, the vector of the vapor liquid equilibrium and the vector of the stoichiometric line must be co-linear. Whether or not this condition is met can be easily checked from a plot of liquid residuum lines and stoichiometric lines. The stoichiometric lines form tangents of the liquid residuum lines. Since in most cases the differences beetween residuum lines and distillations lines are very small (see Section 3.2) the loci of reactive azeotropes can also be determined approximately from a plot of distillation lines.

Figure 5-56 presents the residuum lines of an ideal ternary mixture. For the equimolar reaction $a + b \leftrightarrow 2 \cdot c$ the stoichiometric lines are parallel lines oriented vertically to the a-b-side of the triangle in Fig. 5-56A. The points of tangential contact of stoichiometric lines and residuum lines form a curve of potential reactive azeotropes (dashed dotted line). In most cases this line of co-linearity of vapor liquid equilibrium and stoichiometric lines lies close by the base of the triangle. Whether or not a reactive azeotrope is formed depends on the chemical equilibrium. At large values of the equilibrium constant K_R a reactive maximum azeotrope is formed close by the c-corner of the diagram (see small triangle of Fig. 5-56A). With decreasing values of the equilibrium constant K_R the reactive azeotrope moves to the b-corner of the triangle and vanishes finally. It is important to note that reactive azeotropes exist even in ideal mixtures provided the value of the chemical equilibrium constant K_R is large.

For the equimolar reaction $a + c \leftrightarrow 2 \cdot b$ the stoichiometric lines are oriented vertically to the a-c-side of the triangle, see Fig. 5-56B. The stoichiometric lines intersect the residuum (or distillation) lines with an angle between 30° and 120°. Consequently, no tangential points and, in turn, no reactive azeotropes exist in systems with the stoichiometric pole π at the concave side of the residuum lines. This behavior is very important for the design of reactive distillation processes.

In Fig. 5-56C the equimolar reaction $b + c \leftrightarrow 2 \cdot a$ is considered. In this case the stoichiometric lines are vertical to the base line of the triangle. The points of the tangential contact constitute a line of potential reactive minimum azeotropes. It

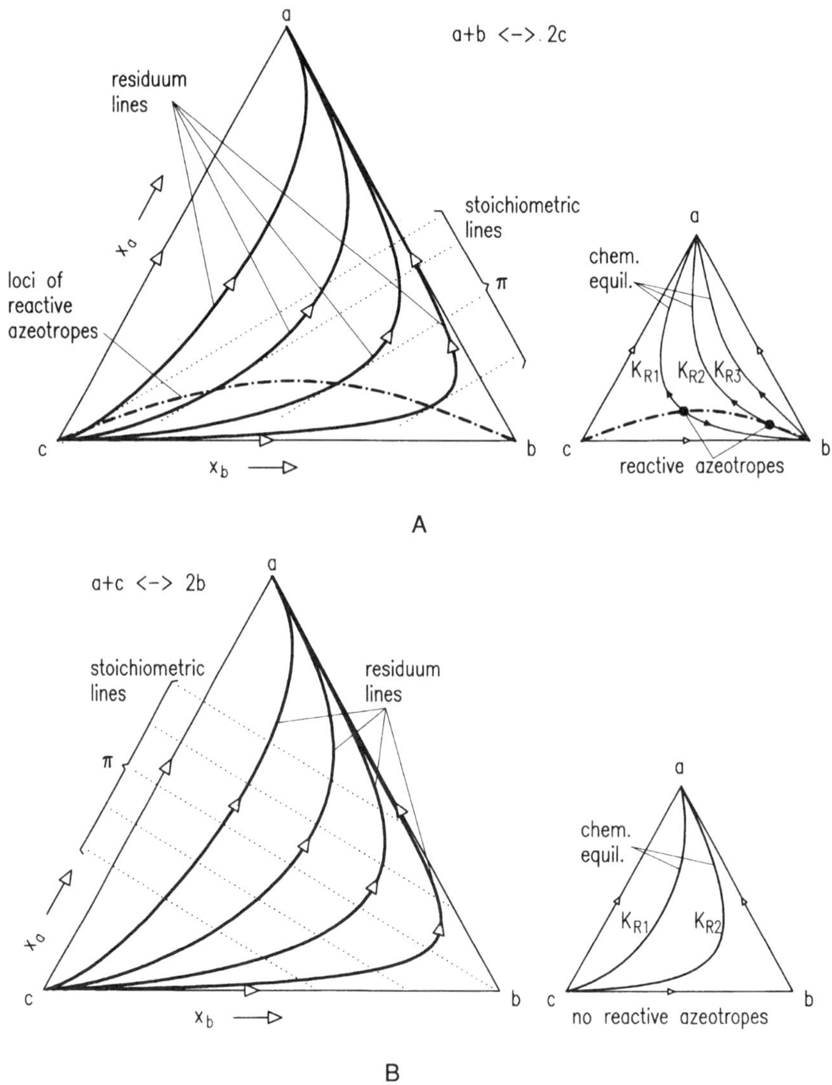

Fig. 5-56. *(A) Determination of reactive azeotropes of the reaction* a + b ↔ 2c. *(B) Determination of reactive azeotropes of the reaction* a + c ↔ 2b. *No reactive azeotropes exist since the stoichiometric pole* π lies at the concave side of the residuum lines.

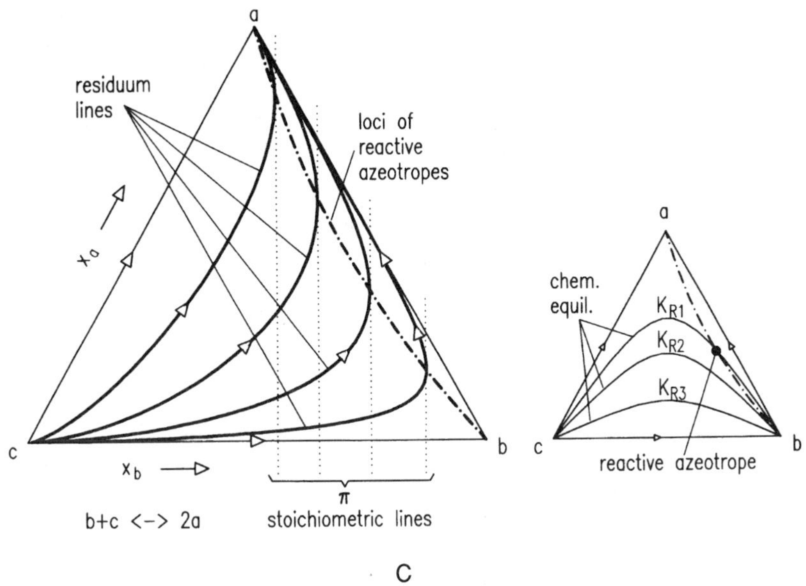

Fig. 5-56. *(C) Determination of reactive azeotropes of the reaction* b + c ↔ 2a.

lies in most cases close by the right hand side of the triangle. The existence of reactive azeotropes and their location in the diagram depends on the value of the equilibrium constant K_R. High values of K_R always result in the formation of a reactive azeotrope, see small triangle of Fig. 5-56C. The value of the equilibrium constant K_R and, in turn, the location of the reactive azeotrope can be manipulated to a certain degree by the operating pressure of the reactive distillation column.

The reaction $a + b \leftrightarrow c$ is considered in a system with a non-reactive a-b-minimum azeotrope, see Fig. 5-57. All residuum (or distillation) lines start at the high boiler c and end at the minimum azeotrope (arrows point to falling boiling temperatures). Reactive azeotropes can be detected by drawing stoichiometric tangents to the residuum lines. Two curves of potential reactive azeotropes exist in this system. One lies close by the base line of the triangle and the other one close by the upper part of the right hand side of the triangle. Reactions with a large value of the equilibrium constant K_R show a reactive maximum azeotrope. Intermediate values of K_R may result in the formation of one or two reactive azeotropes (see small triangle in Fig. 5-57).

A system with a distillation border is considered in Fig. 5-58A in the case of the non-equimolar reaction $a + b \leftrightarrow c$. The stoichiometric lines emerge from the pole π and form tangents to the residuum lines in the lower left section of the diagram only. The curve of potential reactive azeotropes runs between the c-corner and the nonreactive minimum azeotrope. A reaction with a large equilibrium constant K_R

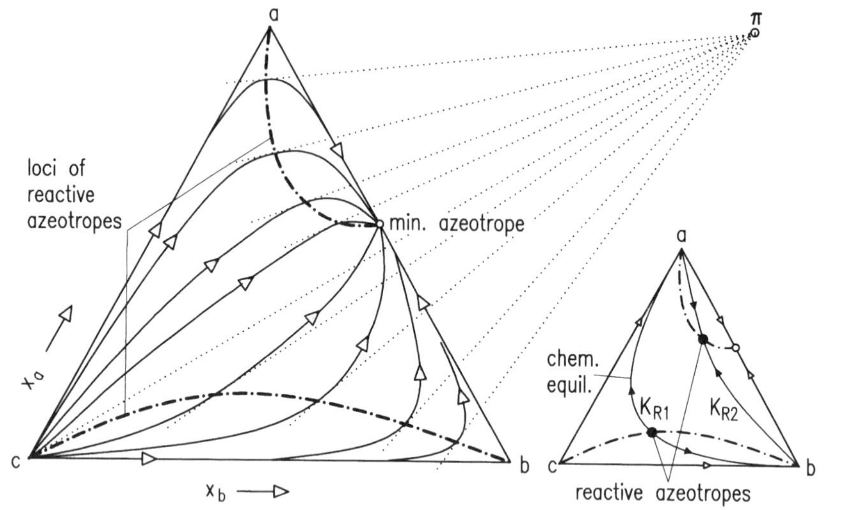

Fig. 5-57. Determination of reactive azeotropes of the reaction a + b ↔ c *in a system with an* a-b-*minimum azeotrope.*

shows two reactive azeotropes, a maximum (left) and a minimum (right) azeotrope. If the value of the equilibrium constant K_R is small no reactive azeotrope exists.

An interesting system is shown in Fig. 5-58B with a distillation border between the *b*-*c*- and the *a*-*b*-azeotrope. Two curves of potential azeotropes exist, one runs between the *c*-corner and the *c*-*b*-azeotrope, the other one between the *a*-corner and the *a*-*b*-azeotrope. In case of a large value of the equilibrium constant K_R two reactive azeotropes exist. In the case of a very low value of K_R just one reactive azeotrope is formed. Surprisingly, no reactive azeotrope exists at intermediate values of the equilibrium constant K_R, see small triangle in Fig. 5-58B.

In all cases studied the lines of potential reactive azeotropes run between singular points of the system, i.e., pure substances or azeotropes. One should be aware that the curves of potential reactive azeotropes may change drastically even by small variations to the residuum (or distillation) line charts. Consequently, a very good knowledge of the vapor liquid equilibrium is neccessary for the localization of reactive azeotropes.

Reactive azeotropes can be more precisely located in a McCabe-Thiele type plot of the vapor and liquid concentrations in transformed coordinates X_i and Y_i (see equation 5.12) [Okasinski and Doherty 1997]

The reactive azeotropes always lie on the chemical equilibrium line. Since in reactive distillation the mixture is at the boiling point the temperature and, in turn, the value of the reaction equilibrium constant K_R are influenced by the operating pressure. As a consequence the location of a reactive azeotrope can be influenced by changing the pressure.

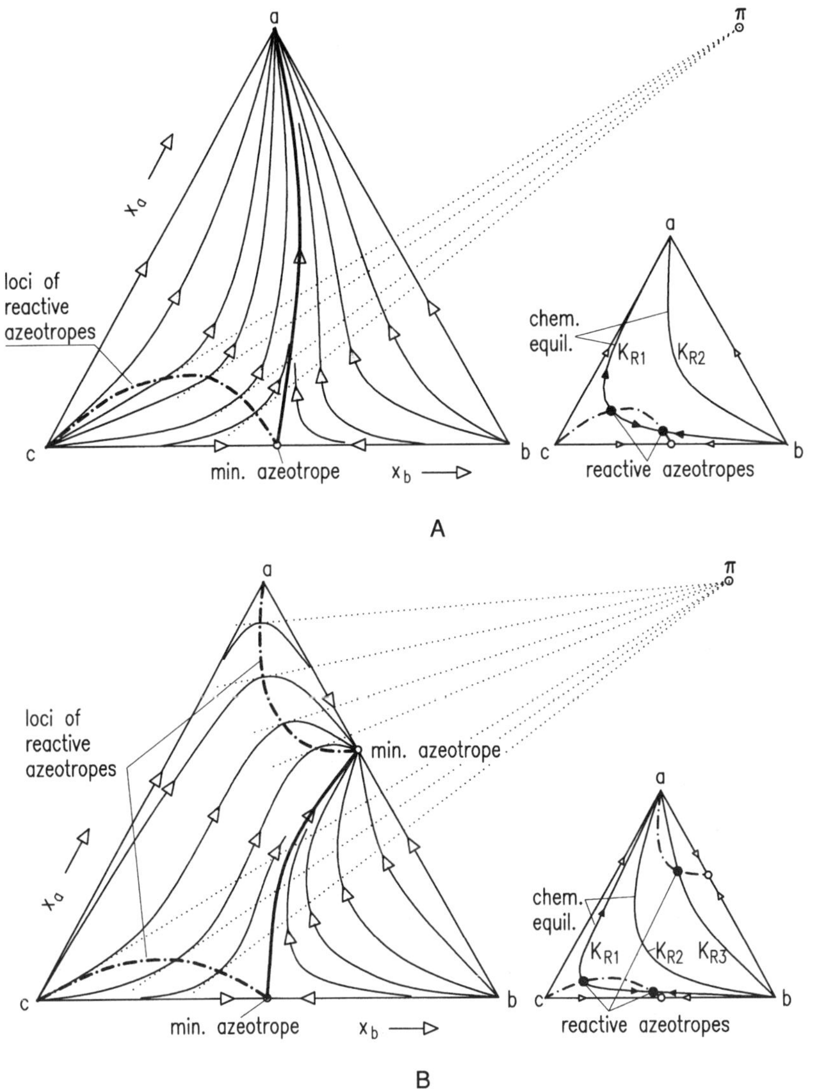

Fig. 5-58. *(A) Determination of reactive azeotropes of the reaction* a + b ↔ c *in a system with a* b-c-*minimum azeotrope and a distillation border. (B) Determination of reactive azeotropes of the reaction* a + b ↔ c *in a system with an* a-b-*and a* b-c-*minimum azeotrope and a border residuum line.*

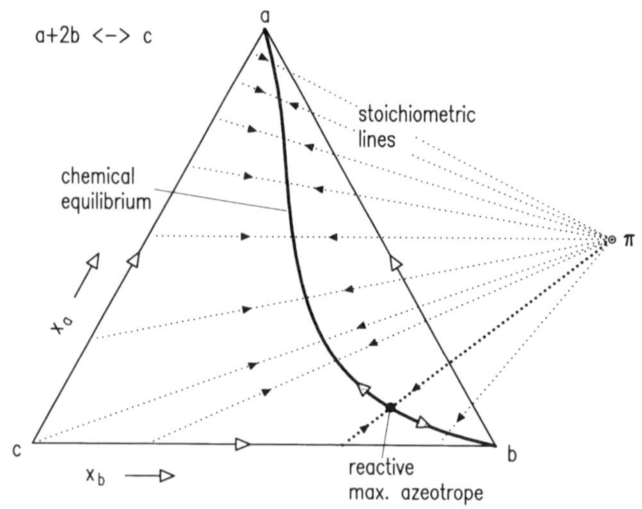

Fig. 5-59. *Reactive distillation line of the reaction* a + 2b ↔ c. *There exists a reactive maximum azeotrope.*

5.7.2 Topology of Reactive Distillation Lines

This section presents some typical examples for reactive distillation lines and reactive azeotropes. Here, too, it is assumed that a reversible and instantaneous reaction takes place in the liquid phase only.

5.7.2.1 Reactions in Ternary Systems Figure 5-59 presents the graphical plot of the following reaction mechanism:

$$a + 2b \leftrightarrow c \quad \text{with } K_R = \frac{x_c}{x_a \cdot x_b^2} = 4 \tag{5.13}$$

Assuming an ideal ternary mixture with the relative volatilities $\alpha_{ac} = 3$ and $\alpha_{bc} = 2$ a reactive maximum azeotrope exists on the reactive distillation line. Above that maximum azeotrope the reactive distillation line runs towards the low boiler and below towards the intermediate boiler b. If the stoichiometric lines are considered as elements of the reactive distillation line, what is often done, then the border is defined by the stoichiometric line through the reactive maximum azeotrope. The reactive distillation lines describe the concentration profile within a column. The reactive azeotrope can not be crossed by reactive distillation (but by non-reactive distillation, see Section 5.7.3).

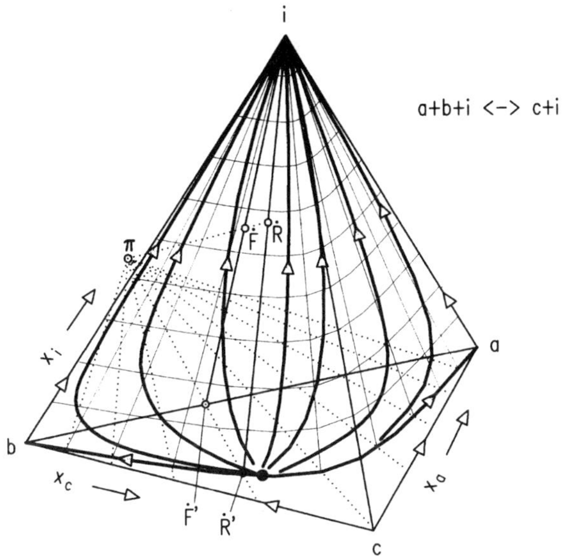

Fig. 5-60. Reactive distillation lines of the reaction a + b ↔ c *with a low boiling inert component* i.

5.7.2.2 Reactions in Ternary Systems with Inerts Sometimes the educts *a* and *b* contain some inert species *i*. Such systems can be graphically presented by a tetrahedron, see Fig. 5-60. The reaction mechanism is:

$$a + b + i \leftrightarrow c + i \quad \text{with } K_R = \frac{x_c}{x_a \cdot x_b} = 10 \tag{5.14}$$

If there is no inert species present the same situation as dealt with in the previous section exists. The triangle *a-b-c* forms the base of a tetrahedron and the chemical equilibrium line runs between *a* and *b*. In the presence of the inerts *i* the chemical equilibrium forms the mantle of a cone constituted by the inert-free chemical equilibrium and the peak of the tetrahedron. The reactive distillation lines lie on the cone mantle. They run from the reactive maximum azeotrope to the peak of the tetrahedron provided the inert component is the lowest boiling species in the mixture. Hence, the chemical equilibrium and the reactive distillation line are no longer identical.

The stoichiometric lines emerge from the pole π that lies in the plane of the tetrahedron base. They allow easy determination of the chemical equilibrium to any concentration of the educts even when the inert *i* is present.

Helpful is the projection of the states on the cone mantle into the plane *i-a-b* along the stoichiometric lines, as is shown in Fig. 5-61. In these transformed coordinates (Equation 5.12) the lever arm rule is valid since the influence of the reaction has been eliminated.

5.7.2.3 Reactions with Side Products A chemical reaction is often accompanied by an unwanted side reaction between one of the educts and the product.

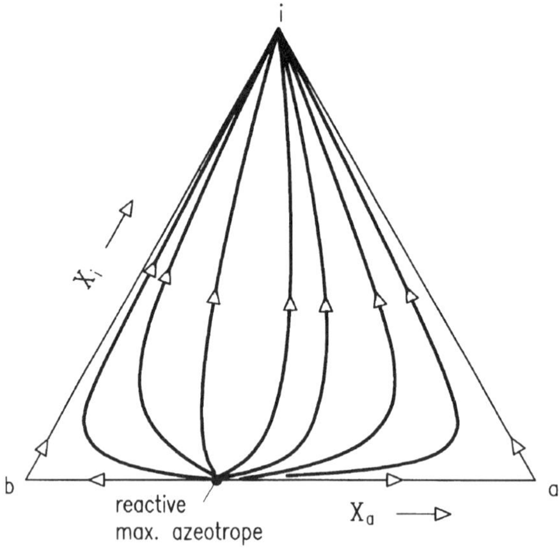

Fig. 5-61. *Reactive distillation lines of Fig. 5-60 presented by transformed coordinates X_i.*

$$a + b \leftrightarrow c \quad \text{with } K_R = \frac{x_c}{x_a \cdot x_b} = 8 \tag{5.15A}$$

$$a + c \leftrightarrow d \quad \text{with } K_R = \frac{x_d}{x_a \cdot x_c} = 4 \tag{5.15B}$$

The chemical equilibrium of the four species can be graphically presented by a tetrahedron. For the main reaction (5.15A) the substance *d* is an inert species. In analogy to the previous section the chemical equilibrium lies on the mantle of a cone that is formed by the chemical equilibrium line of the system without the inert *d* and the *d*-corner of the tetrahedron, see Fig. 5-62.

For the side reaction (5.15B) the species *b* is inert. Consequently, the chemical equilibrium of the side reaction lies on a cone mantle formed by the equilibrium without *b* and the *b*-corner of the tetrahedron. The line of intersection of the two cone surfaces is the chemical equilibrium of the combined reactions. This line is identical with the reactive distillation line. Even here, a reactive azeotrope may exist. The stoichiometric lines that describe pairs of educts and products emerge from two poles π that can be located by the procedure described before.

5.7.2.4 Reactions in Quaternary Systems An important class of reactions is

$$a + b \leftrightarrow c + d \quad \text{with } K_R = \frac{x_c \cdot x_d}{x_a \cdot x_b} = 1 \tag{5.16}$$

The chemical equilibrium is graphically presented in Fig. 5-63A. It forms a saddle

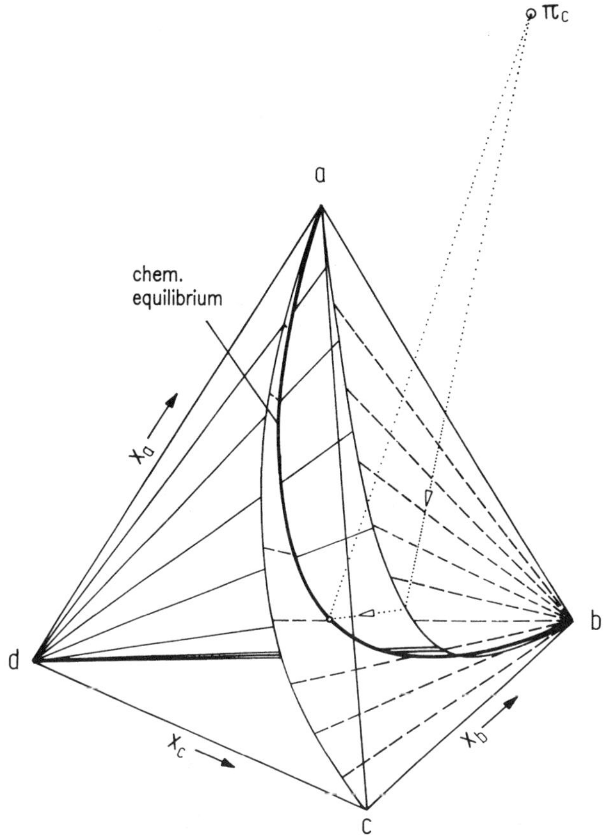

Fig. 5-62. *Reactive distillation line of the reactions* a + b ↔ c *and* a + c ↔ d.

area that extends between the sides *a-c, a-d, b-c* and *b-d* of a tetrahedron. Since the number of moles is not changed by the reaction the stoichiometric lines are all parallel to the main stoichiometric line running from the middle of the *a-b*-side to the middle of the *c-d*-side.

The reactive distillation lines are qualitatively sketched in the saddle area. They originate at the high boiling products *d* and *c* and terminate at the low boiling educts *a* and *b,* respectively. A reactive saddle azeotrope exists as well as two reactive distillation border lines that connect the origins and the termini, respectively, of the distillation lines.

A more abstract but more exact presentation of the reactive distillation lines and the reactive azeotrope offers the use of transformed coordinates X_i (according to equation (5-12)). The states in the chemical equilibrium area are projected along the stoichiometric lines onto a plane oriented vertically to the stoichiometric lines. The tetrahedron transfers into a rhombus with component *a* and *b* at the top and at

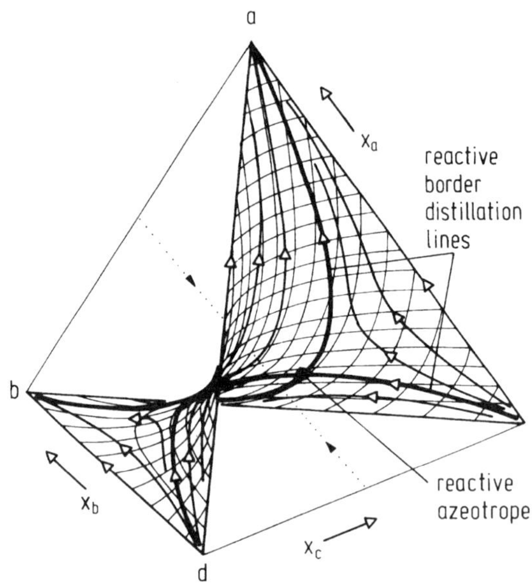

A

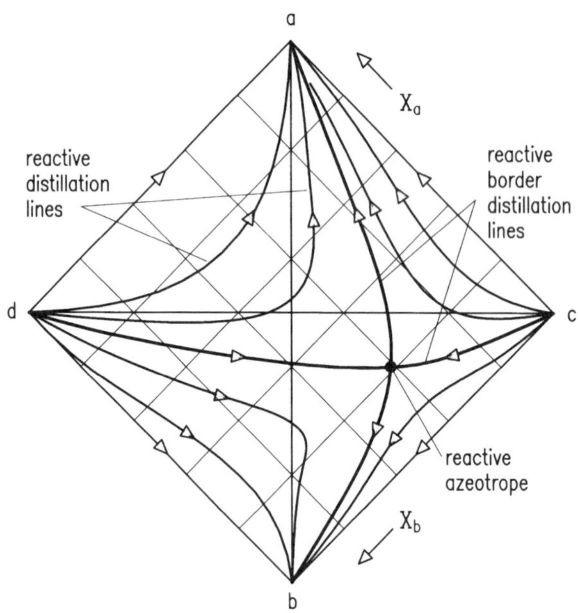

B

Fig. 5-63. *(A) Reactive distillation lines of the reaction* a + b ↔ c + d *with* $K_R = 1$. *(B) Reactive distillation lines of the system of Fig. 5-63A in transformed coordinates.*

the bottom, respectively (see Fig. 5-63B). The reactive saddle azeotrope lies at the intersection of the reactive distillation border lines. They can not be crossed by reactive distillation. Hence, the mixture is divided into four reactive distillation regions.

5.7.3 Topology of Reactive Distillation Processes

Knowledge of the reactive distillation lines is the basis for the conceptual design of reactive distillation processes. The procedure is, in essence, the same as in the design of non-reactive distillation processes described in detail in Section 4.3. The most essential steps are:

- Determination of the feasible products.
- Formulation of the material balance around the column.
- Identification of separation barriers, e.g., azeotropes, border distillation lines or border distillation areas.

The feasible product fractions are determined from a knowledge of the distillation lines. Feasible pure products of distillation columns are either origins or termini of distillation lines. This principle holds also for reactive distillation.

The material balance around the column is graphically represented by a straight line. The balance is met in non-reactive distillation if the feed $\dot{F}$, the distillate $\dot{D}$ and the bottoms $\dot{B}$ lie on the same straight line. In reactive distillation this principle no longer holds since some of the species is converted into other species by the reaction. Hence, the states $\dot{F}$, $\dot{D}$ and $\dot{B}$ are only co-linear after transformation by equation (5.12) along the stoichiometric lines.

Barriers for the separation can be azeotropes, border distillation lines, border distillation areas etc. The effectiveness of reactive distillation is based on the fact that reactive distillation lines can cross non-reactive distillation borders. A disadvantage of reactive distillation is the eventual formation of reactive azeotropes and reactive distillation borders, even in ideal systems.

Reactive distillation is especially effective in catalytic reactions since non-reactive and reactive sections can be combined at will in a column. In the reactive sections non-reactive barriers can be crossed and vice versa. If possible a heterogeneous catalyst should be favored against a homogeneous catalyst in order to avoid the separation of the catalyst from a homogeneous mixture.

In the next sections the potentialities of reactive distillation are presented in a general way by comparing processes of sequential and simultaneous reaction and distillation.

5.7.3.1 Single Product Reactions The following reaction is considered:

$$a + b \leftrightarrow c \tag{5.17}$$

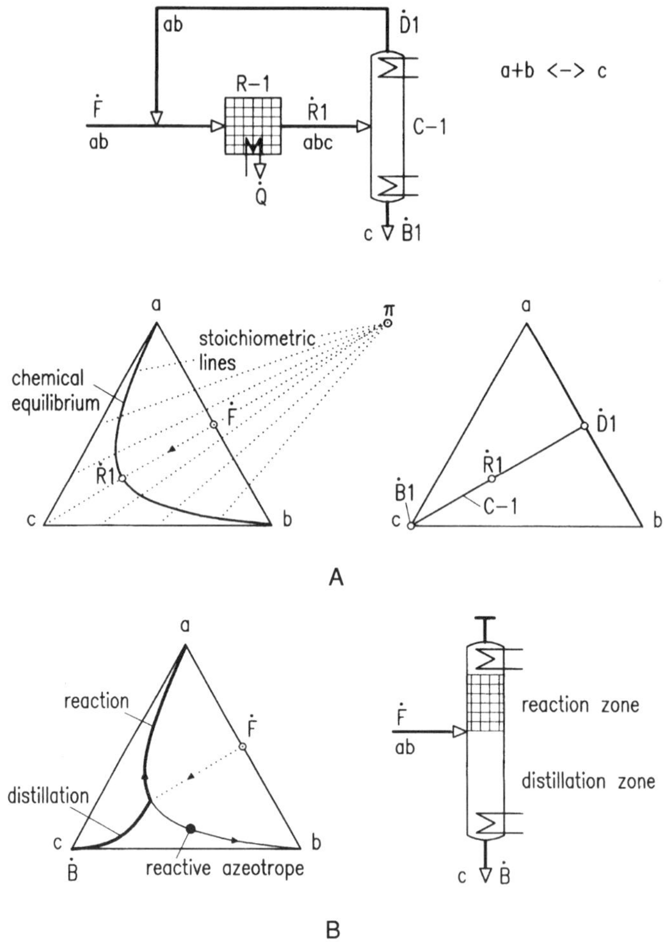

__Fig. 5-64.__ (A) Process with sequential realization of reaction and distillation. (B) Process with simultaneous realization of reaction and distillation.

Since not all of the educt is converted into the product the effluent of the reactor is a ternary mixture that needs to be split into a pure product fraction and an educt fraction. Figure 5-64A presents the conventional process where the reaction and the separation are established sequentially. The stoichiometric educt is fed into the reactor R-1 to be converted. The effluent $\dot{R}1$ consisting of *c* and some unconverted *a* and *b* is subsequently separated in distillation column C-1 into an overhead fraction $\dot{D}1$ and a bottom fraction $\dot{B}1$. The overhead fraction is recycled into the reactor. If the chemical equilibrium lies very close by the educt a large stream must be recycled what makes the process unattractive.

Figure 5-64B shows the process with simultaneous reaction and distillation in a counter currently operated column. The unreacted educts *a* and *b* being the low boilers move upwards in the catalytic column section to be further converted. Since the high boiling reaction product c moves downward in the column a total conversion of a stoichiometric educt is finaly reached. The product is purified in the lower, non-catalytic section of the column. The concentration profile within the column is sketched in the triangular diagram of Fig. 5-64B by a bold line. In the reactive section (with catalyst) the liquid states always lie on the chemical equilibrium line. In the non-reactive section (without catalyst) they follow the concentration profile of a non-reactive distillation line that originates from the high boiler. The endpoints of the internal concentration profile do not lie on a straight line through the feed $\dot{F}$. However, a projection of the bottoms $\dot{B}$ along the stoichiometric line to the a-b side of the triangle gives the feed $\dot{F}$ proving that the material balance is met.

The advantage of reactive distillation is that no external recycle from the distillation column to the reactor is required since there exists a very effective internal recycle. In case of an exothermic reaction the heat of reaction is automatically utilized for distillation.

The merits of reactive distillation are even larger in systems with non-reactive azeotropes. It is supposed in Fig. 5-65A that the chemical equilibrium lies very close by a distillation border. Hence, only a small part of the product c can be separated and, in turn, the recycle stream becomes very large. In reactive distillation, the distillation border is no longer a barrier and, consequently, the educts are completely converted into the desired product c, see Fig. 5-65B.

5.7.3.2 Decomposition Reactions

The following type of decomposition reactions is considered:

$$c \leftrightarrow a + b \tag{5.18}$$

The products a and b have to be completely separated.

A conventional process is shown in Fig. 5-66A. The feed is partially converted in the reactor R-1. The ternary mixture is fed into column C-1 to be split into a bottom fraction $\dot{B}1$ that is recycled and a binary overhead fraction $\dot{D}1$ that is separated into pure a and b, respectively, in column C-2. Two distillation columns are required for the downstream processing.

In a reactive distillation process, Fig. 5-66B, the reaction and the separation are established in a single, fully catalysed column. The condition for such a simple process is that no reactive azeotrope exists. This condition is met when the chemical equilibrium lies very close by the right hand side of the triangle in Fig. 5-66B, see Section 5.7.1.4.

The advantages of reactive distillation are especially large if non-reactive azeotropes exist in the system that hinder the separation of the products. Such an example is presented in Fig. 5-67A. The reaction is

$$c \leftrightarrow a + b \tag{5.19}$$

The three species form two binary azeotropes that are connected by a border distilla-

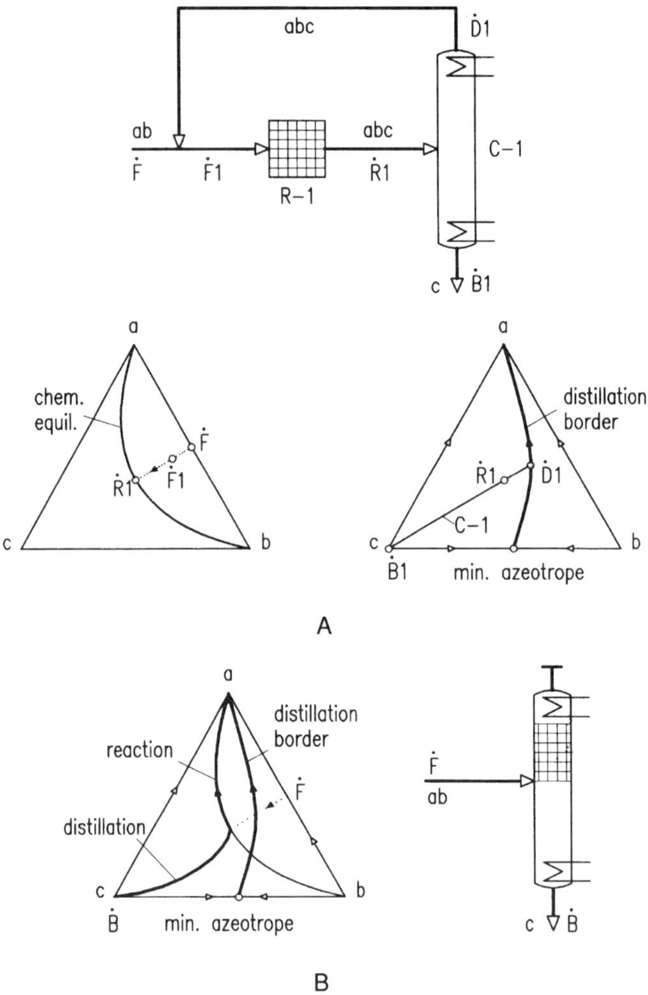

Fig. 5-65. *(A) Process with sequential realization of reaction and distillation in a system with a distillation border. (B) Realization of the process of Fig. 5-65A by reactive distillation.*

tion line. In a conventional process the chemical equilibrium is reached in the reactor R-1. The ternary effluent is fed into column C-1 to be split into an overhead fraction $\dot{D}1$ (mixture of *a* and *b*) and a bottom fraction $\dot{B}1$ (pure high boiler *c*) that is recycled into the reactor. The fraction $\dot{D}1$ is fed into column C-2 where the low boiler *a* is separated as bottoms. The overhead fraction $\dot{D}2$ has the concentration of the *a*-*b*-azeotrope. To enable the separation it is mixed with a low boiling entrainer *e* (fraction $\dot{D}4$) to give a mixture $\dot{M}$ that lies on the left hand side of the distillation border. The stream $\dot{M}$ is split in column C-3 into the bottom fraction $\dot{B}3$ (pure *b*) and an

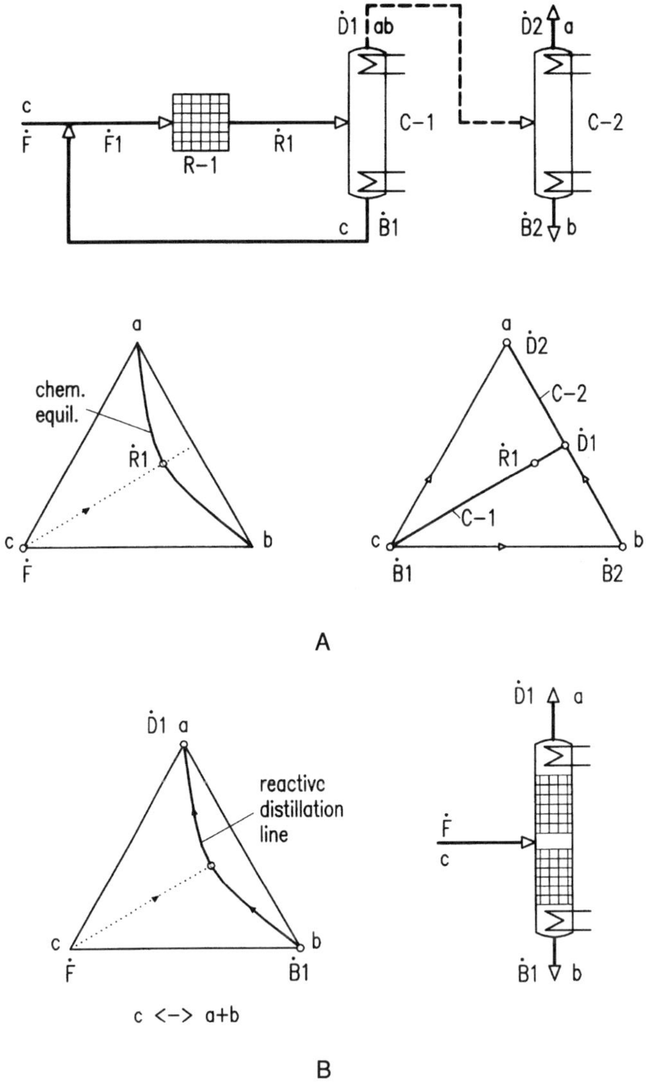

Fig. 5-66. (A) Conventional process for a decomposition reaction c ↔ a + b *and separation of the products* a *and* b. *(B) Realization of the process of Fig. 5-66A by reactive distillation.*

overhead fraction $\dot{D}3$ that lies at the concave side of the distillation border. Fraction $\dot{D}3$ is finally separated in column C-4 into fraction $\dot{D}4$ and $\dot{B}4$. The two fractions are recycled to column C-3 and column C-2, respectively. The process effected in columns C-2, C-3 and C-4 is in essence identical with the general process for the separation of azeotropic mixtures developed in Section 5.6.2.2. Depending on the

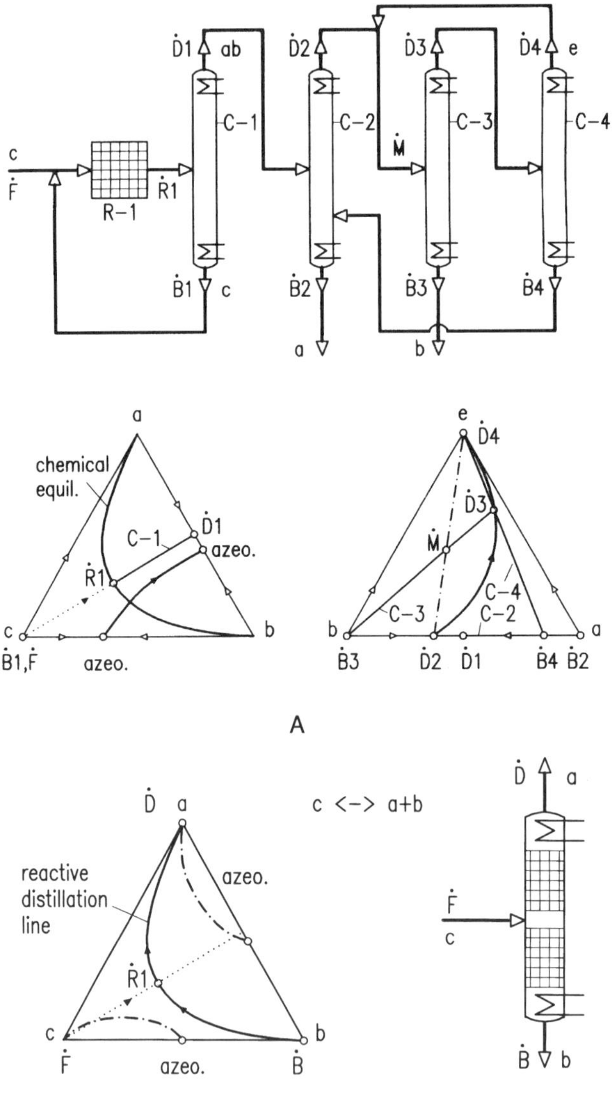

Fig. 5-67. *(A) Conventional process for the decomposition reaction* c ↔ a + b. *The separation of the products* a *and* b *is very complex due to the existence of an* a-b-*minimum azeotrope. (B) Realization of the process of Fig 5-67A by reactive distillation.*

system properties the separation of the azeotropic mixture *ab* can also be effected by azeotropic or extractive distillation. In each version of the process one reactor and four distillation columns are required.

Figure 5-67B presents the process with reactive distillation. Since no reactive azeotrope is formed the internal concentration profile runs without any barrier from the intermediate boiler *b* to the low boiler *a*. No non-reactive section is neccessary since the desired products are an origin and a terminus, respectively, of the reactive distillation line. The educt *c* is completely converted and separated into the pure products *a* and *b* in a single column.

5.7.3.3 Side Reactions In the simultaneous realization of reaction and distillation in a countercurrently operated column the products of the reaction are steadily removed from the reaction zone. This is, in essence, the reason why the educts are completely converted into the products even in reversible reactions. The steady removal of the reaction products can be utilized, furthermore, for an effective suppression of unwanted side reactions. The following reaction mechanism is assumed:

$$a + b \leftrightarrow c \tag{5.20A}$$

$$a + c \rightarrow d \tag{5.20B}$$

The substance *c* is the desired product and the substance *d* the unwanted byproduct. By instantaneous reaction performed in a totally mixed reactor the chemical equilibrium of all four species will be inevitably reached, i.e., a significant amount of the unwanted species *d* will be present in the effluent, see Fig. 5-62. The process with sequential realization of reaction and distillation is shown in Fig. 5-68A. A reactor and two distillation columns are required and two streams are recycled to the reactor.

Figure 5-68B presents the process with simultaneous reaction and distillation. The educts *a* and *b* are fed into the column. It is important to feed the high boiling educt *b* above and the low boiling educt *a* below the catalyzed section of the column. This leads to the favorable situation that in the reaction zone the concentration of *b* in the liquid phase is very high. Most of the low boiler *a* concentrates in the vapor phase. The small percentage of low boiler *a* that is dissolved in the liquid is converted into the product *c*. This situation obviously favors the main reaction. The high boiling product *c* is steadily removed from the reactive section and reaches high concentrations in the lower uncatalyzed zone only. Hence, the main reaction (5.20A) with the desired product *c* is enhanced and the side reaction (5.20B) with the unwanted product *d* is effectively suppressed.

5.7.4 Industrial Processes of Reactive Distillation

The simultaneous realization of reaction and distillation in a counter currently operated column is a very promising alternative to sequential reaction and distillation. This so called reactive distillation offers some important advantages:

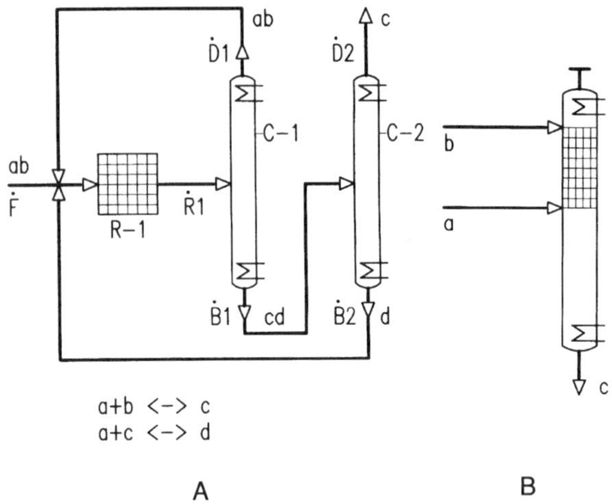

Fig. 5-68. *(A) Conventional process for the reaction* a + b ↔ c *and the side reaction* a + c ↔ d. *Two distillation columns are neccessary to regain the main product* c *in pure form. (B) Realization of the process of Fig. 5-68A by reactive distillation. The side reaction is effectively suppressed by feeding the low boiling educt* a *below the high boiling educt* b.

- Utilization of an exothermic heat of reaction for distillation.
- Total conversion of the educts even in reversible reactions.
- Simple down stream processing of the reactor effluents.
- Overcoming of separation barriers like azeotropes or distillation border lines.

A large number of reactions favorably performed by reactive distillation are described in the literature. Table 5-10 lists some selected examples. A few industrially important processes are described in detail in the next sections.

5.7.4.1 Synthesis of MTBE Methyl-tertiary-butyl-ether (MTBE, $C_5H_{12}O$) is a very effective antiknock additive to gasoline that is produced in amounts larger than $20 \cdot 10^6$ tons per year [Rehfinger and Hoffman 1990]. The educts for MTBE synthesis are mainly i-butene (C_4H_8) and methanol (CH_3OH). The reaction is catalysed either by a homogeneous (e.g., sulfuric acid) or a heterogeneous (strong acidic macroporous ion exchange resin) catalyst. The reaction mechanism is:

$$C_4H_8 + CH_3OH \leftrightarrow C_5H_{12}O \qquad (5\text{-}21)$$

In most cases there is some n-butane present in the feed mixture that is inert for this reaction.

Figure 5-69 shows some important properties of the system that forms three binary minimum azeotropes. The inert species, n-butane, forms the peak of the tetrahedron of Fig. 5-69. The existence of the three binary azeotropes leads to a

Table 5-10. Selected Systems for Reactive Distillation After /Doherty and Buzad 1992, Sundmacher et al. 1994, Ung and Doherty 1995/.

Methyl acetate from methanol and acetic acid	Agreda et al. 1990
Ethyl acetate from ethanol and acetic acid	Komatsu 1977
Butyl acetate from butanol and acetic acid	Hartig, Regner 1971
MTBE from methanol and i-butene	Flato, Hoffmann 1992 Smith 1990
ETBE from ethanol and isobutene	Thiel et al. 1997
TAME from methanol and 2-methyl-2-butene	Bravo, Pyhälahti 1992 Thiel et al. 1997
TAME from methanol and 2-methyl-1-butene	Bravo, Pyhälahti 1992
Ethylenglycol from ethylene oxide and water	Gu, Ciric 1992
Isooctane from isobutane and 1-butene	Huss, Kennedy 1990
Ethylene benzene from benzene and ethylene	Smith et al. 1991
Cumene from benzene and propylene	Shoemaker, Jones 1987
Tert.-butyl-alcohol from isobutene and water	Velo et al. 1988
Nylon 6.6 prepolymer from adipic acid and hexamethylenediamine	Jaswal, Pugi 1975
Tert.-amyl alcohol from isoamylene and water	Gonzales, Fair 1997

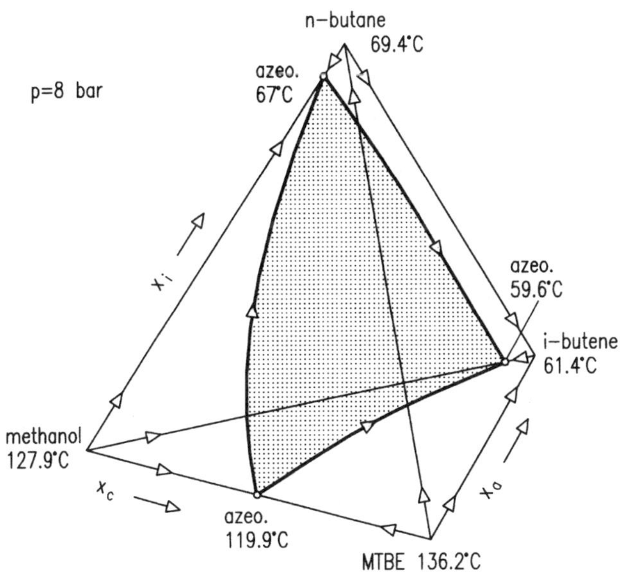

Fig. 5-69. *System properties of the MTBE synthesis process.*

border distillation area that can not be crossed by distillation. Consequently, sequential realization of reaction and distillation is very difficult.

A reactive distillation process for MTBE synthesis is presented in Fig. 5-70. The reaction parameters of the system are well known and, for instance, given by [Espinosa et al. 1995, Thiel et al. 1997]. The chemical equilibrium forms a section of a cone mantle as described in Section 5.7.2.2. All reactive distillation lines run on the cone surface. The origin of the reactive distillation lines is the high boiling educt methanol, the terminus either the n-butane/methanol-azeotrope or the i-butene. There exists a reactive border distillation line that divides the cone mantle in two sections with different termini of distillation lines.

The process is effected at a pressure of 8 bar (or 5 bar) in a column with 14 trays. The upper 8 trays are equipped with a heterogeneous catalyst [Bessling et al. 1997]. The i-butene is totally converted into MTBE that is withdrawn as bottoms from the column. The overhead fraction is the minimum azeotrope formed by n-butane and small parts of methanol. The concentration profile in the column is qualitatively shown in Fig. 5-70A. It follows a reactive distillation line in the upper section and a non-reactive distillation line that originates from pure MTBE in the lower section.

As described earlier the material balance around the column forms a straight line only in the transformed coordinates. Therefore, the state of the product MTBE has to be projected along the stoichiometric line into the n-butane/i-butene/methanol plane. In this plane the feed $\dot{F}$, the overhead $\dot{D}$ and the bottom product $\dot{B}$ are colinear.

The reactive distillation lines in transformed coordinates are shown in Fig. 5-70B. This diagram is more clear but the product MTBE is not visible.

5.7.4.2 Synthesis of Mono-Ethylene Glycol

The synthesis of mono-ethylene glycol ($C_2H_6O_2$) from ethylene oxide (C_2H_4O) and water (H_2O) has been studied [Ciric and Gu 1994]. The main reaction is:

$$C_2H_4O + H_2O \leftrightarrow C_2H_6O_2 \tag{5.22A}$$

A side reaction generates the unwanted by-product di-ethylene glycol ($C_4H_{10}O_3$):

$$C_2H_4O + C_2H_6O_2 \leftrightarrow C_4H_{10}O_3 \tag{5.22B}$$

The boiling points of the four involved species increase from the low boiler ethylene oxide (a) to water (b), ethylene glycol (c) and di-ethylene glycol (d). Hence, the situation is the same as described in section 5.7.3.3. The reaction is sufficiently fast even without a catalyst.

Figure 5-71 shows the process that is established in a single column with 10 equilibrium stages. The low boiling educt ethylene oxide is fed into the middle and the high boiling water into the top of the column. By rigorous simulation the authors proved that in the upper column section inspite of stoichiometric amounts of the educts the concentration of water in the liquid phase is about 200 times higher than the concentration of ethylene oxide. The ethylene oxide is present in the vapor phase

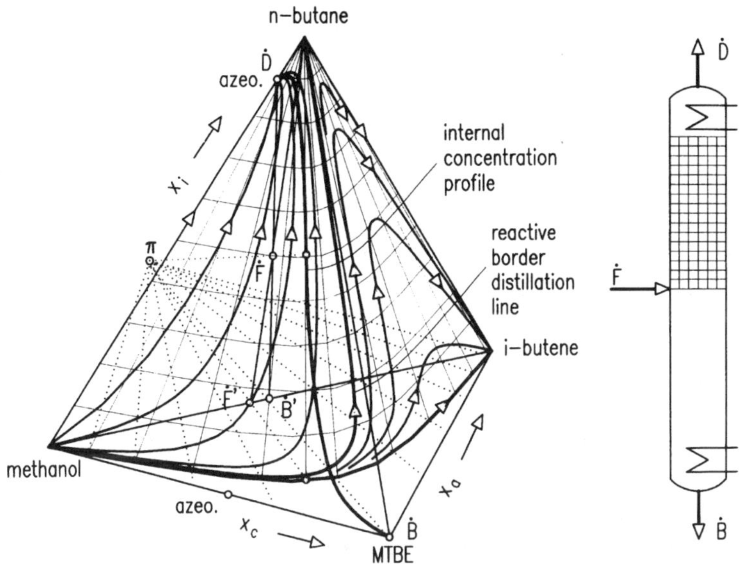

A

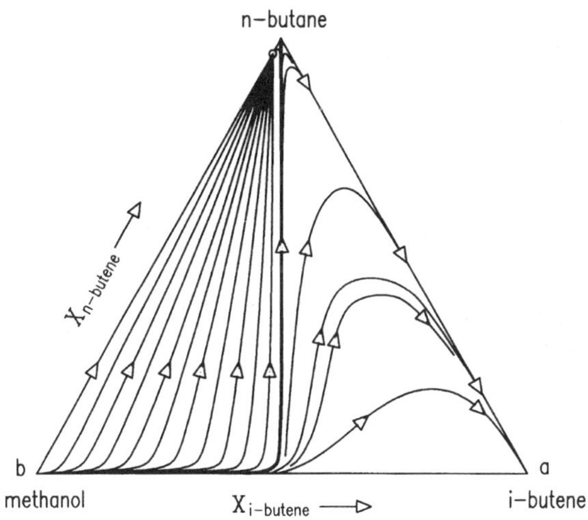

B

Fig. 5-70. *(A) Reactive distillation lines for MTBE synthesis. (B) Reactive distillation lines for MTBE synthesis in transformed coordinates [Espinosa et al. 1995].*

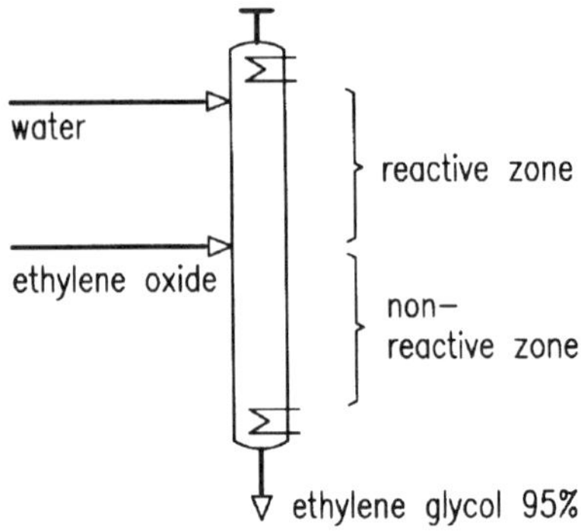

Fig. 5-71. Synthesis of ethylene glycole in a reactive distillation column.

mainly. These conditions favor the main reaction (5.22A). Liquid concentrations of ethylene glycol are lower than 5% in the reactive section of the column. In the lower column section the ethylene glycol is concentrated up to 95%. The rest is diethylene glycol with some unconverted water. The authors state that an excess of water as high as 60% would be neccessary to get the same selectivity in a plug flow reactor. However, this excess water has to be separated by an additional distillation column.

5.7.4.3 Synthesis of TAME Since there is a shortage of isobutene for MTBE synthesis tertiary-amyl-methyl-ether (TAME) is used as an antiknock additive to gasoline in increasing amounts. The advantage of TAME is its synthesis from C_5-olefines.

A modern process for TAME synthesis is presented in Fig. 5-72 [Hoffmann et

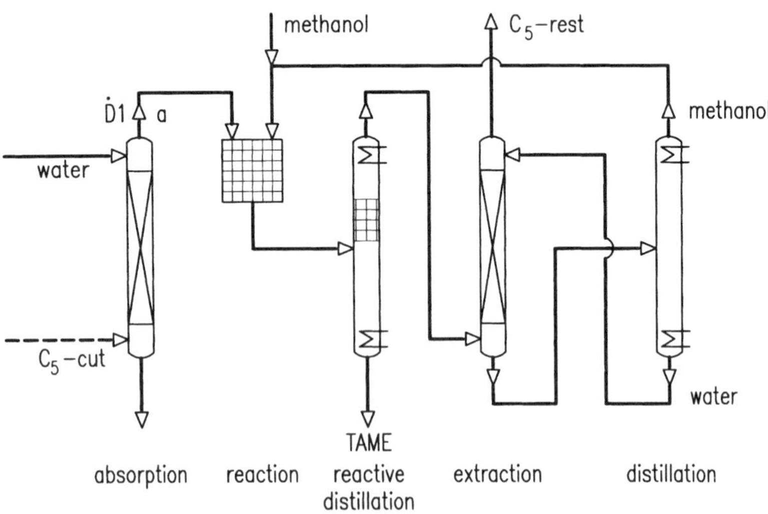

Fig. 5-72. Process for synthesis of TAME.

al. 1997]. The C_5-olefine cut contains between 15 and 50 wt.-% isoamylene that reacts with methanol. In the first step of the process impurities that may poison the catalyst are removed by a water scrubber. The major part (70%) of the conversion takes place in the reactor R-1 with a sulfonic acid ion exchange resin bed. The final conversion is established in a reactive distillation column with a catalytic section above the feed. The high boiler TAME is the bottom fraction of the column. The overhead fraction is the minimum azeotrope formed by C_5-olefines and methanol. The methanol is separated from the olefine by extraction with water. The raffinate is split by distillation into an overhead fraction methanol and a bottom fraction water. Both fractions are recycled within the process to the reactor and the extractor, respectively.

The physical and chemical properties of the systems are described in detail [Hoffmann et al. 1997]. There exists a MeOH/C_5-olefine and a methanol/TAME minimum azeotrope that are connected by a border distillation line. The reactive distillation column is, in the reactive section, equipped with a catalytic active random packing. Below and above of the catalytic section a inert random packing is installed.

LITERATURE

Abu-Eishuh, S. I.; and Luyben, W. L.: ''Design and Control of a Two-Column Azeotropic System,'' *Ind. Eng. Chem. Process Des. Dev.* 24 (1985) No. 1, pp. 132/140.

Agrawal, R.; Woodward, D. W.; Ludwig, K. A.; and Bennett, D. L.: ''Impact of Low Pressure Drop Structured Packing on Air Distillation Preprint Distillation and Absorption 1992,'' *Institution of Chemical Engineers,* Rugby, 1992.

Agreda, V. H.; Parin, L. R.; and Heise, W. H.: ''High-purity methyl acetate via reactive distillation,'' *Chem. Eng. Progr.* 86 (1990) 2, pp. 40/46.

Baker, R. W.: ''Pervaporation,'' *Membrane Separations Systems* No. DE-AC01-88ER30133 (1990) U.S. Department of Energy.

Baldus, H.; Baumgaertner, K.; Knapp, H.; and Streich, M.: ''Verfluessigung und Trennung von Gasen,'' *Winnacker Kuechler: Chemische Technologie* Bd. 3, Carl Hauser Verlag München Wien 1983, pp. 567/650.

Barbosa, D.; and Doherty, M. F.: ''The Influence of Equilibrium Chemical Reactions on Vapor-Liquid Phase Diagrams,'' *Chem. Eng. Sci.* 43 (1988), pp. 529/540.

Berg, L.: ''Selecting the Agent for Distillation,'' *Chem. Eng. Progress* 65 (1969) No. 9, pp. 52/57.

Berg, L.; Harrison, J. M.; and Montgomery, C. W.: ''Azeotropic Dehydration of Pyridine and its Homologs,'' *Ind. Eng. Chem.* 37 (1945) No. 6, pp. 585/587.

Bessling, B.; Schembecker, G.; and Simmrock, K. H.: ''Design of Processes with Reactive Distillation Line Diagrams,'' *Ind. Eng. Chem. Res.* 36 (1997) No. 8, pp. 3032/3042.

Bindal, R. C.; and Misra, B. M.: ''Separation of Binary Liquid Systems by Sorption–A Comparison with Pervaporation,'' *Separation Science and Technology* 21 (1986) No. 10, pp. 1047/1058.

Bittrich, H. J.: *Trennung von Kohlenwasserstoffen mit Selektivloesungsmitteln.* VEB Verlag, Leipzig, 1987.

Black, C.: ''Distillation Modeling of Ethanol Recovery and Dehydration Processes for Ethanol and Gasohol,'' *Chem. Eng. Progress* 76 (1980) No. 9, pp. 78/85.

Blaß, E.: *Entwicklung Verfahrenstechnischer Prozesse.* Verlag Salle + Sauerländer, Frankfurt, 1989.

Bravo, J. L.; and Pyhälahti, A.: ''Investigations in a catalytic distillation plant: VLE, kinetics and mass transfer issues,'' AIChE Annual Meeting, Miami, FL, 1992.

Bril, Z. A.; Mozzhukhin, A. S.; and Petlyuk, F.B.: ''Investigations of Optimal Conditions of Heteroazeotropic Rectification,'' *Theor. Found.* 11 (1977) No. 6, pp. 675/680.

Cairns, B. P.; and Furzer, L. A.: ''Three-Phase Azeotropic Distillation, Experimental Results,'' *I. Chem. E. Symposium Series* 104 (1987) pp. B505/B519.

Carton, A.; Gonzales, G.; de la Torre, A. I.; and Cabezas, J. L.: ''Separation of Ethanol/Water Mixtures Using 3A Molecular Sieve,'' *J. Chem. Tech. Biotechnol* 39 (1987) pp. 125/132.

Ciric, A. R.; and Gu, D.: ''Synthesis of Nonequilibrium Reactive Distillation Processes by MINLP Optimization,'' *J. AIChE* 40 (1994) No. 9, pp. 1479/1487.

Coogler, W. W.: ''Butadiene Recovery Process Employs New Solvent System,'' *Chemical Engineering* 7 (1967) No. 7, pp. 70/72.

Cooper, D.: ''Do You Value Steam Correctly?'' *Hydrocarbon Processing,* July 1989, pp. 44/47.

Doherty, M. F.; and Buzad, G.: ''Reactive Distillation by Design'' *Trans I Chem E.* Vol. 70 (1992) Part A, pp. 448/458.

Doherty, M. F.; and Caldarola, G. A.: ''Design and Synthesis of Homogeneous Azeotropic Distillations,'' *Ind. Eng. Chem Fund.* 24 (1985) pp. 474/485.

Drew, J. W.: ''Solvent Recovery,'' *Handbook of Separation Techniques for Chemical Engineers,* P. A. Schweitzer, McGraw-Hill, New York, 1979.

Drew, J.; and Probst, M.: *Tall Oil* Pulp Chemicals Association, New York 1981

Dutta, B. K.; and Sikdar, S. K.: ''Separation of Azeotropic Organic Liquid Mixtures by Pervaporation,'' *AIChE Journal* 37 (1991) No. 4, pp. 581/588.

Ellerbe, R. W.: ''Steam Distillation/Strippen,'' *Handbook of Separation Techniques for Chemical Engineers* P. A. Schweitzer, McGraw-Hill, New York, 1979.

Espinosa, J.; Aguirre, P.; and Perez, G.: ''The Product Composition Regions of Single-Feed Reactive Distillation Columns,'' *Ind. Eng. Chem. Res.* 34 (1995) pp. 853/861.

Gonzales, J. C.; and Fair, J. R.: ''Preparation of Tertiary Amyl Alcohol in a Reactive Distillation Column, Part 1,'' *Ind. Eng. Chem. Res.* (1997) No. 36, pp. 3833–3844.

Gonzales, J. C.; Subawalla, H.; and Fair, J. R.: ''Preparation of Tert-Amyl Alcohol in a Reactive Distillation Column, Part 2,'' *Ind. Eng. Chem. Res.* (1997) No. 36, pp. 3845–3853.

Flato, J.; and Hoffmann, U.: ''Development and Start-up of a Fixed Bed Reaction Column for Manufacturing the Antiknock Enhancer MTBE,'' *Chem. Eng. Technol.,* 15 (1992) pp. 193.

Fritzweiler, R.: ''Verfahren zur Herstellung von wasserfreiem Spiritus,'' *VDI Zeitschrift* 48 (1938) No. 11, pp. 1373/1378.

Geriche, D.: ''Konzentrieren von Salpetersäure mit Schwefelsäure,'' *Schott Information,* Schott GmbH, Mainz, Germany, 1973.

Gmehling, J.; and Kolbe, B.: *Thermodynamik.* Georg Thieme Verlag Stuttgart, New York, 1988.

Grewer, T.: "Trennung von Chlorwasserstoff und Wasser durch Extraktivdestillation mit Schwefelsäure," *Chem.-Ing.-Tech.* 43 (1971) No. 11, pp. 655/658.

Hafslund, E. R.: "Propylene-Propane Extractive Distillation," *Chemical Engineering Progress* 65 (1969) No. 9, pp. 58/64.

Happe, J.; Cornell, P. W.; and Eastman, D.: "Extractive Distillation of C4-Hydrocarbons Using Furfural," *AIChE Journal* (1946) No. 4, pp. 189/214.

Hebel, H.; and Kaspareit, M.: "Die Gewinnung von Butadien nach dem Difex-Verfahren," *Chemische Technik* 23 (1971) No. 7, pp. 419/423.

Hegner, B.; Hesse, D.; and Wolf, D.: "Möglichkeiten der Berechnung bei heteroazeotroper Destillation," *Chem.-Ing.-Tech.* 45 (1973) No. 14, pp. 942/945.

Herfurth, H.; Meirelles, A.; and Weiß, S.: "Azeotropdestillation von Ethanol/Wasser mit Cyclohexan als Schleppmittel," *Chem. Techn.* 39 (1987) No. 8, pp. 331/334.

Hoffmann, E. J.: *Azeotropic and Extractive Distillation.* Interscience, New York 1964.

Hoffmann, U.; Krummradt, H.; Rapmund, P.; and Sundmacher, K.: "Erzeugung des Kraftstoffethers TAME durch Reaktivdestillation," *Chemie-Ingenieur-Technik 69* (1997), No. 4, pp. 483/487.

Hunsmann, W.; and Simmrock, K. H.: "Trennung von Wasser, Ameisensäure und Essigsäure durch Azeotrop-Destillation," *Chem.-Ing.-Tech.* 38 (1966) No. 10, pp. 1053/1059.

Huss, A.; and Kennedy, C. R.: "Hydrocarbon processes comprised of catalytic distillation using Lewis acid promoted inorganic oxide catalyst systems," US Patent No. 4.935.557 (1989).

Jaswal, I.; and Pugi, K.: "Preparation of Polyamides by Continous Polymerization," US Patent 3.900.450 (1975).

Johannisbauer, W.; Jeromin, L.: "Structured Column Packings in the Oleochemical Industry," *Distillation Symposium,* Birmingham, 1992, pp. B77/B83.

Keyes, D. B.: "The Manufcture of Anhydrous Ethylalcohol," *Ind. and Eng. Chem.* 21 (1929) No. 11, pp. 998/1001.

Komatsu, H.: "Application of the relaxation method for solving reacting distillation problems," *J. Chem. Eng. Japan* 10 (1977) pp 200/205.

Lackner, K.: "N-formyle-morpholine, the Solvent for Atomatics Recovery and its Ranges Of Applicability *Petrochem* 34, (1981), No. 1, pp. 26/30, Erdoel, Kohle, Erdgas.

Mueller, E.: "Gewinnung von Aromaten durch Extraktivrektifikation," *vt "Verfahrenstechnik"* 14 (1980) No. 9, pp. 551/555.

Okasinski, M. J.; and Doherty, M. F.: "Thermodynamic Behavior of Reactive Azeotropes," *J. AIChE* 43 (1997) No. 9, pp. 2227/2238.

Othmer, D. F.: "Azeotropic Distillation Part I: Background Theory," *Verfahrenstechnik* 8 (1974) No. 3, pp. 94/99.

Othmer, D. F.: "Azeotropic Distillation Part II: Uses, Examples, History," *Verfahrenstechnik* 8 (1974) No. 4, pp. 118/123.

Perry, R. H.; Green, D. W.; and Maloney, J. O.: *Perry's Chemical Engineer Handbook.* McGraw-Hill, New York, 1984.

Peters, W. D.; and Rogwers, R. S.: "Improved Furfural Extraction Process," *Hydrocarbon Processing* 47, (1968) No. 11, pp. 131/134.

Pilhofer, T.: "Energiesparende Alternativen zur Rektifikation bei der Rückgewinnung organischer Stoffe aus Lösungen," *vt "Verfahrenstechnik"* 17 (1983) No. 9, pp. 547/549.

Pucci, A.; Mihitenho, P.; and Asselineau, L.: ''Three-Phase Distillation. Simulation and Application to the Separation of Fermentation Products,'' *Chem. Eng. Sci.* 41 (1986), No. 3, pp. 485/494.

Rautenbach, R.; Albrecht, R.: *Membrane Processes.* John Wiley, New York, 1989.

Rautenbach, R.; Knauf, R.; Struck, A.; and Vier, J,: ''Simulation and Design of Membrane Plants with AspenPlus,'' *Chem. Eng. Technol.* 19 (1996) No. 5, pp. 391/397.

Rehfinger, A.; and Hoffmann, U.: ''Kinetics of Methyl Butyl Ether Liquid Phase Synthesis Catalysed by Ion Exchange Resin,'' *Chem. Eng. Sci.* 45 (1990) pp 1605/1617.

Rizvi, S. S. H.; and Heidemann, R. A.: ''Vapor-Liquid Equilibria in the Ammonia/Water System,'' *Ing. Chem. Eng.* Data 32 (1987) No. 2, pp. 183/191.

Ruckenstein, E.; and Park, J. S.: ''The Separation of Water/Ethanol Mixtures by Pervaporation through Hydrophilic-Hydrophobic Composite Membranes,'' *Journal of Applied Polymer Science,* Vol. 40, (1990) pp. 213/220.

Sattler, K.: *Thermische Trennverfahren.* VCH Verlagsgesellschaft, Weinheim, 1988.

Seader, J. D.: ''Distillation,'' *Perry's Chemical Engineer Handbook,* Perry, R. H.; Green, D. W.; Maloney, J. O., McGraw-Hill, New York, 1984.

Schoenmakers, H.: ''Alternativen zur Aufarbeitung desorbierter Lösungsmittel,'' *Chem.-Ing.-Tech.* 56 (1984), No. 3, pp. 250/251.

Schweitzer, P. A.: *Handbook of Separation Techniques for Chemical Engineers.* McGraw-Hill Book Company, New York, 1979.

Shoemaker, J. D.; and Jones, E. M.: ''Cumene by catalytic distillation,'' *Hydrocarbon Processing* (1987) 6, pp. 57/58.

Smith, L. A.: ''Method for the preparation of methyl tertiary butyl ether,'' US Patent No. 4.978.807, 1990.

Smith, L. A.: Jones, E. M.; and Hearn, D.: ''Catalytic distillation—a new chapter in unit operations,'' AIChE Spring Meeting, Houston, TX, 1991.

Sorensen, J. M.; and Arlt, W.: ''Liquid-Liquid Equilibrium Data Collection,'' *DECHEMA,* Frankfurt, since 1979 (Continuing Series).

Sowerby, B.; and Crittenden, B. D.: ''A Vapor Phase Adsorption and Desorption Model for Drying the Ethanol/Water Azeotrope in Small Columns,'' *Trans I Chem E* 69 (1991) Part A, pp. 3/13.

Stage, H.: ''Richtlinien für Einsatz, Auslegung und Betrieb von Fallfilmrohrverdampfern zur Aufarbeitung thermisch sowie oxidationsempfindlicher hochsiedender Flüssigkeitsgemische,'' *Fat. Sci. Technol* 90 (1988) No. 10, pp. 399/411.

Stichlmair, J.: ''Chlorwasserstoffabsorption,'' *Chemie-Anlagen + Verfahren* (1979) No. 1, pp. 40/46.

Stichlmair, J.: ''Distillation and Rectification,'' *Ullmann's Encyclopedia of Industrial Chemistry,* Vol. B3, pp. 4-1/4-94, VCH Verlagsgesellschaft, Weinheim, 1988.

Stichlmair, J.; Herguijuela, J. R.: ''Separation Regions and Prozesses of Zeotropic and Azeotropic Ternary Distillation,'' *AIChE Journal* 38 (1992) No. 10, pp. 1523/1535.

Streich, M.: ''Exergieverluste bei thermischen Trennungen,'' *vt ''Verfahrenstechnik''* 9 (1975) No. 5, pp. 240/243.

Strigle, R. F.: *Random Packings and Packed Towers, Design and Application.* Gulf Publishing Company, Book Division, Houston, 1987.

Tassios, D. P.: ''Extractive and Azeotropic Distillation,'' *Adv. Chem. Ser.* 115, (1972).

Thiel, Ch.; Sundmacher, K.; and Hoffmann, U.: ''Residue curve maps for heterogeneously catalysed reactive distillation of fuel ethers MTBE and TAME,'' *Chem. Engi. Sci.* 52 (1997) No. 6, pp. 993/1005.

Udriot, H.; Araque, A.; and von Stockar, U.: ''Azeotropic Mixtures may be Broken by Membrane Distillation,'' *Chem. Eng. J and the biochem. Eng. J., Laus.* 54 (1994) No. 2, pp. 87/93.

Ung, S.; and Doherty, M. F.: ''Vapor-Liquid Phase Equilibrium in Systems with Multiple Chemical Reactions,'' *Chem. Eng. Sci.* 50 (1995) No. 1, pp. 23/48.

Ung, S.; and Doherty, M. F.: ''Synthesis of Reactive Distillation Systems with Multiple Equilibrium Chemical Reactions,'' *Ind. Eng. Chem. Res.* 34 (1995) No. 8, pp 2555/2565.

Ushio, S.: ''Extract Isoprene with DMF,'' *Chemical Engineering* 79 (1972) No. 5, pp. 82/83.

Velo, E.; Puigjaner, L.; and Rescasens, F.: ''Inhibition by-product in the liquid phase hydration of isobutene to tert-butyl alcohol: kinetics and equilibrium studies,'' *Ind. Eng. Chem. Res.* 27 (1988) pp. 2224/2231.

Volkamer, K.; Schneider, K. J.; and Lindner, A.: ''Entwicklungsarbeiten am Butadienverfahren der BASF,'' *Petrochem* 34 (1981) No. 8, pp. 343/346, Erdoel, Kohle, Erdgas.

Watkins, R. N.: *Petroleum Refinery Distillation.* Gulf Publishing Company, Houston, 1979.

Westphal, G.: *Kombiniertes Adsorptions-Rektifikationsverfahren zur Trennung eines Flüssigkeitsgemisches,* German Patent DE 3712291 A1, 1987.

Widago, S.; and Seider, W. D.: ''Azeotropic Distillation,'' *AIChE Journal* 42 (1996), No. 1, pp. 96–130.

Wunder, R.: ''Abtrennung und Rückgewinnung von anorganischen Stoffen durch Absorption,'' *Stofftrennverfahren in der Umwelttechnik,* Preprints, GVC-Gesellschaft für Verfahrenstechnik und Chemieingenieurwesen, Düsseldorf, 1990.

6

Multistage Batch Distillation

6.1 Batch Distillation of Binary Mixtures
- 6.1.1 Operation with Constant Reflux
  - 6.1.1.1 Infinite Number of Stages
- 6.1.2 Operation with Constant Distillate Composition
  - 6.1.2.1 Infinite Number of Stages
  - 6.1.2.2 Mixtures with Partial Immiscibility in the Liquid Phase
- 6.1.3 Operation with Minimum Energy Input
  - 6.1.3.1 Influence of Number of Equilibrium Stages on the Hamiltonian Function
- 6.1.4 Comparison of Energy Requirement for Different Modes of Distillation

6.2 Batch Distillation of Ternary Mixtures
- 6.2.1 Zeotropic Mixtures
- 6.2.2 Azeotropic Mixtures

6.3 Batch Distillation of Multicomponent Mixtures

6.4 Influence of Column Liquid Holdup on Batch Distillation

6.5 Processes for Separating Zeotropic Mixtures by Batch Distillation
- 6.5.1 Total Slop Cut Recycling
- 6.5.2 Binary Distillation of Accumulated Slop Cuts
- 6.5.3 Recycling of Slop Cuts at the Appropriate Time

6.6 Processes for Separating Azeotropic Mixtures by Batch Distillation
- 6.6.1 Processes in One Distillation Field
- 6.6.2 Processes in Two Distillation Fields
- 6.6.3 Process Simplifications
- 6.6.4 Hybrid Processes
  - 6.6.4.1 Azeotropic Batch Distillation
  - 6.6.4.2 Extractive Batch Distillation

Literature

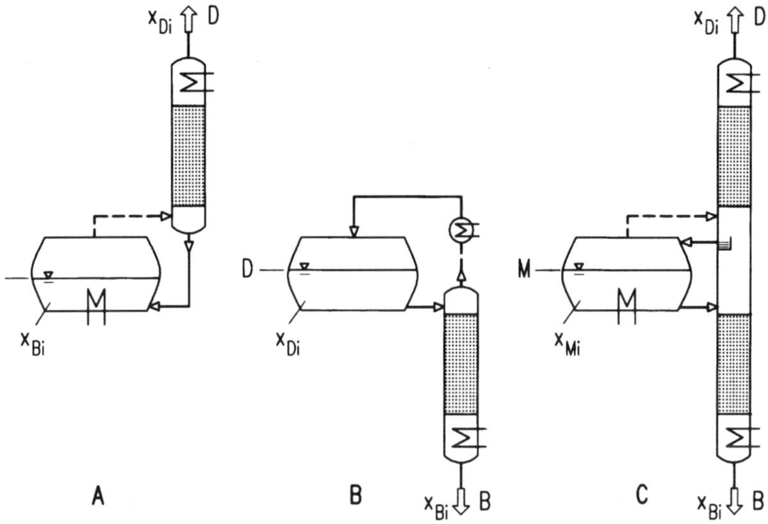

Fig. 6-1. *Schematics of batch distillation devices: (A) Bottom vessel with rectifying column, (B) top vessel with stripping column, (C) middle vessel with rectifying and stripping column.*

Multistage batch distillation of liquid mixtures is of great importance; it is used extensively in laboratories and in small production units. The main advantage of batch distillation over continuous distillation is that a single apparatus can process many different liquid mixtures. A batch distillation column is, in essence, a multiple purpose unit. The starting mixture at hand and different product requirements can be taken into account by simply changing the reflux ratio. Even multicomponent mixtures can be separated by batch distillation in a single column when the constituents are separately collected in different receivers. A disadvantage of batch distillation is the long time the mixture is exposed to high temperatures. This increases the risk of thermal degradation or decomposition of the substances. Furthermore, energy requirement is generally higher in batchwise than in continuous distillation.

A setup for multistage batch distillation is shown in Fig. 6-1A, e.g., [Diwekar 1995]. The still is filled with the liquid mixture F and heated. Vapor flows upwards in the column and condenses at the top. Usually, the entire condensate is initially returned to the column as reflux (i.e., $R_L \rightarrow \infty$). This countercurrent contacting of vapor and liquid considerably improves the separation (cf. Section 4.1). After some time, a part of the overhead condensate is withdrawn continuously as distillate, and the other part is recycled into the column as reflux. The liquid in the still is increasingly depleted of the more volatile components. As the amount of liquid in the still decreases, the concentration of the high boiling constituents increases.

In principle, a setup for batch distillation according to Fig. 6-1B is also feasible, e.g., [Sorensen and Skogestad 1996]. Here, the vessel is located at the top of the column that is operated as a stripping column. During operation, the high boiling constituents are primarily separated from the system. The liquid in the vessel depletes in high boiling constituents and enriches in low boiling ones. This mode of batch distillation is very seldom applied to industrial processes. However, as will be shown later on, it is necessary for separating mixtures with minimum azeotropes.

A third feasible setup for batch distillation is shown in Fig. 6-1C. The column consists of both a rectifying and stripping section, and the vessel is located by the middle of the column. This and similar configurations have recently found much interest in literature, e.g., [Davidyan et al. 1994, Diwekar 1995, Barolo et al. 1996], but are scarcely used in practice. In theoretical studies, such setups proved to be more efficient with respect to energy utilization than a conventional setup, e.g., [Meski and Morari 1995].

6.1 BATCH DISTILLATION OF BINARY MIXTURES

The calculation of multistage batch distillation of binary mixtures is based on the Rayleigh equation, derived in Section 3.2.1 for single-stage batch distillation. Application of this equation to multistage distillation requires the use of x_D, the mole fraction of the more volatile component in the distillate, instead of equilibrium vapor concentration y^* in Equation (3.25):

$$\frac{dB}{B} = \frac{dx_B}{x_D - x_B} \tag{6.1}$$

Here, B denotes the actual amount of liquid mixture in the still.

The concentration x_D of the distillate depends on the vapor-liquid equilibrium (i.e., the relative volatility α), the separation efficiency of the column (i.e., the number of equilibrium stages n or number of transfer units N_{OG}) and the reflux ratio R_L:

$$x_D = f(x_B, \alpha, n, R_L) \tag{6.2}$$

For binary mixtures, the function f is usually determined graphically on the y-x diagram. Calculations depend on the mode of operation.

6.1.1 Operation with Constant Reflux

For operation with constant reflux, the separation efficiency of the column (e.g., number of equilibrium stages n), the reflux ratio R_L and the starting concentration x_F are specified. The amount of product D, the mean composition x_{Dm} of the product and the energy requirement for distillation are then determined. Integration of Equation (6.1) gives:

$$\ln \frac{B_e}{F} = \int_{x_F}^{x_{Be}} \frac{dx_B}{x_D - x_B} \tag{6.3}$$

where F is the initial amount of liquid. Integration of the right-hand side of this equation requires determination of the relationship $x_D = f(x_B, \alpha, n, R_L)$. In a binary system, this is usually done numerically or graphically on the y-x diagram, as shown in Fig. 6-2A. A distillation composition x_{D1} is set and an operating line is drawn, followed by the usual determination of the concentration x_{B1} in the still, e.g., by stepping off the stages. Subsequently, another distillate concentration x_{D2} is set, and the corresponding concentration x_{B2} is determined in an analogous manner. The operating lines are all parallel because the reflux ratio R_L is supposed to be constant. In this way, the correlation between distillate concentration x_D and still concentration x_B is determined point-by-point. The integral in Equation (6.3) is solved by using an auxiliary diagram in which the term $1/(x_D - x_B)$ is plotted on the ordinate and the actual bottoms concentration x_B on the abscissa, see Fig. 6-2B. The area *A* under the curve, determined point-by-point from x_F to x_{Be} represents the solution to the integral. The following equation is obtained:

$$B_e = F\, e^A \tag{6.4}$$

With D = F − B follows:

$$D_e = F\,(1 - B_e/F) = F\,(1 - e^A) \tag{6.5}$$

The mean distillate concentration x_{Dm} results from the material balance:

$$x_{Dm} = \frac{x_F - x_{Be}\, B_e/F}{(1 - B_e/F)} \tag{6.6}$$

The amount of vapor G generated in the reboiler is higher than the amount of overhead product D since some of the distillate is recycled into the column as reflux. With $G = D\,(R_L + 1)$ the heat Q required for the separation is:

$$\frac{Q}{F\,r} = (1 - B_e/F)\,(R_L + 1) \tag{6.7}$$

Here, r is the latent heat of vaporization of the mixture.

6.1.1.1 *Infinite Number of Stages* For ideal mixtures and infinite number of stages in the column, Equation (6.3) can be solved analytically for a preset value of the reflux ratio R_L. Two steps of the process have to be considered.

In the first step, the concentration of the distillate is, under the assumption of an infinite number of stages, always $x_D = 1$ if the reflux ratio is high. This holds as long as the concentration in the still is higher than x_B^*. The value of x_B^* can be calculated with relation (4.35) derived in Section 4.2.4 for the determination of mimimum reflux ratio. Here, the same equation applies, but reflux ratio is preset

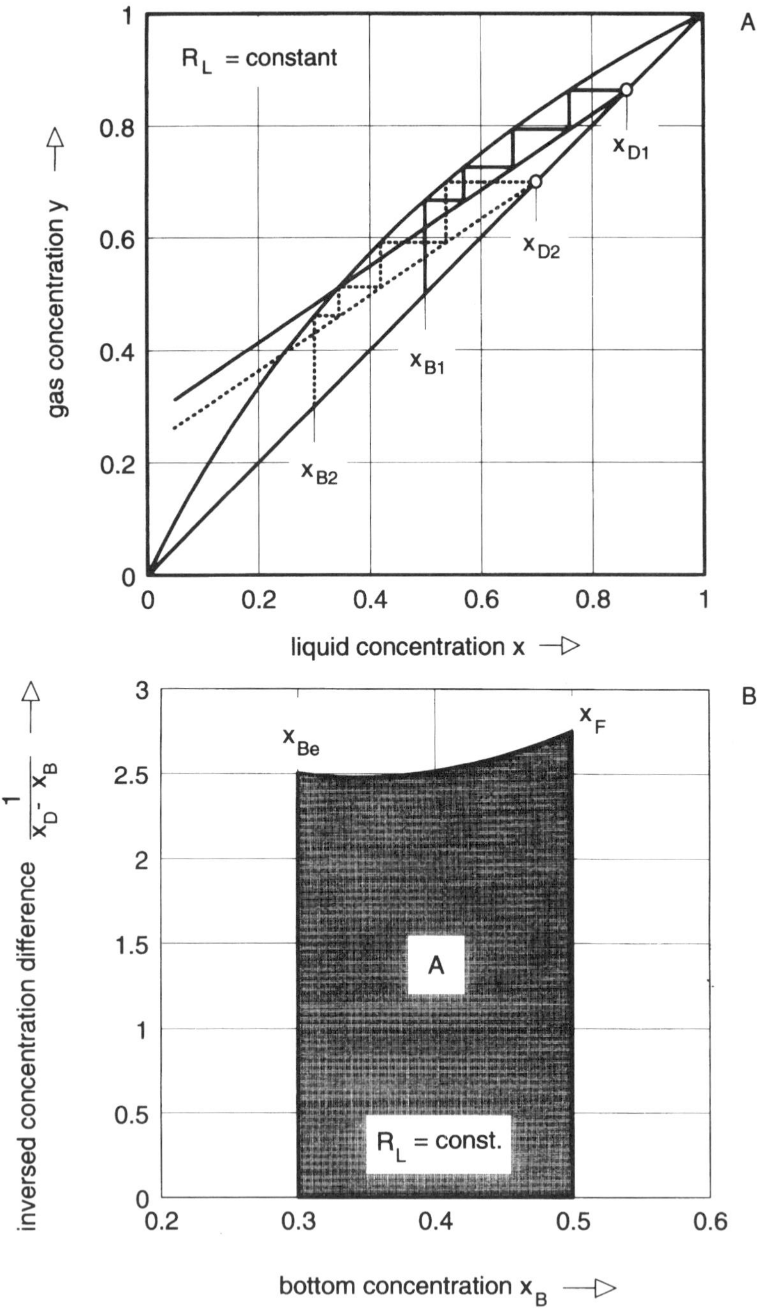

Fig. 6-2. *Batch distillation of a binary mixture at constant relux policy: (A) Graphic determination of pairs of distillate and bottom concentrations, (B) auxiliary diagram for graphical integration of Equation (6.3).*

and feed concentration x_F (i.e., bottom concentration x_B) is the unknown quantity. After transformation we get:

$$x_B^* = \frac{1}{(\alpha - 1) R_L} \tag{6.8}$$

For $x_F > x_B^*$, integrating Equation (6.3) gives:

$$B^*/F = \frac{x_{Dm1} - x_F}{x_D - 1/((\alpha - 1) R_L)} \qquad \text{with } x_{Dm1} = 1 \tag{6.9}$$

Here, B^* denotes the amount of liquid in the still at the end of the first step of the process.

When the concentration in the still falls below x_B^*, pure low boiler is no longer separated from the mixture, i.e., $x_D < 1$, and the second step of the process begins. During this step, both the bottoms and the overhead concentrations, x_B and x_D, decrease continuously. The relation between x_D and x_B is now formulated by Equation (4.34). After transformation we get:

$$x_D = \frac{\alpha x_B}{1 + (\alpha - 1) x_B} (R_L + 1) - R_L x_B \tag{6.10}$$

From Equations (6.3) and (6.10) follows:

$$\ln\left(\frac{B_e}{B^*}\right) = \frac{1}{R_L + 1} \int_{x_B^*}^{x_{Be}} \frac{dx_B}{\alpha x_B/(1 + (\alpha - 1) x_B) - x_B} \tag{6.11}$$

After integration:

$$\frac{B_e}{B^*} = \left(\left(\frac{x_{Be}}{x_B^*}\right)^{1/(\alpha - 1)} \left(\frac{1 - x_B^*}{1 - x_{Be}}\right)^{\alpha/(\alpha - 1)}\right)^{1/(R_L + 1)} \tag{6.12}$$

For a reflux ratio $R_L = 0$, i.e., no reflux, this equation transfers into Equation (3.29) derived for single-stage distillation in Section 3.2.

The mean distillation concentration x_{Dm2} of the second step is calculated from:

$$x_{Dm2} = \frac{x_B^* - x_{Be} B_e/B^*}{(1 - B_e/B^*)} \tag{6.13}$$

The mean distillate concentration of the total process is:

$$x_{Dm} = \frac{1 - B^*/F + (B^*/F - B_e/F)\, x_{Dm2}}{1 - B_e/F}$$
$$\text{with } B_e/F = B_e/B^* \cdot B^*/F \qquad (6.14)$$

The energy requirement follows from:

$$\frac{Q_{min}}{F\, r} = (R_L + 1)(1 - B_e/F) \qquad (6.15)$$

Figure 6-3 displays a plot of these equations for a mixture with a feed concentration $x_F = 0.5$ and a relative volatility $\alpha = 2$. Presented is the minimum energy requirement over the relative amount of distillate for several mean distillation concentrations x_{Dm}. The energy requirement increases sharply when nearly all low boiler has been separated from the mixture in the still. At the point $x_B = 0$, i.e., $D/F = x_F/x_{Dm}$, the energy requirement becomes infinite. Additionally, lines of constant reflux ratio R_L are presented in the diagram. If, for instance, 60% of the feed F has

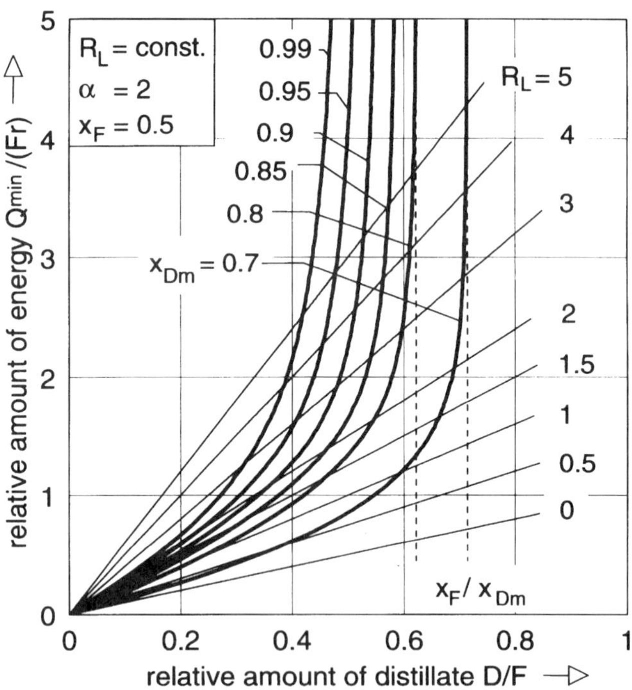

Fig. 6-3. *Minimum (at infinite number of stages) energy demand of batch distillation at constant reflux policy. When the still concentration of the low boiler approaches zero, the energy requirement increases infinitely.*

to be separated as overheads D with a mean concentration $x_{Dm} = 0.80$, the reflux ratio must be as high as $R_L = 3.1$ and the energy demand $Q_{min}/(F\ r)$ is 2.5. Hence, the feed F must be evaporated 2.5 times during the whole process. If only 40% of the feed must be separated from the mixture, the appropriate reflux ratio is $R_L = 1.3$ (at same value of the mean distillate concentration). Here, the energy demand is with $Q_{min}/(F\ r) = 0.95$ much lower.

Operation at constant reflux ratio R_L is relatively simple. However, the distillate concentration x_D decreases with time, as illustrated in Fig. 6-2. Thus, the desired product composition is achieved by having a higher starting purity than specified. However, if a high purity product is required, this is possible only to a very limited extent. Hence, the reflux ratio R_L is often increased several times during operation in order to prevent a steep drop in product purity.

6.1.2 Operation with Constant Distillate Composition

A decrease of distillate concentration x_D over time can be avoided by continuously increasing the reflux ratio R_L throughout the run. The relation between still concentration x_B and the corresponding reflux ratio R_L is, in turn, determined graphically on the y-x diagram. An initial reflux ratio R_L, and thus the position of the operating line, is specified, and the corresponding still concentration x_B is determined graphically. This is illustrated in Fig. 6-4 for two different cases. The correlation obtained depends on the separation efficiency of the column, i.e., number of equilibrium stages n and value of reflux ratio R_L.

Operation with constant distillate concentration x_D permits direct integration of the Rayleigh equation (Equation 6.1) to give:

$$B_e = F \frac{x_D - x_F}{x_D - x_{Be}} \tag{6.16}$$

The quantity of distillate is calculated from:

$$D_e = F \frac{x_F - x_{Be}}{x_D - x_{Be}} \tag{6.17}$$

The energy requirement depends on the reflux ratio R_L, which is variable. Hence, only a differential equation can be given:

$$dQ = dD\ (R_L + 1)\ r \tag{6.18}$$

By using the relation $dD = -dB$, the following relationship is obtained after differentiating Equation (6.16) and inserting it into Equation. (6.18):

$$\frac{dQ}{F\ r} = (x_D - x_F) \frac{R_L + 1}{(x_D - x_B)^2}\, dx_B \tag{6.19}$$

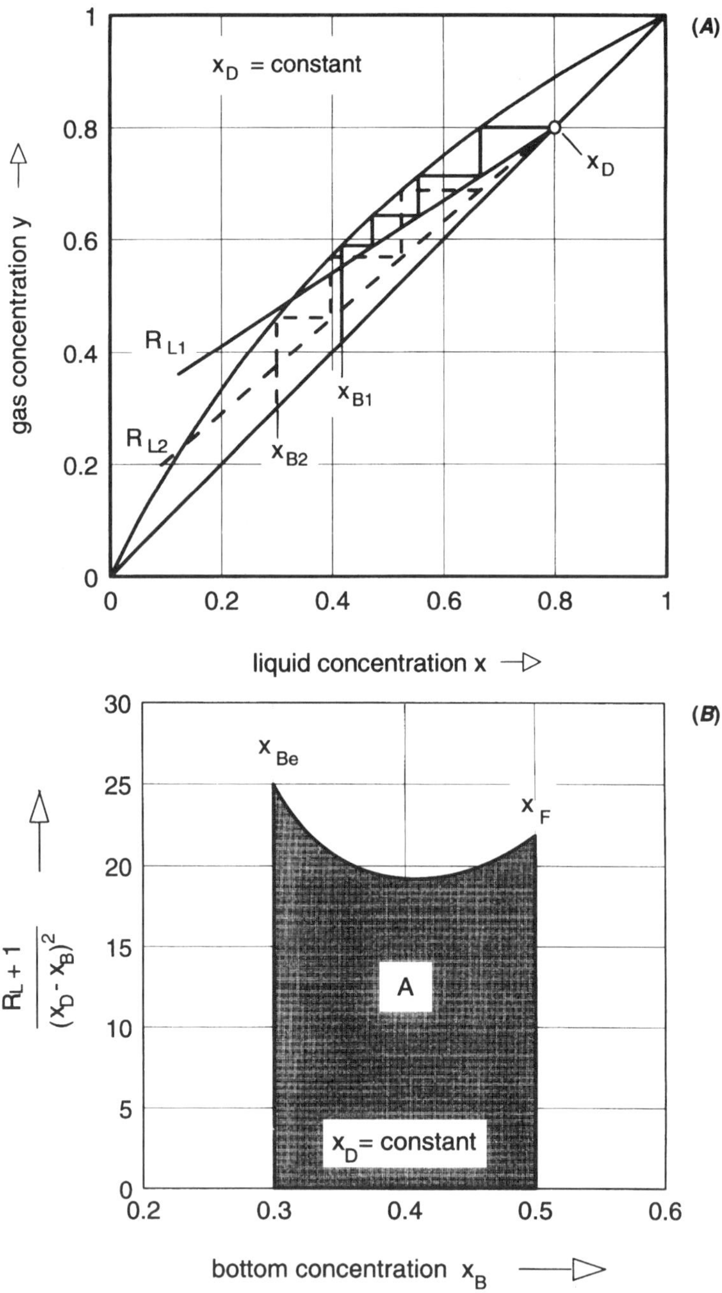

Fig. 6-4. *Batch distillation of a binary mixture at constant distillate reflux policy: (A) Graphic determination of pairs bottom concentrations and reflux ratio, (B) auxiliary diagram for graphical integration of Equation (6.20).*

Integration gives:

$$\frac{Q}{F\,r} = (x_D - x_F) \int_{x_F}^{x_{Be}} \frac{R_L + 1}{(x_D - x_B)^2}\, dx_B \tag{6.20}$$

The term on the right-hand side of Equation (6.20) is integrated graphically by using an auxiliary diagram, see Fig. 6-4B. The still concentration x_B is plotted on the abscissa and the term $(R_L + 1)/(x_D - x_B)^2$ on the ordinate. The area A under the curve between x_F and x_{Be} gives the value of the integral. Hence, the heat requirement is:

$$\frac{Q}{F\,r} = (x_D - x_F)\, A \tag{6.21}$$

6.1.2.1 *Infinite Number of Stages* For ideal mixtures and infinite number of stages in the column, the minimum energy requirement can be analytically calculated. The relation between still concentration x_B and reflux ration R_L at constant distillate concentration x_D is given by Equation (4.34):

$$R_L = \frac{1}{(\alpha - 1)}\left(\frac{x_D}{x_B} - \alpha\frac{1 - x_D}{1 - x_B}\right) \tag{6.22}$$

Inserting into Equation (6.20) yields:

$$\frac{Q_{min}}{F\,r} = (x_D - x_F)\int_{x_F}^{x_{Be}} \frac{1}{(x_D - x_B)^2}\left(\frac{1}{(\alpha - 1)}\left(\frac{x_D}{x_B} - \alpha\frac{1 - x_D}{1 - x_B}\right) + 1\right) dx_B \tag{6.23}$$

After solving the integral of the right-hand side:

$$\frac{Q_{min}}{F\,r} = \frac{x_D - x_F}{(\alpha - 1)x_D(1 - x_D)}\left(\alpha\, x_D \ln\frac{1 - x_F}{1 - x_{Be}} - ((\alpha - 1)\, x_D + 1)\ln\frac{x_D - x_F}{x_D - x_{Be}} + (1 - x_D)\ln\frac{x_F}{x_{Be}}\right) \tag{6.24}$$

A plot of this equation for a mixture with $\alpha = 2$ and $x_F = 0.5$ is presented in Fig. 6-5. As well as in operation with constant reflux, the energy demand increases drastically when nearly all low boiler has been removed from the mixture. Hence, separation with high yield is difficult in both modes of batchwise distillation. A

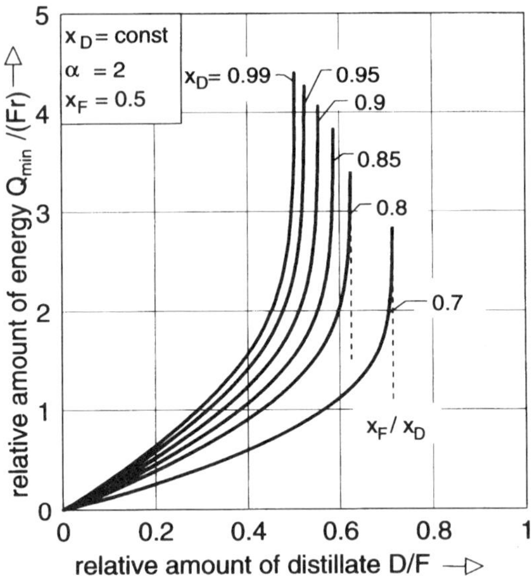

Fig. 6-5. Minimum energy demand of batch distillation at constant distillate concentration reflux policy. The energy demand becomes infinite when the concentration of low boiler a *in the still approaches zero.*

comparison of Fig. 6-3 with Fig. 6-5 reveals that energy requirement is lower at distillation with constant distillate concentration than with constant reflux ratio. This difference is especially high for high purity distillations.

6.1.2.2 Mixtures with Partial Immiscibility in the Liquid Phase In practice, operation at constant distillate concentration x_D is more laborious since the reflux ratio R_L changes (Fig. 6-4) and must continuously be readjusted. On the other hand, this situation occurs automatically during the fractionation of mixtures such as 1-butanol/water which exhibit a region of immiscibility and form a heterogeneous azeotrope (Fig. 6-6). In this special case, part of the separation effect is furnished by the decanter.

The reflux at the top of the column is water saturated with butanol. The concentration of the reflux lies at the left-hand border of the shaded region of immiscibility in the McCabe-Thiele diagram of Fig. 6-6. The state of the vapor lies on the point of intersection of the operating line with the saturation line. Hence, the vapor is enriched in butanol. After condensation the state of the liquid mixture lies in the region of immiscibility. The two liquid phases are separated by a decanter into a water-rich fraction that is recycled as reflux and a butanol-rich fraction that is drawn off as distillate. Thus, the product has the same concentration during the entire process.

The rates of product and distillate flow depend on the concentration of the vapor at the top of the column. At the beginning, the slope of the operating line is small

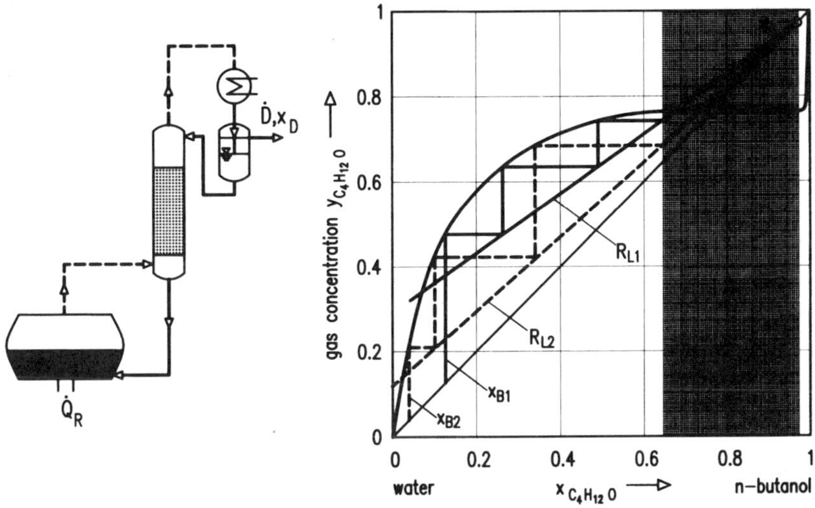

Fig. 6-6. Batch distillation of the mixture 1-butanol/water which exhibits a region of immiscibility with a heterogeneous azeotrope. The shaded area denotes the range of immiscibility. Part of the separation is performed by the decanter.

and, as is seen from Fig. 6-6, the concentration of the overhead vapor is high. This results in a large flow rate of the distillate D and, since D + L = const., in a small flow rate of reflux L. At the end of the process, the reflux is much higher and the concentration of the overhead vapor is smaller. After condensation, the overall liquid concentration lies very close to the left-hand border of the region of immiscibility. According to the lever arm rule, the rate of the product D is much smaller and, consequently, the rate of the reflux L is much higher. Thus, the operation with constant distillate concentration and variable reflux is automatically established without any external action.

6.1.3 Operation with Minimum Energy Input

The two operating modes described in Sections 6.1.1 and 6.1.2 require different quantities of heat Q to obtain the same amount and composition of the product. Generally, less energy is required for the operation at constant distillate concentration. However, in the energetically optimal operation mode the distillate concentration decreases steadily and the reflux ratio increases steadily. Pontryagin's maximum principle [Pontryagin et al. 1962] can be used to calculate the relation between reflux ratio R_L, distillate concentration x_D and still concentration x_B [Coward 1967, Robinson 1969].

The basis of Pontryagin's maximum principle is a system of differential equations that describes the time dependence of the state variables of the process. His principle is, in essence, a generalization of the Lagrange equations developed for easy description of the dynamic behavior of kinematic systems, e.g., [D'Souza and Garg 1984]. Lagrange's equations allow for the derivation of the differential equations of motion, dz/dt, by performing the partial derivatives of the total energy E with respect to the momentum m.

$$\frac{dz}{dt} = \frac{\partial E}{\partial m} \tag{6.25}$$

The time dependence of momentum m is, in turn, formulated by the partial derivatives of the energy E with respect to locus z.

$$\frac{dm}{dt} = -\frac{\partial E}{\partial z} \tag{6.26}$$

Equations (6.25) and (6.26) are essential characteristics of potential functions. The total energy E of a kinematic system is only one example of such a potential function.

In order to determine the potential function for non-kinematic systems, Pontryagin begins with the differential equations for the time dependence of the process variables and defines a new potential function, called the Hamiltonian function H. Function H represents the total driving force for any unsteady process. Alike total energy E in kinematic processes, the potential function H must be kept constant during the process. The higher the value of H the faster the process proceeds. Selection of the maximum value H_{max} at the beginning gives the time-optimum variation of the variables.

Application of Pontryagin's maximum principle to batch distillation starts with the physical description of the unsteady state behavior. The variation of the amount of liquid in the still, dB, with time, dt, follows from a material balance:

$$\frac{dB}{dt} = \frac{-\dot{G}}{R_L + 1} \tag{6.27}$$

Here, the vapor stream generated by the heating system is considered to be constant, i.e., $\dot{G} \neq f(t)$. The component balance is described by the modified Rayleigh Equation (6.1):

$$\frac{dB}{B} = \frac{dx_B}{x_D - x_B} \tag{6.28}$$

Substitution of dB by Equation (6.27) gives after transformation:

$$\frac{dx_B}{dt} = \frac{-\dot{G}}{(R_L + 1)\, B}\,(x_D - x_B) \tag{6.29}$$

From Equations (6.27) and (6.29) follows the Hamiltonian function:

$$H = \frac{-\dot{G}}{R_L + 1} \pi_1 + \frac{-\dot{G}}{(R_L + 1)B} (x_D - x_B) \pi_2 \tag{6.30}$$

The function H and the parameters π are analoga to the energy E and the momentum m, respectively, of a kinematic system. Hence, the derivatives $\partial H/\partial \pi_1$ and $\partial H/\partial \pi_2$ give the time dependence of the variables B and x_B of batch distillation (see Equations (6.28) and (6.29)).

A relation for the time dependence of the parameters π_1, π_2 is, according to Equation (6.26), derived by the partial derivatives $-\partial H/\partial B$ and $-\partial H/\partial x_B$. Hence:

$$\frac{d\pi_1}{dt} = \frac{-\dot{G}}{(R_L + 1)\, B^2} (x_D - x_B)\pi_2 \tag{6.31}$$

and

$$\frac{d\pi_2}{dt} = \frac{+\dot{G}}{(R_L + 1)\, B} \left(\frac{\partial x_D}{\partial x_B} - 1\right) \pi_2 \tag{6.32}$$

Generally, the integration of the system of differential Equations (6.27), (6.29), (6.31) and (6.32) must be performed numerically, e.g., by Simpson's rule or by a Runge-Kutta procedure. At each time step dt the value of the Hamiltonian function H has to be constant. At varying variables B, x_B, π_1 and π_2 this is only possible by additional time variation of the reflux ratio R_L. If, for instance, in the beginning the reflux ratio has been chosen to maximize the Hamiltonian function, the time-optimum variation of the process variables is found. Hence, the resulting variation of reflux ratio represents the time-optimum reflux policy. At operation with constant energy input (i.e., $\dot{Q}$ = const.) the time-optimum also represents the energy minimum.

For numerical integration the starting values of the parameters π_1 and π_2 have to be estimated. The starting value of π_1 may always be set 1 without any restrictions for the solution. Then, the value of π_2 has to be negative. Since the term $(x_D - x_B)$ is also a function of the reflux ratio R_L the Hamiltonian function H exhibits, when plotted versus the reflux ratio R_L, a maximum. Typically, the value of π_1 increases monotonously with time whereas the value of π_2 exhibits a maximum.

The numerical integration starts with that value of R_L that maximizes the Hamiltonian function. For each time step dt the reflux ratio that maximizes the function H must be calculated. The value of H is constant during the process. If the end values x_{Be} and B_e do not meet the material balance, the first estimate of π_2 is wrong and the procedure has to be repeated. After several iterations the solution will be found. A good check for the time- and energy-optimum is the constancy of the value of the Hamiltonian function during the last iteration.

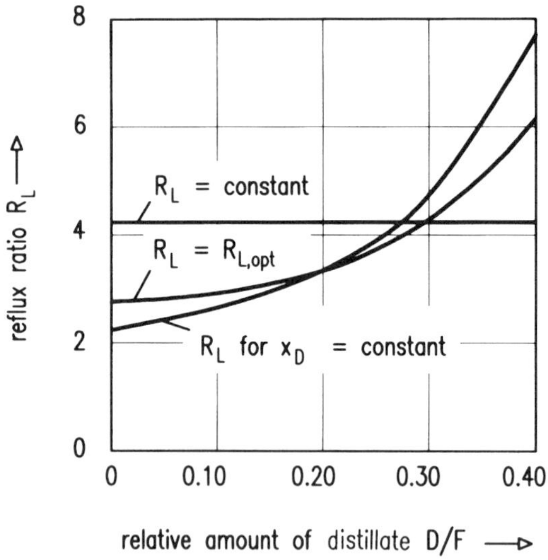

Fig. 6-7. *Variation of the reflux ratio R_L versus relative amount of distillate D/F for all three reflux policies. Typically, the optimum reflux policy lies between R_L = constant and x_D = constant.*

Figure 6-7 presents the course of the optimum reflux ratio, R_{Lopt}, versus the relative amount of distillate D/F [Robinson 1969]. The reflux R_{Lopt} steadily increases during operation. However, the increase is not as drastic as in the operation mode with constant product purity. In the beginning, the reflux is much lower than in the operation mode with constant reflux. The energy demand is lowest in operation with R_{Lopt}. Energy savings of up to approximately 15% can be achieved [Coward 1967], which justify this more laborious method only in special cases. If a product with very high purity is demanded, only operation with constant distillate composition is feasible.

Numerical integration of Equations (6.27), (6.29), (6.31) and (6.32) is tiresome, as it involves the numerical evaluation of the relations $(x_D - x_B)$ and $\partial x_D/\partial x_B$ (e.g., by tray-to-tray calculations. Furthermore, the search of that reflux ratio R_L that maximizes the Hamiltonian function H is very time consuming since the maximum flattens with increasing time or may, even worse, not exist at all (see next section).

6.1.3.1 Influence of Number of Equilibrium Stages on the Hamiltonian Function The basis for the determination of the optimum reflux policy of batch distillation by Pontryagin's maximum principle is the existence of a maximum in the Hamiltonian function H when plotted versus the reflux ratio R_L. However, such a maximum does only exist if the number of stages is small (approximately 2 to 8, depending on vapor-liquid equilibrium). This is outlined on the example of an ideal mixture in the following. The limiting cases $n \to \infty$ and $n = 1$ are considered. If the number of stages is high (approximately more than 8) the distillate concentrations x_D follows from Equation (6.10):

$$x_D = \frac{\alpha\, x_B}{1 + (\alpha - 1)\, x_B}\,(R_L + 1) - R_L\, x_B \qquad (6.33)$$

Here, distillate concentration x_D is a linear function of reflux ratio R_L. Combining Equations (6.30) and (6.33) yields the Hamiltonian function H for $n \rightarrow \infty$:

$$H_\infty = \frac{-\dot{G}}{R_L + 1} \pi_1 + \frac{-\dot{G}}{B} \left(\frac{\alpha \, x_B}{1 + (\alpha - 1)\, x_B} - x_B \right) \pi_2 \tag{6.34}$$

The structure of this function is:

$$H_\infty = \frac{C_1}{R_L + 1} + C_2 \tag{6.35}$$

Function $H\infty$ increases or decreases monotonously with increasing reflux ratio R_L. No extremum exists in such a function. Hence, the optimum reflux policy is identical to operation with constant distillate concentration.

If the number of equilibrium stages is 1 the distillate concentration x_D is always in equilibrium state with the still concentration x_B. Hence:

$$x_D = \frac{\alpha \, x_B}{1 + (\alpha - 1)\, x_B} \tag{6.36}$$

The distillate concentration does not depend on reflux at all. Inserting Equation (6.36) in (6.30) gives the Hamiltonian function for $n = 1$:

$$H_1 = \frac{1}{R_L + 1} \left(-\dot{G}\, \pi_1 + \frac{-\dot{G}}{B} \left(1 + \frac{\alpha \, x_B}{(\alpha - 1)\, x_B} - x_B \right) \pi_2 \right. \tag{6.37}$$

The structure of the function is:

$$H_1 = \frac{C_1}{R_L + 1} \tag{6.38}$$

The Hamiltonian function increases or decreases (depending on the sign of constant C_1) monotonously over the reflux ratio R_L, i.e., no maximum exists. Here, the optimum reflux policy is identical with the operation mode R_L = constant and x_D = variable, which is the only feasible operation mode in this case.

6.1.4 Comparison of Energy Requirement for Different Modes of Distillation

A comparison of minimum energy requirement for different modes of distillation is presented in Fig. 6-8. An ideal mixture with relative volatility $\alpha = 2$, feed concentration $x_F = 0.5$ and infinite number of stages has been considered. Continuous distillation is also included in the comparison. The relevant equation is (see Section 4.2.5):

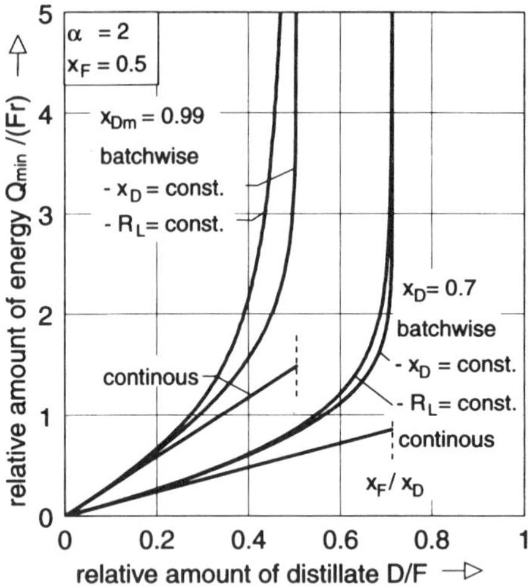

Fig. 6-8. *Comparison of minimum (at infinite number of stages) energy requirement for continuous and batchwise distillations with either R_L = constant or x_D = constant. Especially in high yield separations, continuous distillation is superior to all modes of batch distillation.*

$$\left(\frac{Q_{min}}{F\,r}\right)_{cont} = \left(\frac{1}{\alpha - 1}\left(\frac{x_D}{x_F} - \alpha\,\frac{1 - x_D}{1 - x_F}\right) + 1\right) D/F \tag{6.39}$$

The diagram of Fig. 6-8 makes clear that all modes of batchwise distillation require more energy than continuous distillation. The difference is especially high when the concentration of the low boiler in the still approaches zero. At this point, given by $D/F = x_F/x_{Dm}$, the term $Q_{min}/(F\,r)$ increases infinitely at batchwise distillation. At continuous distillation, however, the minimum energy requirement is a linear function of D/F even at the limiting value $D/F = x_F/x_{Dm}$, where the low boiler concentration in the bottoms is zero. Hence, batchwise distillation is especially disadvantageous when all low boiler has to be separated, i.e., at high yield separations.

6.2 BATCH DISTILLATION OF TERNARY MIXTURES

The application of the fundamental Equation (6.1) to ternary mixtures gives:

$$\frac{dB}{B} = \frac{dx_{Ba}}{x_{Da} - x_{Ba}} \quad \text{and} \quad \frac{dB}{B} = \frac{dx_{Bb}}{x_{Db} - x_{Bb}} \tag{6.40}$$

These equations are similar to Equation (3.39) but distillate concentration x_{Di} replaces equilibrium concentrations y_i^*. In general, determination of distillate concentrations x_{Di} requires the rigorous simulation of the column as described in Section

4.4. However, the results of such a tiresome simulation are only reliable if the liquid holdup in the column is much lower than the liquid holdup in the vessel. If this condition is not met, the dynamic behavior of the column has to be simulated (see Section 6.4).

Batch distillation of ternary (and multicomponent) mixtures is preferentially performed with constant reflux ratio. However, operation with periodically constant distillate concentration is also possible, [Block et al. 1978]. The reflux ratio must be carefully controlled over the entire time of operation. In most cases, this mode of operation does not offer enough advantage to justify the cost of the needed complex control equipment.

6.2.1 Zeotropic Mixtures

Figure 6-9 displays the results of a numerical evaluation of Equation (6.40) for the system methanol/ethanol/2-propanol. Distillate concentrations x_{Di} and bottoms concentrations x_{Bi} are plotted versus the relative amount of distillate D/F in Fig. 6-9A and 6-9B, respectively. In case of high number of equilibrium stages n and high reflux ratio R_L pure low boiler *a* is regained as overhead product in the first step of the process. Hence, the concentration of low boiler in the vessel x_{Ba} decreases. Simultaneously, the concentrations of the higher boiling constituents *b* and *c* increase. After removal of the low boiler *a* the intermediate boiler *b* is the more volatile constituent in the vessel and is, consequently, regained as distillate in the second step of the process. At this point ($D/F = x_{Fa}$), the distillate concentration sharply switches from $x_{Da} = 1$ to $x_{Da} = 0$ and $x_{Db} = 0$ to $x_{Db} = 1$. Now, the concentration of the intermediate boiler *b* decreases in the vessel. After complete removal of *b* (at $D/F = x_{Fa} + x_{Fb}$), the concentration x_{Db} sharply falls to zero and pure high boiler *c* is in the overhead product.

All three constituents of the starting mixture are regained in pure form by collecting the overhead fractions in different receivers. The width of the zone of transition between different pure overhead products depends primarily on reflux ratio R_L and the number n of equilibrium stages.

Figure 6-9 is well suited to display the result of a batch distillation process. However, a plot of top and bottom concentrations in a triangular diagram provides a deeper understanding.

Eliminating dB/B from Equations (6.40) gives:

$$\frac{dx_{Ba}}{dx_{Bb}} = \frac{x_{Da} - x_{Ba}}{x_{Db} - x_{Bb}} \tag{6.41}$$

This linear relation is represented by a straight line in the triangular diagram of Figure 6-10. According to Equation (6.41), distillate concentration x_{Di} and bottoms concentration x_{Bi} have to lie on a straight line that forms a tangent to the residuum curve of the liquid in the still x_{Bi}. This behavior is analogous to the liquid residuum curve of Figure 3-14. If the column provides a sharp separation (i.e., high reflux

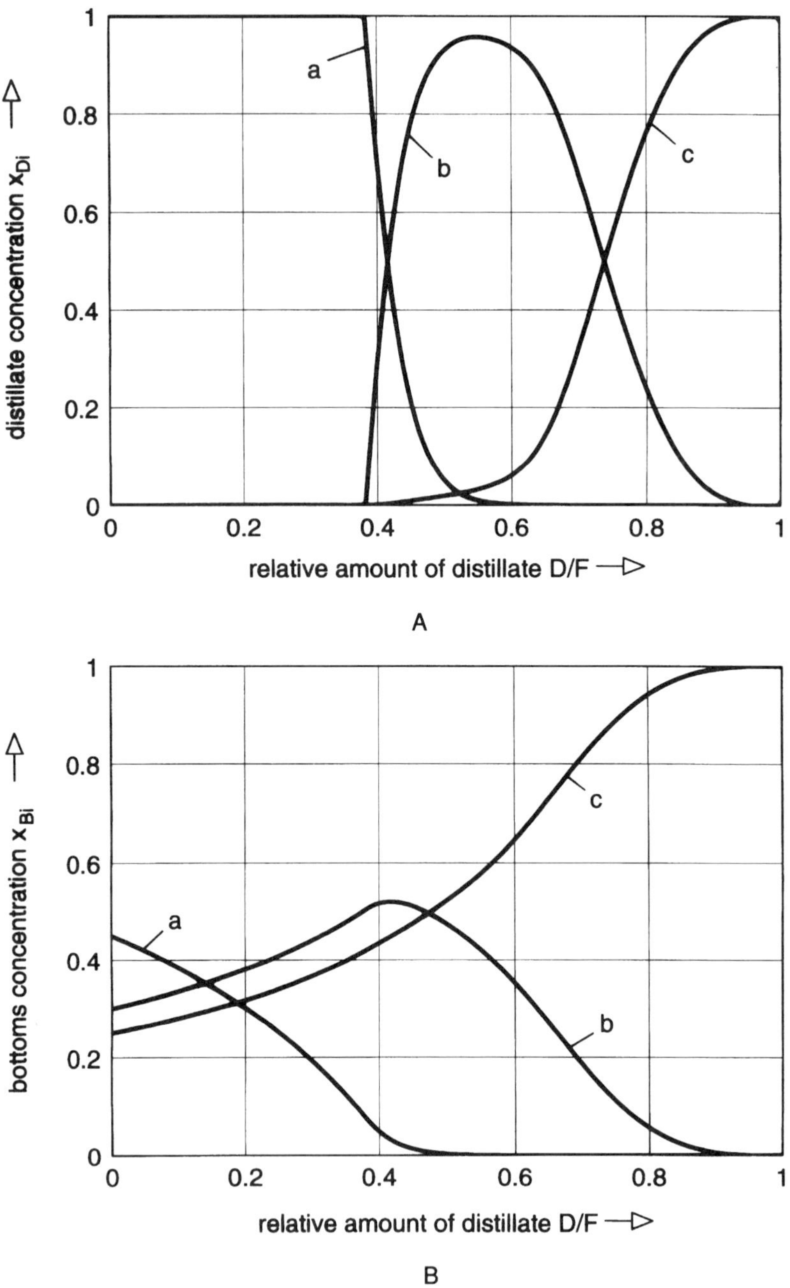

Fig. 6-9. *Concentration profiles for batch distillation of the zeotropic ternary mixture methanol/ethanol/2-propanol calculated with reflux ratio $R_L = 20$ and number of equilibrium stages $n = 30$: A) Distillate concentrations, B) Bottoms concentrations.*

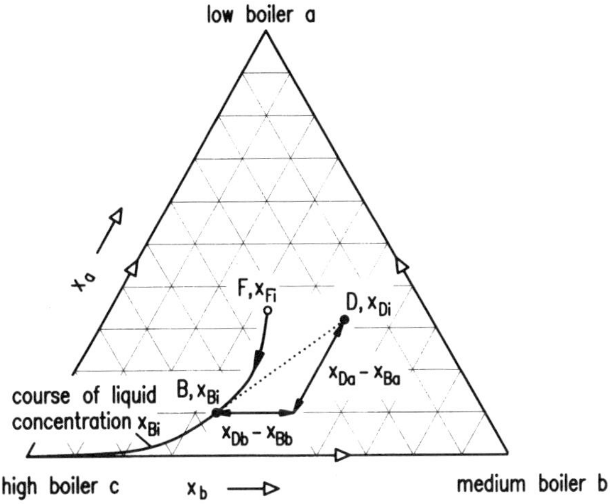

Fig. 6-10. *Graphical interpretation of Equation (6.41). The distillate concentration x_{Di} has to lie on a straight line that forms a tangent to the curve of concentration change x_{Bi} of the liquid in the still.*

ratio and high number of equilibrium stages), the distillate concentration x_{Di} takes an extreme position on the tangent, e.g., at the side of the triangular diagram.

Vice versa, when pure substance *a*, for instance, is removed from the system by batch distillation, the concentrations x_{Bi} of liquid in the still have to change along a straight line through the *a*-corner. The direction of concentration change points away from the *a*-corner. This is indicated by an arrow in Figure 6-11. After the state of the liquid in the still, x_{Bi}, has reached the base of the diagram (i.e., $x_{Ba} = 0$), the intermediate boiler *b* is separated from the system and the concentrations change along the straight line through the *b*-corner moving towards the *c*-corner Figures 6-9 and 6-11 represent the same batch distillation process of a zeotropic mixture.

The intermediate fraction between two succeeding pure products is of prime interest for batch distillation processes. The influence of the reflux ratio on the course of distillate concentrations is presented in Figure 6-12. The number of equilibrium stages is n = 10. It is clearly seen that the distillate quality improves with increasing reflux ratio. The higher the reflux ratio, the sharper the separation. The intermediate fractions formed during the transition from one to the other pure product are smaller at high values of the reflux ratio.

Figure 6-13 shows the influence of the number of equilibrium stages n on the distillate quality at a reflux ratio $R_L = 10$. The separation is significantly improved by increasing the number of equilibrium stages from 5 to 10. A further increase from 10 to 20 is, however, of minor influence. An increase of the number of stages is always connected with an increase of the liquid holdup in the column which has been neglected here. However, a high liquid holdup has, in most cases, a negative influence on the separation performance of a batch distillation column (see Section

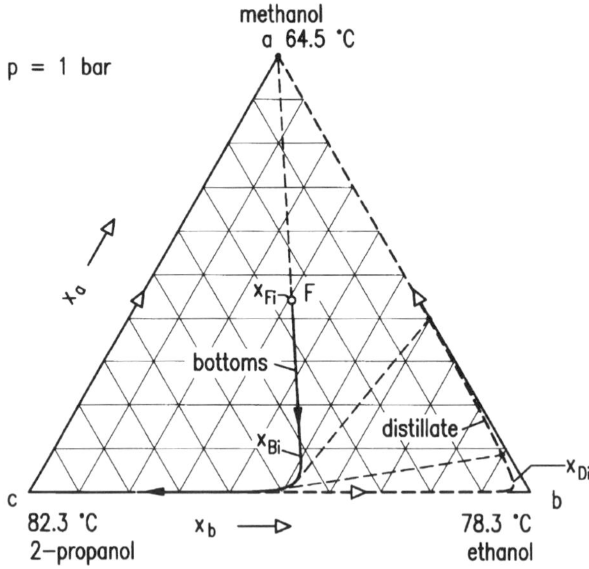

Fig. 6-11. *Presentation of the concentration profiles of Fig. 6-9 in a triangular diagram. As long as a pure component is separated from the mixture, the change of liquid concentrations x_{Bi} has to follow a straight line through this pure component.*

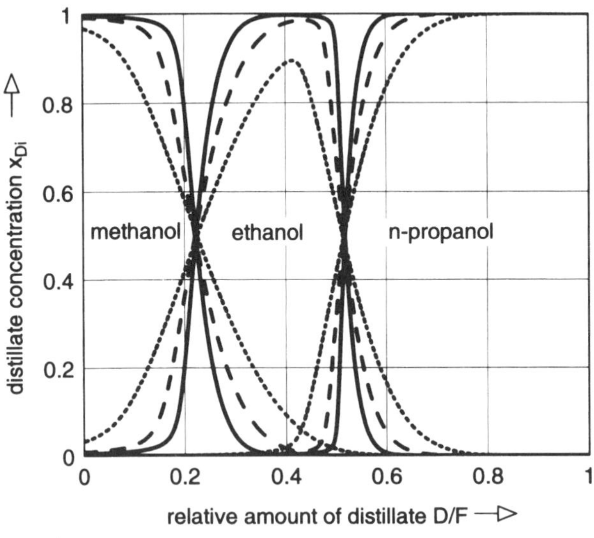

Fig. 6-12. *Influence of the reflux ratio R_L on distillate quality at 10 equilibrium stages. ······· $R_L = 5$; ----- $R_L = 10$; —— $R_L = 20$.*

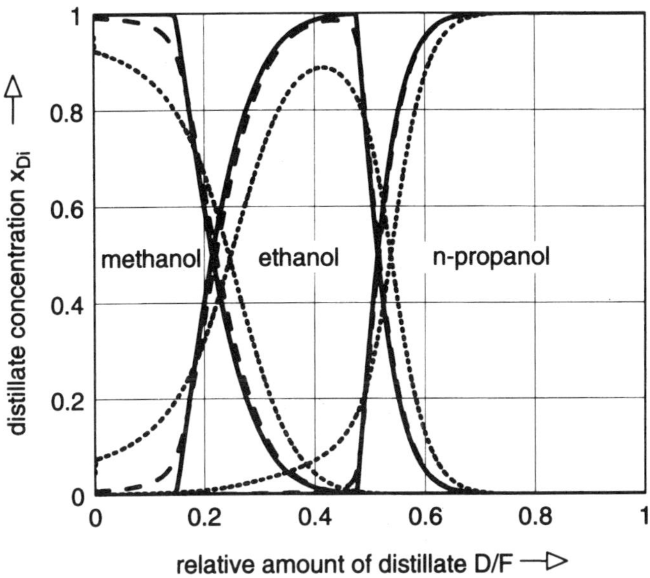

Fig. 6-13. Influence of the number of equilibrium stages n on distillate quality at $R_L = 10$. ······· $n = 5$; ----- $n = 10$; —— $n = 20$.

6.4). Thus, increasing the reflux ratio provides the most effective means of improving product quality.

6.2.2 Azeotropic Mixtures

A more complex situation arises in azeotropic mixtures that exhibit distillation borders (see Section 4.3.2). Such a system is displayed in Fig. 6-14. Here, a distillation border runs from the high boiler benzene to the binary maximum azeotrope and divides the system in two distillation fields. If the starting mixture F lies in the upper field, pure acetone is regained as overhead product during the first step of the batch distillation. The state of the bottoms B moves along a straight line away from the acetone corner. At the intersection of this straight line with the distillation border line, the concentration of acetone in the distillate sharply decreases to a value that lies at the tangent to the distillation border at the point of intersection (dashed line). During the second step of the process, the state of the bottom liquid changes along the distillation border, and due to its curvature, the distillate concentration x_{Da} continues to slowly fall, e.g., [Ewell and Welch 1945; Reinders and De Minjer 1940]. Figure 6-14B displays the distillate concentrations x_{Di} versus the relative amount of distillate D/F. This concentration profile can only be understood with the help of Fig. 6-14A.

Figure 6-15A displays the course of the residuum of batch distillation of the same system but with a starting mixture located in the lower distillation field. Here, the

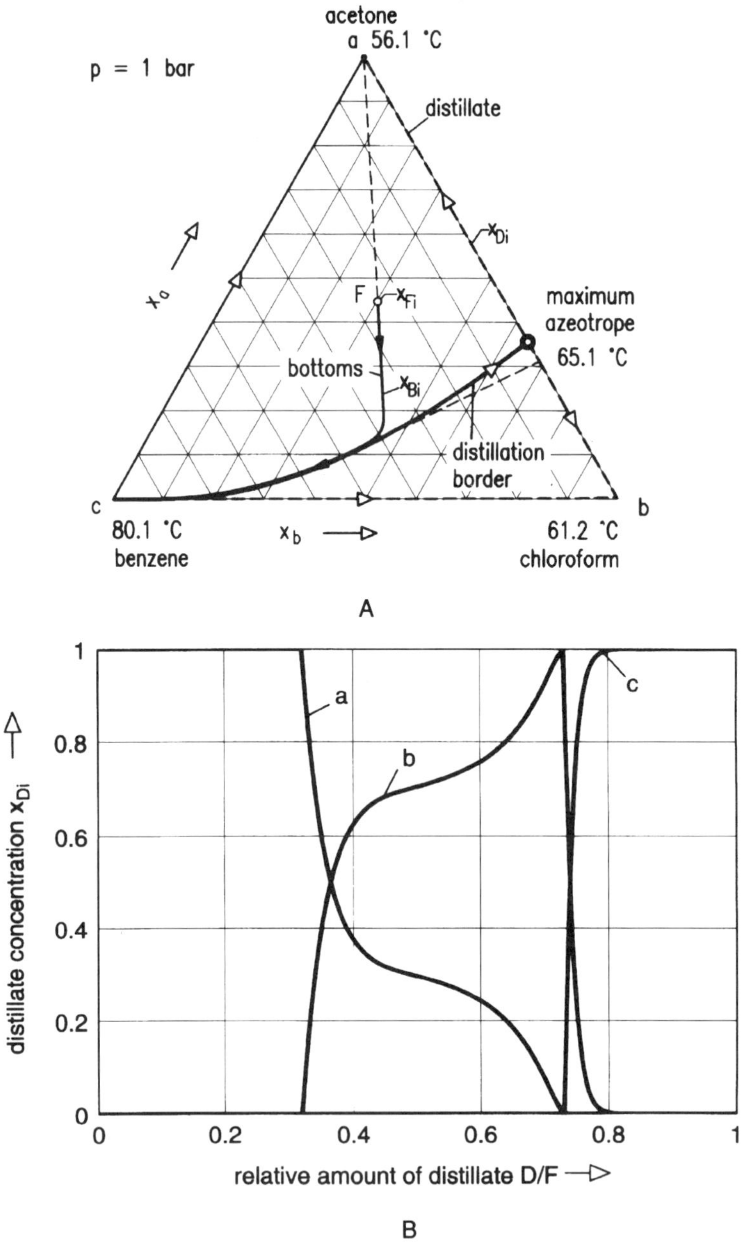

Fig. 6-14. *Concentration profiles for batch distillation of the ternary system acetone/-chloroform/benzene. This system exhibits a maximum azeotrope and a distillation border that divides the mixture in two distillation fields: (A) Change of the state of the bottoms, (B) profiles of distillate concentrations x_{Di} plotted versus the relative amount of distillate D/F.*

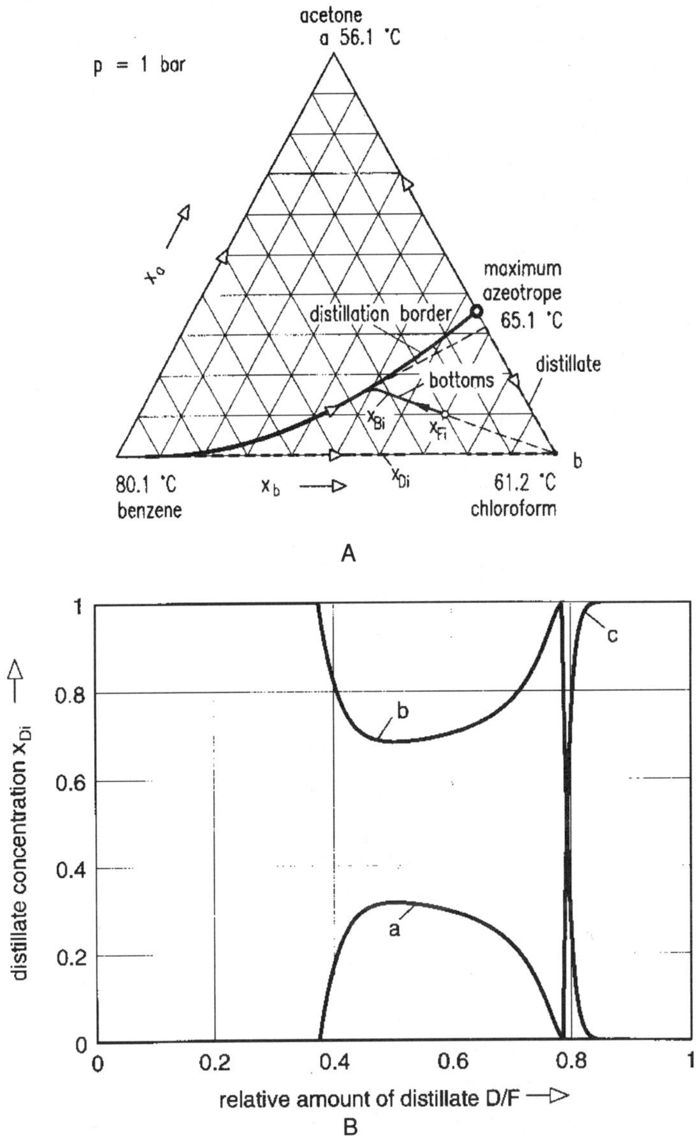

Fig. 6-15. Concentration profiles for batch distillation of the system acetone/chloroform/benzene. The feed is located in that distillation field where the intermediate boiler chloroform is first separated as overhead fraction. When the state in the still reaches the distillation border, chloroform is no longer separated in pure form: (A) Change of the state of the bottoms, (B) profiles of distillate concentrations x_{Di} plotted versus the relative amount of distillate D/F.

intermediate boiler *b* is separated first as overhead product. Hence, the state of the liquid in the vessel follows a straight line through the *b*-corner. At the point of intersection of this straight line with the distillation border, pure substance *b* can no longer be separated. Hence, its concentration decreases sharply to a value that is marked at the side of the triangular diagram by the tangent to the distillation border (dashed line). In the second step of operation, the state of the bottoms changes along the distillation border. According to its curvature the concentration of *b* in the distillate increases again.

This behavior can be seen more clearly in Fig. 6-15B. During the first step of batch distillation, pure intermediate boiler *b* is recovered as distillate. In the second step, the concentration x_{Db} decreases to approximately 0.7 and increases later on again. The concentration of the low boiler *a* shows a corresponding behavior. In the third phase of the process pure high boiler *c* is separated. Thus, only high boiler *c* and part of the intermediate boiler *b* are separated in pure form. The low boiler *a* cannot be regained as a pure product.

Systems with a minimum azeotrope behave alike. However, the vessel has to be located at the top of a stripping column (see Section 6.5).

6.3 BATCH DISTILLATION OF MULTICOMPONENT MIXTURES

Multicomponent batch distillation is also governed by the Rayleigh equation. For each component *i* in the mixture we get:

$$\frac{dB}{B} = \frac{dx_{Bi}}{x_{Di} - x_{Bi}} \tag{6.42}$$

This set of equations can only be solved numerically. The distillate concentrations x_{Di} that correspond with the bottoms concentrations x_{Bi} are determined by rigorous column simulation (see Section 4.4.1).

The result of the evaluation with a constant reflux ratio ($R_L = 5$ and $n = 20$) for a quaternary system is shown in Fig. 6-16. The changes in distillate concentrations x_{Di} clearly indicate that, as distillation progresses, the individual components of the feed mixture are successively obtained in highly pure states in the distillate. The pure substances are collected in order of decreasing volatilities by changing the receiver at the appropriate time. The still is gradually depleted from the low-boiling component *a* early in distillation; concentrations of the higher boiling substances *b*, *c*, and *d* increase simultaneously. When most of substance *a* has been separated, the still is depleted from substance *b*, followed by substance *c*. Finally, the highest boiling substance *d* is obtained in pure form in both the boiler and the distillate.

Batchwise single-stage distillation of the same mixture is shown in Fig. 3-20. Comparison of Figs. 3-20 and 6-16 shows clearly that multistage distillation produces a much sharper separation than single-stage distillation. Furthermore, multistage distillation can be considerably improved by using a larger number of equilibrium stages n or a higher value of reflux ratio R_L.

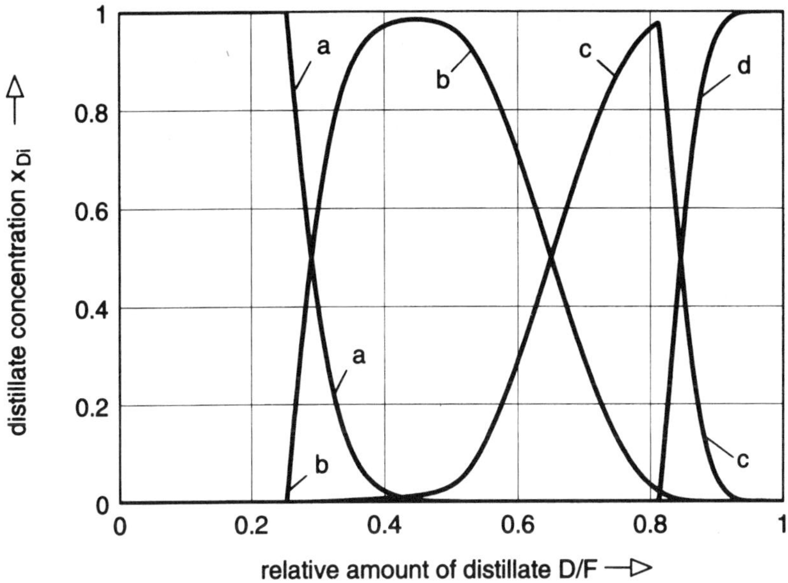

Fig. 6-16. Batch distillation of an ideal quaternary mixture with relative volatilities $\alpha_{ad} = 10$, $\alpha_{bd} = 3$ and $\alpha_{cd} = 2$. Number of stages $n = 20$ and reflux ratio $R_L = 5$. The distillate concentrations x_{Di} are plotted versus the relative amount of distillate D/F. Each constituent of the mixture is recovered in pure form in consecutive periods of time.

To very difficult separations a special mode of operation may be effectively applied. Here, the column is operated with total reflux for a certain period of time. Then, a small amount of top product is withdrawn. Both steps are repeated several times. This alternate operation yields a very pure distillate at the expense of a high energy demand [Heck et al. 1984].

The operating mode described in Section 6.1.3 for minimum energy requirements can, in principle, also be extended to multicomponent mixtures. The calculation procedure is described in [Robinson 1969]. In addition, a chemical reaction in the liquid can be taken into account [Reuter et al. 1989]. The calculation procedures contain a number of simplifications. For instance, a straight operating line is assumed in each case, but this is valid only when the vapor and liquid loads in the column are constant (see Section 4.2.2).

6.4 INFLUENCE OF COLUMN LIQUID HOLDUP ON BATCH DISTILLATION

The liquid holdup H_L, always present in a distillation column, can greatly influence both the yield and the purity of the products of batch distillation. The yield is affected by the column liquid holdup since the sharp decrease and increase of distillate concentrations is hindered. Hence, the intermediate fractions that are off specifications are larger. The influence of column liquid holdup on distillate purity is not so obvious. Information in literature is ambiguous [Mayur et al. 1971, Stewart et al. 1973, Luyben 1971]. Some authors find that a high liquid holdup in the column may even enhance product purity, e.g., [Zuiderweg 1953].

In order to study the influence of liquid holdup H_{Lj} on batch distillation, the dynamic behavior of the column must be considered. The fundamental equations are:

Material balances for $i = 1 \cdots k$ components:

$$L_{j-1} \cdot x_{i,j-1} + G_{j+1} \cdot y_{i,j+1} + F_j \cdot z_{i,j} - (L_j + S_{L,j}) \cdot x_{i,j} - (G_j + S_{G,j}) \cdot y_{i,j} = \frac{\mathbf{dH_{Lj}\,x_{i,j}}}{\mathbf{d\tau}} \tag{6.43}$$

Equilibria for $i = 1 \cdots k$ components:

$$y_{i,j} - K_{i,j} \cdot x_{i,j} = 0 \quad \text{with } K_{i,j} = f(t_j, p_j, x_{i,j}, y_{i,j}) \tag{6.44}$$

Summation of mole fractions in liquid and in vapor phase for $i = 1 \cdots k$ components

$$\sum_{i=1}^{k} x_{i,j} - 1.0 = 0 \tag{6.45}$$

$$\sum_{i=1}^{k} y_{i,j} - 1.0 = 0 \tag{6.46}$$

Heat balances:

$$L_{j-1} \cdot h_{L,j-1} + G_{j+1} \cdot h_{G,j+1} + F_j \cdot h_{F,j} - (L_j + S_{L,j}) \cdot h_{L,j} - (G_j + S_{G,j}) \cdot h_{G,j} + Q_j = \frac{\mathbf{dH_{L,j}\,H_{L,j}}}{\mathbf{d\tau}}$$

$$\text{with } h_{L,j} = f(t_j, p_j, x_{i,j}) \text{ and } h_{G,j} = f(t_j, p_j, y_{i,j}) \tag{6.47}$$

This system of equations contrasts from that for steady-state column operation only by the bold terms that account for the fact that both material and heat may be accumulated in the column at nonsteady state operation. Equations (6.43) to (6.47) represent a nonlinear system of mixed differential and algebraic equations. Mass and energy balances are given by differential equations, further relations between the variables, e.g., vapor-liquid equilibria and molefraction constraints, being described by algebraic equations. Numerous calculation procedures for the solution of Equations (6.43) to (6.47) have been developed. They can be classified into two types.

The first type of computer program uses a rigorous calculation algorithm for the integration of the differential equations. Numerical integration has been performed using the Euler method [Luyben 1971], the method of difference approximation [Meadows 1963], the second-order Runge-Kutta method [Robinson 1970], the fourth-order Runge-Kutta and fifth-order Milne methods [Goldman 1970] and the fifth-order Hamming method [Keith and Brunet 1971]. In all these methods, accurate and stable performance of the numerical integration requires very short time intervals dt and, hence, a very long computation time [Sadotomo and Miyahara 1983].

The second type of computer program simulates the nonsteady state behavior of a distillation column by a sequence of a finite number of steady states of short duration. For each time interval dt the solution is carried out using a mathematical model corresponding to continuous distillation (see Chapter 4.4.1). Holdup is taken into account [Rose 1979]. The robustness of the procedure is the same as for a continuous distillation model and is not influenced by the time integration procedure [Galindez and Fredenslund 1988]. The computer time needed for the simulation is much shorter than with more rigorous simulations without any significant loss of accuracy [Nad and Spiegel 1987].

Figure 6-17 shows the result of evaluating Equations (6.43) to (6.47) for the system methanol/ethanol/2-propanol. Of prime interest is the influence of liquid holdup in the column on the variation of distillate concentrations [Block et al. 1978]. A liquid holdup that is high compared to the amount of liquid in the vessel is very disadvantageous. The holdup smooths the changes in distillate concentration, thus impeding the recovery of highly concentrated top fractions in the distillate receivers.

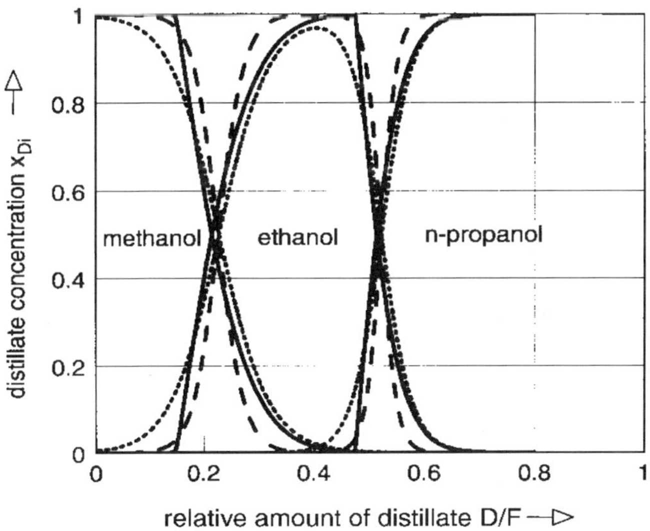

Fig. 6-17. Course of distillate concentration x_{Di} versus D/F of the mixture methanol/ethanol/propanol. The variations of distillate concentrations are significantly influenced by the liquid holdup H_L in the column referred to the holdup in the vessel. ······· H_L = 20%; ----- H_L = 10%; —— H_L = 0%.

However, the recovery of the intermediate boiler *b* (ethanol) is favored by a moderate relative liquid holdup in the column.

6.5 PROCESSES FOR SEPARATING ZEOTROPIC MIXTURES BY BATCH DISTILLATION

Batch distillation has the advantage of being able to completely separate multicomponent mixtures into their pure constituents in a single column. The pure constituents are drawn off as overhead fractions in the order of increasing boiling points. However, during the transition from one to the other pure component, an intermediate fraction that is off specification is formed. This intermediate fraction is often called slop cut or off-cut. The amount and concentration of the slop cuts depend on the separation efficiency and the holdup of the column as has been demonstrated in Sections 6.2.1 and 6.4 (see Figs. 6-12, 6-13 and 6-17). In high efficiency columns the slop cuts are small and approximately binary. This is seen, for instance, in Fig. 6-16. Thus, a slop cut is a prepurified fraction that does not meet the product specifications. However, its quality is much better than that of the feed.

A complete distillation process for separating a zeotropic mixture consists of a sequence of operating cycles. Each of the cycles can be divided in several steps:

- Start up at total reflux
- Withdrawal of the lightest component *a*
- Removal of an intermediate *a*-*b* fraction (slop cut)
- Withdrawal of the second lightest component *b*
- Removal of an intermediate *b*-*c* fraction (slop cut)
- Withdrawal of the third lightest component *c*
- etc.

The performance of a batch distillation process is significantly influenced by the way of slop cut handling. Several strategies of slop cut handling have been developed [Luyben 1988, Quintero-Marmol and Luyben 1990]. These strategies include total slop cut recycling, binary distillation of the accumulated slop cuts and recycling of the slop cuts at the appropriate time.

6.5.1 Total Slop Cut Recycling

All the slop cuts of the previous batch are combined, recycled to the still and mixed with the fresh feed of the next batch. Hence, both the amount and the concentrations of the succeeding batch are changed. That cyclic operation does converge to a steady-state balance after only a few cycles as has been proven by [Chiotti et al. 1993]. Obviously, this mode of slop cut handling is not the most efficient one since the preconcentrated slop cuts are remixed with the feed. The recycling of the binary slop cuts results in a multicomponent mixture. However, this is certainly the most simple and most widely used strategy in practice.

6.5.2 Binary Distillation of Accumulated Slop Cuts

In this mode of slop cut handling, all the slop cuts of the same mixture, e.g., *a-b*, *b-c*, *c-d*, etc., are saved in a segregated tankage and, when enough material has accumulated, each of the binary slop cuts is distilled in a batch distillation column. The resulting binary slop cuts are further processed with the next accumulated slop cut of the same mixture.

6.5.3 Recycling of Slop Cuts at the Appropriate Time

Here, each of the individual slop cuts of the previous cycle is fed into the column at an appropriate time and tray during the next batch distillation cycle. In optimal operation, the state of the liquid at the selected tray should be identical with the state of the slop cut. Thus, any thermodynamic irreversibilities due to mixing are effectively avoided. A modification of this strategy is to fill the *a-b* slop cut into the reflux vessel and the *b-c* slop cut into the column for the start up of the next cycle. The aim of all these modifications is not to spoil the prepurified slop cuts and to avoid mixing of fractions with different composition. However, these slop cut handling strategies require a highly sophisticated system of process control.

6.6 PROCESSES FOR SEPARATING AZEOTROPIC MIXTURES BY BATCH DISTILLATION

The separation of azeotropic mixtures by distillation is very difficult since the difference between vapor and liquid concentration, y-x, is zero at the azeotrope. Hence, there exists no longer a driving force for mass transfer in a distillation column. Separation can be accomplished only by more complex processes. The principles of such processes are outlined in detail in Section 5.6 for steady-state distillation processes. Especially important and effective are those processes that involve a separating agent, called entrainer *e*. As is explained in Section 5.6, two, or even three distillation columns are required for separating a binary azeotropic mixture with the help of an entrainer.

In batchwise operation, different separations are performed in a single distillation column at different periods of time. Thus, a sequence of columns of a continuous process is replaced by a sequence of operation cycles of a single column in a batch process. Hence, only a single column, operated with different operation conditions that vary with time is required.

Three classes of processes can be distinguished:

- Processes in one distillation field
- Processes in two distillation fields

- Hybrid processes that include, in addition to distillation, other separation unit operations (e.g., decantation, absorption, desorption, extraction, adsorption and membrane separations)

6.6.1 Processes in One Distillation Field

This class of processes has been studied extensively by Doherty and coworkers, e.g., [Bernot, et al. 1991]. The starting mixture is supposed to form either a binary minimum or a binary maximum azeotrope. The entrainer *e* that enables separation must be selected so that both products, *a* and *b,* lie in the same distillation field. This condition is most simply met if the boiling point of the entrainer lies between the boiling points of the constituents *a* and *b* of the mixture at hand. Another possibility is the selection of an entrainer that forms either an intermediate boiling maximum azeotrope with the low boiler *a* or an intermediate boiling minimum azeotrope with the high boiler *b* (see Table 5-6).

Figure 6-18 presents the process for the case of an *a*-*b* maximum azeotrope. The feed F consisting of a mixture of *a* and *b* is fed together with the entrainer *e* into the still. From the mixture $B1_\alpha$ the low boiler *a* is separated as pure overhead fraction D1 of a rectifying column.

The bottoms depletes from the low boiler during operation. Hence, the state of the liquid in the still, B1, moves on a straight line from $B1_\alpha$ to $B1_\omega$. When all low boiler has been separated, the intermediate boiling entrainer *e* forms the overhead fraction D2 during the second period of a complete cycle. The state of the liquid in the still, B2, moves on the base of the triangular diagram towards the *b*-corner. When all entrainer has been removed, pure high boiler *b* is left in the still. After removal of all the bottoms fraction $B2_\omega$ from the still a new process cycle is started with a fresh batch.

An analogous process for the separation of a binary minimum azeotrope is presented in Fig. 6-19. Here, the vessel is located at the top of a stripping column. Mixing the feed with the intermediate boiling entrainer *e* gives the mixture $D1_\alpha$ from which the high boiler *b* is removed as bottoms B1 in a stripping column during the first part of a complete cycle. In the second part, which begins when all high boiler *b* has been removed, the entrainer *e,* now being the highest boiling constituent, forms the bottom fraction. At the end of this part the low boiler *a* remains in the vessel.

The processes depicted in Figs. 6-18 and 6-19 can advantageously be performed in a column with a middle vessel, see Fig. 6-20. Here, the column consists of a rectifying and a stripping section, and the low boiler *a* and the high boiler *b* are simultaneously separated from the mixture. During operation, the residuum in the middle vessel changes along a straight line into the direction of the entrainer *e*. Thus, the entrainer concentrates in the vessel. In the next batch just the starting mixture F is fed into the vessel. The entrainer remains in the vessel and need not be handled externally.

This class of processes for separating azeotropic mixtures is very simple and

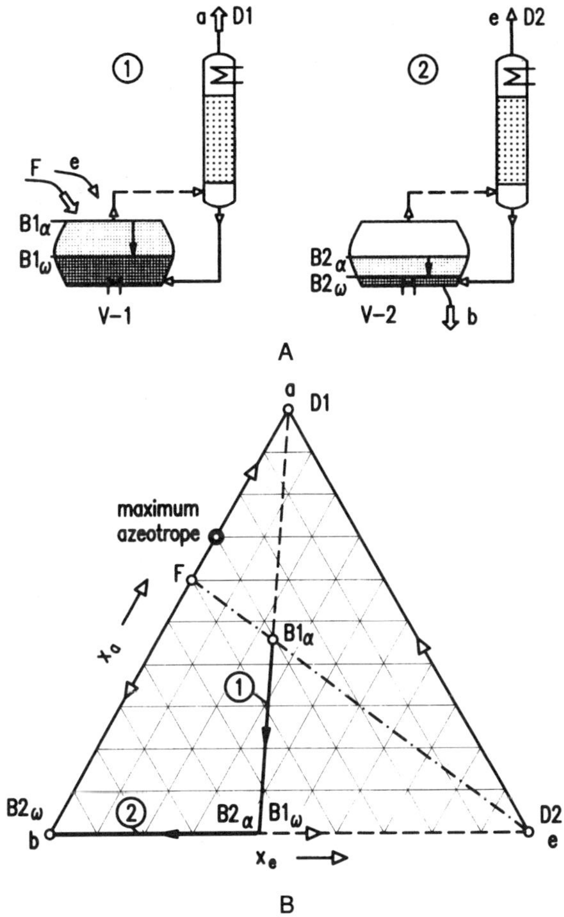

Fig. 6-18. Batch process for separating a binary mixture a-b *that exhibits a maximum azeotrope with the help of an intermediate boiling entrainer* e. *A full cycle consists of two parts performed in consecutive periods of time: (A) Operation sequence, (B) concentration change of the residuum in the vessel.*

elegant. However, application to practical problems is seldom possible since the entrainer *e* has to be the intermediate boiling constituent. Azeotropes are most likely formed by close boiling mixtures. Hence, most of the azeotropic mixtures to be separated by distillation are close boiling ones. Therefore, it might be difficult or even impossible to find an entrainer that boils between. The schemes presented in Fig. 6-18 and 6-20 can only be applied to mixtures a-b with very large boiling point differences.

6.6.2 Processes in Two Distillation Fields

In this class of processes, a border distillation line begins or ends at the a-b azeotrope to make the constituents a and b lie in different but adjoining distillation fields. The

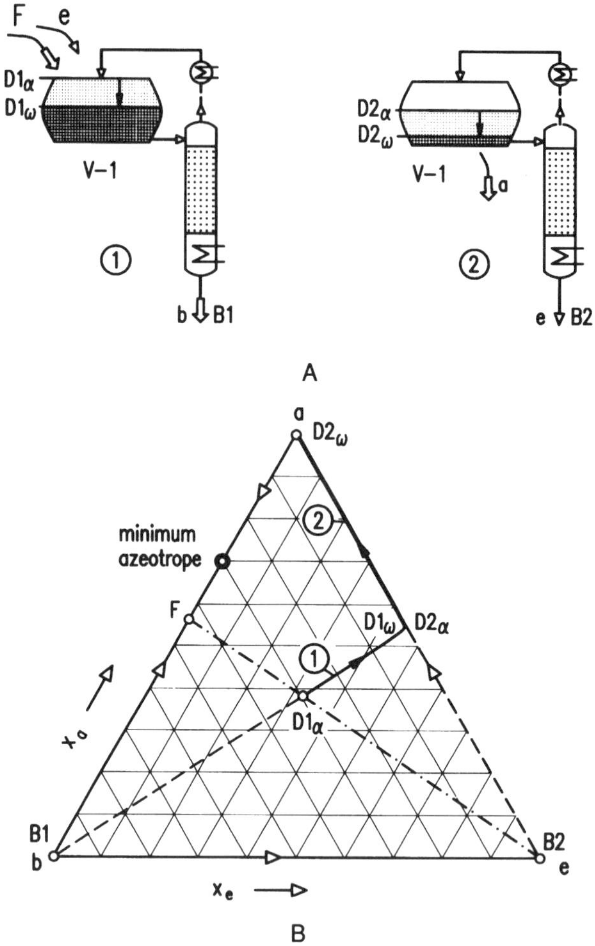

Fig. 6-19. Batch process for separating a binary mixture that exhibits a minimum azeotrope. The vessel must be located at the top of the column: (A) Operation sequence, (B) concentration changes of the residuum in the vessel.

criteria for entrainer selection are developed in detail in Section 5.6. For a minimum azeotrope the entrainer has to be the low boiler or has to form low boiling binary azeotropes. For a maximum azeotrope, a high boiler or a compound that forms binary maximum azeotropes is suited as entrainer (see Table 5.7).

A process that separates an *a*-*b* maximum azeotrope with the help of a high boiling entrainer *e* is presented in Figure 6-21 [Duessel and Stichlmair 1995]. Here, a border distillation line runs from the entrainer *e* to the *a*-*b* azeotrope. The mixture is divided into two distillation fields with the constituents *a* and *b* lying in different fields.

The process needs two vessels located at the bottom of the column and a distilla-

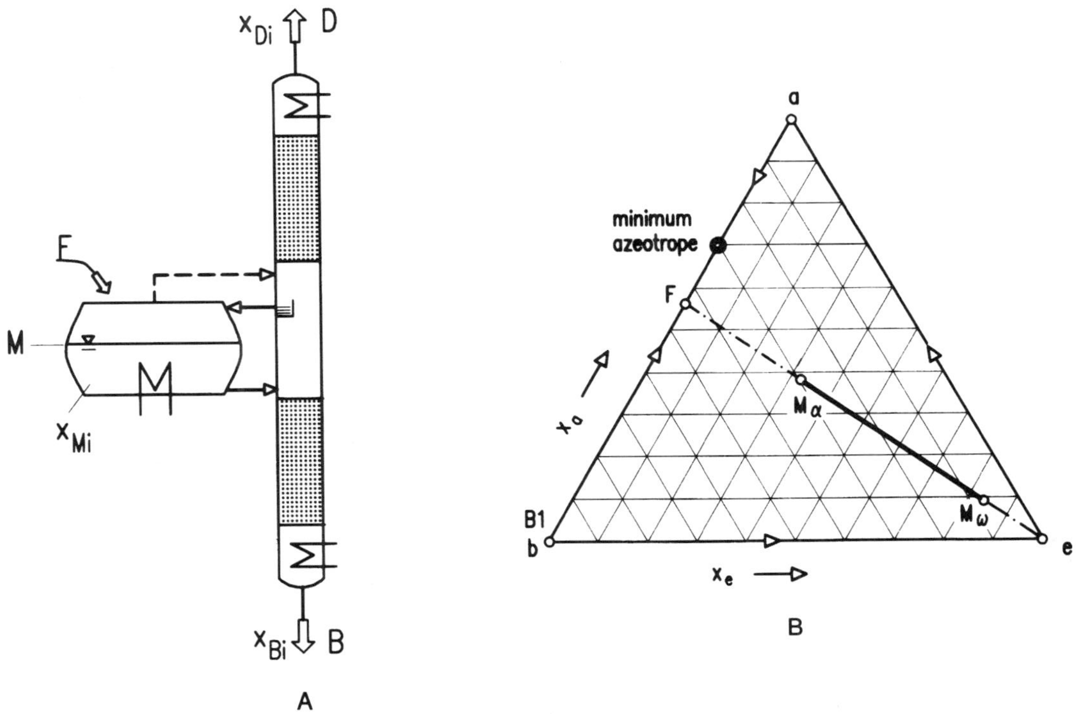

Fig. 6-20. *Process for separating a binary azeotropic mixture by batch distillation with a middle vessel.*

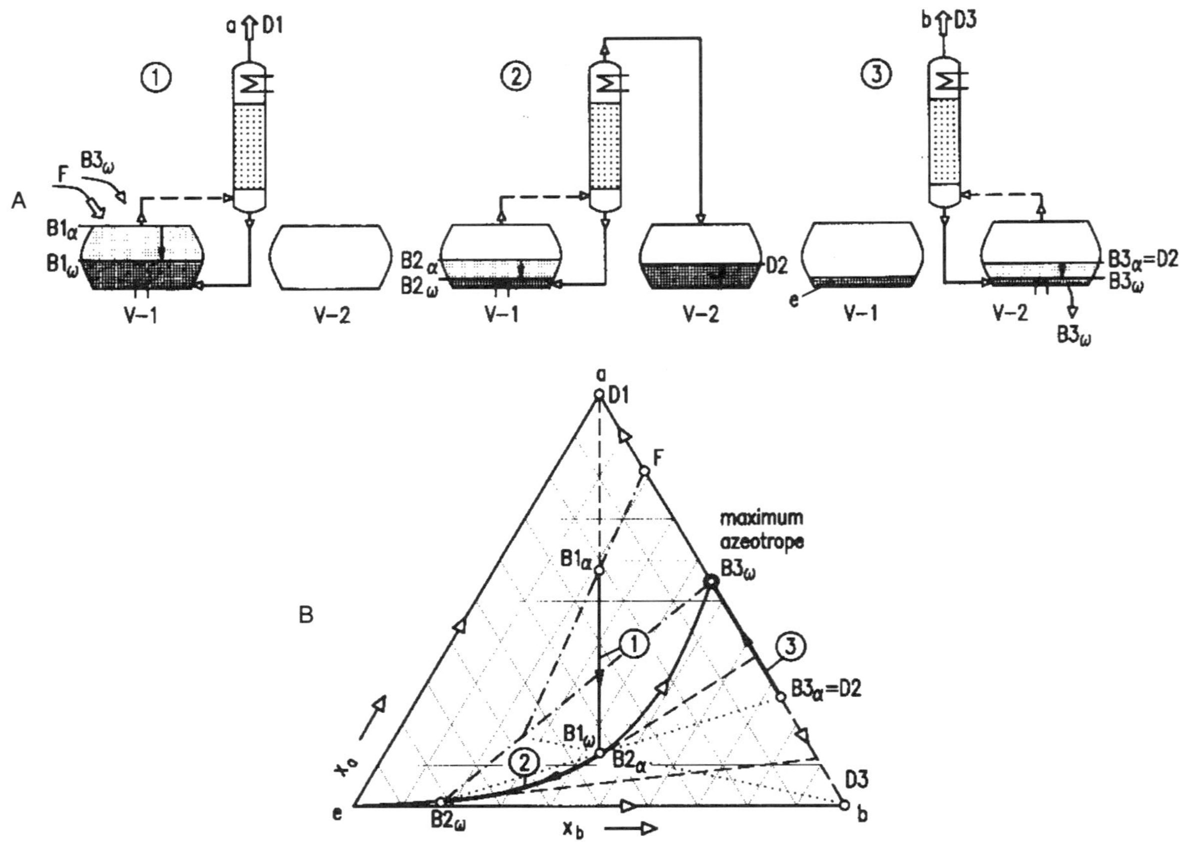

Fig. 6-21. *Batch process for separating a binary mixture with a maximum azeotrope. The entrainer* e *is the high boiler in the system. A full cycle consists of three time-consecutive parts. Two vessels and one column are required: (A) Operation sequence, (B) concentration changes of the residuum in the vessel.*

tion column that consists of a rectifying section only. The entire process cycle consists of three parts. At the beginning of the first part, the starting mixture F is fed into the left vessel and completely mixed with the residua $B2_\omega$ and $B3_\omega$ of the previous cycle. The mixture $B1_\alpha$ lies in that distillation field in which the compound *a* has the lowest boiling temperature and is, therefore, separated in pure form as overhead fraction of the rectifying column. The state of the liquid in the vessel changes continuously from $B1_\alpha$ to $B1_\omega$. All states lie on a straight line through $B1_\alpha$ and D1. The first part of the cycle is finished when the state $B1_\omega$ reaches the distillation border.

In the second part of the cycle, the state of the liquid in the still moves along the distillation border from $B2_\alpha$ to $B2_\omega$. The overhead fraction lies on the tangent to the actual state of B2 to the distillation border. If the separation is high, the actual state of the distillate is found by the intersection of the tangent with the right-hand side of the triangular diagram of Figure 6-21B. No pure constituent is separated during the second part of the cycle. Hence, the distillate D2 has to be collected in the vessel V-2. The material balance for the second part requires that the state of the distillate D2 lies on a straight line through $B2_\alpha$ and $B2_\omega$ (dotted line in Fig. 6-21B). At the end of the second part an entrainer-rich fraction $B2_\omega$ remains in the vessel V-1.

In the third part, the liquid in the vessel V-2 is boiled and the constitutent *b* is separated in pure form as overhead fraction of the column. The state of the liquid B3 in the vessel moves from $B3_\alpha$ to $B3_\omega$, which is the azeotrope. At the end of part 3 of the cycle, the fraction $B3_\omega$ is pumped from vessel V-2 into vessel V-1 and mixed with the residuum $B2_\omega$ of part 2. A new cycle starts by feeding the starting mixture F into the vessel V-1.

An analogous process allows the separation of a binary mixture *a*-*b* that exhibits a minimum azeotrope. In this case, the two vessels V-1 and V-2 are located at the top of the column that is operated as a stripping column (see Fig. 6-22). The starting liquid F is mixed with the residuum of part 2 of the previous cycle $D2_\omega$ (= the entrainer) and the residuum of part 3 $D3_\omega$ (= the binary azeotrope) to give fraction $D1_\alpha$. This fraction lies in that distillation field where high boiler *b* is removed as bottoms. The state of the liquid D1 in vessel V-1 moves on a straight line from $D1_\alpha$ to $D1_\omega$. During part 2 of the cycle, the state of the liquid in vessel V-1 moves on the distillation border from $D2_\alpha$ to $D2_\omega$, and a binary mixture of B2 is removed that lies in the other distillation field. This fraction is collected in the vessel V-2 to be separated in part 3 of the cycle in pure low boiler *a* and the fraction $D3_\omega$.

The process is very similar to the continuous process presented in Fig. 5-37. Here too, three separation steps are required. In the continuous process, the separations are performed in three different loci (i.e., columns) at the same time. In the batch process, all three separations are performed at the same locus (i.e., a single column) at three different periods of time.

6.6.3 Process Simplifications

The processes presented in Figs. 6-21 and 6-22 require three separation steps that are performed in the same column at different periods of time. In special cases, the

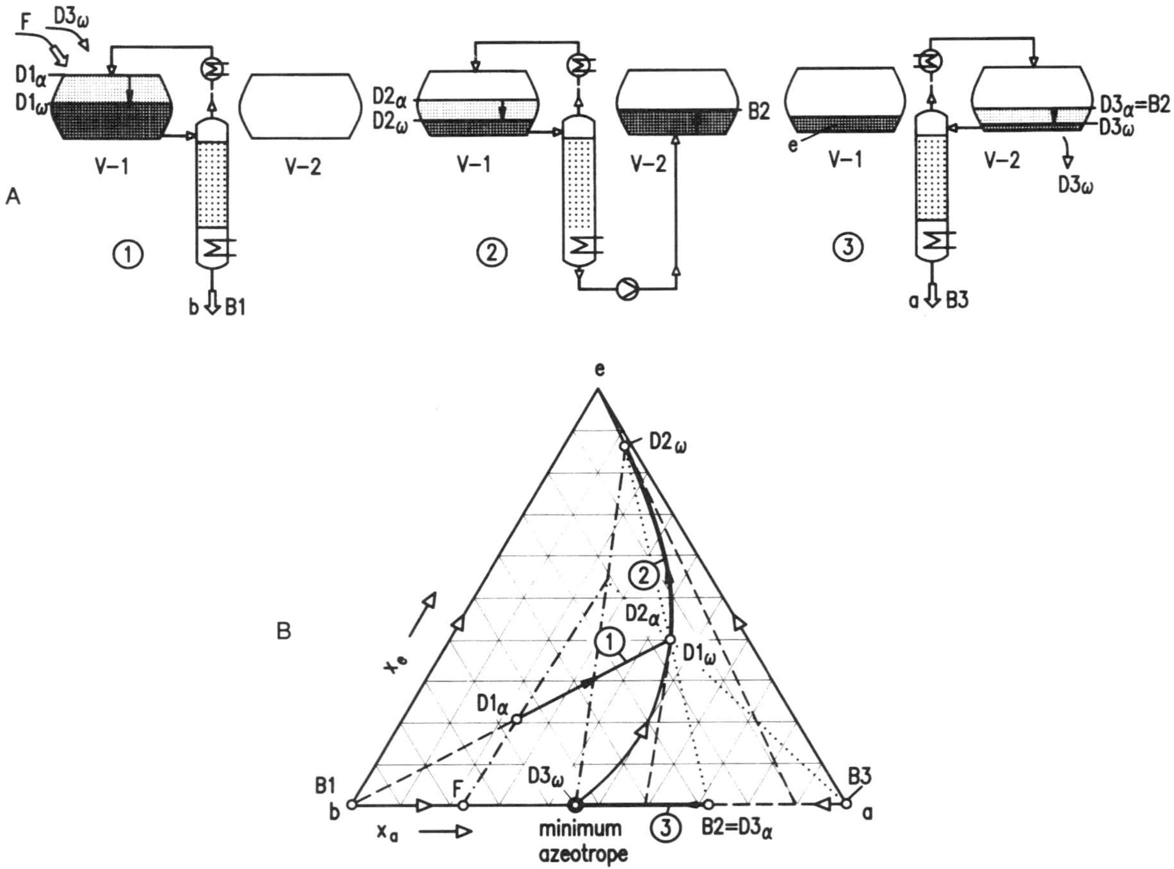

Fig. 6-22. *Batch process for separating a binary mixture with a minimum azeotrope. The entrainer* e *is the low boiler in the system. The two vessels collect the overhead fractions and must be located at the top of the column.*

process can be simplified. The methods of process simplification are analogous to continuous processes dealt with in Section 5.6.2.3. The prerequisites for process simplifications are outlined in detail in this section.

An example of great industrial importance is the separation of the mixture HCl-H_2O that exhibits a maximum azeotrope. An appropriate entrainer is sulfuric acid. A H_2SO_4 concentration of approximately 75 wt-% is sufficient. The starting mixture F is fed into the vessel that contains the entrainer H_2SO_4 as residuum of the previous cycle (Fig. 6-23). The mixture $B1_\alpha$ lies in that distillation field where HCl can be separated as overhead fraction. During operation, the state of the residuum B1 moves from $B1_\alpha$ to $B1_\omega$. The bottom fraction $B1_\omega$ is essentially free of HCl since the distillation border closely joins the base of the triangular diagram. During the second part of a cycle, water is separated as overheads. The cycle is finished when the state of the bottoms $B1_\omega$ in the vessel reaches a concentration high enough to serve as entrainer. Here, the entire cycle consists of only two parts.

This very effective mode of process simplification can often be applied. The only disadvantage is, in most cases, the large amount of the entrainer. However, the entrainer remains in the vessel and automatically performs the separation of the azeotropic mixture into the two pure constituents. The process is operated like a process for separating a zeotropic mixture.

6.6.4 Hybrid Processes

Process simplifications are often possible by replacing one of the distillation steps by a non-distillative separation operation. Some examples are outlined in Section 5.6.3 for continuous separation processes. Here, only the application to batch processes will be demonstrated. For details, please see Section 5.6.3.

6.6.4.1 Azeotropic Batch Distillation A very interesting example for a hybrid process is the separation of the mixture ethanol/water with toluene as entrainer *e*. This type of process is called azeotropic distillation. All the relevant features of the system are discussed in Section 5.6.3.1. The continuous process presented in Fig. 5-45 consists of two distillation columns and one decanter. In the batchwise process, a cycle consists of two distillation and one decantation steps. The separations are performed at consecutive periods of time rather than in consecutive devices (columns and decanter).

In the first part of a cycle (see Fig. 6-24), the ethanol/water mixture and the fraction S1 are fed into vessel V-1 located at the top of the column. The mixture $D1_\alpha$ lies in that distillation field where pure water can be separated as bottoms of a stripping column. The state of the liquid D1 in the vessel moves along a straight line from $D1_\alpha$ to $D1_\omega$, which lies very close by the border distillation line.

At the beginning of the second part of a cycle, the residuum $D1_\omega$ is fed into vessel V-2 and the fraction S2 is added to give fraction $B2_\alpha$. From this mixture the ternary azeotrope, i.e., fraction D2, is removed by a rectifying column. The state of the residuum in the vessel moves from $B2_\alpha$ to $B2_\omega$ along a straight line, see Fig. 6-24B. During this mode of operation the state in the vessel crosses a distillation

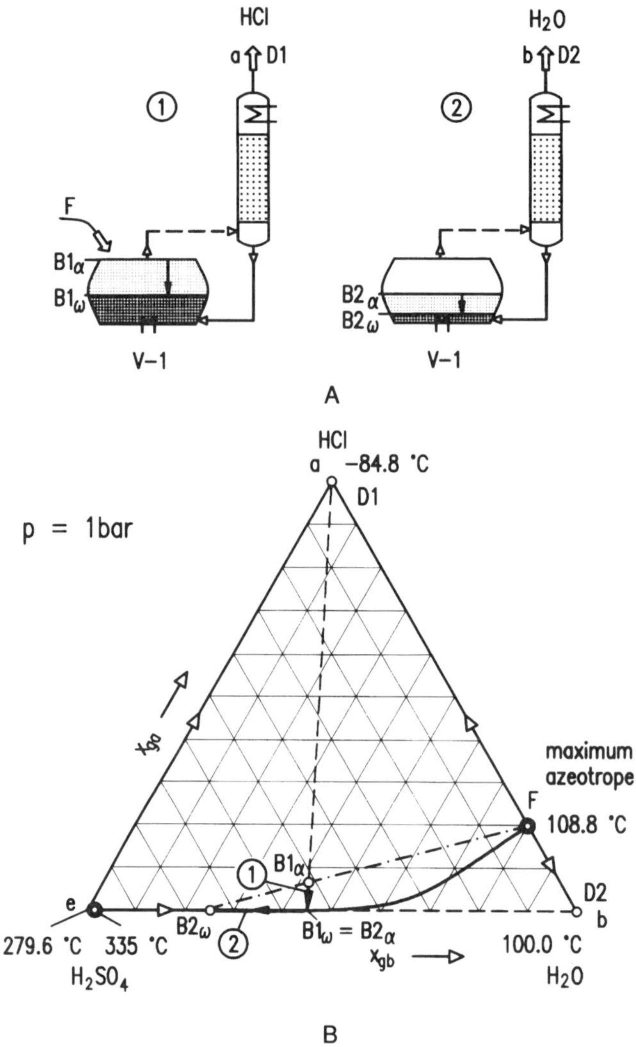

Fig. 6-23. *Batch process for separating the mixture HCl and H_2O by H_2SO_4. A full cycle requires only two parts, since the distillation border approaches the base of the triangular diagram at high sulfuric acid concentrations: (A) Operation sequence, (B) concentration changes of the residuum in the vessel.*

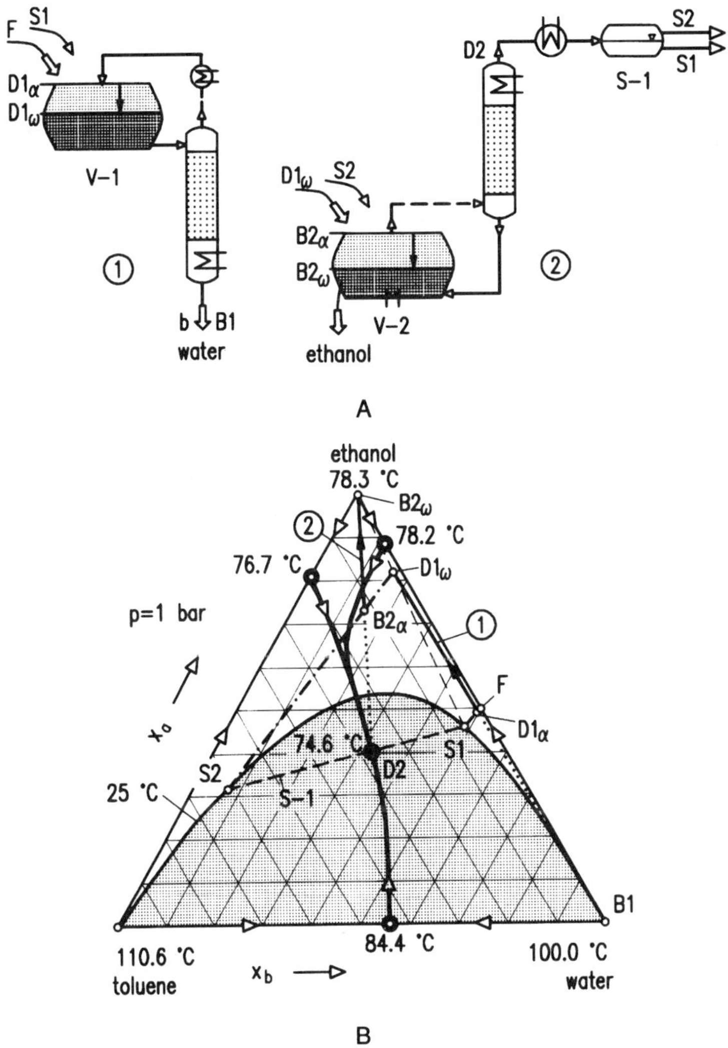

Fig. 6-24. Batch process for separating the mixture ethanol/water by toluene. A decantation step is involved in the second part of a full cycle. This process is referred to as azeotropic distillation: (A) Operation sequence, (B) concentration changes of the residuum in the vessel.

border line. This is possible, since in each distillation field the ternary azeotrope, fraction D2, is the low boiler. The overhead fraction D2 is continuously split into a water-rich fraction S1 and a toluene-rich fraction S2 that are recycled into vessels V-1 and V-2, respectively.

The process requires two vessels, two distillation columns and one decanter. It is referred to as azeotropic distillation.

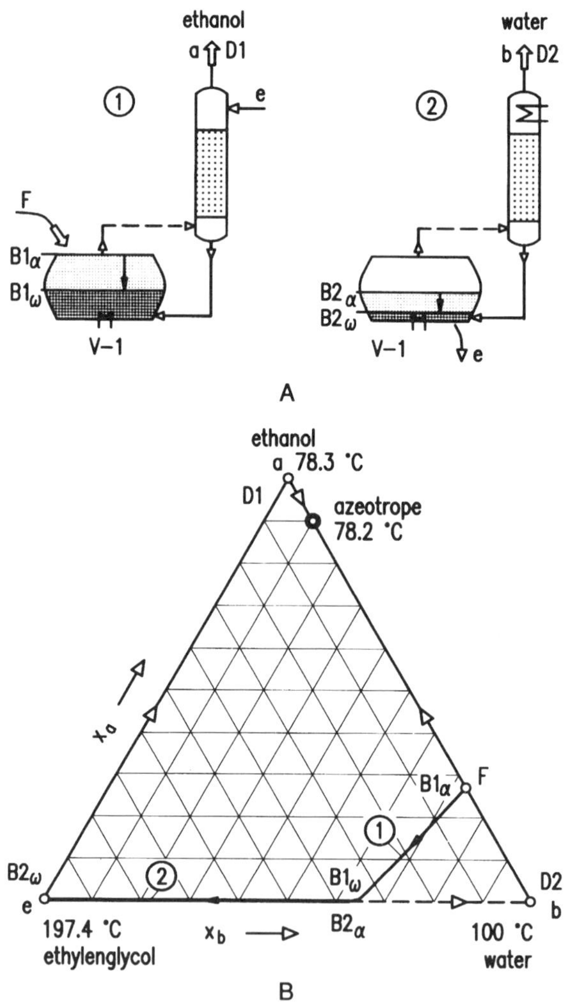

Fig. 6-25. *Batch process for separating the mixture ethanol/water by ethylene glycol. In the first part of a cycle pure ethanol is separated by an absorption process. This process is referred to as extractive distillation: (A) Operation sequence, (B) concentration changes of the residuum in the vessel.*

6.6.4.2 Extractive Batch Distillation Extractive distillation can also be established by batchwise operation, e.g., [Safrit and Westerberg 1996]. The process is explained with the example ethanol/water that exhibits a minimum azeotrope. In the first part of a cycle, the distillation step is replaced by an absorption step. The absorbens has to be a hygroscopic liquid. A good choice is ethylene glycol.

The starting mixture F (ethanol/water) is fed into vessel V-1, see Fig. 6-25. The vapor emerging from the boiling liquid in the still is enriched with ethanol but still contains some water. The water is absorbed by the ethylene glycol by countercurrent contacting in the column. Hence, pure (i.e., water-free) ethanol is the top product. The entrainer is collected in the vessel. The state of the bottoms moves from state $B1_\alpha$ to $B1_\omega$ during the first part of a cycle.

When all ethanol has been removed, the second part of the cycle begins. The binary mixture water-ethylene glycol is separated by batch distillation. Water is the overhead fraction and the residuum in the vessel at the end of the cycle is pure ethylene glycol that is used as entrainer in the next cycle.

The processes presented here for separating azeotropic mixtures by batch distillation prove the efficiency of batch processes. They are as effective as continuous processes but more complex in operation and less complex in device. The significant disadvantage of batch processes to continuous processes is the higher energy demand (see Section 6.1.4).

LITERATURE

Barolo, M.; Guarise, G. B.; Ribon, N.; Rienzi, S.; Treotta, A.; and Macchietto, S.: "Running Batch Distillation in a Column with a Middle Vessel," *Ind. Eng. Chem. Res.* 35 (1996) pp. 4612/4618.

Bernot, C.; Doherty, M. F.; and Malone, M. F.: "Feasibility and Separation Sequencing in Multicomponent Batch Distillation," *Chem. Eng. Science* 46 (1991) No. 5/6, pp. 1311/1326.

Block, U.; Schoenmakers, H.; and Wolf, D.: "Entwurf und Fahrweise diskontinuierlicher Destillationen," *vt"Verfahrenstechnik"* 12 (1978) No. 10, pp. 678/681.

Chiotti, O. J.; Salomone, H. E.; and Iribarren, O. A.: "Selection of Multicomponent Batch Distillation Sequences," *Chem. Eng. Comm.* 119 (1993) pp. 1/21.

Coward, I.: "The Time-Optimal Problem in Binary Batch Distillation," *Chem. Eng. Science* 22 (1967) pp. 503/516.

Davidyan, A.G.; Kiva, V. N.; Meski, G. A.; and Morari, M.: "Batch Distillation in a Column with a Middle Vessel," *Chem. Eng. Sci.* 49 (1994) No. 18, pp. 3033/3051.

Diwekar, U. M.: *Batch Distillation-Simulation, Optimal Design and Control.* Taylor & Francis, Washington, 1995.

D'Souza, A. F.; and Garg, Y. K.: *Advanced Dynamics, Modeling and Analysis.* Prentice-Hall, Englewood Cliffs, 1984.

Duessel, R.; and Stichlmair, J.: "Separation of Azeotropic Mixtures by Batch Distillation Using an Entrainer," *Computers Chem. Eng.* 19 (1995) Suppl. pp. S113/S118.

Ewell, R. H.; and Welch, L.M.: ''Rectification in Ternary Systems Containing Binary Azeotropes,'' *Industrial and Engineering Chemistry* 37 (1945) No. 12, pp. 1224/1231.

Galindez, H.; and Fredenslund A.: ''Simulation of Multicomponent Batch Distillation Processes,'' *Computers Chem. Eng.* 12 (1988) No. 4, pp. 281/288.

Goldman, M. R.: ''Simulating Multicomponent Batch Distillation,'' *Brit. Chem. Eng.* 15 (1970) No. 11, pp. 1450/1453.

Heck, G.; Arnold, D.; and Piening, K.: ''Entscheidungskriterien für die optimale Auslegung und Betriebsweise diskontinuierlicher Destillationsanlagen,'' *Chem.-Ing.-Tech.* 56 (1984) No. 5, pp. 414/415.

Keith, F. M.; and Brunet, J.: ''Optimal Operation of a Batch Packed Distillation Column,'' *Canad. J. Chem. Eng.* 49 (1971), pp. 291.

Luyben, W. L.: ''Some Practical Aspects of Optimal Batch Distillation Design,'' *Ind. Eng. Chem. Process Des. Develop.* 10 (1971) No. 1, pp. 54/59.

Luyben, W. L.: ''Multicomponent Batch Distillation 1. Ternary Systems with Slop Recycle,'' *Ind. Eng. Res.* 27 (1988) No. 4, pp. 642/647.

Mayur, D. N.; and Jackson, R.: ''Time-Optimal Problems in Batch Distillation for Multicomponent Mixtures and for Columns with Holdup,'' *Chem. Eng. Journal* 2 (1971) pp. 150/163.

Meadows, E. L.: ''Multicomponent Batch Distillation Calculations on a Digital Computer,'' *Chem. Eng. Progr.*, Symp. Ser. 59 (1963) pp. 48/55.

Meski, A. G.; and Morari, M.: ''Design and Operation of a Batch Distillation Column with a Middle Vessel,'' *Computers Chem. Eng.* 19 (1995) Suppl. pp. S597/S602.

Nad, M.; Spiegel, L.: ''Simulation of Batch Distillation by Computer and Comparison with Experiment Proc. The Use,'' *Computers Chem. Eng.*, EFCE (1987), pp. 737/742.

Pontryagin, L. S.; Boltyanskii, V. G.; Gamkrelidze, R. V.; and Mishchenko, E.F.: *The Mathematical Theory of Optimal Processes.* Interscience Publishers, New York, 1962.

Quintero-Marmol, E.; and Luyben, W. L.: ''Multicomponent Batch Distillation 2. Comparison of Alternative Slop Handling and Operation Strategies,'' *Ind. Eng. Res.* 29 (1990) No. 9, pp. 1915/1921.

Reinders, W.; and De Minjer, C. H.: ''Vapor-Liquid Equilibria in Ternary Systems, III. The Course of the Distillation Lines in the System Acetone-Chloroform-Benzene,'' *Rec. Trav. Chim.* 59 (1940) pp. 391/406.

Reuter, E.; Wozny, G.; and Jeromin, L.: ''Modeling of Multicomponent Batch Distillation Processes with Chemical Reaction and their Control Systems,'' *Computers Chem. Eng.* 13 (1989) No. 4/5, pp. 499/510.

Robinson, E. R.: ''The Optimisation of Batch Distillation Operations,'' *Chem. Eng. Science* 24 (1969) pp. 1661/1668.

Robinson, E. R.: ''The Optimal Control of an Industrial Batch Distillation Column,'' *Chem. Eng. Science* 25 (1970) pp. 921/928.

Rose, L. M.: ''A Simulation Approach for the Design of Multicomponent Batch Distillation,'' *Chimia* 33 (1979) No. 2, pp. 59/64.

Sadotomo, H.; and Miyahara, K.: ''Calculation Procedure for Multicomponent Batch Distillation,'' *International Chemical Engineering* 23 (1983) No. 1, pp. 56/64.

Safrit, B. T.; and Westerberg, A. W.: ''Improved Operational Policies for Batch Extractive Distillation Columns.'' Engineering Research Center, Carnegie Mellon University, EDRC 06-222-1996

Sorensen, E.; and Skogestad, S.: "Comparison of Regular and Inverted Batch Distillation," *Chem. Eng. Sci.* 51 (1996) No. 22, pp. 4949/4962.

Stewart, R. R.; Weisman, E.; Goodwin, B. M.; and Speight, Ch. E.: "Effect of Design Parameters in Multicomponent Batch Distillation," *Ind. Eng. Chem. Process. Des. Develop.,* 12 (1973) No. 2, pp. 130/136.

Zuiderweg, F. J.: "Absatzweise Destillation, Einfluss der Bodenzahl, des Rücklaufverhältnisses und des Holdup auf die Trennschärfe," *Chem.-Ing.-Techn.* 25 (1953) No. 6, pp. 297/308.

7

Energy Economization in Distillation

7.1 Single Columns
7.1.1 Reduction of Heat Requirement
7.1.2 Reduction of Exergy Losses
7.2 Optimal Separation Sequences for Multicomponent Mixtures
7.2.1 Path a
7.2.2 Path c
7.2.3 Path a-c
7.3 Column Coupling
7.3.1 Direct Coupling of Columns
7.3.2 Indirect Coupling of Columns (Thermal Coupling)
7.3.2.1 Binary Mixtures
7.3.2.2 Ternary Mixtures
7.4. Design of Heat Exchanger Networks
7.4.1 Optimum Heat Exchanger Networks
7.4.2 Modifying the Optimum Heat Exchanger Network
7.4.3 Dual Flow Heat Exchangers Networks
7.4.4 Process Modifications
Literature

The energy requirement W for separation of a mixture into its pure components is calculated [King 1980] from:

$$W = -RT \sum x_{Fi} \ln(\gamma_i \, x_{Fi}) \qquad (7.1)$$

This relationship is independent of the separation method and merely describes the reversal of the mixing procedure (see Equation 2.36).

Energy required for distillation is supplied in the form of heat. The exergy E describes the ability of heat to be converted into work:

$$E = Q \frac{T_1 - T_0}{T_1} \tag{7.2}$$

where Q is the amount of heat, T_1 the heat temperature, and T_0 the ambient temperature. The work performed W is calculated from the exergy difference between the heat input temperature T_1 and the heat output temperature T_2:

$$W = Q \cdot T_0 \left(\frac{1}{T_2} - \frac{1}{T_1} \right) \tag{7.3}$$

Thus, the energy demand depends not only on the quantity of heat Q but also on the temperature levels T_1 and T_2.

In a distillation column, a stream of heat $\dot{Q}_R$ is required at the bottom of the column at temperature T_R and a stream of heat $\dot{Q}_C$ is removed at the top at temperature T_C. The following relationship holds:

$$\dot{Q}_R + \dot{Q}_C \cong 0 \quad \text{with } T_R > T_C \tag{7.4}$$

The temperature difference $T_R - T_C$ is usually in the range from 5 to 20°C, so exergy losses are normally small. However, the heat required for distillation is usually supplied by stream at a temperature $T_1 > T_R$ and is removed by cooling water at a temperature $T_2 < T_C$. Therefore, larger amounts of exergy are lost during separation than are actually necessary.

Means of saving energy during distillation include reducing the heat requirement $\dot{Q}_R$ and minimizing the temperature difference $(T_1 - T_2)$ between heat input and output.

7.1 SINGLE COLUMNS

7.1.1 Reduction of Heat Requirement

The heat requirement $\dot{Q}_R$ of a single distillation column can be minimized best by effective heat exchange between the hot products and the cold feed, and also by efficient insulation of the column. Furthermore, the operational reflux ratio R_L should be kept as small as possible; optimum ratios are in the range of $R_L = (1.05 - 1.1) \cdot R_{Lmin}$ [Luedeker 1985].

The thermal state of the feed (caloric factor q_F) greatly influences the heat require-

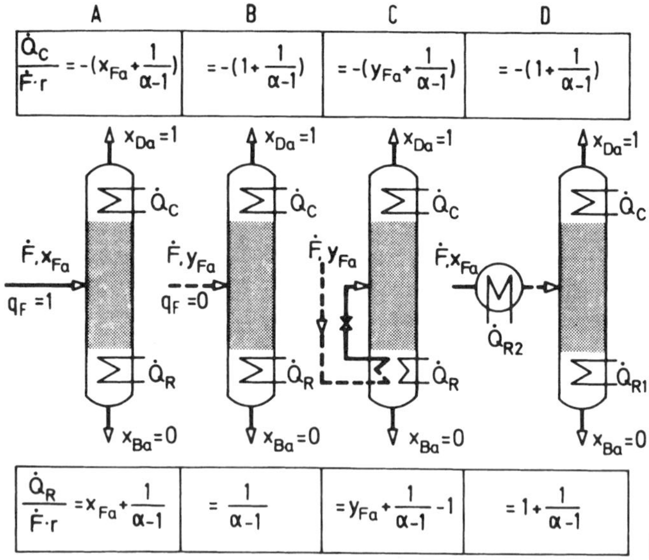

Fig. 7-1. Influence of thermal state q_F of the feed on minimum heat turnover in complete separation of an ideal binary mixture: (A) Liquid feed $q_F = 1$, (B) vapor feed $q_F = 0$, (C) condensation of vapor feed in the reboiler, (D) prevaporization of liquid feed.

ment of a column. The theoretical minimum values of the energy demand $\dot{Q}_{Rmin}/(\dot{F} \cdot r)$ for ideal binary mixtures are given in Fig. 7-1.

The heat requirement of a liquid feed (at boiling point) is greater by the additive term x_{Fa} than that of a vapor feed (see Figs. 7-1A and 7-1B). However, in a vapor feed, the latent heat is utilized only in the upper section of the column. Hence, the vapor feed should be condensed in the reboiler before introducing it into the column (Fig. 7-1C). In this way, the latent heat is utilized in both sections of the column, which leads to a lower external heat requirement $\dot{Q}_R$. The column can be operated without any external heat source if the feed concentration y_{Fa} is low and the relative volatility α is high. This mode of operation is often employed in low-temperature processes, but the feed pressure must be higher than the operating pressure of the column.

Hence, a vapor feed is preferable to a liquid feed. However, a liquid should not be prevaporized because then the total energy requirement is higher than with a liquid feed (Fig. 7-1D).

When the feed is composed of several streams with different composition, multiple feeding should be considered (see Fig. 4-10). Optimal choice of feed points and careful column control are very important.

7.1.2 Reduction of Exergy Losses

The use of intermediate reboilers and condensers (see Fig. 4-11) has frequently been proposed. Although they do not lower the heat requirement of a column, part of the heat required can be supplied at a lower temperature and removed at a higher temperature, i.e., exergy losses are reduced. This method offers practical advantages only if different external heat sources (e.g., steam nets with different pressure) or several heat sinks are available, or if thermal coupling is possible (Section 7.3.2).

Exergy losses can effectively be reduced by utilizing waste heat obtained at the top of the column to produce low-pressure steam (Fig. 7-2A) [Kleinhenz 1980]. No steam is "consumed" during distillation, just high-pressure steam is converted into low-pressure steam. Thus, exergy, and not energy, is consumed. If a low-pressure steam net is not available, steam is returned to the high-pressure net after compression (Fig. 7-2B).

Systematic development of this operation mode leads to incorporation of a heat pump with supplementary circulation (Fig. 7-2C). Here, the question of finding an appropriate working medium for the heat pump always arises. The substances separated in the distillation column can often be used to advantage as a working medium for the heat pump. Consequently, the heat pump has to be integrated into the distillation process. Utilization of the overhead product as a heat pump medium (Fig. 7-3A) is usually more favorable than utilization of the bottom product (Fig. 7-3B).

The integrated heat pump is used in many industrial low-temperature plants, e.g., ethylene plants [Kreuter 1973]. In high-temperature processes, however, a problem occurs because the high molecular substances present have a very small value of the heat capacity ratio $\kappa = c_p/c_v$. Figure 7-4 presents a plot of the heat capacity ratio versus the number of atoms in a molecule. The higher the number of atoms in the molecule the lower is the heat capacity ratio. Consequently, the vapor temperature barely rises on compression of high molecular substances; hence, the dew point may

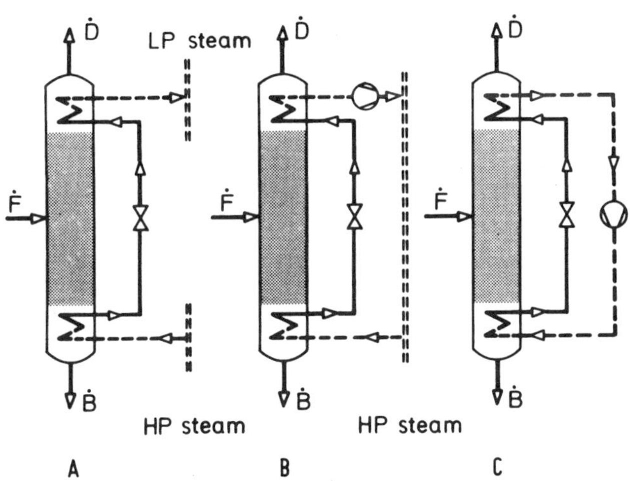

Fig. 7-2. *Reduction of exergy losses during distillation: (A) Generation of low-pressure steam from condensate, (B) compression and recycling of low-pressure steam, (C) heat pump with supplementary circulation.*

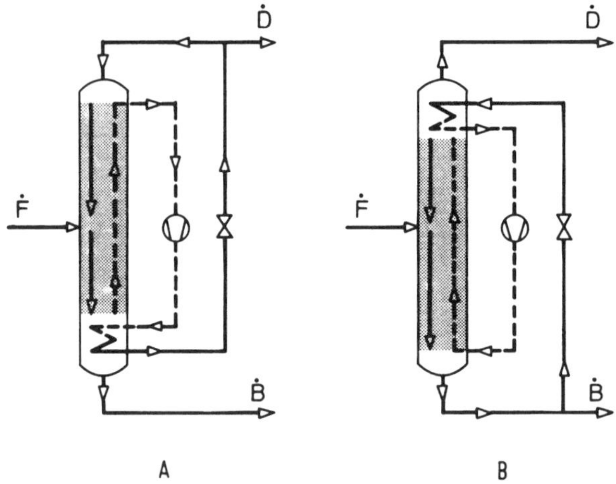

Fig. 7-3. *Schematic of integrated heat pumps that use the products of the distillation column as working medium: (A) The overhead product is the working medium, (B) the bottom product is the working medium.*

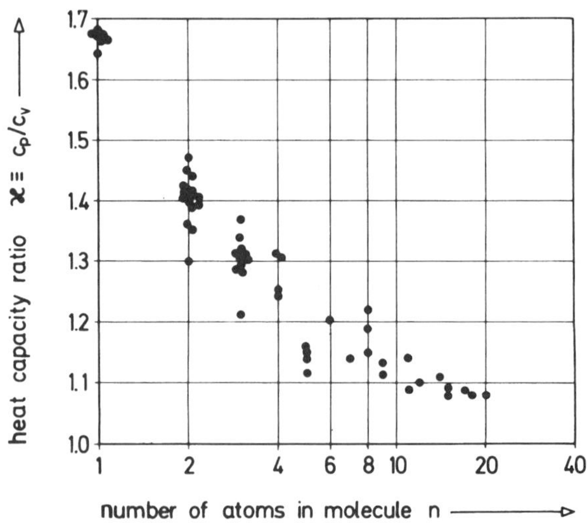

Fig. 7-4. *Values of heat capacity ratios versus number of atoms in a molecule. The data are taken from [Perry, Green et al. 1984].*

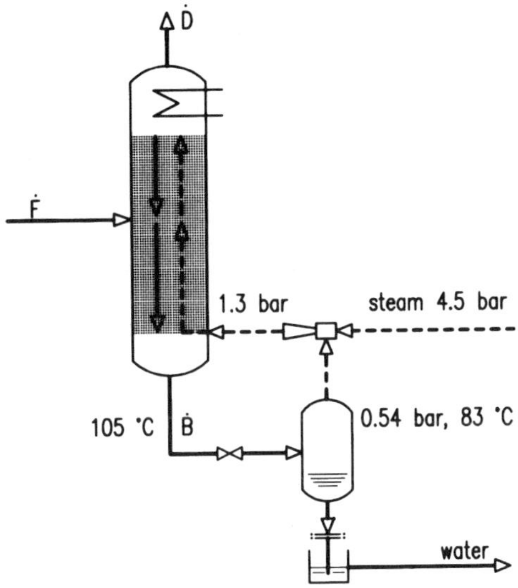

Fig. 7-5. Modification of a heat pump applicable to columns that produce water as bottom product.

be crossed during compression which causes machine damage due to condensation [Kleinhenz 1982]. In some cases the problem can be met by using a compressor with a low efficiency. An example is the separation of 1,2-dichlorethane from high boilers [Dummer and Schmidhammer 1991].

A problem of columns with integrated heat pumps is the start up. In most cases a special start-up reboiler that is operated by steam is installed. However, it is sometimes possible to start column operation with air or nitrogen. These gases have a very high heat capacity ratio κ ($\kappa = 1.4$). As a consequence, the temperature rises very high in the compressor and the operating temperature increases gradually up to the boiling point of the mixture.

A modification of a heat pump that is often applied to columns with a pure water bottom fraction is presented in Fig. 7-5. The bottom liquid (water) is expanded by a live steam-driven ejector. At the lower pressure, the bottom liquid is in superheated state and, consequently, partly evaporated. The emerging vapor is mixed in the ejector with life steam and fed into the column. Such a process reduces the demand of live steam significantly. At the operating conditions given in Fig. 7-5, the demand of life steam is reduced by approximately 30%.

A good example for the reduction of exergy losses is the multiple-stage-flash (MSF) process for the recovery of desalinated water from sea water, e.g., [Greig 1987]. In principle, desalinated water can be recovered by single-stage distillation as shown in Section 3.1 since the vapor is free of salt. However, energy consumption of such a process would be very high because the total heat of vaporization has to be covered by an external heat source.

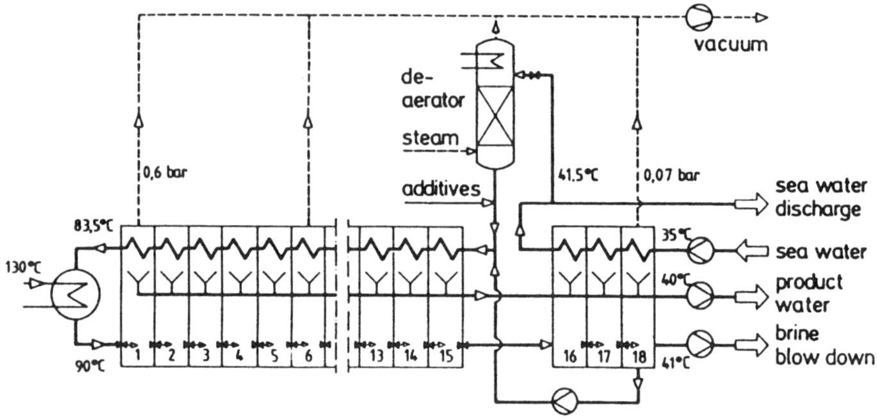

Fig. 7-6. Multiple Stage Flash process for the recovery of desalinated water from sea water.

Figure 7-6 presents the simplified flow diagram of a typical MSF plant. The fresh sea water is first used as cooling agent in the evaporation stages, and is hereby heated to approximately 83°C. It is then heated up to 90°C in an external heat exchanger by low pressure steam. The hot brine is flashed to approximately 0.07 bar in a multistage evaporator typically with 18 stages. The steam generated in each stage is condensed and withdrawn as product water. The liquid load of the evaporator is approximately 12.5 times as high as the product water. To reduce the expenses for sea water make up, the greater part of the brine is recycled. Only 30% of the brine is blown down and replaced by fresh sea water.

In stages 1 to 15, the heat recovery section, an internal heat transfer from boiler to condenser takes place with an average temperature difference of approximately 5 to 6°C. The heat supplied by low pressure steam in the heat exchanger is finally removed in the condensers of stages 16 to 18 by additional sea water. This part is called the heat rejection section.

After stage 16, the sea water is deaerated by life-steam stripping in a packed column to avoid the accumulation of inert gases in the evaporator stages. Additives are neccessary to prevent precipitation and foaming and to reduce the corrosivity of the brine.

The energetic effectivness of the process can be expressed by the performance ratio that represents the ratio of external heat demand to the latent heat of vaporization. Technically feasible are values of 7 to 8. Consequently, only one-seventh of the latent heat of vaporization of water (2330 kJ/kg) has to be provided by an external heat source. The energy consumption of the process is as low as 330 kJ (or 0.09 kWh) per kg product water. In addition, about 0.005 kWh/kg of electric energy is required.

Such MSF plants for sea water desalination are built in huge units with a product water flow of more than 1000 m^3/day. The heat exchanger area of such a unit is as high as 80 000 m^2. The dimensions of a single stage are typically 15 m in width, 4 m in length and 4 m in height [Ali El Saie et al. 1989].

7.2 OPTIMAL SEPARATION SEQUENCES FOR MULTICOMPONENT MIXTURES

As described in Section 5.3, a large number of different separating sequences exists for complete fractionation of multicomponent mixtures. Investment and operating costs depend on the separating path selected. Operating costs are determined primarily by energy costs. This is approximately also true for the investment cost since the amount of vapor determines the diameter of the columns. Heat requirements for the different separating paths in a ternary mixture are presented below.

The minimum heat requirements for separating an ideal ternary mixture *a-b-c* into pure substances by the *a*-, *c*-, and *a-c* paths can be formulated in general terms as described in Chapter 4 and are illustrated in Fig. 7-7.

7.2.1 Path a

In path *a* (Fig. 7-7A) the low boiling substance *a* is removed first as overhead fraction. The remaining binary mixture *b-c* is fed into column C-2 and separated into *b* and *c*. If the feed is a boiling liquid, then, according to Section 4.3.4.2, the following expression holds for the minimum heat requirement of column C-1:

$$\frac{\dot{Q}_{R1,min}}{\dot{F} \cdot r} = (R_{La,min} + 1) \cdot x_{Fa} \tag{7.5}$$

where r is the molar latent heat of vaporization of the feed and $R_{La,min}$ the minimum reflux ratio that is defined in Equation (4.88).

The minimum heat requirement of column C-2 is given by:

$$\frac{\dot{Q}_{R2,min}}{\dot{F} \cdot r} = x_{Fb} + \frac{1 - x_{Fa}}{\alpha_{bc} - 1} \tag{7.6}$$

The heat requirement of the entire process is:

$$\frac{\dot{Q}_{R,min}}{\dot{F} \cdot r} = \left(\frac{\dot{Q}_{R1,min}}{\dot{F} \cdot r} + \frac{\dot{Q}_{R2,min}}{\dot{F} \cdot r} \right) \tag{7.7}$$

Values of $\dot{Q}_{R,min}/(\dot{F} \cdot r)$ for the mixture 2,2-dimethylbutane/2-methyl-pentane/hexane are in the range of 3.4 to 6.3. In Fig. 7-7A, they are plotted in parametric form against the feed composition of the mixture on a triangular diagram. The heat requirement increases drastically with increasing concentration of intermediate boiling substance *b*.

7.2.2 Path c

In path c, the high boiler *c* is removed in column C-1 as bottom fraction (Fig. 7-7B). The overhead product of column C-1 is fed in the vapor state into column C-2, where pure products *a* and *b* are regained. For a boiling-liquid feed, the heat required by column C-1 is (see Section 4.3.4.3):

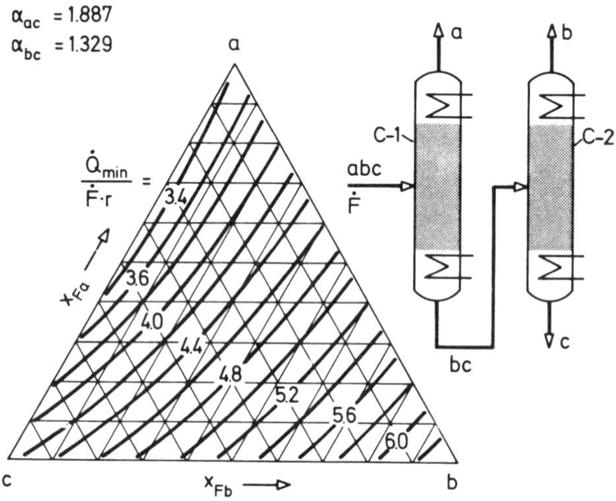

A

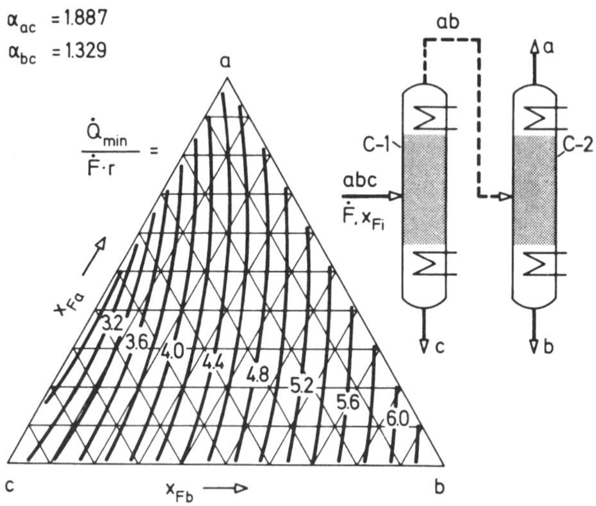

B

Fig. 7-7. *Minimum heat requirement $\dot{Q}_{R,min}/(\dot{F} \cdot r)$ as a function of the feed composition for separating a ternary mixture of 2,2-dimethylbutane (a)/2-methylpentane (b)/hexane (c). Relative volatilities: α_{ac} = 1.887; α_{bc} = 1.329: (A) Path a, (B) Path c.*

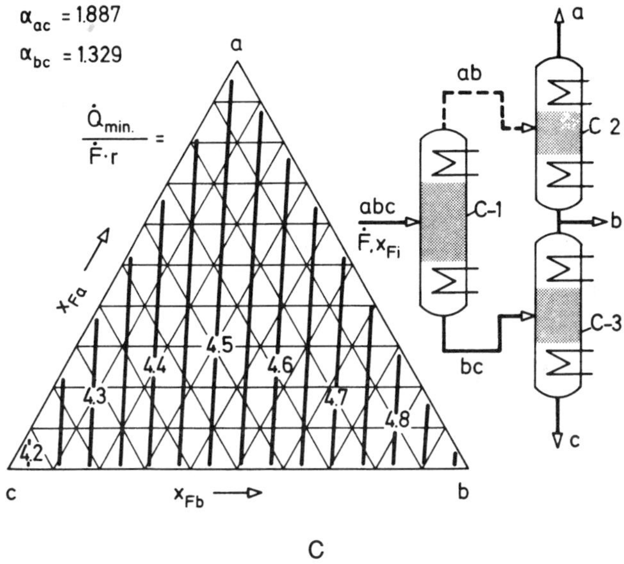

Fig. 7-7. (continued) (C) Path a-c.

$$\frac{\dot{Q}_{R1,min}}{\dot{F} \cdot r} = R_{Lc,min} x_{Fc} \tag{7.8}$$

where $R_{Lc,min}$ is obtained from Equation (4.89). For a liquid feed, the heat required by column C-2 is:

$$\frac{\dot{Q}_{R2,min}}{\dot{F} \cdot r} = x_{Fa} + \frac{1 - x_{Fc}}{\alpha_{ab} - 1} \tag{7.9}$$

For the more favorable vapor feed, the heat required by column C-2 is:

$$\frac{\dot{Q}_{R2,min}}{\dot{F} \cdot r} = \frac{1 - x_{Fc}}{\alpha_{ab} - 1} \tag{7.10}$$

The heat requirement of the entire process is expressed by:

$$\frac{\dot{Q}_{R,min}}{\dot{F} \cdot r} = \left(\frac{\dot{Q}_{R1,min}}{\dot{F} \cdot r} + \frac{\dot{Q}_{R2,min}}{\dot{F} \cdot r} \right) \tag{7.11}$$

The values of $\dot{Q}_{R,min}/(\dot{F} \cdot r)$ for the mixture 2,2-dimethylbutane/2-methylpentane/hexane are in the range of 3.0 to 6.3; here again, the highest values are obtained at high concentrations of the intermediate boiler *b* (Fig. 7-7B).

7.2.3 Path a-c

In path a-c, three distillation columns are required (Fig. 7-7 C). To obtain pure substances, the overhead product from column C-1 must be free of the high boiler *c*. Correspondingly, the bottom product of column C-1 must be free of the low boiler *a*. The overhead product should preferably be fed in the vapor state into column C-2. In the following discussion, the assumption is made that the "preferred" separation is carried out in column C-1 (cf. Section 4.3.4.1). Thus, the reflux ratio and the heat requirement of this column reach an absolute minimum. If the feed is a boiling liquid, Equation (4.94) is applied:

$$\frac{\dot{Q}_{R1,min}}{\dot{F} \cdot r} = \frac{1 + (\alpha_{ac} - 1) \cdot x_{Fa} + (\alpha_{bc} - 1) \cdot x_{Fb}}{\alpha_{ac} - 1} \tag{7.12}$$

The heat requirement of column C-2 for a liquid feed is:

$$\frac{\dot{Q}_{R2,min}}{\dot{F} \cdot r} = x_{Fa} + \frac{(\alpha_{ac} - 1) \cdot x_{Fa} + (\alpha_{bc} - 1) \cdot x_{Fb}}{(\alpha_{ac} - 1) \cdot (\alpha_{ab} - 1)} \tag{7.13}$$

For a vapor feed, the following equation is obtained for column C-2:

$$\frac{\dot{Q}_{R2,min}}{\dot{F} \cdot r} = \frac{(\alpha_{ac} - 1) \cdot x_{Fa} + (\alpha_{bc} - 1) \cdot x_{Fb}}{(\alpha_{ac} - 1) \cdot (\alpha_{ab} - 1)} \tag{7.14}$$

The heat requirement of column C-3 is:

$$\frac{\dot{Q}_{R3,min}}{\dot{F} \cdot r} = x_{Fb} + \frac{\alpha_{ac} - \alpha_{bc}}{\alpha_{ac} - 1} + \frac{\alpha_{ac} - (1 + (\alpha_{ac} - 1) \cdot x_{Fa} + (\alpha_{bc} - 1) \cdot x_{Fb})}{(\alpha_{ac} - 1) \cdot (\alpha_{bc} - 1)} \tag{7.15}$$

For the entire process:

$$\frac{\dot{Q}_{R,min}}{\dot{F} \cdot r} = \frac{\dot{Q}_{R1,min}}{\dot{F} \cdot r} + \frac{\dot{Q}_{R2,min}}{\dot{F} \cdot r} + \frac{\dot{Q}_{R3,min}}{\dot{F} \cdot r} \tag{7.16}$$

In Fig. 7-7C, the minimum heat requirement for the system 2,2-dimethylbutane/2-methyl-pentane/hexane is plotted versus the composition of the starting mixture giving a series of parallel lines. Values for $\dot{Q}_{R,min}/(\dot{F} \cdot r)$ are in the range of 4.2 to 5.0; they increase only slightly with increasing concentration of the intermediate boiling substance *b*.

Comparison of heat requirements of the three separating paths of a ternary mixture shows large variations depending on the composition of the feed. Each separation

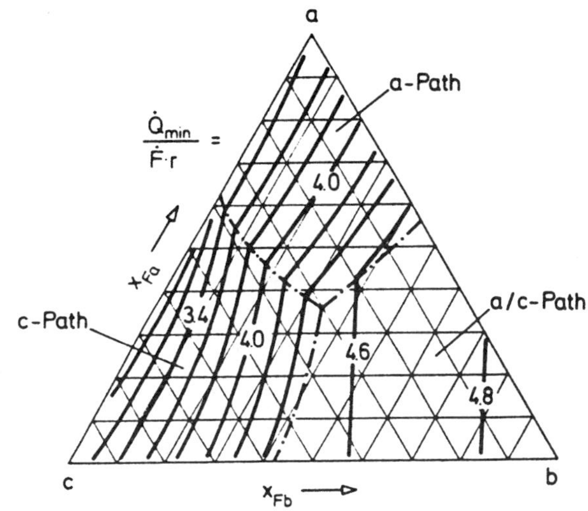

Fig. 7-8. *Comparison of the minimum energy requirement $\dot{Q}_{R,min}/(\dot{F} \cdot r)$ of all three separation paths of a ternary mixture. Each path can be the best choice. At high concentrations of the intermediate boiler* b *the path a-c has the lowest energy demand.*

path can be the best choice (see Fig. 7-8). If the feed contains a high concentration of the low boiler *a,* path a is the most favorable one. However, if the feed contains an excess of the high boiler *c,* path c is preferred. Path a-c is favored when the feed contains an excess of intermediate boiling substance *b*. Path a-c requires three separating columns; i.e., it requires more equipment than the other two paths. However, this sequence is exergetically advantageous because heat is added or removed at several different temperature levels; therefore, its energy saving potential is high. If exploited properly, path a-c gives a more efficient separation than the other two sequences (see Section 7.3.2).

7.3 COLUMN COUPLING

Column coupling may be advantageous in processes with several distillation columns. In direct coupling, columns are coupled by means of their product streams [Schoenmakers 1986]. Indirect or thermal coupling exploits the waste heat from one column to heat a second column. Indirect coupling is a very effective way of substantially reducing the heat requirement of distillation processes and, therefore, is more often used than direct coupling. However, it increases equipment requirements; additional pumps and compressors are usually needed. Direct coupling not only saves energy but also lowers equipment requirements.

7.3.1 Direct Coupling of Columns

Direct coupling of columns C-2 and C-3 is possible in separating a ternary mixture by path a-c, see Fig. 7-7C. Vapors of substance *b* are condensed at the top of column

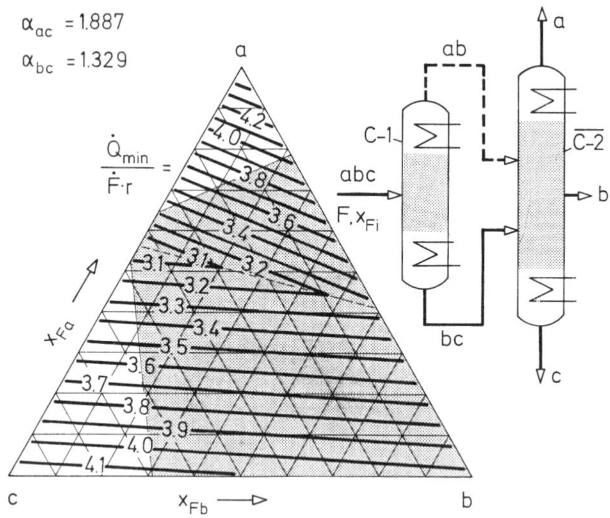

Fig. 7-9. *Minimum heat requirement $\dot{Q}_{R,min}/(\dot{F}\cdot r)$ as a function of the feed composition for the path a-c with direct-column coupling in the separation of a ternary mixture of 2,2-dimethylbutane (a)/2-methylpentane (b)/hexane (c). The shaded area represents the concentration range in which this path is energetically more favorable than path a or path c.*

C-3 to obtain the required reflux. However, substance *b* must also be vaporized at the bottom of column C-2 to establish a countercurrent flow in this column. Hence, columns C-2 and C-3 can be combined by omitting a condenser and a reboiler to give column $\overline{\text{C-2}}$ (Fig. 7-9). The amount of heat converted in the two columns generally differs. In principle, this heat difference can be compensated by an intermediate heat exchanger. However, the equipment requirement is lower if the heat supplied at the bottom of column $\overline{\text{C-2}}$ corresponds to the maximum heat requirement of one of the two original columns C-2 or C-3, whichever is larger. Thus, the heat requirement of the entire process in Fig. 7-9 is:

$$\frac{\dot{Q}_{R,min}}{\dot{F}\cdot r} = \frac{\dot{Q}_{R1,min}}{\dot{F}\cdot r} + \text{Max}\left(\frac{\dot{Q}_{R2,min}}{\dot{F}\cdot r} \text{ or } \frac{\dot{Q}_{R3,min}}{\dot{F}\cdot r}\right) \tag{7.17}$$

The individual amounts of heat are calculated by using Equations (7.12), (7.14) and (7.15). Direct coupling reduces the heat requirements, at best, to approximately two-thirds of the original amount. In Fig. 7-9, Equation (7.17) is applied to the mixture 2,2-dimethylbutane/2-methylpentane/hexane. The heat requirement is presented in parametric form versus the feed composition. The values for $\dot{Q}_{R,min}/(\dot{F}\cdot r)$ are in the range of 3.1 to 4.4, i.e., substantially lower than the values obtained without direct coupling, see Fig. 7-7C. The lowest heat requirements are obtained in the intermediate concentration range in which $\dot{Q}_{R2,min}/(\dot{F}\cdot r)$ and $\dot{Q}_{R3,min}/(\dot{F}\cdot r)$ are almost equal; hence, omission of the smaller value results in the largest energy savings.

As a result of the direct coupling of columns C-2 and C-3, only two columns are neccessary for path a-c. Thus, equipment requirements for all three paths of the separation of a ternary mixture are roughly the same. With regard to the heat requirement, however, path a-c is more efficient over a wider concentration range than the other two paths. The shaded part of Fig. 7-9 represents the optimal region. Comparison with Fig. 7-8 clearly shows that direct coupling is particularly useful for increasing the optimal range of this path.

Another variation in the direct coupling of columns for path a-c is shown in Fig. 7-10B. However, the heat requirement is not lowered because the steam for column C-1 must additionally be generated in column $\overline{\text{C-2}}$. Nevertheless, omission of two heat exchangers of column C-1 reduces the equipment cost.

The column design shown in Fig. 7-10C is still more efficient. Here, columns C-1 and C-2 are combined such to give a single column $\overline{\text{C-2}}$ in which the middle part is divided by a vertical wall [Kaibel 1987]. This design can easily be applied to packed columns and involves lower investment costs. However, its disadvantage is that all the heat is added at the highest temperature and removed at the lowest. In addition, the different separations on the right and left sides of the wall must be carried out at the same packing height. Thus, the amounts of liquid in both parts of the column must be carefully controlled. Furthermore, the heat transfer across the dividing wall must be taken into account in detail engineering [Lestak et al. 1994].

Direct coupling of columns can also be established in the separation of quaternary

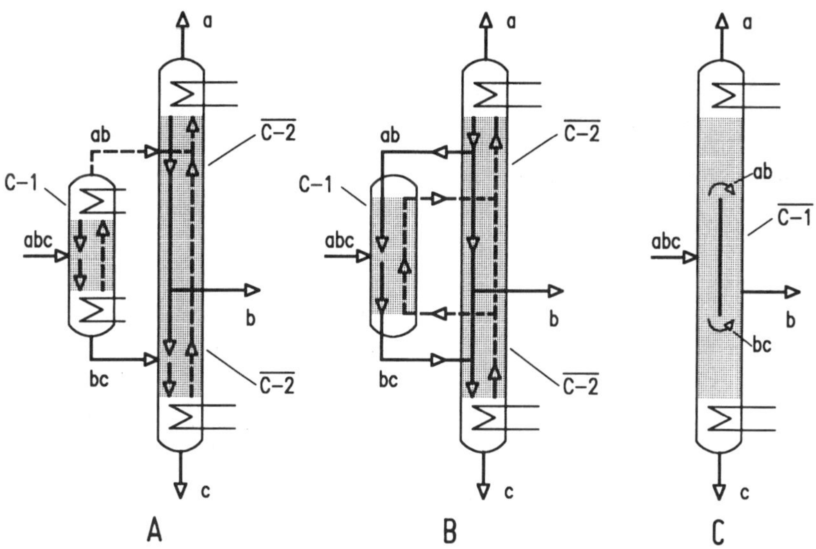

Fig. 7-10. *Simplification of the equipment for separating a ternary mixture by the path a-c with direct-column coupling: (A) Basic flow sheet of Fig. 7-9, (B) omission of the heat exchangers in the first column, (C) combination of two columns to give a single column in which the middle part is divided by a vertical wall.*

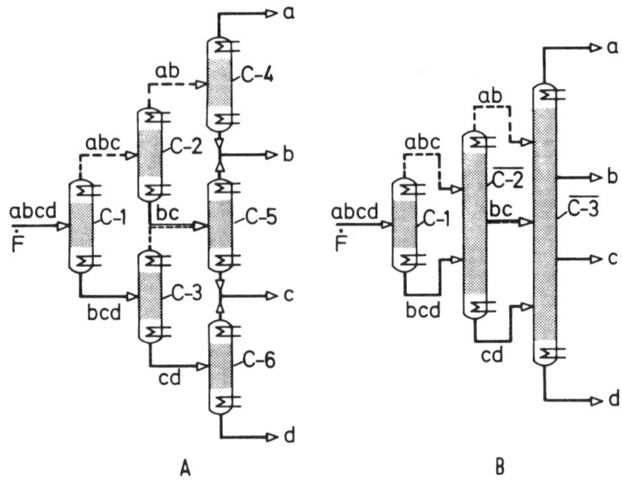

Fig. 7-11. Path a-d for the separation of a quaternary mixture: (A) Process without direct coupling, (B) process with direct coupling.

mixtures. Figure 7-11 shows the path in which the overhead product and the bottom product from each column are free of the high boiler and the low boiler, respectively. If each column is operated at the lowest possible reflux ratio, the preferred separation according to Section 4.3.3.1 is obtained. The amounts and compositions of all material streams are clearly fixed. If a liquid feed is introduced into column C-2, the composition of the bottom product of C-2 is identical to the overhead product of C-3 (for a vapor feed the two products are in equilibrium). Thus, the product streams can be combined and fed into column C-5. Six columns are required for complete separation into pure substances.

Direct coupling can again be carried out between columns C-2 and C-3, C-4 and C-5 and C-5 and C-6 (Fig. 7-11B). Hence, only three columns are required for the process. The heat requirement is:

$$\frac{\dot{Q}_{R,min}}{\dot{F}\cdot r} = \frac{\dot{Q}_{R1,min}}{\dot{F}\cdot r} + \mathrm{Max}\left(\frac{\dot{Q}_{R2,min}}{\dot{F}\cdot r} \text{ or } \frac{\dot{Q}_{R3,min}}{\dot{F}\cdot r}\right) + \mathrm{Max}\left(\frac{\dot{Q}_{R4,min}}{\dot{F}\cdot r} \text{ or } \frac{\dot{Q}_{R5,min}}{\dot{F}\cdot r} \text{ or } \frac{\dot{Q}_{R6,min}}{\dot{F}\cdot r}\right) \quad (7.18)$$

Direct coupling reduces the heat requirement of a four-component mixture by as much as one-half of the original value.

Another example for the direct coupling of two columns is the separation of an acetone/water/1-butanol mixture by using a decanter as shown in Fig. 7-12A. The vapor at the bottom of column C-1 has approximately the same concentration as the vapor at the top of column C-2. Hence, the two columns can be combined

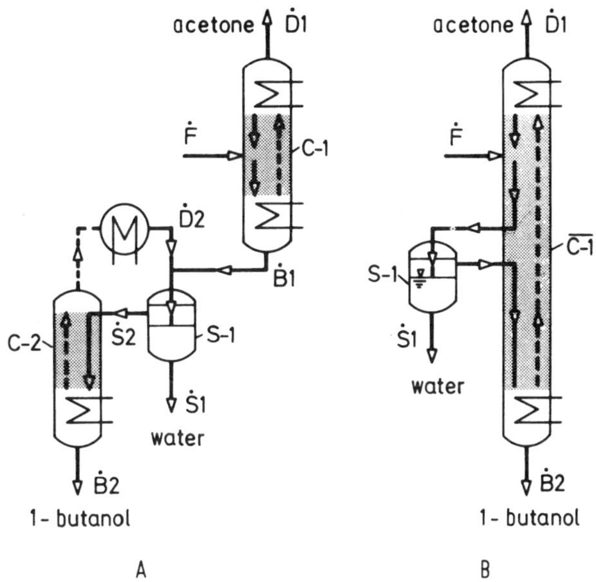

Fig. 7-12. Direct-column coupling in the separation of acetone/water/1-butanol. The top fraction of column C-2 and bottom fraction of column C-1 have approximately the same concentration: (A) Basic process, (B) process with direct column coupling.

[Umeda et al 1979], as shown in Fig. 7-12B. One condenser and one reboiler can be omitted, and the heat requirement of the new column $\overline{\text{C-1}}$ corresponds to the value for either column C-1 or column C-2, whichever is larger.

These examples show that direct-column coupling is particularly useful and always leads to substantial reduction of the heat requirement. It also simplifies equipment because considerably fewer columns and heat exchangers are required. This, in turn, lowers investment costs.

7.3.2 Indirect Coupling of Columns (Thermal Coupling)

In indirect or thermal coupling, waste heat from one column is used to heat another column. In most cases, the columns must be operated at different pressures to obtain the necessary temperature levels. Since the vapor pressure of liquids increases exponentially with temperature, relatively large changes of pressure are needed to produce a sufficient change of temperature. Hence, indirect coupling is of particular interest in close boiling systems.

7.3.2.1 Binary Mixtures If a process contains several columns, indirect coupling should always be considered. Separation in two thermally-coupled columns is sometimes preferable even when the process requires a single column only. Figure

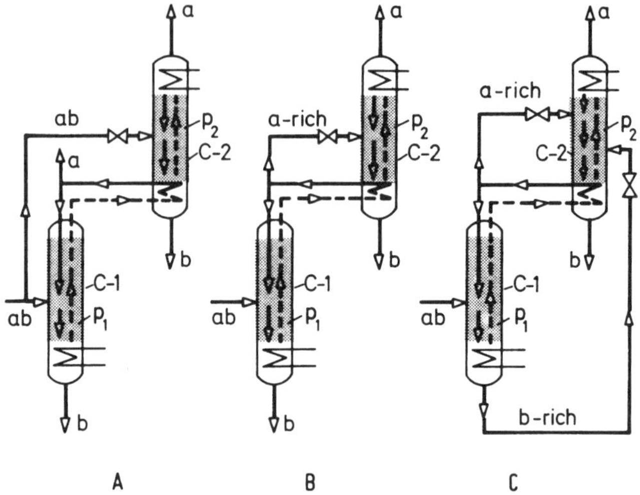

Fig. 7-13. *Indirect-(thermal) column coupling in the separation of binary mixtures. Column C-1 operates at a higher pressure than column C-2: (A) Two columns operated in parallel with complete separation in each column, (B) separation of the* a-rich *overhead product in column C-1, (C) prefractionation in column C-1 followed by complete fractionation in column C-2.*

7-13A shows two columns operating in parallel. Each column separates the binary feed completely. However, the pressure p_2 of column C-2 is much lower than the pressure p_1 of column C-1. Hence, the overhead temperature of column C-1 is higher than the bottom temperature of column C-2, so waste heat from column C-1 can be utilized for heating column C-2. Thus, the heat requirement of distillation is halved. The feed, however, must be compressed to the higher pressure p_1.

A variation of this process (Fig. 7-13B) permits indirect coupling of the two columns at a lower pressure difference. Pure substance *b* is taken from the bottom of column C-1, and an *a*-rich fraction containing a significant amount of *b* is taken from the top of the column. The vapor is used to heat column C-2. The condensate thus obtained is partially returned to column C-1 as reflux and partially fed into column C-2, where it is completely separated into pure substances *a* and *b*.

In Fig. 7-13C, column C-1 works as a prefractionator from which an *a*-rich overhead fraction and a *b*-rich bottom fraction are fed to column C-2. Separation into pure products takes place in column C-2. The heat requirement of column C-2 is lowered by this double feeding. Indirect coupling of the columns permits prefractionation in column C-2 without additional heat requirement.

Several modifications of the processes presented in Fig. 7-13 are employed in industry. For instance, the air-separation process described in Section 5.2.4 is comparable to the process shown in Fig. 7-13C. Exceptional features of air separation also permit omission of the reboiler in column C-1 and the condenser in column C-2.

Requirements for the indirect coupling of columns are always fulfilled in the

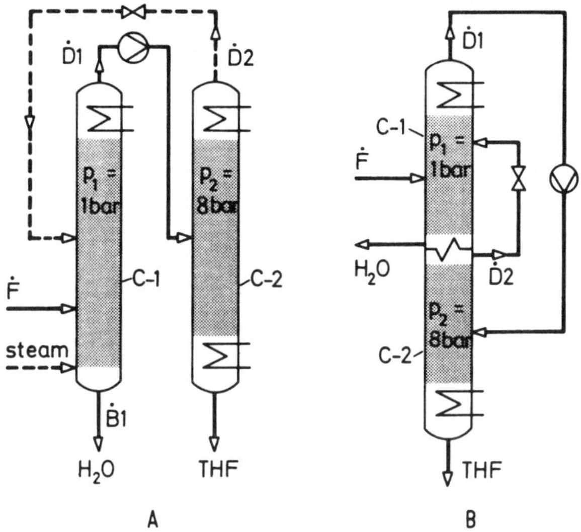

Fig. 7-14. *Indirect (thermal) column coupling in separating a mixture of water and tetrahydrofuran by changing system pressure: (A) Basic process, (B) thermally coupled process.*

separation of azeotropic mixtures by pressure change. Figure 7-14 shows indirect coupling in the process used to separate a mixture of water and tetrahydrofuran, see Section 5.5. The temperature at the top of column C-2 is 136°C, but the temperature at the bottom of column C-1 is only 100°C. Therefore, indirect coupling is possible, and the heat requirement is reduced to approximately 65% of the original value [Abu-Eishah and Luyben 1985]. Equipment requirements are also substantially lower.

7.3.2.2 Ternary Mixtures Separation of ternary mixtures requires at least two distillation columns; therefore, the prerequisite for indirect column coupling is always met.

Indirect coupling of the two columns of path a is shown in Fig. 7-15A. Column C-2 operates at a higher pressure, so its waste heat can be used to heat column C-1. Indirect coupling reduces the heat requirement of the entire process to a value equivalent to the larger value of the two original columns. Hence:

$$\frac{\dot{Q}_{R,min}}{\dot{F} \cdot r} = \mathrm{Max}\left(\frac{\dot{Q}_{R1,min}}{\dot{F} \cdot r} \text{ or } \frac{\dot{Q}_{R2,min}}{\dot{F} \cdot r}\right) \tag{7.19}$$

The quantities $\dot{Q}_{R1,min}/(\dot{F}r)$ and $\dot{Q}_{R2,min}/(\dot{F}r)$ can be calculated by using Equations (7.5) and (7.6). Indirect coupling reduces the heat requirement of the process by as much as one-half, at best.

The triangular diagram in Fig. 7-15A shows the values calculated for $\dot{Q}_{R,min}/(\dot{F} \cdot r)$ from Equation (7.19) for the mixture 2,2-dimethylbutane/2-methylpentane/hexane.

The heat requirement is plotted in parametric form against the feed composition and is in the range of 1.7 to 4.0, i.e., considerably lower than the values obtained without indirect coupling, see Fig. 7-7A. In the intermediate concentration range, the heat required by each of the two columns is about the same; therefore, omission of one quantity of heat results in the largest energy saving. Here, indirect coupling halves the heat requirement.

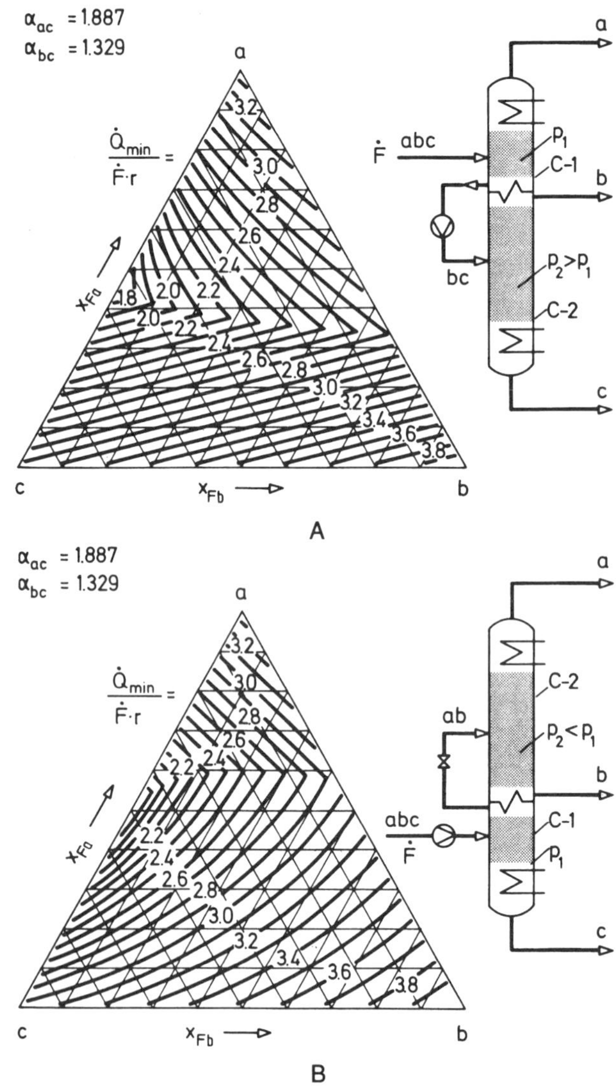

Fig. 7-15. *Minimum heat requirement $\dot{Q}_{R,min}/(\dot{F} \cdot r)$ as a function of feed composition for all three separation paths with indirect coupling in the separation of a ternary mixture of 2,2-dimethylbutane (a)/2-methylpentane (b)/hexane (c). Relative volatilities: α_{ac} = 1.887; α_{bc} = 1.329: (A) Path a, (B) Path c.*

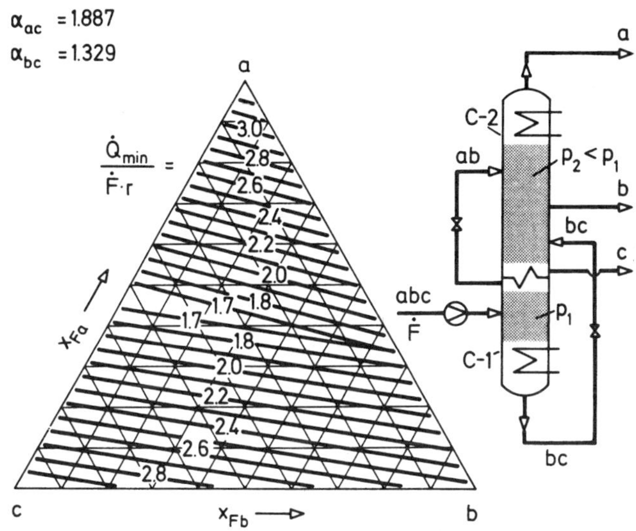

C

Fig. 7-15. (continued) (C) Path a-c with direct and indirect coupling.

Analogously, the indirectly coupled columns used in path c are shown in Fig. 7-15B. The liquid overhead product of column C-1, which operates at a higher pressure, is introduced into column C-2. The heat requirement of the entire process is:

$$\frac{\dot{Q}_{R,min}}{\dot{F}\cdot r} = \text{Max}\left(\frac{\dot{Q}_{R1,min}}{\dot{F}\cdot r} \text{ or } \frac{\dot{Q}_{R2,min}}{\dot{F}\cdot r}\right) \tag{7.20}$$

The individual quantities of heat are calculated from Equations (7.8) and (7.9).

The results of Equation (7.20) are shown on the triangular diagram in Fig. 7-15B. The parametric representation clearly shows the transition between the two quantities of heat. Energy savings are at a maximum in the transition range. However, a 50% reduction is not achieved because the feeding of vapor into column C-2 represents a form of thermal coupling (see Fig. 7-7B), which is, however, less effective than that shown in Fig. 7-15B.

The indirect coupling of path a-c is presented in Fig. 7-15C. Here also, overhead product from the high-pressure column C-1 is fed into column C-2 in the liquid state, e.g., [Triantafyllou and Smith 1992]. The minimum heat requirement of the process is:

$$\frac{\dot{Q}_{R,min}}{\dot{F}\cdot r} = \text{Max}\left(\frac{\dot{Q}_{R1,min}}{\dot{F}\cdot r} \text{ or } \frac{\dot{Q}_{R2,min}}{\dot{F}\cdot r} \text{ or } \frac{\dot{Q}_{R3,min}}{\dot{F}\cdot r}\right) \tag{7.21}$$

The individual quantities of heat are calculated from Equations (7.12), (7.13) and (7.15). The results of Equation (7.21) for the mixture 2,2-dimethylbutane/2-methylpentane/hexane are presented in parametric form on the triangular diagram shown in Fig. 7-15C. A series of parallel lines is obtained in each region. The value $\dot{Q}_{R1,min}/(\dot{F} \cdot r)$ is always smaller than $\dot{Q}_{R2,min}/(\dot{F} \cdot r)$ or $\dot{Q}_{R3,min}/(\dot{F} \cdot r)$ so that only two regions are obtained on the triangular diagram.

Values of the heat requirement $\dot{Q}_{R,min}/(\dot{F} \cdot r)$ range from 1.6 to 3.3 and are lower than those obtained for paths a and path c, see Figs. 7-15A and 7-15B, respectively. The three processes shown in Fig. 7-15 are equivalent only if the mixture has a very low content of intermediate boiler *b*. In all other cases, path *a-c* with direct and indirect coupling is the most favorable separation sequence.

7.4 DESIGN OF HEAT EXCHANGER NETWORKS

Generally, chemical processes consist of a large number of units that are interconnected by material streams. For proper function heat has to be supplied and removed at many places in both different quantities and different temperature levels. From a thermodynamic point of view, any thermal chemical process involves a heat exchanger network.

The simplest design of such a network would apply utility heaters to all streams that have to be heated (cold streams) and utility coolers to all streams that have to be cooled (hot streams). However, a more sophisticated design would try to match cold and hot streams in order to reduce the quantity of external heat required. The objective of heat recovery network design is to select streams for matching such that the network is cheapest in terms of utility and capital cost required [Linnhoff 1983].

7.4.1 Optimum Heat Exchanger Networks

Development of systematic procedures for the design of heat exchanger networks has been an active area of interest in the chemical engineering literature for many years. A review is given by [Nishida, et al. 1981]. Early applications of such methods are in cryogenic processes, e.g., air separation [Baldus et al. 1983]. All these methods are based on the graphical plot of temperature levels versus enthalpy streams similar to the design procedure for a single heat exchanger. These methods have recently been further developed and propagated by Linnhoff and coworkers [Linnhoff and Flower 1978; Linnhoff and Hindmarsh 1983; Linnhoff and Vredeveld 1984; and Linnhoff and Sahde 1989].

The procedure will be explained in the following on the simple example of two hot streams and three cold ones without any phase change, e.g., neither evaporation

Table 7-1. Listing of Temperature Levels and Enthalpies of Five Streams that Have to be Matched. Constant Values of Heat Capacities are Supposed.

Stream	Inlet Temperature	Outlet Temperature	Enthalpy Stream
1 hot	150°C	60°C	200 kW
2 hot	200°C	100°C	100 kW
3 cold	40°C	130°C	120 kW
4 cold	110°C	130°C	140 kW
5 cold	110°C	180°C	60 kW

nor condensation. Temperature levels and heat quantities of the streams are listed in Table 7-1.

With the assumption of constant heat capacities, the streams are represented by straight lines in the temperature/enthalphy stream diagrams of Fig. 7-16. The direction of temperature change is marked by arrows. First, the enthalpies of all hot streams are added at their individual temperature levels resulting in a single line for all hot streams, see Fig. 7-17A. According to [Linnhoff and Flower 1978] the resulting line is called *hot composite curve*. The *cold composite curve* is constructed in the same way, see Fig. 7-17B. In practical applications, the number of streams involved is generally much larger and the individual lines might be curved. However, the design procedure of the composite curves essentially remains the same.

In the next step, the cold and hot composite curves are combined in a single

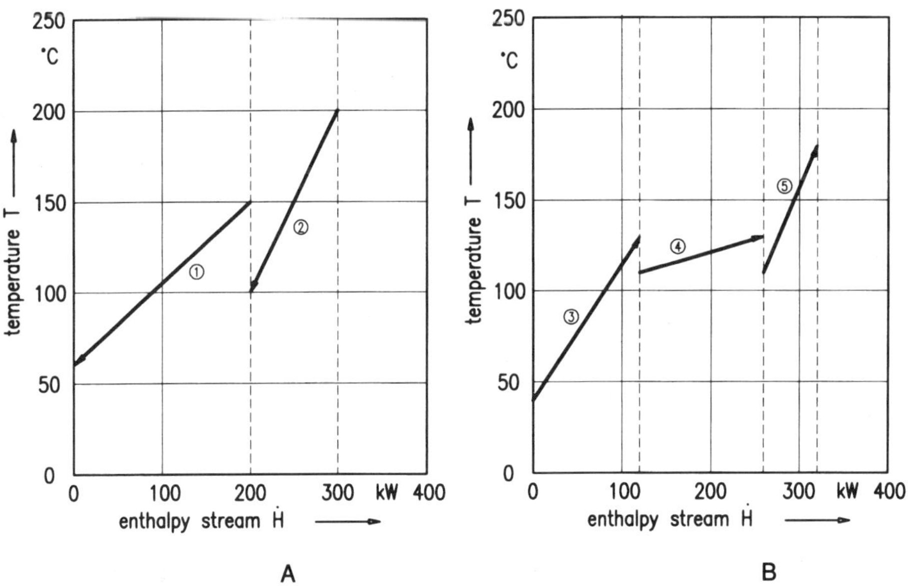

Fig. 7-16. *Graphical plot of temperatures and enthalpies of the individual streams of Table 7-1: (A) Hot streams, (B) cold streams.*

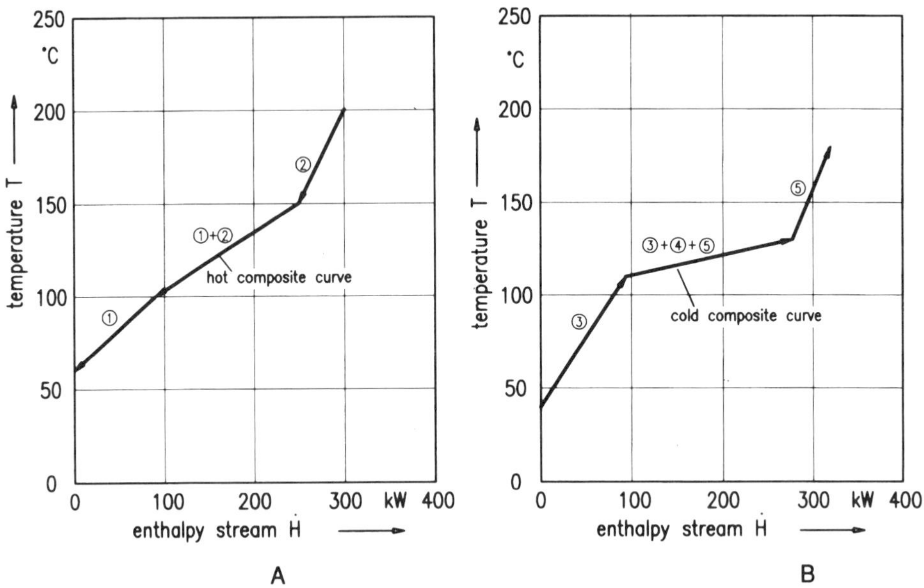

Fig. 7-17. *Composite curves of the hot and cold streams: (A) Hot streams, (B) cold streams.*

diagram. Since the abscissa represents enthalpy streams $\dot{H}$ the composite curves can be shifted in horizontal direction until a minimum temperature difference ΔT_P (e.g. 10°C) between hot and cold curve is reached, Fig. 7-18. The locus of the minimum vertical distance between the composite curves is called *pinch.* The curves overlap along the abscissa. The quantity of heat in the overlapping section can be internally covered by an optimum heat exchanger network. However, at either end there exists an overhang such that the top of the cold composite curve needs external heating $\dot{Q}_H$ and the bottom of the hot composite curve needs external cooling $\dot{Q}_C$. These external heats are referred to as the hot and cold utility targets [Linnhoff and Sahde 1989]. From the diagram in Fig. 7-18, the quantities of external heat streams $\dot{Q}_H$ and $\dot{Q}_C$ are easily determined. In our example we get $\dot{Q}_H = 80$ kW and $\dot{Q}_C = 60$ kW.

However, knowledge of heat quantities $\dot{Q}_H$ and $\dot{Q}_C$ is not sufficient for the design of the optimum heat exchanger network. As important as heat streams $\dot{Q}_H$ and $\dot{Q}_C$, are their temperature levels T_H and T_C. From a thermodynamic point of view, the heat stream $\dot{Q}_C$ should be withdrawn at a temperature level as high as possible and, in turn, the heat stream $\dot{Q}_H$ should be supplied at a temperature level as low as possible. By doing so, the exergy losses of the heat exchanger network are minimized.

It is obvious from Fig. 7-18 that the cold utility $\dot{Q}_C$ need not be withdrawn at the lowest temperature level (here 40°C). The length of $\dot{Q}_C$ can be shifted upwards along the cold composite curve in the T/$\dot{H}$-diagram until the minimum temperature difference ΔT_p between cold and hot composite curve is reached, see Fig. 7-19. The value of the temperature level T_C is most easily determined by shifting the lower

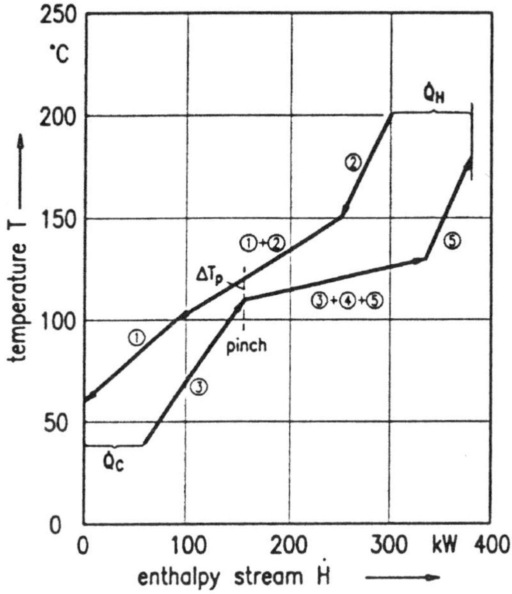

Fig. 7-18. Position of the composite curves at a pinch temperature difference of $\Delta T_p = 10°C$. Heat can be internally recovered in the overlapping section (approximately 240 kW). Utility heats are $\dot{Q}_H = 80$ kW and $\dot{Q}_C = 60$ kW.

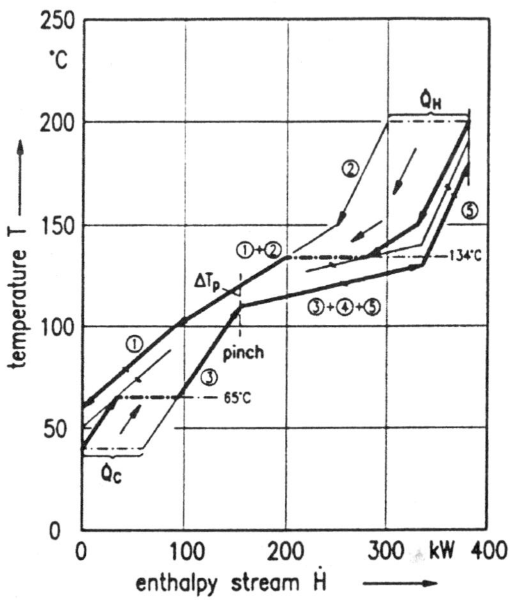

Fig. 7-19. Determination of the temperature levels of cold and hot utilities.

part of the hot composite curve vertically downwards by the distance ΔT_p (here 10°C). The lower part of the cold composite curve is shifted horizontally to the left by the distance of $\dot{Q}_C$ (here 60 kW). The point of intersection of these two auxiliary lines gives the value of T_C (here 65°C) in Fig. 7-19.

An analogous approach allows determination of the lowest value for temperature T_H of the hot utility $\dot{Q}_H$. The upper part of the cold composite curve is shifted vertically upwards by the distance ΔT_P (here 10°C). The upper part of the hot composite curve is horizontally shifted to the right by the distance $\dot{Q}_H$ (here 80 kW). The intersection of both auxiliary lines gives the value of T_H (here 134°C). By implementation of the hot and cold utilities at their optimum temperature levels the final courses of both composite curves result. They are presented in Fig. 7-21.

A different approach for determination of the optimum temperature levels of external duties has been proposed by Linnhoff. [Linnhoff and Flower 1978]. He determines the *grand composite curve* in an auxiliary diagram, see Fig. 7-20. First, the hot composite curve is shifted downwards in vertical direction by the distance $\Delta T_p/2$. Analogously, the cold composite curve is shifted upwards by the same distance until both curves touch. Then, the horizontal distances between the composite curves are plotted in a separate $T^*/\Delta\dot{H}$-diagram. The resulting line is referred to as the *grand composite curve*. The external heat stream $\dot{Q}_H$ has to be supplied above the pinch at the lowest temperature T_H^* where the abscissa of the grand composite curve reaches the value of $\dot{Q}_H$. An analogous procedure yields the temperature T_C^*.

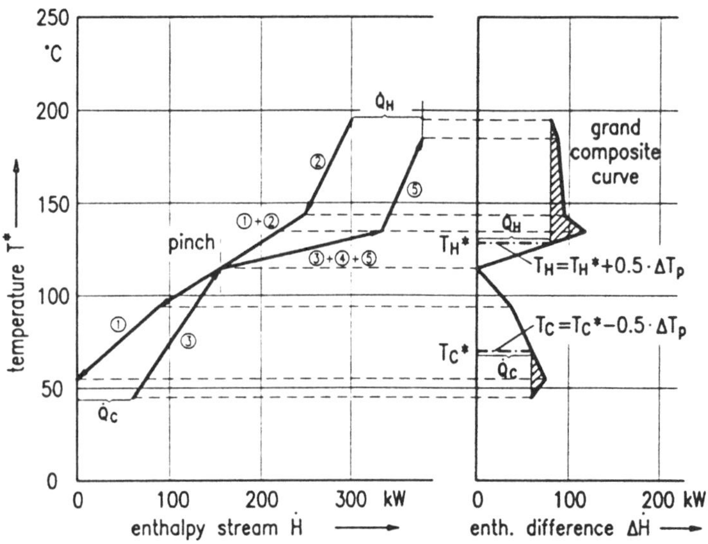

Fig. 7-20. *Construction scheme of the* grand composite curve. *The composite curves of Fig. 7-18 are shifted upwards and downwards, respectively, until they touch. The horizontal distances form the* grand composite curve *(right-hand side) from which the temperature levels T_H and T_C are determined.*

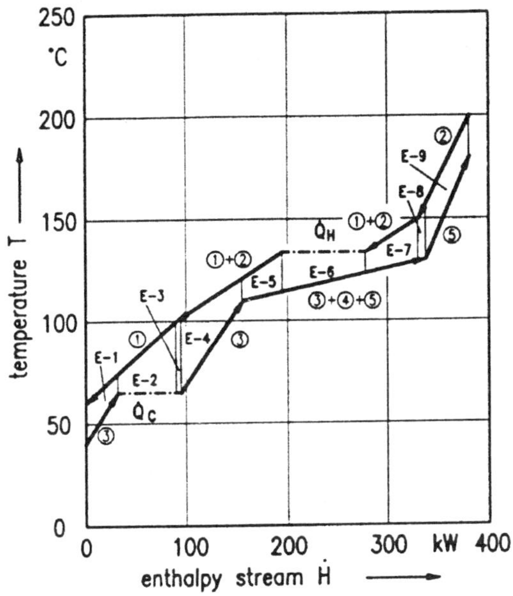

Fig. 7-21. *Temperature-enthalpy plot of streams and utilities. This diagram contains all information neccessary for developing the optimum heat exchanger network. Nine heat exchangers are required. Their duties and mean temperature differences are listed in Table 7-2.*

Since the ordinate of the grand composite diagram is T* and not T, the temperature levels need to be corrected by:

$$T_H = T_H^* + 0.5\Delta T_p \qquad T_C = T_C^* - 0.5\,\Delta T_p \tag{7.22}$$

The same values for the level of hot and cold utilities result from Figs. 7-19 and 7.20. However, the approach of Fig. 7-19 is simpler and easier to understand.

The optimum heat exchanger network is designed from a temperature-enthalpy diagram in which the utilities $\dot{Q}_H$ and $\dot{Q}_C$ are implemented in the hot and cold composite curve, respectively, see Fig. 7-21. Here, the cold and hot composite curves have the same length and, generally, the minimum temperature difference ΔT_p exists several times. This plot contains all the information necessary for the design of the network presented in Fig. 7-22. Each break of one of the composite curves is the end of a heat exchanger and the beginning of the next one because at this point a stream is, in most cases, added or removed. Exceptions to this rule are processes with evaporation and condensation of pure substances. All temperatures and enthalpy differences necessary for heat exchanger dimensioning can be taken from the diagram of Fig. 7-21. The characteristic values of the heat exchangers are listed in Table 7-2. For our simple example with only five streams, as many as nine heat exchangers are required, most of them multiple-path heat exchangers. Such complex heat exchanger networks are difficult to realize in technical plants. However, there are examples for such networks, especially in cryogenic processes (see Fig. 5-9).

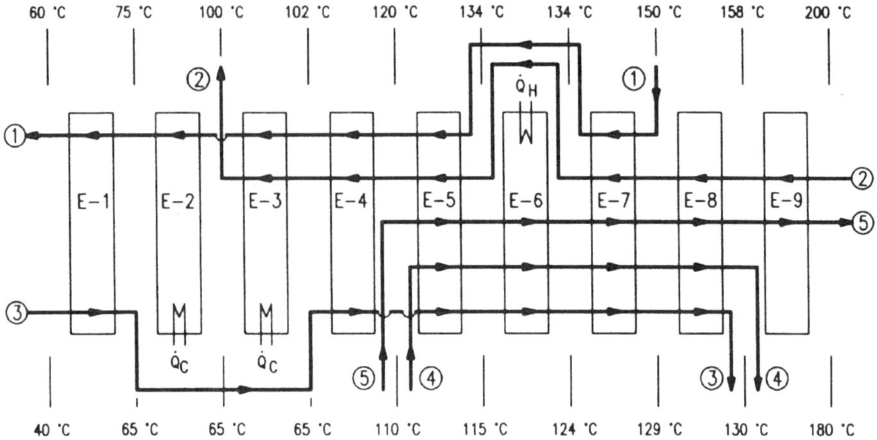

Fig. 7-22. Schematic of the optimized heat exchanger network with temperatures of the cold and hot streams of each heat exchanger. Most of the nine heat exchangers are multiple path exchangers.

In most cases the optimum network cannot be realized and, hence, needs to be simplified.

For modifying the heat exchanger network, a simpler, more abstract graphical presentation of the network is convenient. All streams are represented by equally long, parallel horizontal lines with temperatures increasing to the right direction. The cold streams are arranged below the hot streams. Heat exchangers are symbolized by circles that are connected by vertical lines between all involved streams. The external utilities are represented by circles also. The network of Fig. 7-22 is shown in this more abstract manner in Fig. 7-23. This scheme of the network is very clear and allows for easy modification.

Table 7-2. Heat Duties and Mean Logarithmic Temperature Differences of the Optimized Heat Exchanger Network Presented in Fig. 7-22.

Heat Exchanger	Heat Duty	Mean Logarithmic Temperature Difference
E-1	4 kW	14.4°C
E-2	56 kW	19.9°C
E-3	34 kW	36.0°C
E-4	61 kW	20.6°C
E-5	43 kW	13.6°C
E-6	80 kW	13.6°C
E-7	52 kW	14.8°C
E-8	7 kW	24.3°C
E-9	43 kW	23.7°C

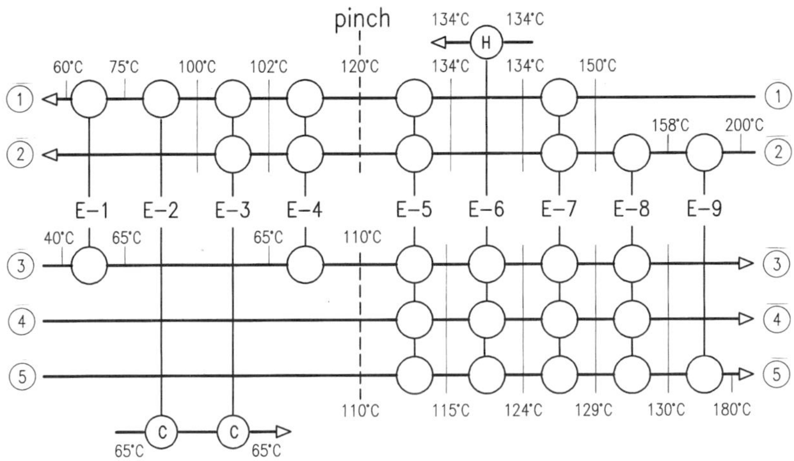

Fig. 7-23. *Grid representation of the heat exchanger network of Fig. 7-22. In this more abstract form all streams are represented by horizontal lines of equal length. This grid presentation allows easy modification of the network.*

7.4.2 Modifying the Optimum Heat Exchanger Network

For modifying the network, the pinch principles developed by Linnhoff [Linnhoff and Hindmarsh 1983] should be applied. The network can be considered as two separate systems—one above the pinch temperature, and one below. The system above the pinch requires only residual heat and is, therefore, a heat sink. The system below the pinch has heat to reject and is, therefore, a heat source. Hence, the following rules can be laid down [Linnhoff and Sahde 1989]:

- Heat must not be transferred across the pinch.
- There must be no outside cooling above the pinch.
- There must be no outside heating below the pinch.

One of the consequences of these rules is, for instance, that a heat pump has to be integrated into a process so that the heat is pumped across the pinch [Townsend and Linnhoff 1983] since it only makes sense to pump heat from a heat source to a heat sink. Many heat pumps installed in industrial plants violate this rule, since only a single unit and not the entire process has been considered in the design procedure.

Modifications of the optimum heat exchanger network are often prompted by the large number of heat exchangers required in such a network. The number can often be reduced by variation of the temperature levels of hot and cold utilities. From Fig. 7-21, for instance, it follows that the exchangers E-2 and E-3 can be united by

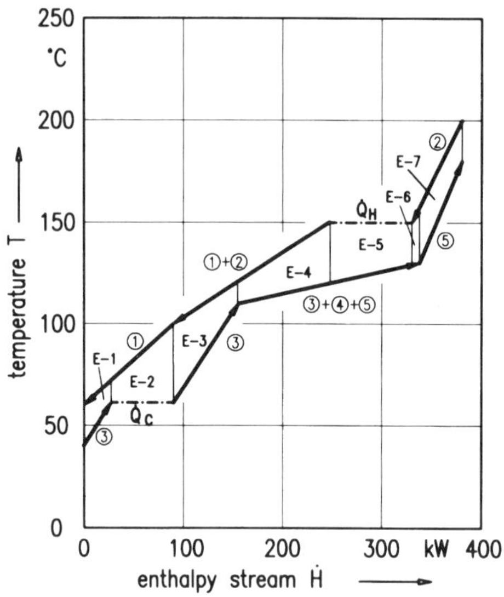

Fig. 7-24. Simplification of the heat exchanger network by decreasing the temperature level of the cold utility and increasing the temperature level of the hot utility to reduce the number of heat exchangers.

simply decreasing the level of outside cooling from 65°C to approximately 62°C. The exchanger E-7 can be discarded by increasing the level for outside heating from 134°C to 150°C. From these modifications the temperature-enthalpy-diagram of Fig. 7-24 results. The modified network is schematically depicted in Fig. 7-25. Only seven heat exchangers are required after changing the temperature levels of the hot

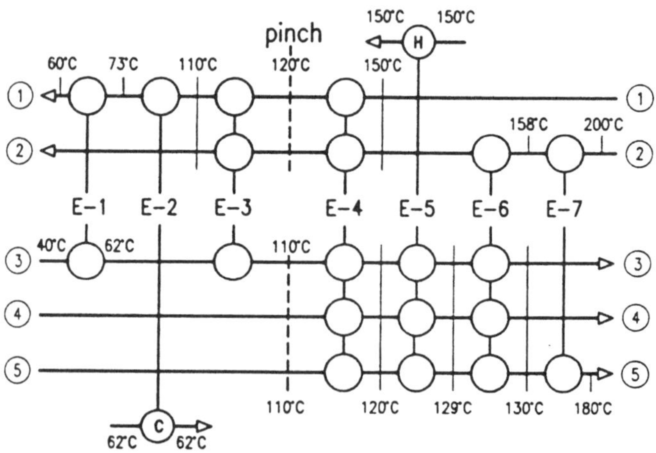

Fig. 7-25. Grid presentation of the simplified heat exchanger network of Fig. 7-24. Only seven heat exchangers are required.

and cold utilities. The feasibility of both measures depends on the temperature levels of external heat sink and heat source. The simplification of the heat exchanger network must, however, be paid by higher exergy losses of the process.

7.4.3 Dual Flow Heat Exchangers Networks

One of the essential drawbacks of the optimum heat exchanger network of Fig. 7-23 is the application of multiple-path heat exchangers with up to five paths. The standard design prefers dual-path heat exchangers because they are cheaper and easier to install and operate.

For developing a heat exchanger network with dual flow heat exchangers only, the rules of Linnhoff are helpful. To avoid heat transfer across the pinch, the networks of exchangers below and above the pinch should be developed seperately. [Linnhoff and Hindmarsh 1983].

First, the network below the pinch is developed because it is the simpler one in our example. In principle, the combined streams of the composite curves of Fig. 7-21 must be replaced by an alternating sequence of small elements of the individual streams. The higher the temperature difference between hot and cold composite curve in Fig. 7-21, the longer the elements of the individual streams can be. In our example the hot composite curve is combined of the streams No. 1 and 2 between 100°C and the pinch (120°C). Hence, in the $T/\dot{H}$-diagram of the individual streams in Fig. 7-26 (left-hand side) only the hot streams No. 1 and 2 appear in this temperature range. The sequence of the streams is ordered such that the distance between hot and cold streams never falls below the value ΔT_p.

The network of dual-path heat exchangers above the pinch is developed in Fig. 7-26 (right-hand side). The cold composite curve between pinch and 130°C is combined of streams No. 3, 4 and 5. Hence, the $T/\dot{H}$-curves of these streams are arranged in consecutive elements. The hot composite curve is combined from streams No. 1 and 2 in the temperature range from the pinch (120°C) to 150°C. Since the temperature differences between hot and cold composite curves in Fig. 7-21 are very small, the $T/\dot{H}$-curves of the individual streams have to be alternatively arranged in small consecutive elements. The hot stream No. 1, for instance, is divided into two parts that are separated by the hot utility $\dot{Q}_H$. The cold streams No. 4 and 5 are also separated in two elements that are arranged in mixed order with stream No. 3 between them.

The resulting $T/\dot{H}$-diagram of the individual streams has the same minimum temperature difference ΔT_p as the overall diagram of Fig. 7-21. However, this minimum temperature difference exists more often. From the diagrams of Fig. 7-26 the heat exchanger network that consists of dual-flow heat exchangers only is easily developed. The network is presented in Fig. 7-27. Now, the number of heat exchangers is ten instead of nine but only dual-flow heat exchangers are involved.

7.4.4 Process Modifications

The energy demand of a chemical process can be drastically reduced by process modifications. The shape of the composite curves is, for instance, changed by varia-

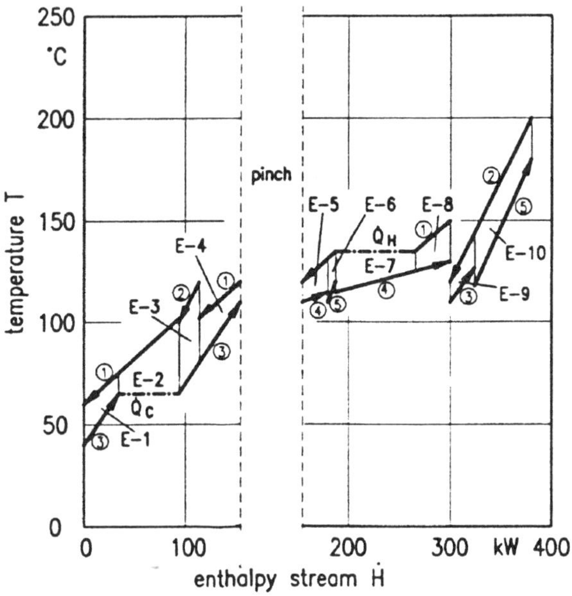

Fig. 7-26. *Temperature-enthalpy diagram of a network with dual-path heat exchangers only. The diagram is divided into sections below the pinch and above the pinch to safely avoid heat transfer across the pinch.*

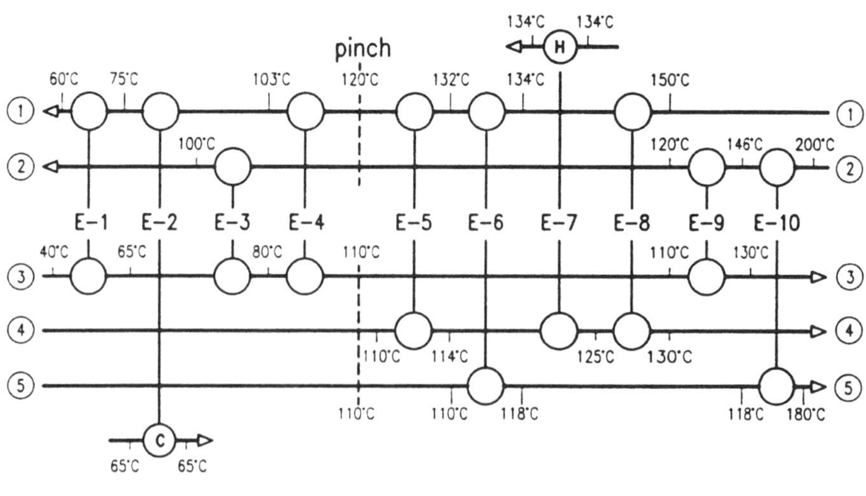

Fig. 7-27. *Grid presentation of the dual-flow heat exchanger network of Fig. 7-26.*

tion of the pressures of evaporators and condensers. Such modifications result in a different location of the pinch and can manage to increase the overlapping region in the enthalpy-temperature plot of Fig. 7-18. In the optimum version the hot and cold composite curves no longer form a sharp pinch. Both curves should have approximately the same distance everywhere.

At a chemical site in Germany, 130 processes were examined in a large energy saving program [Koerner 1988]. The heat exchanger networks of 70 out of these 130 processes were modified, resulting in a reduction of steam consumption of up to $700 \cdot 10^3$ kg/h. Thus, investment of a new $300 million power station has been avoided. The study reveals that the potential for energy savings lies primarily in process modifictions, e.g., change of operation pressure, and not in mere stream matching. However, for a systematic approach to advantageous process modifications, the pinch technology as outlined in this section is the appropriate engineering tool.

LITERATURE

Abu-Eishah, S. I. and Luyben, W.-L.: ''Design and Control of a Two-Column Azeotropic Distillation System,'' *Ind. Eng. Chem. Process Des. Dev.* Vol. 24, (1985) No. 1, pp. 132/140.

Ali El Saie; El Kafrawi and Kamel: ''Study of Operation Conditions for Tree Large MSF Desalination Units,'' *Desalination 73* (1989), pp. 207/230.

Baldus, H.; Baumgaertner, K.; Knapp, H. and Streich, M.: ''Verfluessigung und Trennung von Gasen,'' Winnacker Kuechler: *Chemische Technologie* Bd. 3, pp. 567/650. Carl Hauser Verlag München Wien, 1983.

Dhole, R. V. and Linnhoff, B.: ''Distillation Column Targets,'' *Computers and Chemical Engineering Journal* Vol. 17 (1993) pp. 549/560.

Dummer, G. and Schmidhammer, L.: ''Grosstechnische Nutzung der Thermokompression bei der Destillation von 1,2-Dichlorethan,'' *Chem.-Ing.-Tech.* 63 (1991) No. 1, pp. 74/75.

Greig, H. W.: ''An Economic Comparison of Two 2000 m^3 per Day Desalination Plants,'' *Desalination* 64 (1987) pp. 17/50.

Kaibel, G.: ''Distillation Columns with Vertical Partitions,'' *Chem. Eng. Technol* 10 (1987) pp. 92/98.

King, C. J.: *Separation Processes.* McGraw-Hill, New York, 1980, p. 660.

Kleinhenz, K.: ''Energiesparen beim Verdampfen und Destillieren in der Chemischen Industrie,'' *VDI Berichte* 383 (1980) pp. 43–45.

Kleinhenz, K.: ''Forderungen an Bruedenverdichter bei Wasserdampf und bei anderen Medien,'' *VDI Berichte* 442 (1982) pp. 45–51.

Koerner, H.: ''Optimaler Energieeinsatz in der Chemischen Industrie,'' *Chem.-Ing.-Tech.* 60 (1988) No. 7, pp. 511/518.

Kreuter, W.: ''Optimization of Ethylene/Ethane Separation in the Production of Ethylene,'' *Linde Reports on Science and Technology* 18 (1973) pp. 23–26.

Lestak, F.; Smith, R. and Dhole, V. R.: ''Heat Transfer Across the Wall of Dividing Wall Columns,'' *Trans IChemE* Vol. 72 (1994) Part A, pp. 639/644.

Linnhoff, B.: ''New Concepts in Thermydynamics for Better Chemical Process Design,'' *Chem. Eng. Res. Des.* Vol. 61, July 1983, pp. 207–223.

Linnhoff, B. and Flower, J. R.: ''Synthesis of Heat Exchanger Networks I. Systematic Generation of Energy Optimal Networks,'' *AIChE Journal* 24 (1978) pp. 633/642.

Linnhoff, B. and Hindmarsh, E.: ''The Pinch Design Method for Heat Exchanger Networks,'' *Chem. Eng. Sci.* 38 (1983) No. 5, pp. 745–763.

Linnhoff, B. and Sahde, V.: ''Pinch Technology,'' *Ullmanns Encyclopedia of Industrial Chemistry* Vol. B 3, pp. 13–1/13–6, VCH Verlagsgesellschaft, Weinheim, 1989.

Linnhoff, B. and Turner, J. A.: ''Simple Concepts in Process Synthesis Give Energy Savings and Elegant Designs,'' *The Chemical Engineer, Dez.,* 1980, pp. 742/746.

Linnhoff, B. and Vredeveld, D. R.: ''Pinch Technology Has Come of Age,'' *Chemical Engineering Progress,* July 1984, pp. 33/40.

Luedeker, H.: ''Wirtschaftliche Einsatzbereiche von Wärmepumpen und Mehrstufenschaltungen in der Rektifiziertechnik,'' Dissertation TU Berlin, 1985.

Luedeker, H. and Gelbe, H.: ''Wirtschaftliche Einsatzbereiche von Wärmepumpen- und Mehrstufenschaltungen in der Rektifiziertechnik,'' *Chem.-Ing.-Tech.* 58 (1986) No. 11, pp. 847/855.

Nishida, N.; Stephanopoulos, G. and Westerberg, A. W.: ''A Review of Process Synthesis,'' *AIChE Journal* Vol. 27 (1981) pp. 321/351.

Perry, R. H.; Green, D. W. and Maloney, J. O.: *Chemical Engineer's Handbook,* Sixth Edition McGraw-Hill, New York, 1984, pp. 3–144.

Schoenmakers, H.: ''Internal Coupling of Distillation Columns—A Further Step to Process Integration,'' *Ger. Chem. Eng.* 9 (1986) pp. 199–203.

Smith, R. and Linnhoff, B.: ''The Design of Separators in the Context of Overall Processes,'' *Chem. Eng. Res. Des.* Vol. 66, May 1988, pp. 195/228.

Townsend, D. W. and Linnhoff, B.: ''Heat and Power Networks in Process Design, Part I: Criteria for Placement of Heat Engines and Heat Pumps in Process Networks,'' *AIChE Journal* Vol. 29 (1983) No. 5, pp. 742/771.

Triantafyllou, C. and Smith, R.: ''The Design and Optimization of Fully Thermally Coupled Distillation Columns,'' *Trans IChem E.* Vol. 70 (1992) Part A, pp. 118/132.

Umeda, T. K.; Niida, K.; and Shiroko, K.: ''A Thermodynamic Approach to Heat Integration in Distillation Systems,'' *AIChE Journal* 25 (1979) No. 3, pp. 423–429.

8

Design and Dimensioning Mass Transfer Equipment

8.1 Types of Construction
- 8.1.1 Tray Columns
- 8.1.2 Packed Columns
- 8.1.3 Criteria for Use of Tray or Packed Columns

8.2 Dimensioning Tray Columns
- 8.2.1 Design Parameters of Tray Columns
- 8.2.2 Operating Region of Tray Columns
  - 8.2.2.1 Maximum Gas Load
  - 8.2.2.2 Minimum Gas Load
  - 8.2.2.3 Maximum Liquid Load
  - 8.2.2.4 Minimum Liquid Load
  - 8.2.2.5 Operating Region of Tray Columns
- 8.2.3 Two-Phase Flow in Tray Columns
  - 8.2.3.1 Flow Regimes
  - 8.2.3.2 Relative Liquid Holdup
  - 8.2.3.3 Froth Height
  - 8.2.3.4 Liquid Entrainment
  - 8.2.3.5 Liquid Mixing in the Two-Phase Layer
  - 8.2.3.6 Liquid Maldistribution
  - 8.2.3.7 Interfacial Area
  - 8.2.3.8 Pressure Drop
    - 8.2.3.8.1 Pressure Drop of Dry Sieve Trays
    - 8.2.3.8.2 Pressure Drop of Dry Valve and Bubble Cap Trays
- 8.2.4 Mass Transfer in the Two-Phase Layer on Column Trays
  - 8.2.4.1 Definitions of Mass Transfer Efficiencies
  - 8.2.4.2 Relationship between Point Efficiency and Tray Efficiency

8.2.4.3 Point Efficiency
8.2.4.4 Mass Transfer Coefficients
8.2.4.5 Practical Determination of Tray Efficiencies
8.3 Dimensioning Packed Columns
8.3.1 Design Parameters of Packed Columns
8.3.1.1 Types of Packings
8.3.1.1.1 Dumped Packings
8.3.1.1.2 Structured Packings
8.3.1.2 Model Structures of Packings
8.3.1.2.1 Particle Model Structure
8.3.1.2.2 Channel Model Structure
8.3.1.3 Packing Supplements
8.3.1.3.1 Liquid Distributor
8.3.1.3.2 Liquid Redistributor
8.3.1.3.3 Liquid Collector
8.3.1.3.4 Support Grid
8.3.1.3.5 Hold Down Plate
8.3.1.3.6 Gas Distributor
8.3.2 Operating Region of Packed Columns
8.3.2.1 Flooding
8.3.2.2 Minimum Liquid Load
8.3.3 Two-Phase Flow in Packed Columns
8.3.3.1 Liquid Holdup
8.3.3.1.1 Holdup Below the Loading Point
8.3.3.1.2 Holdup Above the Loading Point
8.3.3.2 Pressure Drop
8.3.3.2.1 Pressure Drop of a Dry Packing
8.3.3.2.2 Channel Model Structure
8.3.3.2.3 Particle Model Structure
8.3.3.2.4 Pressure Drop of an Irrigated Packing
8.3.3.2.5 Channel Model Structure
8.3.3.2.6 Particle Model Structure
8.3.3.2.7 Determination of the Flood Point from the Pressure Drop Equation
8.3.3.3 Maldistribution of Liquid
8.3.3.4 Interfacial Area
8.3.4 Mass Transfer in Packed Columns
8.3.4.1 Height of a Transfer Unit
8.3.4.2 Mass Transfer Coefficients and Effective Interfacial Area
8.3.4.2.1 Onda Model
8.3.4.2.2 Bravo/Fair Correlation
8.3.4.2.3 Billet/Schultes Correlation
8.3.4.2.4 Problems in Predicting Mass Transfer in Packed Columns
8.3.4.3 Practical Determination of the Height of a Transfer Unit
Appendix: Pressure Drop in Packed Beds
Literature

The separation principle of distillation is the mass transfer across the interface of a two-phase gas liquid system. Any distillation equipment has to provide a high mass transfer rate across the interphase. The basic requirements for good mass transfer are best understood from the following fundamental mass transfer equation:

$$N = K_{OG}\, A\, (y^* - y) \tag{8.1}$$

The maximum mass transfer ratė N is obtained when all three terms on the right-hand side of the above equation are as large as possible:

- The mass transfer coefficient K_{OG} depends on the flow condition of the liquid and gas phases; it increases with increasing velocities in both phases. The situation is improved if new fluid elements constantly reach the phase boundary since this results in a nonsteady mass transfer with short transfer times. Columns that allow steady renewal of the interface area are particularly effective.
- The interface area A for mass transfer depends on the geometry of the apparatus and the local velocities of the two phases. It is strongly affected by the column internals.
- The largest driving force $(y^* - y)$ for mass transfer between the phases is achieved when the overall flow pattern at all places in the apparatus allows countercurrent contact between equal streams of gas and liquid without any remixing. Thus, both phases must be distributed uniformly over the entire flow area.

8.1 TYPES OF CONSTRUCTION

Tray columns and packed columns are most often used for distillation [Kunesh et al. 1995] since they guarantee excellent countercurrent flow and permit a large overall height. The internals provide a large mass transfer area, which is constantly renewed, especially in tray columns. The use of mass transfer machines has been proposed repeatedly for distillation. Although such devices have proved useful for extraction, they have not been accepted in distillation [Mersmann et al. 1986].

8.1.1 Tray Columns

Figure 8-1 shows the most important features of a tray column. The gas flows upwards within the column through perforations in the horizontal trays, and the condensed liquid flows countercurrently downwards. However, on the individual trays the two phases exhibit cross flow relative to each other. The ascending gas passes through the perforations into the liquid flowing across the tray, to form a large interfacial area between the two phases. The phases subsequently separate; the gas rises to the tray above, whereas the liquid flows over the exit weir into the downcomer, which is usually segmental in shape. Some of the liquid accumulates in each downcomer. This compensates for the pressure drop caused by the gas as it passes through the tray. The liquid then flows through an inlet onto the tray below. The orifices in the tray floor can have various designs.

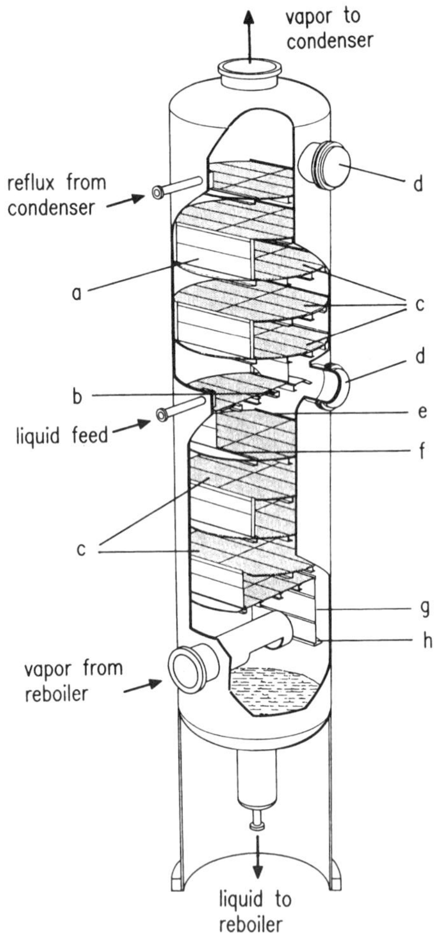

Fig. 8-1. *Cutaway section of a tray column: (a) Downcomer, (b) tray support, (c) sieve trays, (d) man way, (e) outlet weir, (f) inlet weir, (g) side wall of downcomer, (h) liquid seal.*

Bubble cap trays have been used almost exclusively for many years. Their design prevents liquid from leaking downward (weeping) through the tray. However, the complex bubble caps are relatively expensive and have a higher pressure drop than other designs.

Sieve trays have become increasingly important. They are inexpensive, have a high separation efficiency, and produce a low pressure drop across the tray. Previously, sieve trays were used with holes of approximately 1 mm diameter, which were easily blocked by dirt particles. However, hole diameters of 12 and sometimes even 25 mm are now employed.

Valve trays are a relatively new, but widely used, development. The cross section of the orifices in the tray is adjusted to prevent the liquid from leaking at low gas loads and to avoid excessive pressure increase at high gas loads. The gas load can be varied up to a factor 4 or 5 that gives value trays a much larger operational flexibility than any other tray design. Approximately ten vendors of valve trays are on the market. Most of them offer more than ten different types of valves. Many other variations of the sieve, valve, and bubble cap trays are also available, e.g., Performkontakt [Hoppe et al. 1969].

8.1.2 Packed Columns

Figure 8-2 shows the most important features of a packed column. Gas enters at the bottom of the column and flows upwards through the free cross-sectional area of

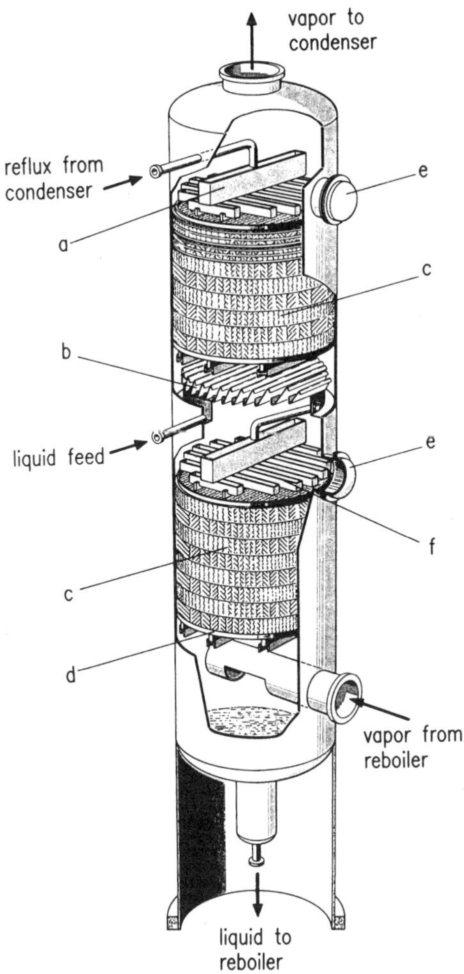

Fig. 8-2. Cutaway section of a packed column with a structured wire gauze packing [Sulzer 1982]: (a) Liquid distributor, (b) liquid collector, (c) structured packing, (d) support grid, (e) manway, (f) liquid redistributor.

the internals. Liquid flows downwards countercurrently to the ascending gas. Thus, a packed column provides countercurrent contacting of the two phases, which is crucial for good mass transfer. The liquid should wet the column packing (random or structured) as completely as possible to maximize the mass transfer contact area between phases. The type of packing greatly influences the efficiency of the column. Random dumped packings made of specially formed particles (rings, saddles, cf. Fig. 8-31) are often used: their surface area should be as large as possible; they should have a low flow resistance; and they should be distributed homogeneously in the column (no large cavities). Structured packings (see Fig. 8-32) comply more readily with these requirements and are widely used.

Uniform distribution of liquid over the entire cross-sectional area is very important. In Fig. 8-2, a cascade distributor is used for this purpose. Although the initial liquid distribution is uniform, liquid maldistribution usually occurs as the height of the packing increases; this can greatly impair separation efficiency. Therefore, the height of individual packed beds is limited to 3 to 6 m. Liquid leaving the first packing bed is collected and fed into an annular channel, where it is thoroughly mixed. It is then redistributed by a second cascade distributor. Smaller columns generally use side deflectors for liquid redistribution.

The internals of the column must offer minimum resistance to gas flow. Modern packings have a relative free-cross-sectional area of more than 90%. The liquid collector and distributor should also reach this value. The support grid of the packing is often a barrier to flow, especially when its free-cross-sectional area is blocked by the lowest particle layer.

8.1.3 Criteria for Use of Tray or Packed Columns

Proper choice of equipment is very important for effective and economical separation. An unambiguous choice is not always possible because both tray and packed columns can be modified to fulfill different requirements.

Tray columns are generally employed in large diameter (e.g., larger than 1 m) towers; several downcomers are often necessary. The gas load in tray columns must be kept within a relatively narrow range; only valve trays allow greater operational flexibility. The liquid load, however, can be varied over a very wide range, see Fig. 8-3. Tray columns can also be operated efficiently at very low liquid loads, e.g., in vacuum operation. However, their relatively high pressure drop (typically 7 mbar per equilibrium stage) is a disadvantage, especially when the separation requires a large number of trays. Tray columns have a relatively high liquid holdup in both the two-phase layer on the tray and the downcomer. This compensates for fluctuations in feed composition. However, the high holdup results in a high liquid residence time in the column, which may lead to the decomposition of thermally unstable substances. The holdup is also significant when a chemical reaction occurs during distillation. Special cap designs allow the liquid on the tray to be retained for short shutdown periods so that column operation can be resumed quickly. A further advantage of tray columns is that heat exchanger coils incorporated on the trays allow

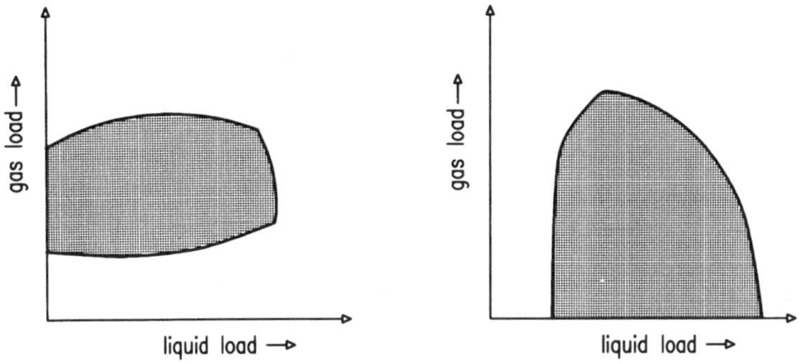

Fig. 8-3. *Sketch of the operating region of a tray column (left) and a packed column (right).*

heat to be added or removed easily. Tray columns are also relatively insensitive to impurities in the liquid.

Packed columns are used almost exclusively in small diameter (e.g., smaller than 0.7 m) towers. However, the development of structured packings permits the use of packed columns with very large diameters. Packed columns are extremely flexible as far as gas load is concerned, but they require a minimum liquid load, see Fig. 8-3, which often excludes their use in vacuum operation. Newly developed structured packings, however, have brought substantial progress to this area, too. The small pressure loss in packed columns, in comparison with tray columns, is a tremendous advantage. It is typically 0.5 mbar per equilibrium stage, *i.e.*, more than one order of magnitude lower than in tray columns. The danger of decomposition of thermally unstable substances is also less in packed columns because liquid holdup is very small. Packed columns are not suitable for the treatment of liquids that contain particulate contaminants or tend to crystallize. They are, however, less sensitive to foaming than tray columns. Ceramic random packings are resistant to corrosion.

8.2 DIMENSIONING TRAY COLUMNS

Schematic cross sections of bubble cap, valve and sieve trays together with their most important dimensions are shown in Fig. 8-4. On each tray the gas emerging from the openings in the plate penetrates into the liquid and, thus, generates a two-phase mixture with a large interfacial area. The larger the interfacial area the higher is the mass transfer rate. Very essential for high mass transfer rate is also a uniform gas flow through all openings in the tray (see Section 8.2.3.6).

8.2.1 Design Parameters of Tray Columns

Figure 8-5 shows some examples of bubble caps and valves. All types of elements are arranged on the plate in a triangular spacing. Typically, valves are smaller than bubble caps. Very essential parameters are plate spacing, weir height and height of

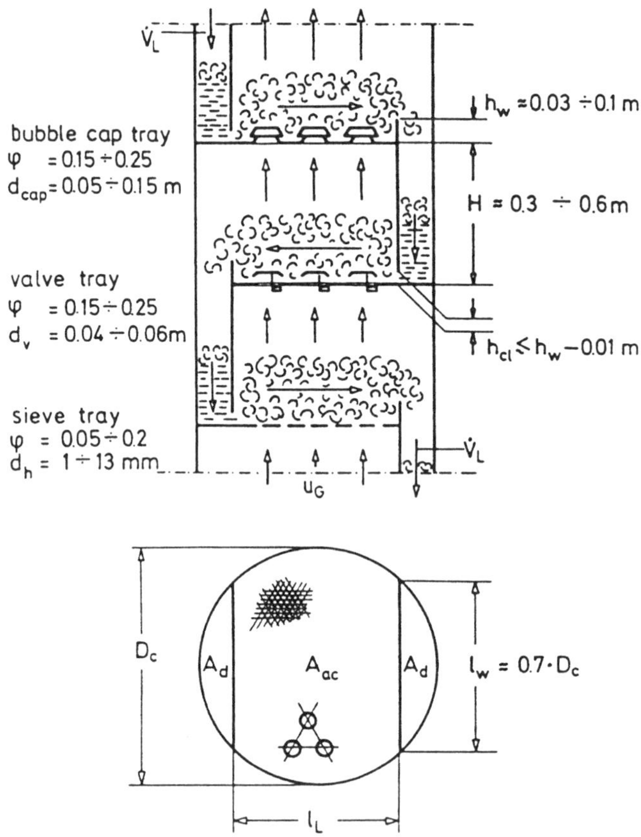

Fig. 8-4. *Schematic of a tray column showing some characteristic parameters.*

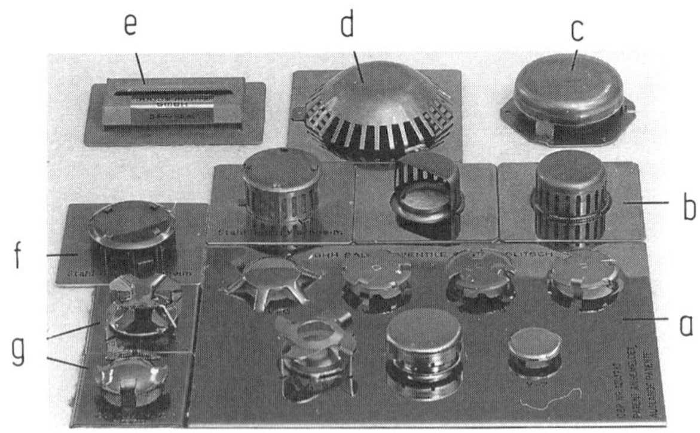

Fig. 8-5. *Examples of bubble caps and valves of several vendors: (a) Glitsch valves, (b) bayer bubble cap, (c) Linde bubble cap, (d) ACV bubble cap, (e) Montz KSG bubble cap, (f) Stahl valve, (g) Koch valves.*

Table 8-1: Standard Geometrical Dimensions of Trays

	Vacuum	Atmospheric Pressure	High Pressure
Tray spacing H (m)	0.4 to 0.6	0.4 to 0.6	0.3 to 0.4
Weir length l_W (m)	(0.5 to 0.6) · D_c	(0.6 to 0.75) · D_c	0.85 · D_c
Weir height h_W (m)	0.02 to 0.03	0.03 to 0.07	0.04 to 0.1
Skirt clearance h_{cl} (m)	0.7 h_W	0.8 h_W	0.9 h_W
Bubble cap diameter d_{cap} (m)	0.08 to 0.15	0.08 to 0.15	0.08 to 0.15
Bubble cap spacing	1.25 d_{cap}	(1.25 to 1.4) d_{cap}	1.5 d_{cap}
Valve diameter d_v (m)	0.04 to 0.05	0.04 to 0.05	0.04 to 0.05
Valve spacing	1.5 d_v	(1.7 to 2.2) d_v	(2 to 3) d_v
Hole diameter d_h (m)	0.004 to 0.013	0.004 to 0.013	0.004 to 0.013
Hole spacing	(2.5 to 3) d_h	(3 to 4) d_h	(3.5 to 4.5) d_h
Relative free area ϕ %	15 to 10	10 to 6	7.5 to 4.5

skirt clearance. Standard geometrical dimensions, developed by practical experience, are given in Table 8-1. These empirical findings will be theoretically founded in the following chapters.

8.2.2 Operating Region of Tray Columns

Careful column dimensioning requires determination of the entire operating region and the proper selection of the operating parameters so that a safety margin can be maintained with respect to the limits for liquid and gas loads. Hence, all the phenomena that limit column operation have to be studied first. From the right, understanding of the limiting factors' relations for calculating the border lines of the operating region can be developed.

8.2.2.1 Maximum Gas Load The gas load in a tray column can be increased up to a point where the gas blows the liquid off the tray in form of fine droplets. The liquid then no longer flows countercurrently to the gas, and proper column operation ends. [Souders and Brown 1934] were the first to develop a systematic approach for calculating the maximum gas load of a tray. They considered a single drop suspended above the tray and formulated the equilibrium of friction force and weight force minus buoyancy:

$$\frac{d^2 \pi}{4} \zeta_{sp} \frac{\rho_G}{2} u_G^2 = \frac{d^3 \pi}{6} (\rho_L - \rho_G) g \tag{8.2}$$

Here, u_G is the superficial gas velocity, d the diameter and ξ_{sp} the friction factor of the drop, ρ_L is the density of the liquid, and ρ_G is the density of the gas. The gas load is best expressed by the F-factor that represents the square root of the kinetic energy of the gas flow. It is defined by:

$$F = u_G \sqrt{\rho_G} \tag{8.3}$$

From Equations (8.2) and (8.3) follows:

Table 8-2: Correlations for the Gas Capacity Factor C_G

Reference	Parameters	Range of Validity
[Souders and Brown 1934]	$C_G = f(H, \sigma)$	bubble caps
[Kirschbaum 1950]	$C_G = f(H^{1/2}, d_{cap})$	bubble caps
[Kirschbaum 1960]	$C_G = f(H^{1/2})$	sieve trays
[Wenzel 1957]	$C_G = f(H)$	cryogenic systems
[Huang and Hodson 1958]	$C_G = f(H, \sigma)$	bubble caps
[Fair and Matthews 1958]	$C_G = f(H, \dot{V}_L/\dot{V}_G \cdot \rho_L/\rho_G)$	bubble caps
[Fair 1961]	$C_G = f(H, \dot{V}_L/\dot{V}_G \cdot \rho_L/\rho_G, \sigma^{0.2}, \phi)$	sieve trays
[Smith et al. 1963]	$C_G = f(H\text{-}h_f, \dot{V}_L/\dot{V}_G \cdot \rho_L/\rho_G)$	all tray design
[Hoppe and Mittelstrass 1966]	$C_G = f(H^{1/2}, \phi^{0.167})$	sieve trays
[Glitsch 1974]	$C_G = f(H, \rho_G)$	valve trays
[Koch 1968]	$C_G = f(H)$	valve trays

$$F = \sqrt{\frac{4\ d\ g}{3\ \xi_{sp}}}\sqrt{\rho_L - \rho_G} \tag{8.4}$$

The first right-hand term contains the drop diameter d that is unknown in most cases. According to [Souders and Brown 1934] this term is replaced by the gas capacity factor C_G. Then Equation (8.4) becomes:

$$F = C_G \sqrt{\rho_L - \rho_G} \tag{8.5}$$

The capacity factor C_G must be determined by experiments. Many correlations have been developed from experimental data. Some of them are listed Table 8-2. The most frequently used correlation is that of Fair presented in Fig. 8-6 [Fair 1961].

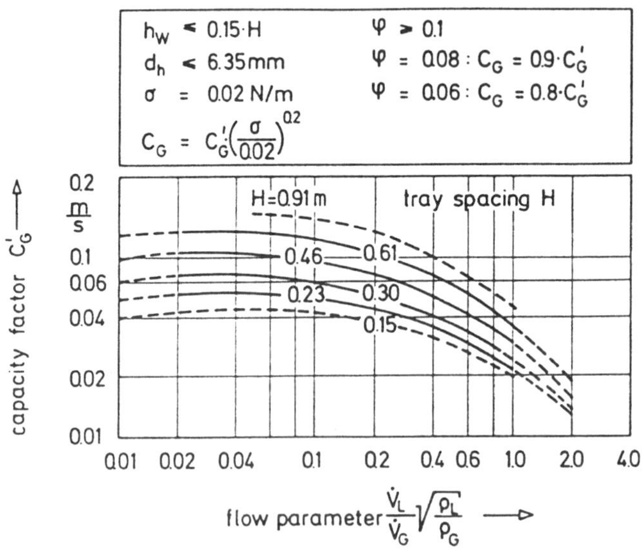

Fig. 8-6. Correlation of the capacity factor C_G for sieve trays according to [Fair 1961]. The capacity factor C'_G has to be corrected with respect to surface tension σ and relative free area ϕ.

He found that C_G is proportional to $\sigma^{0.2}$ (σ denotes the surface tension of the liquid) and dependent on the relative free area ϕ.

A theoretical approach based on first principles has been developed by [Stichlmair 1974]. He assumes that the gas on the tray forms the continuous phase while the liquid is dispersed into droplets (drop regime, see Section 8.2.3.1). The size of the droplets in the two-phase layer on a tray is determined by the gas jets emerging from the holes or slots of a sieve tray and a bubble cap tray, respectively. When the jets penetrate into the two-phase mixture they act alike a jet mill. The liquid drops are entrained into the jets, accellerated due to the friction forces and, simultanously, deformed. After having lost their spherical shape the drops are torn into pieces. Decisive for drop stability in a gas jet is the ratio of friction force to surface force, i.e., the Weber-Number. Drops are stable in a gas jet up to a critical value of the Weber-Number, We_{cr}, which is formed with the gas velocity in the orifices u_{Gh}. From the definition of the Weber-Number, $We = \rho\, u^2\, d / \sigma$, follows:

$$d = We_{cr}\, \sigma / F_h^2 \quad \text{or} \quad d = We_{cr}\, \sigma\, \phi^2 / F^2 \tag{8.6}$$

where σ is the surface tension of the liquid, ϕ is the relative free hole area, F is the gas loading factor defined in Equation (8.3), and F_h is the gas load factor in the holes ($F_h = F/\phi$). Combining equation (8.6) with equation (8.4) yields:

$$F_{max} = \left(\frac{4\, We_{cr}}{3\, \xi_{sp}}\right)^{1/4} (\phi^2\, \sigma\, (\rho_L - \rho_G)\, g)^{1/4} \tag{8.7}$$

According to [Wallis 1960], the value of the critical Weber-Number is approximately 12 for liquids with low viscosity. If the friction coefficient ξ_{sp} is assumed to be 0.4 (a value valid for rigid spheres in turbulent flow), then Equation (8.7) becomes:

$$F_{max} = 2.5\, (\phi^2\, \sigma\, (\rho_L - \rho_G)\, g)^{1/4} \tag{8.8}$$

This theoretically derived equation corresponds very well with published experimental data and with Fair's correlation [Fair 1961]. A proportionality of $\sigma^{0.25}$ exists in Equation (8.8), whereas Fair empirically obtained a proportionality of $\sigma^{0.2}$. Fair's dependence on the relative free area ϕ also corresponds well with the $\phi^{1/2}$ relationship in Equation (8.8).

Equation (8.8) describes the process of blowing the liquid off the tray and is applicable only to columns with very large tray spacings (approximately $H > 0.5$ m). When tray spacing is smaller, the height of the two-phase layer, and not the blow-off process, is decisive for the highest feasible gas load. An equation for predicting the froth heigh is given in Section 8.2.3.3. For safe operation the froth height must always be less than the tray spacing.

8.2.2.2 Minimum Gas Load The gas load of the column tray can be decreased to a point where either the gas no longer flows uniformly through all the tray openings or the liquid leaks through the tray. Both modes of operation should be avoided because they diminish tray efficiency.

According to [Mersmann 1963], gas flows uniformly through all holes of a sieve tray when the Weber-Number $We = \rho\, u_{Gh}^2\, d_h / \sigma$ formed with the hole diameter d_h exceeds a critical value of 2. From this condition follows:

$$F_{h,min} = \sqrt{2\, \sigma / d_h} \quad \text{or} \quad F_{min} = \phi \sqrt{2\, \sigma / d_h} \tag{8.9}$$

Hence, the minimum gas load is proportional to $d_h^{-1/2}$.

According to [Ruff et al. 1976], the liquid is prevented from leaking through a sieve tray if the Froude-Number Fr exceeds a critical value. From this condition follows:

$$F_{h,min} = \sqrt{0.37\, d_h\, g\, (\rho_L - \rho_G)^{1.25} / \rho_G^{0.25}} \quad \text{or} \tag{8.10}$$

$$F_{min} = \phi \sqrt{0.37\, d_h\, g\, (\rho_L - \rho_G)^{1.25} / \rho_G^{0.25}}$$

Here, the minimum gas load is proportional to $d_h^{+1/2}$.

Equations (8.9) and (8.10) are evaluated for three different systems in Fig. 8-7. At small hole diameters up to approximately 2 or 3 mm, the uniform gas flow through all holes is the limiting mechanism. The minimum gas load $F_{h,min}$ decreases with increasing hole diameter. At hole diameters greater than 2 to 3 mm, weeping of liquid through the tray becomes the decisive factor; the minimum gas load then increases with increasing hole diameter.

Equations (8.9) and (8.10) do not take into account the influence of the liquid

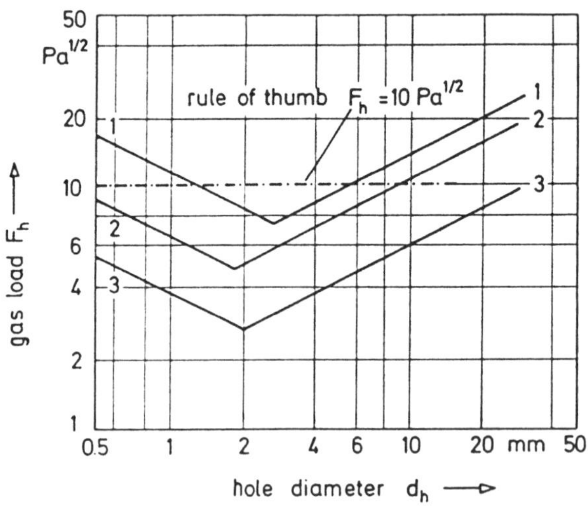

Fig. 8-7. *Minimum gas load of sieve trays F_{hmin} for three different systems: (1) Air/water, (2) benzene/toluene, (3) isobutane/n-butane.*

flow on the tray, although, for instance, large liquid loads require a somewhat higher minimum gas load to prevent leaking of liquid through the tray. However, if some weeping occurs at high liquid loads, it does not affect separation efficiency significantly since there still exists a counter-current flow of the two phases. Thus, some weeping of liquid can be tolerated.

On bubble cap trays the determination of the minimum gas load is more complex, since the bubble caps form a resistance to liquid flow across the tray. The flow resistance is overcome by a hydaulic gradient along the flow path of the liquid. Hence, the height of clear liquid varies on the tray and the risk of non-uniform gas flow increases. A detailed study of minimum gas load of bubble cap trays has been published by [Kemp and Pyle 1949]. They found that the minimum gas load depends significantly on liquid load per weir length and on the number of bubble cap rows to be crossed by the liquid flow across the tray. The values of F_{min} are generally higher on bubble cap trays than on sieve trays.

8.2.2.3 Maximum Liquid Load Liquid flows down the column through the downcomers due to gravity. This limited driving force permits a limited liquid flow rate $\dot{V}_L$ only. The following four empirical rules are often used to determine the maximum liquid flow rate, e.g., [Hoppe and Mittelstrass 1967]:

- The weir load $\dot{V}_L/l_w$ (l_w = weir length) should be less than 60 $m^3/(m\ h)$.
- The liquid velocity in the downcomer should not exceed a value of 0.1 m/s.
- The volume of the downcomer should permit a liquid residence time of more than 5 s.
- The height of the clear liquid in the downcomer should not exceed half of the tray spacing.

These rules have been developed by long-term practical experience. However, they do not permit any insight into the limiting phenomena.

The flow of liquid through the downcomer is, in essence, comparable with the flow of a liquid through the opening of a vessel. This classical flow problem has been solved by Torricelli as early as in 1644. In modern writing Torricellis equation is:

$$u_{Ld} = \alpha \sqrt{2\ g\ H} \tag{8.11}$$

Here, α denotes the discharge coefficient that is approximately 0.61.

Some additional phenomena have to be taken into account for applying equation (8.11) to the downcomer of a tray column.

- The downcomer does not contain clear liquid but a liquid gas mixture with a relative liquid content of ϵ_{Ld}.
- The velocity of clear liquid in the downcomer exit is expressed by $u_{Ld} = \epsilon_{Ld}\ \dot{V}_L/(l_{cl}\ h_{cl})$.

- The density of the gas ρ_G must not be neglected against the density of the liquid ρ_L.
- The liquid flows onto a tray with a clear liquid height h_L.
- Between downcomer inlet and outlet exists a pressure difference due to the pressure loss Δp of a tray. It can be expressed in terms of clear liquid height that is defined by $h_p = \Delta p/(\rho_L\ g)$.

With these phenomena the Torricelli Equation (8.11) turns into [Stichlmair 1978]:

$$\left(\frac{\dot{V}_L}{l_{cl}}\right)_{max} = 0.61\ \epsilon_{Ld}\ h_{cl}\sqrt{2\ g\ H\ (\rho_L - \rho_G)/\rho_L\left(1 - \frac{h_p + h_L}{\epsilon_{Ld}\ H}\right)} \quad (8.12)$$

where l_{cl} is the skirt clearance length and h_{cl} the skirt clearance height of the downcomer. The parameter ϵ_{Ld} is unknown; an approximate value of 0.4 can be employed for non-foaming systems. Further details are given by [Volpert and Stichlmair 1986].

8.2.2.4 Minimum Liquid Load A column tray can, in principle, be operated even with very small liquid loads because the height of the two-phase layer on the tray is kept at a minimum value by the exit weir. However, at extremely low liquid loads, liquid flows unevenly across the tray (maldistribution), which decreases the mass transfer efficiency (see Section 8.2.4). Accordingly, the minimum height of the weir overflow is usually set at approx. 5 mm; this corresponds to a minimum liquid weir load of $(\dot{V}_L/l_w)_{min} \cong 2\ m^3/(m\ h)$. In small diameter columns, however, this value may be considerably lower. Especially at high gas loads, entrainment of liquid by the gas may determine the minimum liquid load (see Section 8.2.2.5).

8.2.2.5 Operating Region of Tray Columns In careful dimensioning of a tray column, the complete operating region has to be determined. Figure 8-8 presents

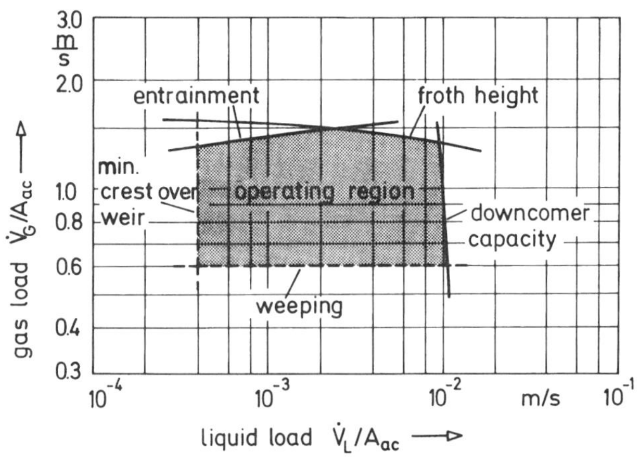

Fig. 8-8. Operating region of a tray column. The upper limits for gas and liquid flow cannot be crossed. The lower limits may be crossed without encountering serious flow problems.

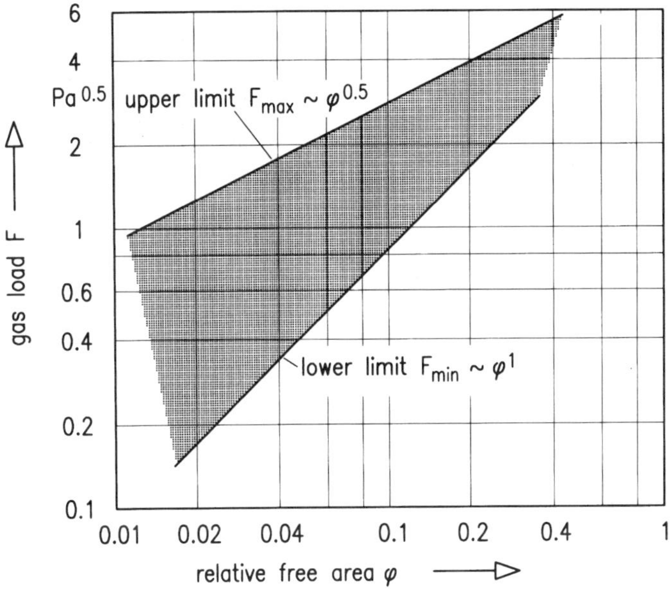

Fig. 8-9. *Influence of the relative free area ϕ on the maximum and minimum gas load.*

a graphical plot of the operating region for the system benzene/toluene at a pressure of 1 bar. The gas load $\dot{V}_G/A_{ac}$ is plotted versus the liquid load $\dot{V}_L/A_{ac}$ according to [Molzahn, and Schmidt 1975]; loads are referred to the active area A_{ac} of a tray. The operating region has been calculated with the equations developed in the previous sections. One should be aware that the upper borders for gas and liquid flow are absolute borders that can never be crossed. The lower borders (dashed lines), however, may be exceeded to a certain extent without encountering any flow problems. However, the mass transfer efficiency may gradually decrease. Generally, the operating conditions should be selected so that a sufficient safety margin to the border lines exists.

It is very interesting to study the influence of some design parameters on the operating region of a column. An example is depicted in Fig. 8-9. Here, the influence of the relative free area ϕ on the upper and lower border for gas flow is shown. According to Equation (8.8) the maximum gas load is proportional to $\phi^{1/2}$, whereas the minimum gas load is proportional to ϕ^1, see Equation (8.10). Hence, the width of the operating region becomes smaller with increasing relative free area ϕ. That is why the relative free area should not be choosen higher than $\phi = 0.15$ (see Table 8-1).

The influence of surface tension σ on the upper and the lower gas load limit is shown in Fig. 8-10. According to Equation (8.8) the upper limit is proportional to $\sigma^{1/4}$. The lower limit, however, shows a behavior that is different for sieve trays with small and with large holes, respectively. Equation (8.9), which is valid for small holes, contains a proportionality of $\sigma^{1/2}$. Hence, the operating region is wide

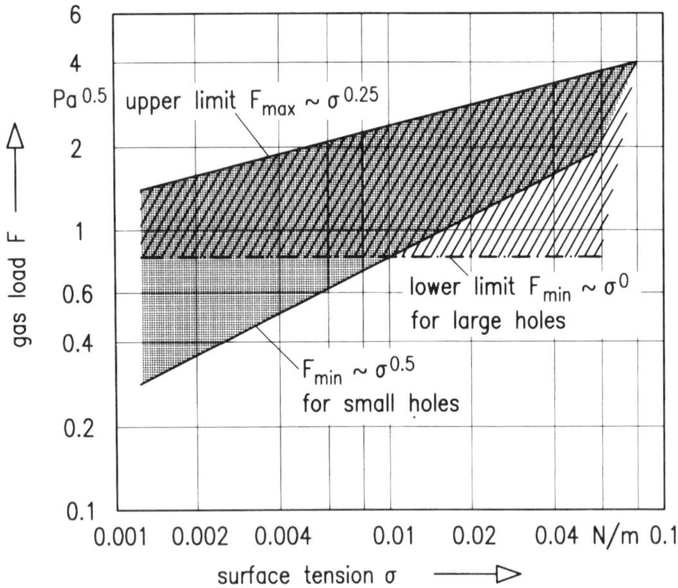

Fig. 8-10. *Maximum and minimum gas load as a function of surface tension* σ of the liquid.

at small surface tensions and small at large surface tensions. For trays with large holes, Equation (8.10) is valid which does not exhibit any influence of the surface tension. In this case, the operating region is small at low surface tensions and large at high surface tensions. Hence, in systems with low surface tensions, trays with small holes should be used. This agrees well with practical experience. For instance, columns for air separation, a system with very low surface tension, are typically equipped with sieve trays with very small holes, e.g., 1 mm or less.

8.2.3 Two-Phase Flow in Tray Columns

After determination of the operating region of the column, the characteristic data of the two-phase flow on the tray must be studied to ensure safe column operation. The behavior and properties of the two-phase layer on the trays are important for the effectiveness of tray columns.

8.2.3.1 Flow Regimes The two-phase layer may have any of three different structures or regimes:

- Bubble Regime: The liquid forms the continuous phase. The gas rises in the liquid in the form of discrete bubbles.
- Drop Regime: The gas forms the continuous gas phase and the liquid is dispersed into fine droplets.

- Froth Regime: This regime represents the intermediate state between the bubble and the drop regime. The gas-liquid layer is agitated intensively and no clearly dispersed phase exists.

The bubble regime can exist only at very high pressures. This statement is based on the following considerations:

The gas load at the holes of a tray should be no lower than $F_h \approx 5$ (see Fig. 8-7); therefore, the gas velocity at the holes is $u_{Gh} = 5 / \sqrt{\rho_G}$. However, the bubbles rise with a velocity of approximately 0.1 to 0.2 m/s. Only if the density of the gas ρ_G is very high (i.e., at high pressure) the gas velocity in the holes does not significantly surpass the rising velocity of the bubbles, as is necessary in a bubble regime. At low or moderate pressures, the superficial gas velocity is much higher than the rising velocity of the bubbles. Thus, bubbles accumulate above the orifices and form a region with very high gas content, which suddenly breaks through the two-phase layer and rises upwards. This jerky mode of operation is called froth regime.

The gas load at which the transition from froth to drop regime (phase inversion) occurs is of special interest [Stichlmair 1978]. Figures 8-11 and 8-12 present some experimental data for phase inversion on a sieve tray [Lockett et al. 1976, Loon et al. 1973, Pinczewski and Fell 1972, Porter and Wong 1969, Pinczewski et al. 1975]. From the data it follows that if the liquid level on the tray h_L is high, phase inversion occurs at $F/F_{max} \cong 0.65$. However, if clear liquid height h_L is very low, phase

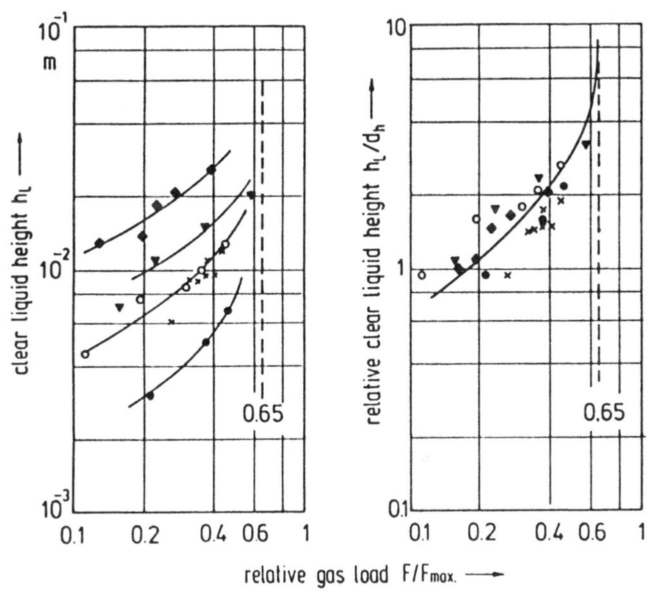

Fig. 8-11. Transition from froth to drop regime (phase inversion). Air/water data from [Lockett et al. 1976]. All the data lie on a single line when the clear liquid height h_L is referred to the hole diameter d_h (right-hand side).

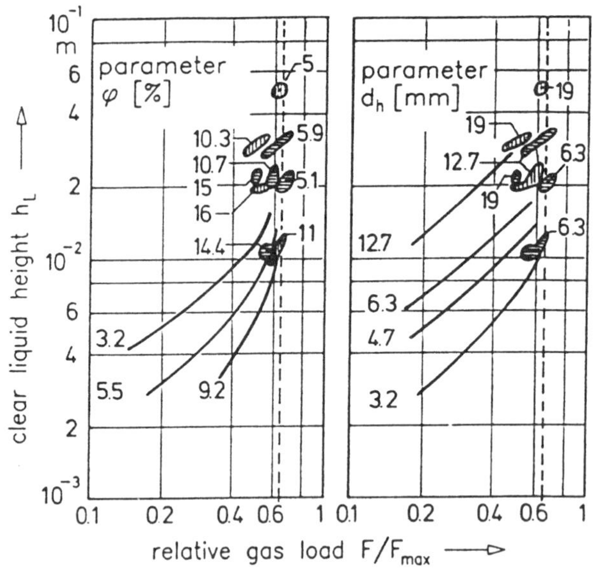

Fig. 8-12. Transition from froth to drop regime (phase inversion). Air/water data from [Loon et al. 1973; Pinczewski and Fell 1972; Porter and Wong 1969; Pinczewski et al. 1975]. Parameter is the relative free area ϕ.

inversion may occur much earlier. Hence, column dimensioning requires a minimum gas penetration depth of approximately 20 mm. In most cases tray columns are operated in the froth region. Drop regime may sometimes exist in vacuum distillation columns.

8.2.3.2 Relative Liquid Holdup The relative liquid holdup ϵ_L is a very important characteristic parameter of the two-phase layer on a tray. It is defined as the ratio of clear liquid height h_L and froth height h_f.

$$\epsilon_L = h_L/h_f \tag{8.13}$$

From a knowledge of the froth height h_f and the relative liquid content ϵ_L, the height of clear liquid h_L on a tray is easily calculated. Clear liquid height h_L is one of the most important parameters for pressure drop of a tray and, moreover, together with the liquid hold-up in the downcomer it determines the residence time of the liquid in the column.

Table 8-3 lists some references for the relative liquid content ϵ_L together with the parameters involved and the range of validity. The correlations have been evaluated for air/water and a sieve tray with a relative free area $\phi = 0.1$ and a weir height $h_W = 0.05$ m. The results are plotted versus the gas load F in Fig. 8-13. Nearly all correlations lie in a narrow range. All the correlations and most of the published data are summarized by the following equation:

Table 8-3: Listing of Some Correlations for the Relative Liquid Content ϵ_L in the Froth

Reference	Parameters	Range of Validity
1 [Kastanek 1970]	u_G	methanol/water
2 [Stichlmair 1978]	F/F_{max}	several
3 [F.R. Michigan 1960]	F	several
4 [Mashelkar 1970]	u_G	several
5 [Crozier 1957]	F	—
6 [Solomakha et al. 1972]	F, h_L, η_L, σ	several
7 [F.R. Delaware 1958]	F, $\dot{V}_L/l_W$, h_W	air/water
8 [Thomas, Campbell 1967]	F, $\dot{V}_L/l_W$, h_W	air/water
9 [Takahashi et al. 1974]	u_G, h_L	air/water
10 [Andrew 1969]	u_G	air/water
11 [Rama Iyer, Murti 1965]	F, $\dot{V}_L/l_W$, h_W, ρ_L	air/water
12 [Thomas, Haq 1976]	F, $\dot{V}_L/l_W$, h_W	air/water

$$\epsilon_L = 1 - (F/F_{max})^{0.28} \tag{8.14}$$

This relationship is valid only if the term $\sqrt{\phi\ \rho_L/\rho_G}$ is larger than 50. Determination of the relative liquid holdup for smaller values of this parameter is described in [Stichlmair 1978].

8.2.3.3 Froth Height A knowledge of the froth height on a tray is very essential since it must be considerably lower than tray spacing. If the froth extends up to the

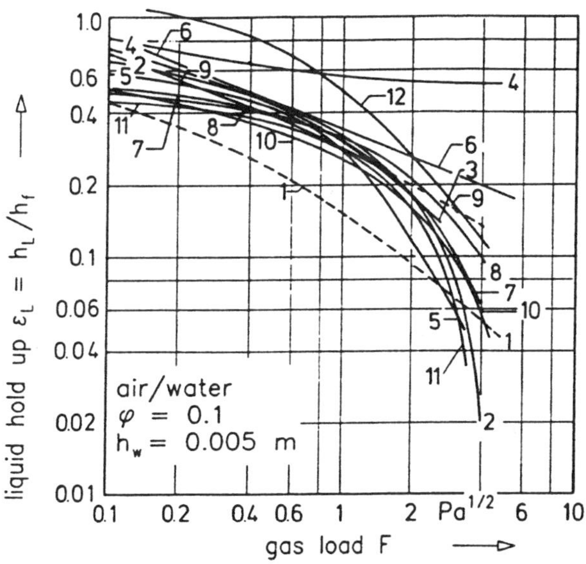

Fig. 8-13. Comparison of published correlations for the relative liquid content ϵ_L in the two-phase layer on a tray (see Table 8-3).

Table 8-4: Listing of Some Correlations for the Froth Height

Reference	Parameters	Validity
1 [F.R. Delaware 1958]	F, $\dot{V}_L/l_W$, h_W	air/water
2 [F.R. Delaware 1958]	F^2, h_W	hexane/heptane acetone/benzene
3 [Bernard and Sargent 1966]	F, h_W	air/water
4 [Thomas, Campbell 1967]	F, $\dot{V}_L/l_W$, h_W	air/water
5 [Stichlmair 1978]	F/F_{max}, $\dot{V}_L/l_W$, h_W	several
6 [Kastanek 1970]	u_G, h_W	methanol/water
7 [Andrew 1969]	u_G, $\dot{V}_L/l_W$, h_w	air/water
8 [Weiss, John 1973]	$F^{1.7}$, h_W,	several
9 [Mukhlenov 1958]	u_G, $\dot{V}_L/l_W$, σ, η_L/ρ_L	several
10 [Rama Iyer, Murti 1965]	F, $\dot{V}_L/l_W$, h_W, ρ_L	air/water
11 [Thomas, Haq 1976]	F, $\dot{V}_L/l_W$, h_W	air/water

next tray it may block the openings in the tray and cause a sharp increase in pressure drop and, finally, flooding.

Published correlations for the froth height are often neither dimensionally correct nor do they adequately represent the limiting conditions. Table 8-4 lists some references for froth height correlations together with their parameters and range of validity. These correlations have been evaluated for a sieve tray ($h_W = 0.05$ m, $\dot{V}_L/l_W = 0.005$ m^3/(m s)). The results are plotted in Figure 8-14 versus the gas load factor F. The correlations differ considerably. Based on our own and published experimental data the following equation is recommended [Stichlmair 1978]:

$$h_f = h_w + \frac{1.45}{g^{1/3}}\left(\frac{\dot{V}_L/l_w}{\epsilon_L}\right)^{2/3} + \frac{125}{(\rho_L - \rho_G)\, g}\left(\frac{F - 0.2\sqrt{\rho_G}}{1 - \epsilon_L}\right)^2 \tag{8.15}$$

where h_w is weir height, l_w weir length, and g the acceleration due to gravity. The influence of the liquid load on the height of the two-phase layer is described primarily by the second term, which represents the overflow height of the two-phase mixture over the exit weir. The last term takes into account upward spraying of liquid caused by the gas, which occurs only in the froth and drop regimes. Therefore, it should be considered only when $F - 0.2\sqrt{\rho_G} > 0$.

Equation (8.15) is valid only when gas is allowed to pass through the liquid up to the exit weir. If the exit weir is preceded by a calming zone, gas liquid separation can take place, with the result that clear liquid flows over the exit weir. The height of the two-phase layer is then expressed by:

$$h_f = (h_w + \frac{1.45}{g^{1/3}}(\dot{V}_L/l_w)^{2/3})/\epsilon_L \tag{8.16}$$

Equation (8.16) usually gives a larger value of the froth height h_f than Equation (8.15) does. Figure 8-15 shows a graphical plot of Equation (8.15). For a relative

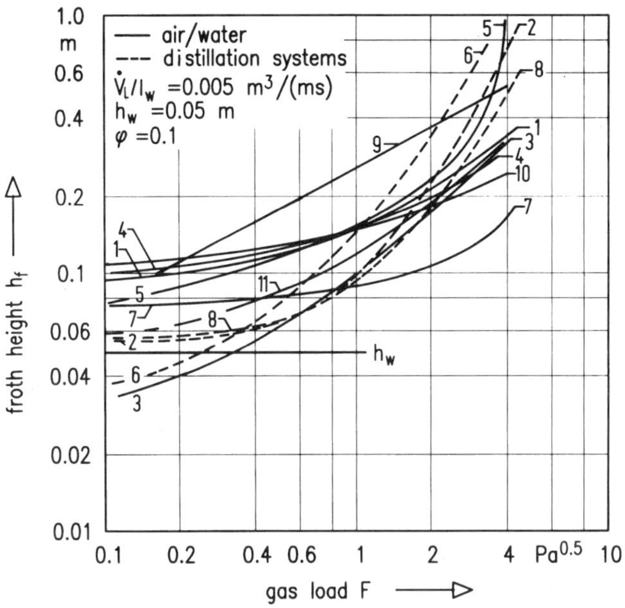

Fig. 8-14. *Comparison of published correlations for the froth height h_f (see Table 8-4).*

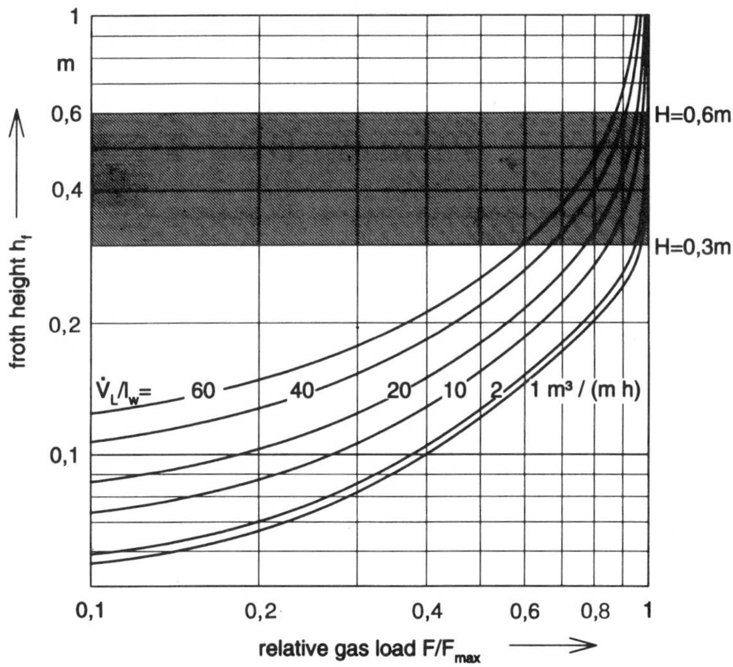

Fig. 8-15. *Froth height h_f versus the relative gas load F/F_{max}. From this plot follows that the optimal tray spacing H lies in the region of 0.3 to 0.6 m.*

gas load $F/F_{max} \rightarrow 1$ the froth height h_f rises drastically, and so does the plate spacing H required to avoid flooding since the condition $H > h_f$ must be met. From the plot follows that depending on liquid load, the favorite values of plate spacings lie between 0.3 m and 0.6 m. Plate spacings higher than 0.6 m permit only a small increase of gas load whereas plate spacings lower than 0.3 m require a drastic reduction of the gas load. Hence, a plate spacing too low leads to a large column diameter to handle a given gas flow.

8.2.3.4 Liquid Entrainment Gas flowing through the two-phase layer on a tray always entrains some liquid in form of fine droplets. This liquid entrainment is unfavorable because it affects the countercurrent flow of gas and liquid within the column and, in turn, decreases mass transfer efficiency of a tray.

The decision mechanism for liquid entrainment is the inevitable generation of liquid droplets in the two-phase layer. Even at very low gas loads (e.g., bubble regime) fine droplets are formed by bursting bubbles. The jerky mode of motion at froth regime results in an increased rate of drop formation. At drop regime all the liquid on the tray is dispersed into fine droplets. But the droplets do not have a uniform drop size as has been supposed in Section 8.2.2.1. There rather exists a drop size distribution that is unsymmetrical, i.e., a swarm contains more small than large droplets.

Liquid drops are entrained by the gas flow if their terminal velocity is lower than gas velocity. However, due to the high-speed motion within the two-phase layer on a tray, even some large drops are thrown upwards to form a layer of drops above the froth. After having reached a certain height, most droplets fall down again into the two-phase layer on the tray. The smaller the drops the higher they rise. Hence, at small tray spacing, even drops with a terminal velocity higher than the gas velocity may be entrained.

Table 8-5: Liquid Entrainment Data Plotted in Fig. 8-16

Reference	Tray Design and Tray Spacing H (m)	System and Surface Tension σ(kg/s^2)	Liquid Load $\dot{V}_L/l_w$(m^3/(mh))
▼ [Peavy, Baker 1937]	caps, 0.15 to 0.3	ethanol/water, 0.02	0.73 to 5.35
■ [Ashraf et al. 1934]	caps, 0.38	air/kerosene, 0.03	–
◆ [Holbrook, Baker 1934]	caps, 0.3 to 0.79	steam/water, 0.062	0.2 to 1
○ [Sherwood, Jenny 1935]	caps, 0.23	air/water, 0.073	–
◇ [Hunt et al. 1955]	sieve, 0.51	air/water, 0.073	–
□ [Pinczewski, Fell 1975]	sieve, 0.91	air/water, 0.073	15
▽ [Calcaterra et al. 1968]	sieve, 0.33 to 0.66	air/water, 0.073	10.9
△ [Friend et al. 1960]	sieve, 0.15 to 0.3	air/water, 0.073	7.18 to 29
△ [F.R. Delaware 1958]	caps, 0.61	acetone/benzene, 0.02	8.7 to 18.6
▲ [F.R. Delaware 1958]	caps, 0.61	steam/water, 0.062	7.45 to 44.7
● [F.R. Delaware 1958]	caps, 0.61	cyclohexane/-n-heptane, 0.012	3.9 to 54.2
⊕ [Piqueur, Verhoye 1976]	valve, 0.3 to 0.5	air/water, 0.073	0.45 to 4.5
⊙ [Lockett et al. 1976	sieve, 0.36	air/water, 0.073	3.6 to 21.6

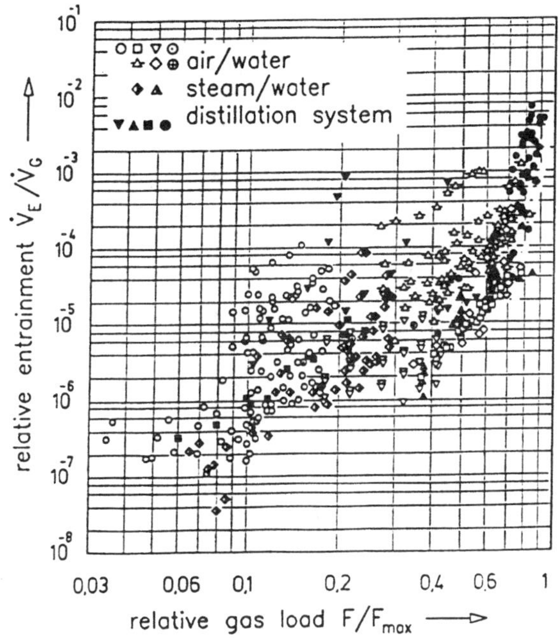

Fig. 8-16. *Plot of published entrainment data versus the relative gas load F/F_{max} (see Table 8-5).*

Many investigations on liquid entrainment have been reported in literature (see Table 8-5). Often used is the correlation of [Fair 1961]. Liquid entrainment rate is a quantity that may vary over several orders of magnitude. All authors agree that the gas load factor F is the dominant factor. An increase of the gas load by a factor of 10 increases liquid entrainment by a factor of 10^3 or 10^4. Another major factor is surface tension σ because it influences the drop size. Liquid viscosity is only of influence in high viscous systems [Barker and Choudhury 1959] seldom encountered in distillation columns. Further important design parameters are the relative free area ϕ [Pinczewsky et al. 1975] and tray spacing H, since it determines the clearance between froth height and next tray, i.e., the distance an entrained droplet has to cover. Figure 8-16 shows a plot of published experimental entrainment data (see Table 8-5). The relative entrainment $\dot{V}_E/\dot{V}_G$ is plotted versus the relative gas load F/F_{max}. According to [Stichlmair and Hofer 1978], the quantity $F_{max} = 2.5\ (\phi^2\ \sigma\ \Delta\rho\ g)^{1/4}$ accounts adequately for the influence of the surface tension σ and the relative free area ϕ of the tray. The data scatter very widely since the important parameter tray spacing H is not considered. Dimensional analysis yields the parameter $(H\text{-}h_f)\ \Delta\rho\ g/F_{max}^2$ as dimensionless clearance.

Figure 8-17 shows the correlation of the data of Fig. 8-16. The relative entrainment $\dot{V}_E/\dot{V}_G$ increases significantly with increasing gas load F/F_{max} (please keep in mind that the abscissa covers 1 decade and the ordinate 7 decades). The sharp bend of

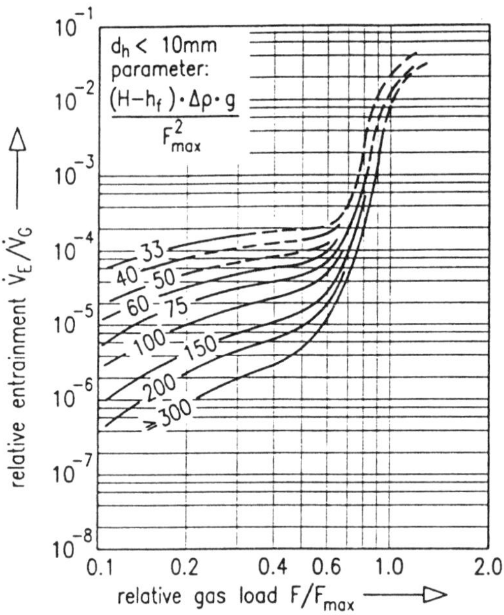

Fig. 8-17. *Correlation of published entrainment data according to [Stichlmair and Hofer 1978]. Parameter is the dimensionless tray clearance.*

the parameter lines at $F/F_{max} = 0.65$ is obviously caused by the transition from froth to drop regime (phase inversion, see Section 8.2.3.1). For large values of the dimensionless clearance $(H\text{-}h_f)\ \Delta\rho\ g/F_{max}^2$ (in air water systems approximately $H - h_f > 0.5$ m) the entrainment no longer depends on the clearance, since only drops with a terminal velocity smaller than the superficial gas velocity are entrained. However, with decreasing clearance the entrainment increases drastically since even large drops may be tossed to the next tray.

The correlation of Fig. 8-17 describes the major factors for entrainment quite well. More than 95% of the data are correlated with an accuracy better than 50% which is very good, keeping in mind the wide variation of the data from 10^{-7} to 10^{-2}. The accuracy of the correlation is limited by the strong influence of gas velocity on entrainment. For instance, a deviation of 10% in the gas velocity causes a deviation of liquid entrainment of approximately 60%.

The data base of the correlation (Table 8-5) contains data of bubble cap, sieve and valve trays with different tray spacings and different liquid loads (accounted for by the froth height h_f). Most data have been collected with the system air/water but steam/water data and hot distillation data are also correlated very well.

Liquid entrainment is inevitable in tray operation. However, too large an entrainment affects mass transfer efficiency since a pure countercurrent flow of gas and liquid no longer exists. A rule of thumb says that an entrainment rate of up to 10% of the liquid flow onto the tray may be tolerated.

8.2.3.5 *Liquid Mixing in the Two-Phase Layer* In tray columns, liquid flows across the tray to the exit weir; i.e., there exists a cross flow of liquid and gas. Mass transfer between the two phases generates a concentration profile in the liquid, which greatly influences tray efficiency. The two-phase layer on the tray is agitated intensely by the gas flow, particularly in the froth and drop regimes. This results in partial mixing of the liquid on the tray, which, in turn, reduces the differences in liquid concentration. According to the dispersion model, the degree of liquid mixing on the tray is represented by the dimensionless Peclet Number which is defined by:

$$Pe = \frac{l_L^2}{D_E \tau_L} \tag{8.17}$$

where l_L is the path length of the liquid, D_E is the dispersion coefficient, and τ_L is the residence time of the liquid in the two-phase layer on the tray. The liquid is totally mixed when Pe = 0; the opposite case (i.e., no mixing, plug flow) is characterized by $Pe \rightarrow \infty$.

A result of liquid mixing experiments performed by [Stichlmair and Weisshuhn 1973] is presented in Fig. 8-18. The test tray, a rectangular sieve tray 6 cm in width and 25.5 cm in length, has been operated with warm water and air. On the tray the air becomes nearly saturated with water due to the high mass transfer rate in the two-phase layer. In this case, mass transfer produces not a change of liquid concentration but of liquid temperature since the high heat of vaporization of water has to be covered by the liquid phase (as in a cooling tower). The temperature profile in the

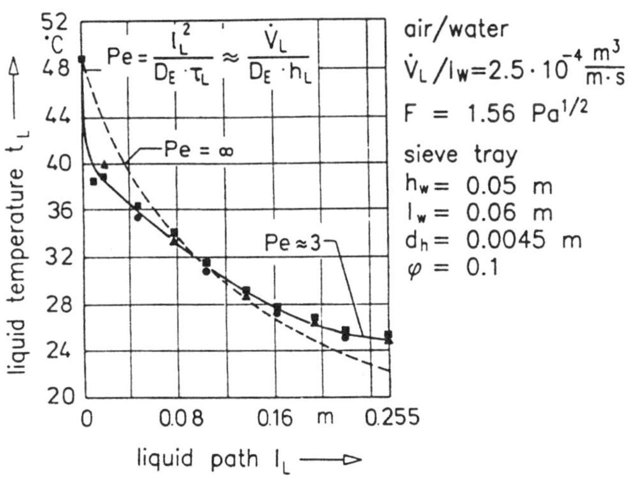

Fig. 8-18. *Liquid mixing in the two-phase layer on a sieve tray. The dashed line has been calculated for plug flow with an efficiency of E_{OG} = 0.7. Data were measured 20 mm (●), 50 mm (▲) and 70 mm (■) above the plate.*

froth has been measured by thermo-couples. The data reveal that the liquid temperatures remain practically unchanged in vertical direction. Hence, the liquid is well mixed vertically. In horizontal direction, however, there exists a significant temperature gradient.

In case of plug flow (i.e., $Pe \rightarrow \infty$) the temperature profile on a tray can easily be calculated from a knowledge of the point efficiency E_{OG}. The result is presented in Fig. 8-18 by a dashed line (calculated with $E_{OG} = 0.7$). The data points differ significantly from the calculated line indicating the degree of liquid mixing. A more detailed analysis of the data points (see Section 8.2.4.2) gave a degree of liquid mixing of $Pe \approx 3$. By a series of such experiments with variation of tray length, gas load and liquid load, a correlation for the dispersion coefficient D_E (i.e., eddy diffusion coefficient) has been developed by [Weisshuhn 1976].

$$D_E = 1.06\, h_f\, F \sqrt{\frac{\epsilon_L}{\rho_L - \rho_G}} \tag{8.18}$$

Equation (8.18) applies only to the froth and the drop regimes. In the bubble regime the degree of liquid mixing is typically so small that it can be neglected.

8.2.3.6 *Liquid Maldistribution* Undisturbed plug flow of liquid provides the highest mass transfer efficiency of any tray design. In small diameter columns plug flow is primarily disturbed by liquid mixing (see previous Section). In large diameter columns, however, liquid mixing is of minor importance. Here, the major mechanism that adversely affects plug flow is maldistribution. Maldistribution denotes a non-uniform flow rate profile. Liquid flow rates vary locally on the tray and, in turn, a wide variation of liquid residence time exists in the froth.

Maldistribution strongly affects tray efficiency of large diameter columns. In sections with high liquid residence time the liquid virtually reaches equilibrium state with the gas phase and, due to the nonexisting driving force, these sections no longer contribute to the mass transfer. The effect of maldistribution on tray efficiency may be much worse than that of liquid mixing since tray efficiency of a large diameter tray may fall below point efficiency. However, knowledge of the degree of maldistribution is very poor and, even worse, no theory exists to quantitatively describe the relation between maldistribution and tray efficiency.

There exist several techniques for maldistribution measurement on trays. The simplest method is to suddenly inject a dye into the incoming liquid and to watch the spreading of the colored liquid across the tray by eye, e.g., [Lockett 1986]. [Bell 1972] used a fluorescing dye and glass fiber detectors to gain more quantitative results. The movement of ping pong balls on the froth has been watched by [Porter and Lockett 1972]. They found that the balls may stop and even circle in some regions of the tray.

A good method for detecting liquid maldistribution on a tray is the measurement of liquid temperature profiles in the froth [Stichlmair 1971] either by thermo-couples or, even better, by an infrared camera (see the front cover). Figure 8-19 shows some typical results of experiments performed by [Stichlmair and Ullrich 1985] on a large

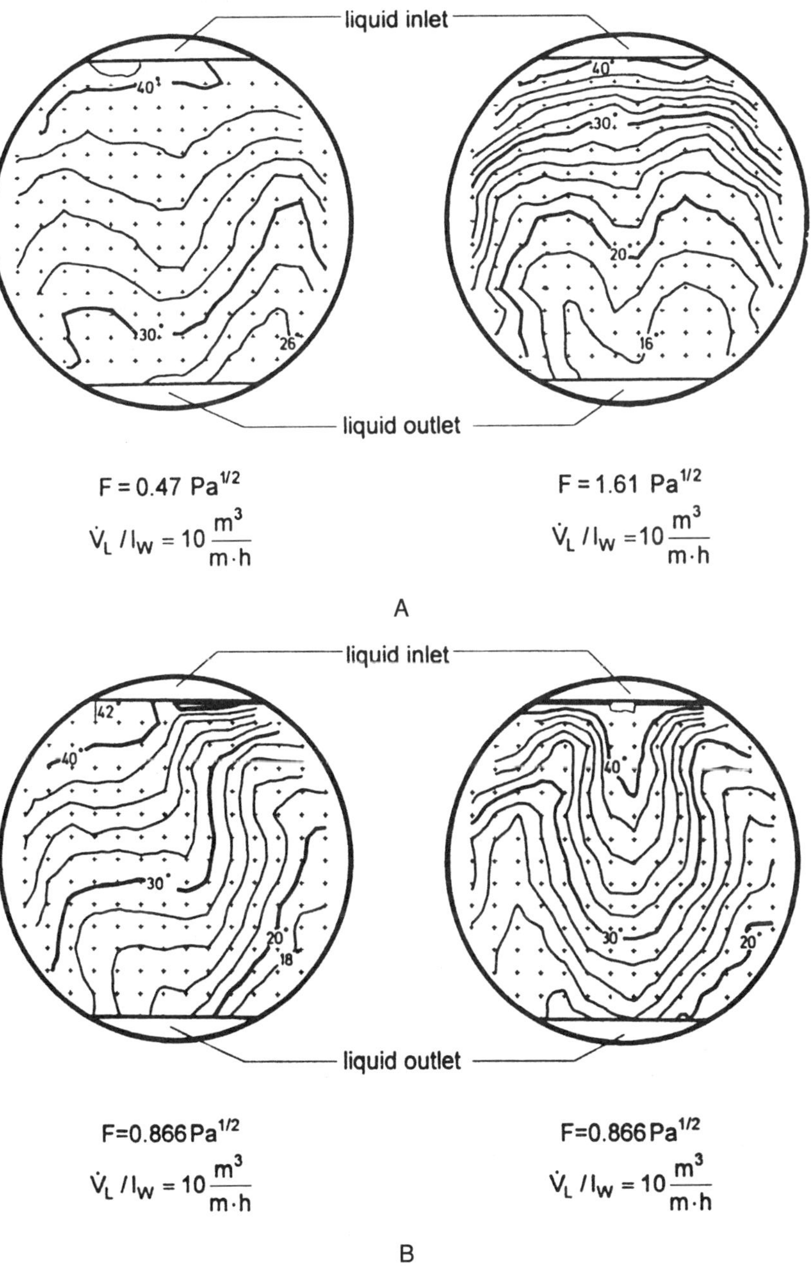

Fig. 8-19. *Liquid isotherms in the two-phase layer on a bubble cap tray of 2.3 m diameter. Deviations from horizontal straight lines are caused by maldistribution: (A) Modifications of the gas load, (B) modifications of the inlet weir.*

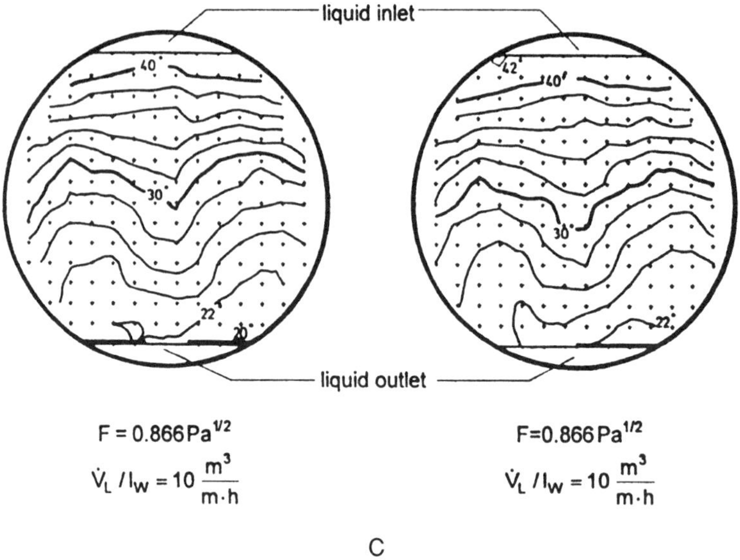

Fig. 8-19. *(continued)* (C) Modifications of the outlet weir.

diameter bubble cap tray that has been operated with warm water and air. The tray had a diameter of 2.3 m and 440 Bayer bubble caps (see Fig. 8-5). The liquid temperatures have been measured by 182 thermo-couples that have been uniformly distributed over the active area of the tray. The positions of the thermo-couples are marked by crosses in Fig. 8-19. From the data, lines of constant liquid temperatures have been evaluated via linear interpolation by a computer and plotted in the active area of the tray. The isotherms can be interpreted as lines of constant residence time of the liquid in the froth. In case of pure plug flow, the isotherms are straight lines that are parallel to the inlet and the outlet weir. Any deviations from horizontal straight lines are caused by a non-uniform liquid flow, i.e., by maldistribution.

Figure 8-19A presents two results with different gas loads. The form of the isotherms reveals a large degree of maldistribution existing on the tray. Near the wall and in the center, liquid flow rate is increased, obviously by a higher cap spacing in these regions. However, variations of cap spacing are inevitable in tray design.

Experimental results with modifications of the inlet weir are depicted in Fig. 8-19B. Blocking of half of the inlet weir (left-hand side) totally changes the liquid flow across the tray. On the blocked side the temperatures are significantly lower than on the free side indicating a lower liquid throughput. It is worthwhile to note that no maldistribution could be detected by eye during these experiments. Central liquid feeding (right-hand side) also dominates the liquid flow across the whole tray.

Modifications of the outlet weir have only little effect on liquid flow rate profiles across the tray as can be seen in Fig. 8-19C. The same modifications of the weir have been made as in Fig. 8-19B but of the outlet instead of the inlet weir. Hence, the installation of zigzag outlet weirs often applied to trays with very low liquid load is no means to improve liquid flow pattern.

More than 80 experiments of this type have been performed with systematical variation of gas and liquid flow, of tray levelness and of the form of liquid inlet and outlet weirs. In some experiments baffles have been installed on the tray and especially near the column wall. Bubble cap trays as well as sieve trays have been tested with this technique. The results can be summarized as follows:

- Even small variations of cap spacing cause severe liquid maldistribution.
- Standard baffles often installed near the column wall are in most cases too small and ineffective.
- Modifications of the form of the outlet weir, e.g. zigzag weir, have little or no effect.
- Modifications of the form of the inlet weir are very effective.
- Levelness of the plate and its horizontal installation is of major importance.

In practice, a certain degree of liquid maldistribution is virtually inevitable. Hence, mass transfer efficiency will never be as high as predicted for plug flow (see Section 8.2.4.2). However, the unfavorable effects of maldistribution are restricted to each tray since in the succeeding downcomer the liquid is totally remixed. The effects of maldistribution do not add up over the height in tray columns as they do in packed columns (see Chapter 8.3.3.3). The permanent liquid mixing in the downcomers compensates for maldistribution to a certain degree and makes trays a favorite candidate for large diameter column internals.

8.2.3.7 *Interfacial Area* A high interfacial area in the two-phase layer on the tray is the most essential prerequisite for good mass transfer. The interfacial area A is conveniently referred to the volume V_f of the froth and expressed by $a = A/V_f$. Present knowledge on the interfacial area is very poor and no satisfactory correlation is available. Published data differ by a factor of up to 10, apparently due to different measuring techniques [Stichlmair 1978].

Therefore, an attempt will be made to estimate the order of magnitude of the interfacial area and the influencing factors. Basis is the following equation, which is valid for a two-phase layer with spherical dispersed particles:

$$a = 6\ \epsilon_d/d \tag{8.19}$$

Here, ϵ_d is the volume fraction of the dispersed phase and d is the particle diameter. Equation (8.19) can be applied to the bubble and drop regime, but not to the froth regime.

In bubble regime the diameter d of the bubbles and their volume fraction ϵ_d is approximately (Equation 8.14):

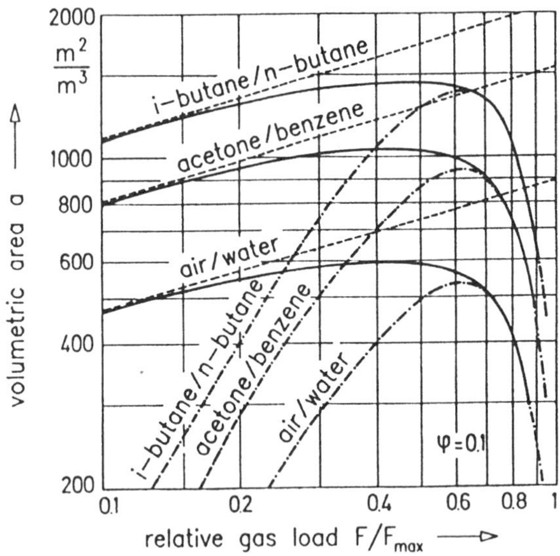

Fig. 8-20. Calculated values for the volumetric interfacial area a of the two-phase layer on column trays: —— Bubble regime, –·–·– Drop regime, — interpolation.

$$d = \left(\frac{6\,\sigma}{(\rho_L - \rho_G)\,g}\right)^{1/2} \quad \text{and} \quad \epsilon_d = 1 - \epsilon_L = (F/F_{max})^{0.28} \tag{8.20}$$

From Equations (8.19) and (8.20) follows the interfacial area in the bubble regime:

$$a = 6\left(\frac{(\rho_L - \rho_G)\,g}{6\,\sigma}\right)^{1/2} (F/F_{max})^{0.28} \tag{8.21}$$

In drop regime the liquid is the dispersed phase. The drop diameter d follows from Equation (8.6) and the volume fraction ϵ_d from Equation (8.14), respectively:

$$d = 12\,\sigma\,\phi^2/F^2 \quad \text{and} \quad \epsilon_d = \epsilon_L = 1 - (F/F_{max})^{0.28} \tag{8.22}$$

Analogously, we get for the drop regime:

$$a = \frac{F^2}{2\,\sigma\,\phi^2}\,(1 - (F/F_{max})^{0.28}) \tag{8.23}$$

where ϕ is the relative free hole area, i.e., ratio of total hole area to the active area.

Equation (8.21) for bubble regime and (8.23) for drop regime have been evaluated and plotted for three systems in Fig. 8.20. The froth regime cannot be evaluated because no univocal dispersed phase exists. However, the interpolated values shown in Fig. 8-20 should approximate the real course. The interfacial area in aqueous systems is approximately a ≈ 500 m^2/m^3, but in organic systems, values of a ≈ 1000 m^2/m^3 or more can be reached due to a lower surface tension. The interfacial tension σ has a considerable effect on the area of the interfacial area. The analysis of the above equations gives a proportionality of $a \approx \sigma^{-0.57}$ which is in excellent accordance with the theory of [Kolmogoroff 1958] that predicts a proportionality of $a = \sigma^{-0.6}$ in dispersed systems with homogeneous turbulence.

The above theoretical approach only permits a rough estimation of the interfacial area and the most important influencing factors. The values obtained are only partially applicable to the calculation of the mass transfer efficiency of a tray column. This is because it is not known whether all parts of the interface contribute in the same extent to mass transfer between phases. Thus, interfacial area is a weak point in predicting tray efficiency.

8.2.3.8 Pressure Drop The permanent pressure loss a gas suffers as it passes through a single tray is illustrated in Fig. 8-21. The pressure drop of a tray is rather high, typically in the order of $3 \cdot 10^2$ to $8 \cdot 10^2$ N/m^2 (i.e., 30 to 80 mm clear liquid height). The major parameters that influence the pressure loss are tray design, gas load F and liquid load per weir length, $\dot{V}_L/l_w$.

The basic equation for calculating the pressure drop of a tray is:

$$\Delta p = \Delta p_d + \Delta p_L + \Delta p_R \qquad (8.24)$$

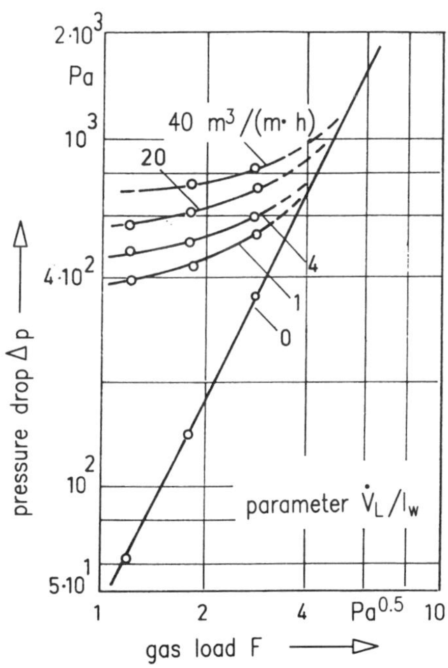

Fig. 8-21. Pressure drop of a sieve tray versus gas load factor F. Parameter is the liquid flow per weir length $\dot{V}_L/l_W$.

The first term Δpd accounts for the pressure drop of a dry tray. The second term Δp_L represents the pressure drop resulting from the clear liquid height h_L and can be formulated by:

$$\Delta p_L = h_L \rho_L g = h_f \epsilon_L \rho_L g \tag{8.25}$$

The froth height h_f and the relative liquid content ϵ_L can be determined by Equation (8.15) and (8.14), respectively.

The last term of Equation (8.24), Δp_R, accounts for several residual factors including bubble formation, liquid mixing and vertical acceleration of the liquid. This term is much smaller than the first two terms and may be neglected in most cases.

8.2.3.8.1 Pressure Drop of Dry Sieve Trays The pressure drop of the gas as it passes through a dry sieve tray Δp_d is given by:

$$\Delta p_d = \zeta \frac{\rho_G}{2} u_{Gh}^2 \quad \text{or} \quad \Delta p_d = \frac{\zeta}{2} F_h^2 \tag{8.26}$$

The unknown quantity in Equation (8.26) is the orifice coefficient ζ. For sieve trays equations for the orifice coefficient can be derived from first principles.

The mechanism of fluid flow through a sieve tray is easily understood by considering the flow through a single hole in a plate. In principle, such a flow consists of two essential parts, i.e., a converging part before the plate and a diverging part after the plate, Fig. 8-22.

In the converging part the fluid velocity increases and, according to Bernoulli's law, the static pressure decreases. No significant pressure loss is generated in the

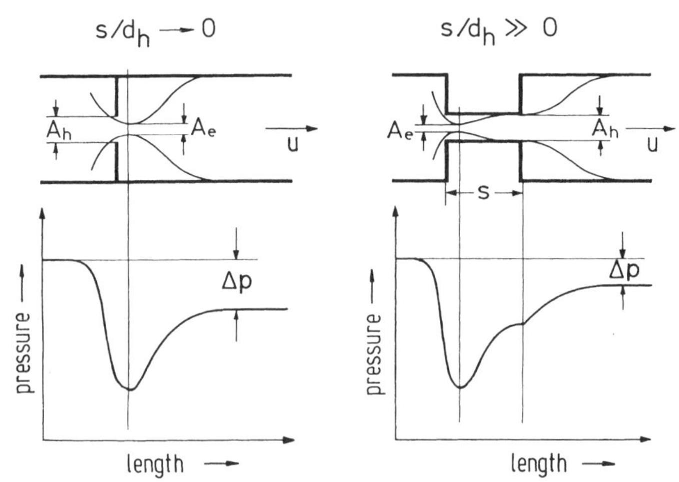

Fig. 8-22. Mechanism of fluid flow through a hole in a thin plate (left) and a thick plate (right).

converging part since fluid flow is directed towards a falling static pressure. Potential energy is transformed into kinetic energy without any significant losses.

In the diverging part after the plate, the fluid velocity decreases and, consequently, the static pressure increases. Flow against an increasing static pressure is accompanied by the generation of eddies that cause severe pressure losses.

From the laws of conservation of momentum and energy follows the very fundamental Borda-Carnot equation for the permanent pressure loss during expansion [e.g., Perry et al. 1984]:

$$\Delta p_d = \frac{\rho_G}{2} (u_{Ge} - u_G)^2 \quad \text{or} \quad \Delta p_d = \frac{\rho_G}{2} u_{Gh}^2 \left(\frac{u_{Ge}}{u_{Gh}} - \frac{u_G}{u_{Gh}} \right)^2 \tag{8.27}$$

A comparison with Equation (8.26) gives for the orifice coefficient:

$$\zeta = \left(\frac{u_{Ge}}{u_{Gh}} - \frac{u_G}{u_{Gh}} \right)^2 \tag{8.28}$$

The continuity equation allows calculating the velocity ratios u_{Ge}/u_{Gh} and u_G/u_{Gh} from a knowledge of the flow cross sections.

At the end of the converging part the fluid flow contracts to an area A_e (vena contracta) smaller than the hole area A_h. The ratio A_e/A_h is described by the discharge coefficient α:

$$\alpha = A_e/A_h \tag{8.29}$$

The continuity equation leads to:

$$u_{Ge}/u_{Gh} = 1/\alpha \tag{8.30}$$

The ratio of the hole area A_h and the total area A is described by the relative free area ϕ:

$$\phi = A_h/A \tag{8.31}$$

With the continuity equation follows:

$$u_G/u_{Gh} = \phi \tag{8.32}$$

At thin plates with large holes the flow diverges in a single step from A_e to A (Fig. 8-22). Combining Equations (8.28), (8.30) and (8.32) yields:

$$\zeta = \left(\frac{1}{\alpha} - \phi \right)^2 \quad \text{for } s/d_h \rightarrow 0 \tag{8.33}$$

This is the well known discharge formula.

At thick plates with small holes the flow diverges in two steps, first from A_e to A_h and second from A_h to A (Fig. 8-22). The pressure drops of the two expansions are added to give:

$$\zeta = \left(\frac{1}{\alpha} - 1\right)^2 + (1 - \phi)^2 \quad \text{for } s/d_h \gg 1 \tag{8.34}$$

In above equation the friction losses in the hole are neglected.

At high Reynolds Numbers and sharp edged holes the discharge coefficient has a value of $\alpha = \pi/(\pi + 2) = 0.611$. For a single hole in a large diameter plate the relative free area ϕ approaches zero, i.e. $\phi \rightarrow 0$. Evaluation of Equations (8.33) and (8.34) for $\phi = 0$ yields:

$$\zeta_o = \left(\frac{1}{0.611} - 0\right)^2 = 2.67 \quad \text{for } s/d_h \rightarrow 0 \tag{8.35}$$

and

$$\zeta_o = \left(\frac{1}{0.611} - 1\right)^2 + (1)^2 = 1.41 \quad \text{for } s/d_h \gg 1 \tag{8.36}$$

Hence, the orifice coefficient ζ_o of thin plates is higher than that of thick plates. This can easily be understood keeping in mind that flow expansion in two small steps causes smaller losses than in one large step. In a uniformly diverging cross section, for instance, the permanent pressure drop approaches zero since the expansion takes place in many small steps.

In practice, values of the ratio of plate thickness and hole diameter s/d_h lie typically between the two special cases considered until now. Hence, the orifice coefficient

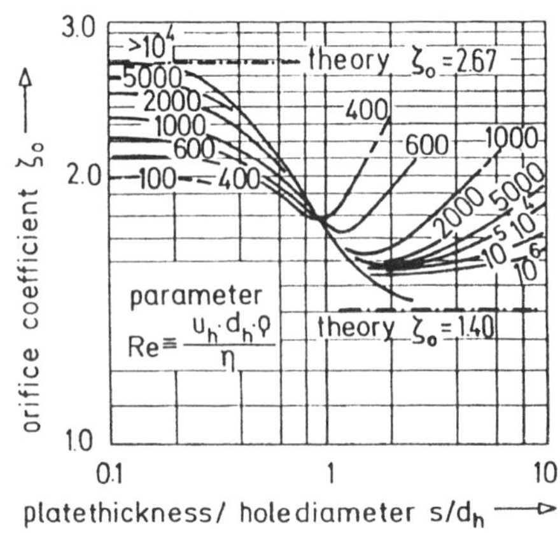

Fig. 8-23. Dependence of the orifice coefficient ξ_o ($\phi \rightarrow 0$) of sieve trays with sharp-edged holes on the ratio of plate thickness to hole diameter s/d_h. Parameter is the Reynolds-Number in the holes.

has to be determined by experiments. A correlation of published data is presented in Fig. 8-23. The orifice coefficient ζ_o, i.e., for $\phi \rightarrow 0$, is plotted versus the ratio of plate thickness and hole diameter s/d_h. Parameter is the Reynolds-Number $Re_h = u_{Gh}\, d_h\, \rho_G/\eta_G$ in the hole. The increase of ζ_o with increasing Reynolds-Numbers at small values of s/d_h results from the variation of the discharge coefficient α. The decrease of ζ_o with increasing Reynolds-Numbers at large values of s/d_h is caused by the friction losses in the hole that have been neglected in the theoretical considerations.

The relationship between the orifice coefficient ζ and the relative free area ϕ differs for thin and thick plates. For thin plates follows from Equation (8.33):

$$\zeta = \zeta_o + \phi^2 - 2\,\phi\,\sqrt{\zeta_o} \quad \text{for } s/d_h \rightarrow 0 \tag{8.37}$$

For thick plates, the following is obtained from Equation (8.34):

$$\zeta = \zeta_o + \phi^2 - 2\,\phi \quad \text{for } s/d_h \gg 1 \tag{8.38}$$

In practice, the differences between Equations (8.37) and (8.38) are very small because the value of ϕ does not exceed 0.15. Therefore, Equation (8.37) can be used with sufficient accuracy for most tray thicknesses.

8.2.3.8.2 Pressure Drop of Dry Valve and Bubble Cap Trays

Bubble caps and valves are so complex in design that prediction of the pressure drop is very difficult or almost impossible. Hence, the pressure drop must be measured. Both the orifice coefficient ζ of a single element and the number of elements that can be accommodated per unit area depend on the design; therefore, the comparison is made with a tray orifice coefficient ζ_t, whose definition is:

$$\zeta_t = 2\,\Delta p_d/F^2 \tag{8.39}$$

Here, the gas load is referred to the total active area A_{ac} of the tray and not, as usual, to the area of the slots or holes.

A compilation of the orifice coefficients for different types of trays is shown in Fig. 8-24 [Stichlmair 1978]. The symbols mark the closest installation of the elements as recommended by the vendors. The lines allow estimation of the orifice coefficient when less elements have been installed per unit area. For comparison, the transformed orifice coefficients of a sieve tray with $\zeta_o = 1.0$ and $\zeta_o = 1.7$ are also shown. The orifice coefficients of valve trays and bubble cap trays are always larger than that of sieve trays. However, the relative free area ϕ of bubble cap and valve trays can be larger than in sieve trays because the lower risk of liquid leaking through the tray. Thus, differences between the three types of trays can be minimized.

8.2.4 Mass Transfer in the Two-Phase Layer on Column Trays

The difficulty of a distillative separation is expressed either by the number of *equilibrium stages* or the number of *transfer units* (see Section 4.1). The higher the number

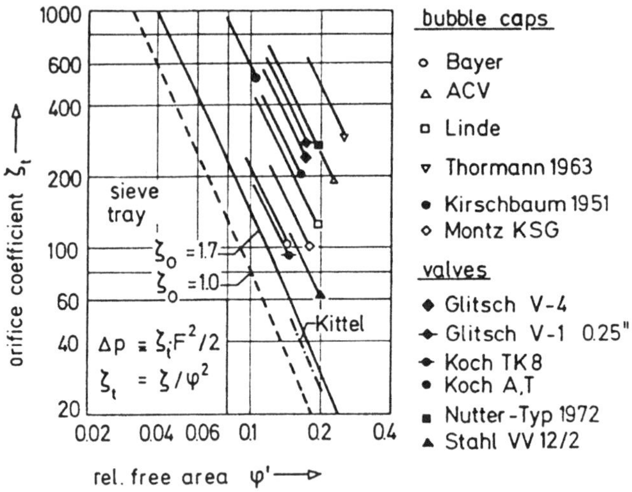

__Fig. 8-24.__ Dependence of the orifice coefficient ζ_t of bubble cap and valve trays on the relative free area ϕ. The symbols show the closest arrangement of the elements.

of these quantities the more difficult it is to achieve the separation. Both quantities are defined from a thermodynamic point of view, i.e., the type and details of the equipment are not accounted for. However, it is obvious that a difficult separation requires a tall column.

A major objective of column dimensioning is the transformation of equilibrium stages or transfer units into column height. To tray columns the concept of equilibrium stages is nearly exclusively applied. However, one should be aware that the separation achieved by a tray is not identical with that of an equilibrium stage. In most cases the concentration changes achieved by an actual tray are significantely smaller. State of the art is the use of mass transfer efficiencies that are defined as the ratio of the concentration changes of an actual tray and an equilibrium stage. Mass transfer efficiencies can be defined either as gas side efficiency E_{OG} by considering the concentration changes in the gas phase or as liquid side efficiency E_{OL} by considering the concentration changes in the liquid phase.

In distillation, gas side mass transfer efficiencies are almost exclusively used which describe the ratio of the actual concentration change of the gas to a theoretical one. Two definitions of gas side mass transfer efficiencies have to be distinguished: the point efficiency and the tray efficiency (see next Section).

From a knowledge of the mass transfer efficiency of a tray the number of trays required to achieve the specified products is determined by drawing the effective gas concentration line in the McCabe-Thiele diagram, see Fig. 8-25. The number of actual trays is found by drawing steps between operating line and effective gas concentration line. The rule, that the number of actual trays can be determined by dividing the number of equilibrium stages by the tray efficiency, is correct only if

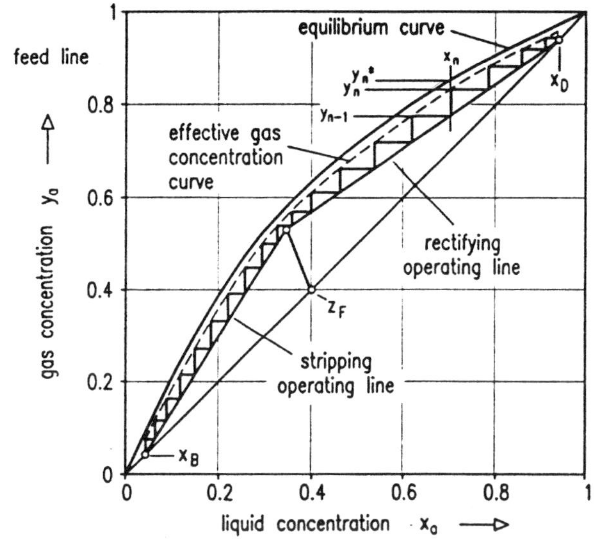

Fig. 8-25. Determination of the number of actual trays from a knowledge of tray efficiency in the McCabe-Thiele diagram.

operating and equilibrium lines are parallel. The more the slopes of the two lines differ the more the number of actual trays is underestimated by this rule.

8.2.4.1 Definitions of Mass Transfer Efficiencies The different definitions of mass transfer efficiencies are best explained by Fig. 8-26. In point efficiency E_{OG}, the concentration change along an individual stream line of the gas flow is considered. The definition is:

$$E_{OG} = \frac{y - y_{n-1}}{y^*(x) - y_{n-1}} \qquad (8.40)$$

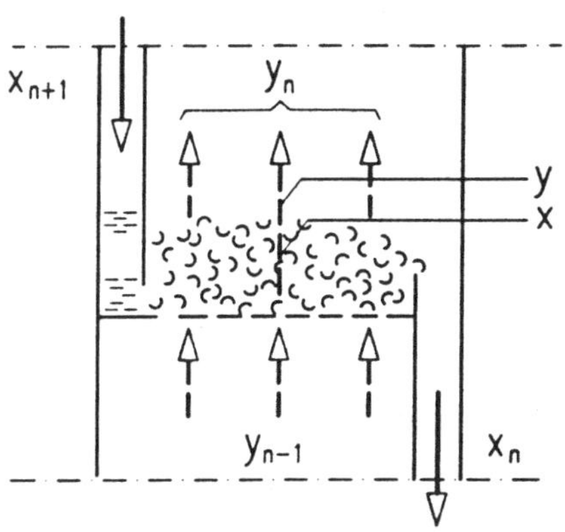

Fig. 8-26. Sketch of a single sieve tray with notations for point efficiency and tray efficiency definitions.

The subscript OG expresses that the overall resistance has been formally placed into the gas phase. The actual change in concentration of the gas $(y - y_{n-1})$ as it flows through the two-phase layer on tray n is related to the maximum concentration change attainable; i.e., $y^*(x) - y_{n-1}$. Here, y^* is the gas concentration in equilibrium with liquid concentration x at the point on the tray where the gas stream line passes through the two-phase layer. Point efficiency can reach a maximum value of 1 provided that the liquid is completely mixed throughout the height of the two-phase layer. Thus, point efficiency is a very useful theoretical quantity but it has no immediate practical significance.

In practice, tray efficiency E_{OGM} (Murphree Efficiency) is used which is defined by:

$$E_{OGM} = \frac{y_n - y_{n-1}}{y^*(x_n) - y_{n-1}} \tag{8.41}$$

Here, the overall gas flow through the column is considered. The actual concentration change of the total gas flow $(y_n - y_{n-1})$ is related to the maximum possible concentration change $(y^*(x_n) - y_{n-1})$. Here, y^* is the gas concentration in equilibrium with the state of the liquid (concentration x_n) leaving tray n. This definition is arbitrary because the liquid concentration usually is different at each point on the tray. Thus, efficiencies higher than 100% may be obtained in some cases. Although the theoretical definition of tray efficiency E_{OGM} is unsatisfactory, it is of practical interest because it describes the separation efficiency of a real tray, see Fig. 8-25. Equations (8.40) and (8.41) define the gas side mass transfer efficiency; analogous equations are also sometimes used for the liquid side mass transfer efficiency. However, they have no practical importance because the main resistance to mass transfer in distillation is attributed to the gas.

8.2.4.2 *Relationship between Point Efficiency and Tray Efficiency*

The relationship between point and tray efficiency can be easily formulated when the liquid flows across the tray without mixing (i.e., plug flow). According to [Lewis 1936]:

$$\frac{E_{OGM}}{E_{OG}} = \frac{\dot{L}/\dot{G}}{m} \cdot \frac{1}{E_{OG}} \left(\exp\left(\frac{m}{\dot{L}/\dot{G}} E_{OG} \right) - 1 \right) \tag{8.42}$$

The gas under each tray is assumed to be completely mixed. As a rule, the liquid in the two-phase layer on column trays is at least partially mixed (see Section 8.2.3.5).

For complete liquid mixing, tray efficiency and point efficiency have the same value:

$$E_{OGM}/E_{OG} = 1 \tag{8.43}$$

The normal condition lies between the two extreme cases described by Equations (8.42) and (8.43). The degree of liquid mixing, which is expressed in terms of a

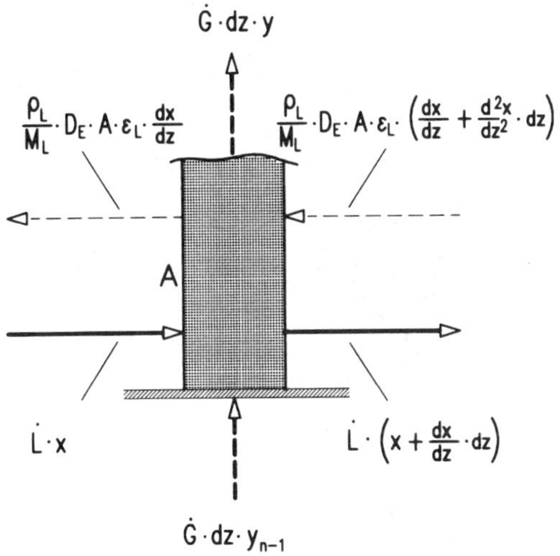

Fig. 8-27. *Application of the dispersion model to liquid mixing on a tray.*

Peclet number Pe (see Section 8.2.3.5), must then be considered. Liquid mixing on a tray is adequately described by the dispersion model. Here, the process of mixing is formulated in analogy to the process of molecular diffusion with the eddy diffusion coefficient D_E instead of the molecular diffusion coefficient D_L. An eddy diffusion stream is countercurrently superimposed on the liquid flow across the tray, see Fig. 8-27. A mass balance at an element of the two-phase layer of differential length dz yields:

$$\frac{1}{Pe}\frac{d^2x}{dz^2} - \frac{dx}{dz} + \frac{m}{(\dot{L}/\dot{G})} E_{OG}\,(x - y_n/m) = 0 \quad \text{with } dx/dz = 0 \text{ for } z = 1 \tag{8.44}$$

Here, z denotes the dimensionless tray length.

From a knowledge of the concentration profile in the liquid phase, the mass transfer efficiency of the tray, E_{OGM} can be determined. The solution is graphically represented in Fig. 8-28. [Final Report, University of Delaware 1958] with the Peclet-Number as parameter. In plug flow (no mixing) the value of the Peclet-Number becomes infinite, Pe $= \infty$. The case of complete liquid mixing is expressed by Pe $= 0$. When both the parameter Pe and the abscissa $E_{OG}\, m/(\dot{L}/\dot{G})$ are large, the ratio of the tray efficiency to the point efficiency E_{OGM}/E_{OG} becomes very large. However, for the abscissa to be very large, the liquid flow must be small. The high residence time at low liquid flow rates causes considerable remixing of liquid on the tray so that the parameter Pe in Fig. 8-28 becomes very small. Thus, the value of tray efficiency is very seldom much higher than point efficiency.

In practice, tray efficiency is affected adversely by some other processes such as

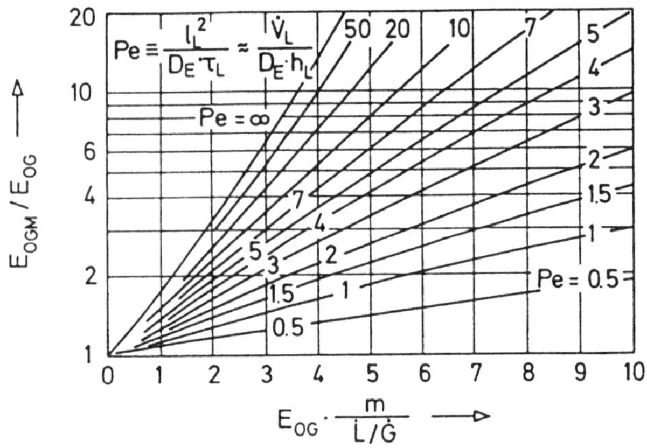

Fig. 8-28. Relation between the ratio of overall tray efficiency E_{OGM} to point efficiency E_{OG} and the degree of liquid mixing represented by the Peclet number Pe.

liquid leakage through the tray, entrainment of liquid to the next tray, and maldistribution. In careful column design, a tray efficiency that is only slightly higher than the point efficiency should be used.

8.2.4.3 Point Efficiency The prediction of mass transfer efficiency is based on the following fundumental equation see, for instance, [Taylor and Krishna 1993]:

$$E_{OG} = 1 - \exp(-N_{OG}) \tag{8.45}$$

This equation formulates a relation between two definitions only, i.e., gas side point efficiency E_{OG} and the gas side overall transfer units N_{OG}. The overall number of gas phase transfer units N_{OG} can be calculated from the number of transfer units in the gas N_G and in the liquid N_L by using the following equation:

$$\frac{1}{N_{OG}} = \frac{1}{N_G} + \frac{m}{\dot{L}/\dot{G}} \frac{1}{N_L} \tag{8.46}$$

The term $m/(\dot{L}/\dot{G})$ is known as the stripping factor. The number of transfer units in the gas phase is given by:

$$N_G = \frac{\beta_G \, a \, h_f \, A_{ac}}{\dot{V}_G} \tag{8.47}$$

Here, β_G is the mass transfer coefficient in the gas, a is the relative interfacial area (Fig. 8-20), h_f is the height of the two-phase layer (Equation 8.15), A_{ac} is the active area, and $\dot{V}_G$ is the volumetric gas flow. Analogously, for the liquid phase:

$$N_L = \frac{\beta_L \, a \, h_f \, A_{ac}}{\dot{V}_L} \tag{8.48}$$

Inserting Equations (8.48) and (8.47) into Equation (8.46) and replacing $\dot{V}_G/A_{ac}$ by the superficial gas velocity u_G gives:

$$N_{OG} = \frac{a \, \beta_G \, h_f/u_G}{1 + m \, \beta_G/\beta_L \, M_L/M_G \, \rho_G/\rho_L} \sim h_f/u_G \tag{8.49}$$

Here, M_L and M_G are the molar masses of gas and liquid phase and ρ_L and ρ_G the densities of the liquid and gas, respectively. The numerator in this equation represents the mass transfer rate in the gas phase, while the denominator takes into account the additional resistance to mass transfer in the liquid phase. Equation (8.49) is generally applicable and contains four unknown quantities that are specific to the apparatus: the relative interfacial area a, the height of the two-phase layer h_f and the mass transfer coefficients in the gas β_G and in the liquid β_L.

8.2.4.4 Mass Transfer Coefficients

The mass transfer coefficients in the gas, β_G, and in the liquid, β_L, can be estimated by assuming nonsteady mass transfer:

$$\beta_G = \sqrt{\frac{4}{\pi}\frac{D_G}{\tau}} \quad \text{and} \quad \beta_L = \sqrt{\frac{4}{\pi}\frac{D_L}{\tau}} \tag{8.50}$$

where D_G and D_L are the diffusion coefficients of the gas and liquid, respectively. Determination of the contact time τ is difficult. According to [Stichlmair 1978], it is equivalent to the lifetime of the interface in the two-phase layer. Since the interface is produced by the penetration of gas into the liquid, lifetime of the interface is identical with the residence time of the gas in the two-phase layer. Thus, the same value is obtained for the contact time τ in both the liquid and the gas phases. Hence, the mass transfer coefficient in the gas is:

$$\beta_G = \sqrt{\frac{4}{\pi}\frac{D_G}{h_f}\frac{u_G}{\epsilon_G}} \tag{8.51}$$

Analogously, the mass transfer coefficient in the liquid is:

$$\beta_L = \sqrt{\frac{4}{\pi}\frac{D_L}{h_f}\frac{u_G}{\epsilon_G}} \tag{8.52}$$

These relationships apply to the froth regime and, to a close approximation, to the drop regime, but not to the bubble regime since here the surface is renewed by the rising bubbles. Inserting Equations (8.51) and (8.52) into Equation (8.49) gives:

$$N_{OG} = \frac{a\sqrt{4/\pi \, D_G/u_G \, h_f/\epsilon_G}}{1 + m\sqrt{D_G/D_L} \, M_L/M_G \, \rho_G/\rho_L} \sim (h_f/u_G)^{1/2} \tag{8.53}$$

The denominator in this equation takes into account the additional mass transfer resistance in the liquid phase. It contains system properties only.

Whereas in the basic relationship (Equation 8.49) the overall gas mass transfer units N_{OG} are directly proportional to the height of the two-phase froth layer h_f and inversely proportional to the gas velocity u_G, a proportionality to the square root of these parameters is obtained in Equation (8.53); this agrees well with experimental data.

8.2.4.5 Practical Determination of Tray Efficiencies The theory of mass transfer on trays is, as has been shown in the previous Chapters, well developed. However, present knowledge of some of the most important parameters is very poor; e.g., interfacial area, mass transfer coefficients in the gas and liquid phases, liquid mixing and liquid maldistribution. Hence, at the present state of the art mass transfer efficiency cannot be predicted rigorously. Empirical values of tray efficiencies are used in practice and, in most cases, quite large safety factors are included to reduce the risk of malfunction of the separation column.

Empirical tray efficiency data lie in the order of 70% for distillation systems under normal conditions. However, for wide boiling systems or high viscous systems much lower values have been found. In wide boiling systems often encountered in absorption and desorption, values of tray efficiencies are reported to be as low as 1% (e.g., absorption of carbon dioxide in water) or even 0.1% (e.g., desorption of oxygen from water).

In order to put the empirical findings into the frame set by mass transfer theory, some experimental data have been plotted in the diagram of Fig. 8-29. [Stichlmair and Weisshuhn 1973]. The structure of the diagram results from Equations (8.45) and (8.46) with the gas side point efficiency as ordinate, the stripping factor as abscissa and the number of transfer units N_G and N_L as parameter. The regions where empirical data lie are marked in the diagram. The parameter lines have been determined by evaluating Equations (8.45) and (8.46) with different sets of gas side and liquid side transfer units, N_G and N_L, respectively. From the comparison of the parameter lines and the empirical data, follows that the gas side transfer units lie in the range of $N_G \approx 0.5$ to $N_G \approx 2$, whereas the liquid side transfer units vary from $N_L \approx 1$ up to $N_L \approx 20$.

The most important parameter on gas side tray efficiency is the stripping factor $m/(\dot{L}/\dot{G})$. It represents the ratio of the slope of the equilibrium line, m, and the slope of the operating line, $\dot{L}/\dot{G}$, in the McCabe-Thiele diagram. The stripping factor varies over up to 4 decades. In distillation, however, the values of the stripping factor very seldom significantly deviate from $m/(\dot{L}/\dot{G}) \approx 1$, i.e., the equilibrium line and the operating line are nearly equidistant in the McCabe-Thiele diagram. Under these conditions the tray efficiency is quite high. Fig. 8-29 confirms the rule of thumb that tray efficiency of close boiling systems has a value of approximately 70%. The

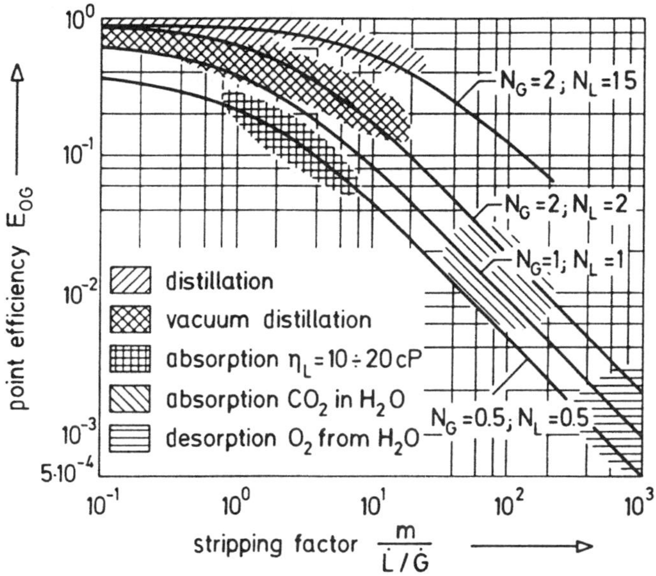

Fig. 8-29. Typical experimental values of the gas side point efficiency E_{OG} versus the stripping factor.

very low efficiencies of absorption and desorption are primarily caused by the large value of the slope of the equilibrium line in such systems. It is important to note that Fig. 8-29 is valid for the gas side definitions of mass transfer efficiencies only.

In practice, empirical correlations are sometimes used for calculating the number of gas side and liquid side transfer units N_G and N_L, respectively. Well proven correlations are presented in the AIChE Bubble Tray Design Manual [Gerster et al. 1958]:

$$N_G = (0.776 + 4.57\,h_w - 0.238\,F + 104.8\,\dot{V}_L/l_w)\cdot(\rho_G\,D_G/\eta_G)^{1/2} \tag{8.54}$$

and

$$N_L = 19.7\cdot 10^3\,D_L{}^{1/2}(0.4\,F + 0.17)\,\frac{A_{ac}\,h_L}{\dot{V}_L} \quad \text{for sieve trays} \tag{8.55}$$

$$N_L = 20.3\cdot 10^3\,D_L{}^{1/2}(0.21\,F + 0.15)\,\frac{A_{ac}\,h_L}{\dot{V}_L} \quad \text{for bubble cap trays} \tag{8.56}$$

These correlations are dimensionally incorrect. The length must be expressed in m (metre), the time in s (second) and the mass in kg (kilogram). Their evaluation is demonstrated in Example 8-1. The use of these correlations is more reliable than the use of empirical tray efficiency data since the strong influence of the stripping factor is considered by evaluating Equations (8.46) and (8.45) to find the value of the efficiency.

Example 8-1: DIMENSIONING A SIEVE TRAY COLUMN

Determine the diameter and the tray efficiency of a sieve tray column for distilling the mixture benzene/toluene at a pressure $p = 1$ bar. The internal flows are $\dot{G} = 0.05$ kmol/s and $\dot{L} = 0.034$ kmol/s.

The system properties are:

	Liquid Phase	Gas Phase
Molar mass	$M_L = 85.8$ kg/kmol	$M_G = 85.8$ kg/kmol
Density	$\rho_L = 800$ kg/m^3	$\rho_G = 2.6$ kg/m^3
Surface tension	$\sigma = 0.02$ kg/s^2	
Viscosity		$\eta_G = 8.9 \cdot 10^{-6}$ kg/(m s)

Solution:

1. Column diameter: The diameter of a column is determined by the gas flow. Firstly, a gas flow rate must be chosen that lies between maximum and minimum gas load.

1.1. Maximum gas load—Equation (8.8): $F_{max} = 2.5\,(\phi^2\,\sigma\,(\rho_L - \rho_G)\,g)^{1/4}$. Selected value for the relative free area: $\phi = 0.1$; $F_{max} = 2.5\,(0.1^2\,0.02\text{kg/s}^2\,(800\text{kg/m}^3 - 2.6\text{kg/m}^3)\,9.81\text{m/s}^2)^{1/4} = 2.8\ \text{Pa}^{1/2}$.

1.2. Minimum gas load—Equation (8.10):

$$F_{min} = \phi\sqrt{0.37\,d_h\,g\,(\rho_L - \rho_G)^{1.25}/\rho_G{}^{0.25}}.$$

Selected value for the hole diameter: $d_h = 8$ mm;

$$F_{min} = 0.1\sqrt{0.37\,0.008\text{m}\,9.81\text{m/s}^2\,(800\text{kg/m}^3 - 2.6\,\text{kg/m}^3)^{1.25}/(2.6\text{kg/m}^3)^{0.25}}$$

$$= 0.984\,\text{Pa}^{1/2}.$$

1.3. Design value of the gas load and column diameter—Design value of the gas load: $F = 1.6\ \text{Pa}^{1/2}$. Superfical gas velocity: $u_G = F/\sqrt{\rho_G} = 1.6\text{Pa}^{1/2}/\sqrt{2.6\text{kg/m}^3} = 1$ m/s. Volumetric gas flow: $\dot{V}_G = \dot{G}\,M_G/\rho_G = 0.05\text{kmol/s}\ 85.8\text{kg/kmol}/2.6\text{kg/m}^3 = 1.65\ \text{m}^3/\text{s}$. Active area: $A_{ac} = \dot{V}_G/u_G = 1.65\ \text{m}^3/\text{s}/1\ \text{m/s} = 1.65\ \text{m}^2$. Cross section: $A_c = A_{ac}/(1 - 2\,A_d/A_c)$. Selected value for the relative weir length: $l_w/D_c = 0.7$. From geometric tables follows: $A_d/A_c = 0.09$; $A_c = 1.65\text{m}^2/(1 - 2 0.09) = 2\ \text{m}^2$. Column diameter: $D_c = \sqrt{4\,A_c/\pi} = \sqrt{4\,2\text{m}^2/\pi} = 1.6$ m.

1.4. Preliminary check of liquid flow—Weir length: $l_w = 0.7\ D_c = 0.7$ 1.6m = 1.12 m. Volumetric liquid flow: $\dot{V}_L = \dot{L}\ M_L/\rho_L = 0.034$ kmol/s 85.8kg/kmol/800kg/m^3 = $3.65 \cdot 10^{-3}$ m^3/s. Liquid flow per weir length: $\dot{V}_L/l_w = 3.65 \cdot 10^{-3}$m^3/s/1.12m = $3.26 \cdot 10^{-3}$ m^3/(m s) = 11.72 m^3/(m h). This value is higher than the minimum value 2 m^3/(m h) and lower than the maximum value 60 m^3/(m h).

1.5. Liquid holdup in the froth—Equation (8.14):

$$\epsilon_L = 1 - (F/F_{max})^{0.28} = 1 - (1.6/2.8)^{0.28} = 0.145.$$

1.6. Froth height—Equation (8.15):

$$h_f = h_w + \frac{1.45}{g^{1/3}}\left(\frac{\dot{V}_L/l_w}{\epsilon_L}\right)^{2/3} + \frac{125}{(\rho_L - \rho_G)\ g}\left(\frac{F - 0.2\sqrt{\rho_G}}{1 - \epsilon_L}\right)^2.$$

Selected value for the weir height: h_w = 50 mm;

$$h_f = 0.05\text{m} + \frac{1.45}{(9.81\text{m/s}^2)^{1/3}}\left(\frac{3.26 \cdot 10^{-3}\text{m}^3/(\text{m s})}{0.145}\right)^{2/3}$$
$$+ \frac{125}{(800\text{kg/m}^3 - 2.6\text{kg/m}^3)\ 9.81\ \text{m/s}^2}$$
$$\times \left(\frac{1.6\text{Pa}^{1/2} - 0.2\sqrt{2.6\text{kg/m}^3}}{1 - 0.145}\right)^2;$$

$$h_f = 0.05\text{m} + 0.054\ \text{m} + 0.036\ \text{m} = 0.14\ \text{m}.$$

1.7. Entrainment: The entrainment rate is determined from Fig. 8-17. Abscissa: F/F_{max} = 1.6/2.8 = 0.57. Selected value for tray spacing: H = 0.5 m. Parameter:

$$\frac{(H - h_f)\ \Delta\rho\ g}{F_{max}^2} = \frac{(0.5\text{m} - 0.14\text{m})\ 797.4\text{kg/m}^3\ 9.81\ \text{m/s}^2}{(2.8\text{Pa}^{1/2})^2} = 355.$$

From Fig. 8-17 follows:

$$\dot{V}_E/\dot{V}_G = 1 \cdot 10^{-5};$$
$$\dot{V}_E = \dot{V}_G\ 1 \cdot 10^{-5} = 1.65 \cdot 10^{-5}\ \text{m}^3/\text{s};$$
$$\dot{V}_E/\dot{V}_L = \frac{1.65 \cdot 10^{-5}\ \text{m}^3/\text{s}}{3.65 \cdot 10^{-3}\ \text{m}^3/\text{s}} = 0.005.$$

This value is smaller than the highest tolerable value of 0.1.

1.8. Pressure drop—Equation (8.26): $\Delta p_d = (\zeta/2)\,F_h^2$. The value of ζ is determined from Fig. 8-23 and Equation (8.37). Selected value for plate thickness: s = 2 mm. Abscissa: s/d_h = 0.002m/0.008m = 0.25. Parameter: Re = $u_{Gh}\, d_h\, \rho_G/\eta_G = u_G/\phi\, d_h\, \rho_G/\eta_G$ = 1m/s/0.1 0.008m 2.6kg/m^3/(8.9·10^{-6}kg/(m s)) = 2.34 · 10^4. From the diagram in Fig. 8-23 follows: ζ_0 = 2.60 for $\phi \rightarrow 0$.
Equation (8.37): $\zeta = \zeta_0 + \phi^2 - 2\,\phi\,\sqrt{\zeta_0}$ = 2.60 + 0.1^2 − 20.1 $\sqrt{2.60}$ = 2.29; $\Delta p_d = (\zeta/2)\,F_h^2 = (\zeta/2)\,(F/\phi)^2$ = (2.29/2) (1.6 Pa$^{1/2}$/0.1)^2 = 293 N/m^2; $\Delta p_L = h_L\,\rho_L\,g = h_f\,\epsilon_L\,\rho_L\,g$ = 0.140m 0.145 800kg/m^3 9.81m/s^2 = 159 N/m^2. Total pressure drop of one tray: $\Delta p = \Delta p_d + \Delta p_L$ = 293 N/m^2 + 159 N/m^2 = 452 N/m^2 or 46 mm clear liquid height.

1.9. Check of downcomer capacity—Equation (8.12):

$$\left(\frac{\dot{V}_L}{l_{cl}}\right)_{max} = 0.61\;\epsilon_{Ld}\,h_{cl}\sqrt{2\,g\,H\,\frac{(\rho_L - \rho_G)}{\rho_L}\left(1 - \frac{h_p + h_L}{\epsilon_{Ld}\,H}\right)}.$$

Assumption for non foaming systems: ϵ_{Ld} = 0.4; $h_p = \Delta p/(\rho_L\,g)$ = 452N/m^2/(800kg/m^3 9.81m/s^2) = 0.058 m. Selected value for skirt clearance: h_{cl} = 0.8 h_w = 0.8 0.05m = 0.04 m;

$$\left(\frac{\dot{V}_L}{l_{cl}}\right)_{max} = 0.61 \cdot 0.4 \cdot 0.04$$

$$\times \sqrt{29.81\text{m/s } 0.5\text{m}\,\frac{(800\text{kg/m}^3 - 2.6\text{kg/m}^3)}{800\text{kg/m}^3}\left(1 - \frac{0.058\text{m} + 0.02\text{m}}{0.4\;0.5\text{m}}\right)}$$

$$= 23.8 \cdot 10^{-3}\ \text{m}^3/(\text{m s}) = 86\ \text{m}^3/(\text{m h}).$$

The actual liquid load is much lower than the maximum downcomer capacity.

2. Mass transfer efficiency

2.1. Gas side point efficiency—Assumption: Drop regime. Interfacial area from Equation (8.23):

$$a = \frac{F^2}{2\,\sigma\,\phi^2}\,(1 - (F/F_{max})^{0.28}) = \frac{(1.6\text{Pa}^{1/2})^2}{2\;0.02\text{kg/s}^2\;0.1^2}\,(1 - 0.57^{0.28})$$

$$= 932\ \text{m}^2/\text{m}^3.$$

This value agrees very well with Fig. 8-20.
Number of transfer units from Equation (8.53):

$$N_{OG} = \frac{a\sqrt{4/\pi\; D_G/u_G\; h_f/\epsilon_G}}{1 + m\,\sqrt{D_G/D_L}\;M_L/M_G\;\rho_G/\rho_L}.$$

Values of the diffusion coefficients: D_G = 2.66·10^{-6} m^2/s (from equation of [Gilliland 1934]); D_L = 2.2·10^{-8}m^2/s (from equation of [Wilke and Chang 1955]);

$$N_{OG} = \frac{932\text{m}^2/\text{m}^3 \sqrt{\frac{4}{\pi}\frac{2.66\cdot 10^{-6}\text{m}^2/\text{s}}{1\text{m/s}}\frac{0.14\text{m}}{(1-0.145)}}}{1 + 1\sqrt{\frac{2.66\cdot 10^{-6}\text{m}^2/\text{s}}{2.2\cdot 10^{-8}\text{m}^2/\text{s}}\frac{85.8\text{kg/kmol}}{85.8\text{kg/kmol}}\frac{2.6\text{kg/m}^3}{800\text{kg/m}^3}}} = \frac{0.7}{1.04} = 0.68.$$

Point efficiency from equation (8.45): $E_{OG} = 1 - \exp(-N_{OG}) = 1 - \exp(-0.68) = 0.49$.

Alternative determination of the point efficiency:

Equation (8.54):

$$N_G = (0.776 + 4.57\ h_w - 0.238\ F + 104.8\ \dot{V}_L/l_w)\left(\frac{\rho_G\ D_G}{\eta_G}\right)^{1/2}$$

$$= (0.776 + 4.57\ 0.05\text{m} - 0.238\ 1.6\text{Pa}^{1/2}$$

$$+ 104.8\ 3.26\cdot 10^{-3}\text{m}^2/\text{s})\left(\frac{2.6\text{kg/m}^3\ 2.66\cdot 10^{-6}\text{m}^2/\text{s}}{8.9\text{kg/s}^2}\right)^{1/2} = 0.85$$

Equation (8.55):

$$N_L = 19.7\cdot 10^3\ D_L^{1/2}\ (0.4\ F + 0.17)\ \frac{A_{ac}\ h_L}{\dot{V}_L}$$

$$= 19.7\cdot 10^3\ (2.2\cdot 10^{-8}\text{m}^2/\text{s}^{1/2}\ (0.4\ 1.6\text{Pa}^{1/2} + 0.17)\ \frac{1.65\text{m}^2\ 0.02\text{m}}{3.65\cdot 10^{-3}\text{m}^3/\text{s}} = 21.4$$

Equation (8.46):

$$\frac{1}{N_{OG}} = \frac{1}{N_G} + \frac{m}{\dot{L}/\dot{G}}\frac{1}{N_L} = \frac{1}{0.85} + \frac{1}{0.68}\frac{1}{21.4} = 1.2452\ N_{OG} = 0.8031$$

Equation (8.45):

$$E_{OG} = 1 - \exp(-\ N_{OG}) = 1 - \exp(-0.8031) = 0.55$$

2.2. Gas side tray efficiency—Degree of liquid mixing from Equation (8.17): $Pe = l^2{}_L/D_E\ \tau_L \approx \dot{V}_L/D_E\ \epsilon_L\ h_f$. Eddy diffusion coefficient from Equation (8.18):

$$D_E = 1.06\ h_f\ F\sqrt{\frac{\epsilon_L}{\rho_L - \rho_G}}$$

$$= 1.06\ 0.14\text{m}\ 1.6\text{Pa}^{1/2} \sqrt{\frac{0.145}{800\text{kg/m}^3 - 2.6\text{kg/m}^3}} = 3.2{\cdot}10^{-3}\ \text{m}^2/\text{s};$$

$$\text{Pe} = \frac{3.65{\cdot}10^{-3}\text{m}^3/\text{s}}{3.2{\cdot}10^{-3}\text{m}^2/\text{s}\ 0.145\ 0.14\text{m}} = 54, \quad \text{i.e, plug flow (see Fig. 8-28)}$$

Equation (8.42):

$$\frac{E_{OGM}}{E_{OG}} = \frac{\dot{L}/\dot{G}}{m}\frac{1}{E_{OG}}\left(\exp\left(\frac{m}{\dot{L}/\dot{G}}E_{OG}\right) - 1\right)$$

$$\dot{L}/\dot{G} = \frac{0.034\text{kmol/s}}{0.050\text{kmol/s}} = 0.68$$

$$\frac{E_{OGM}}{E_{OG}} = \frac{0.68}{1}\frac{1}{0.49}\left(\exp\left(\frac{1}{0.68}0.49\right) - 1\right) = 1.46$$

Tray efficiency: $E_{OGM} = 1.46\ E_{OG} = 1.46\ 0.49 = 0.71.$

8.3 DIMENSIONING PACKED COLUMNS

In packed columns a countercurrent flow of gas (upwards flow) and liquid (downwards flow) exist throughout the column. This feature in principle, makes, a packed column more effective for mass transfer than a tray column, which is characterized by a cross flow of gas and liquid on each tray. A countercurrent flow of the two phases utilizes the existing driving force for mass transfer more efficiently. However, the countercurrent flow of gas and liquid is not perfect in a packed column since the liquid flow is not uniform over the cross section. There is a strong tendency for maldistribution of liquid caused by a non-homogeneous structure of the packing. Well known flow mechanisms of maldistribution are liquid channeling and wall flow.

A good packing design should have a high capacity as well as a high separation efficiency. As important is a large flexibility to gas and liquid throughput combined with a nearly constant separation efficiency. Figure 8-30 shows a rough comparison of capacity and separation efficiency of several metal packings. Significant differences in the performance of different types of packings exist. Structured packings are generally superior to random packings in both capacity and separation efficiency. Modern random packings perform better than the standard Pall ring packing. Even the standard corrugated sheet packing is surpassed by modern developments.

8.3.1 Design Parameters of Packed Columns

In order to provide a high interfacial area for mass transfer between gas and liquid, the column is filled with a bed of solid material with high porosity and large volumet-

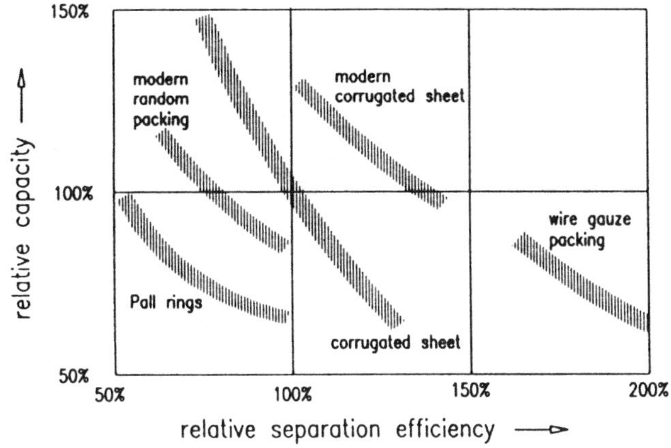

Fig. 8-30. *Qualitative comparison of capacity and separation efficiency of several metal packings. Reference point is a corrugated metal sheet packing with 250 m^2/m^3 [Eiden, private communication 1997].*

ric area. The liquid trickles downward in the bed in form of thin films, small rivulets and even small droplets. Ideally, all the surface of the bed is wetted by the liquid, and liquid flow through the bed is uniform. Very important is a low flow resistance of the bed against gas flow. Hence, the bed should have a high voidage and a shape that reduces pressure losses of the gas flow.

8.3.1.1 Types of Packings The types of packings presently in industrial use can be distinguished into dumped packings (random packings) and ordered packings (structured packings).

8.3.1.1.1 Dumped Packings The majority of packed columns use random packings composed of a large number of individual particles. Presently, some 50 different types of packings, most of them in up to 7 sizes, are offered on the market. Some examples of the more important packing elements are shown in Fig. 8-31. The flow rate and the separation efficiency of the column are directly dependent on particle shape. Randomly dumped packings should be as homogeneous as possible; bridging and large cavities between particles should be avoided. The particles should be sufficiently stable and should form a large packing surface with a low pressure drop of gas flow. The feasible particle size depends on column diameter; depending on particle shape, it should not exceed 1/10 to 1/30 of the column diameter. Most particles are made of metal or ceramics, but plastics are increasingly being used. Characteristic data of the most important packings currently in use are compiled in Table 8-6. The data are average values with limited accuracy. If available, data published by the vendors should be used instead of the values given in Table 8-6.

The oldest packing element is the Raschig ring patented as early as in 1907. The characteristic feature is that the length of the ring is equal to its diameter. This

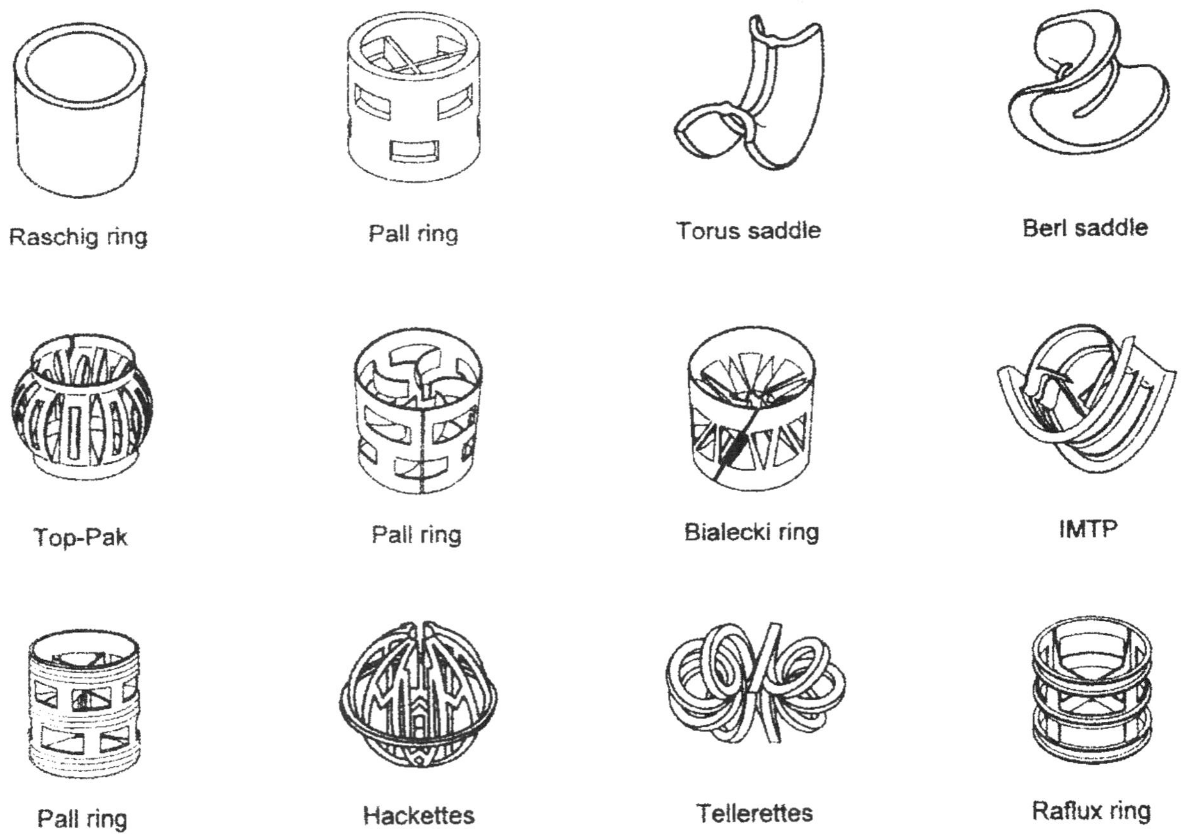

Fig. 8-31. *Selected particles used in random packings: Upper row: ceramic, Middle Row: metal, Bottom Row: plastic.*

Table 8-6: Typical Values of Specific Area a and Porosity ϵ for Several Random Column Packings

		Ceramic		Metal		Plastic	
Particle	Diameter d_n mm	Specific Area a, m^2/m^3	Porosity ϵ	Specific Area a, m^2/m^3	Porosity ϵ	Specific Area a, m^2/m^3	Porosity ϵ
Spheres	2	1800	0.4				
	4	900	0.4				
	6	600	0.4				
	8	450	0.4				
	10	360	0.4				
	15	240	0.4				
	20	180	0.4				
	25	144	0.4				
	30	120	0.4				
	35	100	0.4				
	50	72	0.4				
Raschig rings	5	1000	0.56	1000	0.87		
	10	440	0.65	500	0.89		
	15	330	0.70	350	0.92	350	0.86
	20	240	0.72			290	0.87
	25	195	0.73	220	0.92	220	0.87
	35	140	0.76	150	0.93	150	0.90
	50	98	0.77	110	0.95	110	0.91
	80	60	0.77	65	0.96	65	0.91
Pall rings	10			515	0.92		
	15			360	0.93	350	0.88
	25	220	0.73	215	0.94	220	0.91
	35	165	0.76	145	0.94	160	0.93
	50	120	0.77	105	0.95	110	0.93
	80	75	0.77	78	0.96	100	0.94
Bialecki rings	25			225	0.95		
	35			155	0.95		
	50			110	0.96		
	80			68	0.97		
Torus saddles	15	450	0.71				
	20	355	0.72				
	25	255	0.74				
	35	166	0.76				
	50	120	0.79				
	75	92	0.80				
Hiflow	20	280	0.72				
	25			214	0.92	218	0.92
	35	128	0.76				
	50	102	0.78	99	0.98	90	0.94
	75	70	0.80				
Tellerettes	no. 1					180	0.87
	no. 2					125	0.93
	no. 3					98	0.92
Cascade minirings	no. 0			338	0.97	321	0.89
	no. 1			249	0.96	183	0.92
	no. 2	89		114	0.97	141	0.94
	no. 3	66	0.76	105	0.98	75	0.96
	no. 4	49	0.78	75	0.98		
	no. 5			49	0.99		
	no. 7	46	0.78				

feature makes the particles form quite a homogeneous bed structure during pouring into the column. Raschig rings have a rather high pressure drop since the walls of those rings in horizontal position block the gas flow.

The Pall ring avoids this disadvantage since parts of the wall are punched out and deflected into the inner part of the ring. A Pall ring has the same porosity and the same volumetric area as a Raschig ring but a considerably lower pressure drop. For many years Pall rings have been the standard random packing element. Modern developments of random packing elements are characterized by the transition from plate structures to lattice structures [Billet 1993]. This reduces the pressure drop of gas flow and, in turn, enhances surface renewal of the liquid phase by increased drop formation.

8.3.1.1.2 Structured Packings In any random packing a certain degree of inhomogeneity is unavoidable. These inhomogeneities cause liquid maldistribution that adversely affects mass transfer efficiency of the column. Thus, for years, attempts have been made to develop ordered packing structures [Fair 1988].

The first idea in developing an ordered structure is a concentrical or rolled up sheet structure that fits perfectly into the cylindrical mantle of a column. However, all these structures failed in large diameter columns since lateral mixing of gas and liquid was poor. Even slots or holes in the sheets to enhance radial mixing could not help. A break-through was the development of the corrugated sheet structure by Sulzer Company in the mid sixties [Sulzer 1982]. Here, corrugated sheets are assembled parallel in vertical direction with alternating inclinations of the corrugations of neighboring sheets (see Fig. 8-32). Lateral mixing of liquid is enhanced parallel to the sheets but suppressed vertical to the sheets. Hence, the orientation of the sheets is changed in subsequent layers (approximately 21 cm height) by 90 degrees to

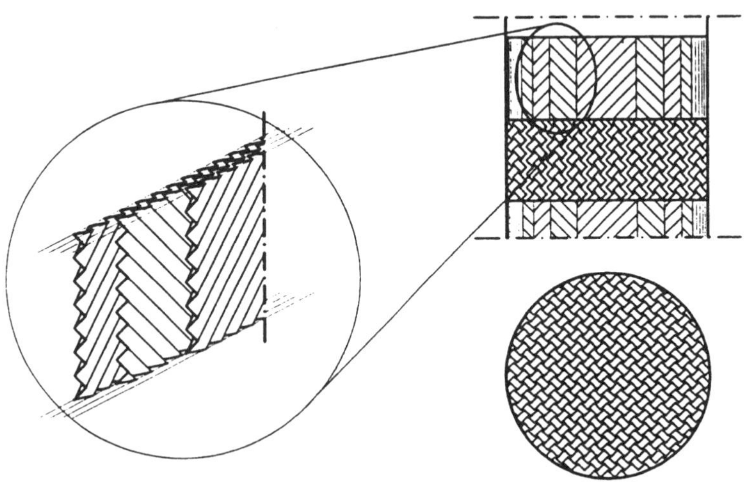

Fig. 8-32. *Structured packing assembled by corrugated sheets.*

Table 8-7: Typical Values of Specific Area a and Porosity ϵ for Several Structured Packings

Vendor	Designation	Specific Area a, m^2/m^3	Porosity ϵ
Sulzer (metal gauze)			
	BX	500	0.90
	CY	700	0.85
Sulzer (metal sheet)			
	Mellapak 250.X	250	0.988
	Mellapak 250.Y	250	0.988
	Mellapak 350.Y	350	0.975
Montz (metal sheet)			
	B1 100	100	0.99
	B2 200	200	0.98
	B3 300	300	0.97
Glitsch (metal sheet)			
	Gempak 1A	131	0.96
	Gempak 2A	223	0.95
	Gempak 3A	394	0.92
	Gempak 4A	525	0.91
Koch (metal sheet)			
	Flexipak 4	69	0.98
	Flexipak 3	135	0.96
	Flexipak 2	223	0.95
	Flexipak 1	558	0.91
Norton (metal sheet)			
	Intalox 2T	231	0.95
	Intalox 1T	315	0.95
Kühni (metal sheet)	Rombopak	754	

enhance lateral mixing of the liquid in all directions. Since the packing does not fit perfectly into the cylindrical column shell, additional tightening strips have to be installed between packing and column wall.

These packings provide a homogeneous bed structure and, due to the vertical orientation of the sheets, a low pressure loss of the gas flow. The first field of application of such structures with wire gauze elements was the vacuum distillation. Meanwhile, such structures are widely used in low pressure distillation with excellent results. Sheet metal and plastic packings are increasingly being used. Characteristic data of some important structured packings are given in Table 8-7. The packings of different vendors differ mainly in the microstructure of the sheet surface. Ordered packings with a different geometric structure are marked by [Bühlmann 1997, Sulzer Chemtech 1996]. These packings provide a small scale lateral mixing of gas and liquid in all directions.

8.3.1.2 Model Structures of Packings

The structure of any packing is too complex for modeling due to the large number of geometrical dimensions. Thus, at the present state of the art, the complex real structure of the packing is substituted

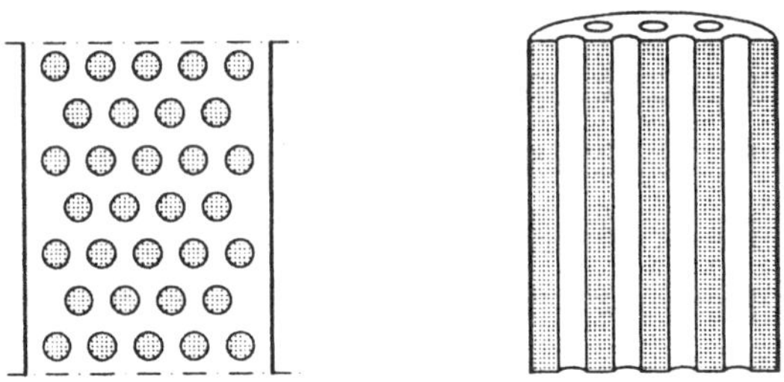

Fig. 8-33. *Models of packing structure.*

by a simple model structure with the same specific area a and the same porosity ϵ. Two simplified model structures are feasible: the particle model structure and the channel model structure.

8.3.1.2.1 Particle Model Structure Here, the real packing is substituted by a collective of solid particles (i.e., spheres). Thus, the solid material of the packing is considered as the dispersed phase, see Fig. 8-33. At high porosities ($\epsilon > 0.4$) the spheres do not touch each other, i.e., the model structure is very similar to a fluidized bed. For the porosity follows:

$$\epsilon = \frac{V_G}{V} = 1 - \frac{z\, d_p^3 \pi/6}{V} \tag{8.57}$$

With z the number of spheres and V the volume of the bed. The specific area of the spheres is:

$$a = \frac{A}{V} = \frac{z\, d_p^2 \pi}{V} \tag{8.58}$$

From both equations follows after elimination of z/V:

$$d_p = \frac{6\,(1 - \epsilon)}{a} \tag{8.59}$$

The diameter d_p characterizes the structure of the **solid** material of the packing.

The particle model structure is quite a realistic one since the solid phase is considered as dispersed phase. Hence, this model structure is used in the following chapters.

8.3.1.2.2 Channel Model Structure Here, the real packing element is replaced by a system of parallel channels or pipes, see Fig. 8-33, with the equivalent diameter d_{eq}. The equivalent diameter is chosen so that the porosity ϵ and the specific area a of the model structure have the same values as the real packing. For the porosity follows:

$$\epsilon = \frac{V_G}{V} = \frac{z\ H\ {d_{eq}}^2 \pi/4}{V} \tag{8.60}$$

Here, z denotes the number of channels. The specific area of the model structure is:

$$a = \frac{A}{V} = \frac{z\ H\ d_{eq}\ \pi}{V} \tag{8.61}$$

Combining Equations (8.60) and (8.61) yields:

$$d_{eq} = \frac{4\ \epsilon}{a} \tag{8.62}$$

The diameter d_{eq} characterizes the *open* structure of the packing. Hence, the definitions and meanings of d_p and d_{eq} are totally different. The two quantities must not be confused.

The equivalent diameter d_{eq} of the channel can be expressed by the particle diameter d_p. Eliminating the volumetric area a from Equation (8.62) and (8.62) yields:

$$d^*_{eq} = \frac{2}{3} \cdot \frac{\epsilon}{1 - \epsilon} \cdot d_p \tag{8.63}$$

Equation (8.63) in most cases is used without the factor 2/3:

$$d_{eq} = \frac{\epsilon}{1 - \epsilon} d_p \tag{8.64}$$

Equation (8.64) formulates a relation between the particle diameter d_p and the channel diameter d_{eq}. It is, in principle, a nonsense since first, the particle diameter is determined by Equation (8.59) from a knowledge of a and ϵ, and second, the equivalent diameter is determined from the particle diameter. Equation (8.62) allows a more straightforward determination of the equivalent channel diameter d_{eq}.

The channel model is very often used in modeling of packings. However, it is unrealistic since the solid material of the packing is treated as the continous phase. In most real packings the solid material is the dispersed phase. Moreover, at high values of the porosity (i.e. $\epsilon > 0.907$) the cross sections of the channels intersect each other making the channel model structure geometrically unfeasible.

8.3.1.3 Packing Supplements Obviously, the most important element of packed column internals is the packing itself. However, some supplementary ele-

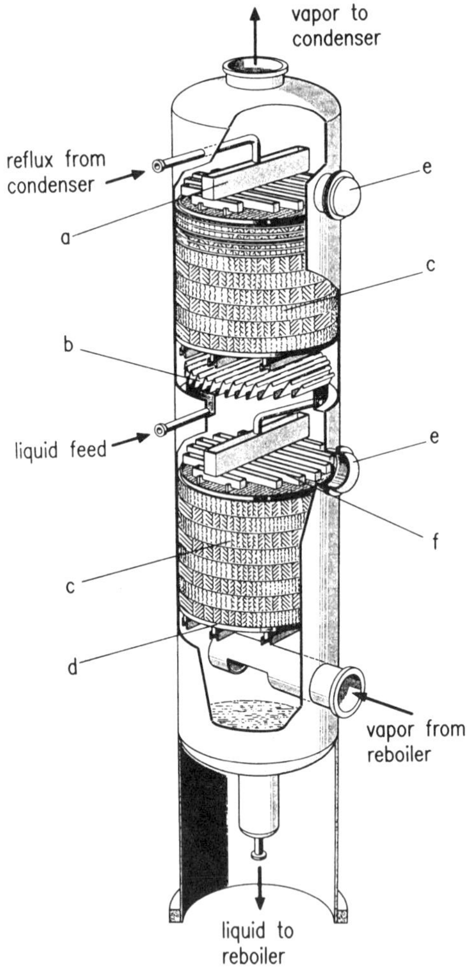

Fig. 8-34. *Cutaway section of a packed column showing column iternals [Sulzer 1982]: (a) Liquid distributor, (b) liquid collector, (c) structured packing, (d) support grid, (e) manway, (f) liquid redistributor.*

ments are necessary for proper column operation. These supplementary internals are schematically depicted in Fig. 8-34. The most important elements are:

- Liquid distributor and redistributor
- Liquid collector
- Wall wiper
- Support grid
- Hold down plate
- Gas distributor

All these elements need careful adjustment to the packing. Packing and supplements form a complex and highly integrated system. Even a high efficiency packing may fail when, for instance, the liquid distributor is not properly designed. Hence, most vendors offer sets of packings and supplements.

The fundamental requirement to all packing supplements is that they must not hinder the countercurrent flow of gas and liquid through the column. Ideally, they should have a free area that is as high as the porosity of the bed. This demand is hard to meet since the porosity of a modern structured packing can be as high as 95%. Often the packing supplements give rise to an increased pressure drop and, even worse, to flooding below the flooding point of the bed.

8.3.1.3.1 Liquid Distributor A key element of any packed column design is the liquid distributor. Especially in vacuum distillation which is characterized by a low liquid load, the liquid distributor is decisive for separation efficiency of the whole column. The requirements to a liquid distributor are:

- Uniform distribution of liquid across the whole cross section of the column
- Large flexibility to liquid flow rates
- Low resistance against gas flow rate (low pressure drop)
- Low sensitivity to fouling

High performance distributors provide a flow rate variation per irrigation point of less than 5 to 6% of average flow [Strigle 1987]. Larger deviations between neighboring irrigation points can be tolerated since lateral liquid mixing flattens the flow rate profile. Particularly unfavorable are large scale flow rate variations because they are not compensated for by lateral mixing within the bed.

Industrial liquid distributors use nozzles, weirs or perforated plates and troughs as distributing elements. Nozzles demand a high liquid pressure ($\Delta p > 1$ bar) for good operation and imply the risk of entrainment of small droplets [Ullrich et al. 1983]. Weirs are often equipped with dripping elements (noses) or zigzag edges. Plate distributors have simple holes or short pipes. A rule of thumb requires 65 $1/m^2$ [Moore and Rukovena 1987] or 100 $1/m^2$ [Strigle 1987] irrigation points per square meter. Decisive for good operation is a precise leveling of weirs and trays, respectively.

Up to a column diameter of 0.8 m the standard design is an orifice distributor that consists of a plate with small holes for downward liquid flow and chimneys for upward gas flow. Columns with a diameter larger than 0.8 m require a more complex distributing system with pipes or troughs. Distributors are made up either of metal or plastics, in very special cases of ceramics. Important is the transportability of the elements through the rather small man way in the column shell.

8.3.1.3.2 Liquid Redistributor After a certain height of the packing the liquid needs redistribution to avoid a high degree of liquid maldistribution. Depending on the type of packing and the liquid load, the maximum feasible distance for redistribution is 6 m or even 8 m. A rule of thumb says that not more than 20 equilibrium stages should be realized in one section. Additionally, the ratio of packing height and column diameter should meet the condition $H/D_c < 6$. In small-diameter columns maldistribution is mainly caused by wall flow. Here, simple wall wipers are used to improve the liquid flow profile. In large diameter columns, however, a system of liquid collector and liquid distributor has to be installed for redistribution as is shown, for instance, in Fig. 8-34. The height of such a system of liquid collector and redistributor is approximately 2.5 m, significantly increasing the column height.

8.3.1.3.3 Liquid Collector Liquid collectors have to be installed for the withdrawal of side stream products and for pump arounds (see Chapter 5.3.2.2). In high efficiency packings an effective liquid collector should be installed above each liquid redistributor. A liquid collector has to interrupt the liquid flow without any blocking of gas flow. A good design is shown in Fig. 8-34.

An essential additional task of a liquid collector is the effective mixing of the liquid before the redistribution since high differences of liquid concentrations may exist across the flow cross section (see Fig. 8-51).

8.3.1.3.4 Support Grid A grid or perforated plate is installed within the column to support the packing and, even more important, the liquid holdup of the packing. The support grid must pass both the downwardly flowing liquid phase and the upwardly flowing gas phase [Strigle 1987] over the whole range of operation of the packing. Hence, the free area of the support grid should be as high as the porosity of the bed. Especially in dumped packings a high risk exists in that the packing elements (e.g. Raschig rings) block some of the open area of the support grid. This increases the pressure drop of the column and may even cause flooding (see Chapter 8.3.3.2.1). Plane support grids are used in small diameter columns (e.g., $D_c < 0.6$ m). Profiled support grids have a higher stability and are, therefore, used in columns with a larger diameter.

8.3.1.3.5 Hold Down Plate The primary function of hold down plates (or bed limiters) is to prevent expansion of the packed bed as well as to maintain a horizontal bed surface level [Strigle 1987]. Since the height of a dumped packing may shrink during long time operation, the hold down plate is not fixed at the column wall, but acts by its own weight. The hold down plate should not restrict the gas and liquid flow through the column. Hence, the free area of these plates should be very high. The installation of hold down plates is especially important above low weight packings as random metal and plastic packings. At structured packings the liquid distributor is often immediately installed on top of the packing itself. Hence, there is no need for a hold down plate.

8.3.1.3.6 Gas Distributor Uniform gas flow across the column cross section is as important as uniform liquid flow. The risk of gas maldistribution is, however, much lower than that of liquid maldistribution since the gas flow rate profile is flattened by the pressure drop. However, modern structured packings in large diameter columns require careful initial gas distribution. For column diameters up to $D_c = 2.5$ m a simple inlet pipe is sufficient (see Fig. 8-34). For a diameter larger than $D_c = 2.5$ m a distributing system is recommended [Moore and Rukovena 1987].

8.3.2 Operating Region of Packed Columns

The feasible operating region of packed columns is depicted in Fig 8-35; the gas load $\dot{V}_G/A_c$ is plotted against the liquid load $\dot{V}_L/A_c$ (A_c is the cross-sectional area of the column). The operating region of packed columns differs considerably from that of tray columns (see Chapter 8.2.2.5). In packed columns, the gas load can be reduced to extremely small values, but the liquid load must be maintained within an upper and a lower limit. This difference influences the choice of apparatus. In most cases, the flexibility of the liquid distributor is much smaller than the flexibility of the packing itself. Two principal mechanisms limit the gas and liquid flow rates of the packing: flooding and dewetting.

8.3.2.1 Flooding The upper capacity limit of a packed column is set by flooding. At the flood point the pressure drop of the gas flow through the bed increases so drastically that the liquid is no longer able to flow downward against the gas flow. Hence, the countercurrent flow of gas and liquid breaks down. Often, a two-phase layer develops on top of the packing.

Experimental determination of the flood point is best performed by increasing the gas load at constant liquid load. The maximum feasible gas load is easily found from a plot of the pressure drop versus the gas load, see, for instance, Fig. 8-44.

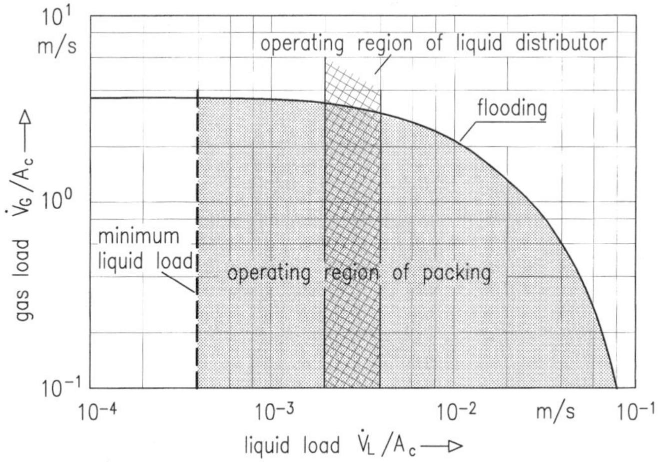

Fig. 8-35. Operating region of a packing. In most cases the flexibility of the liquid distributor is much lower than that of the packing itself.

This method only becomes inaccurate at operation with very low liquid load. One should be aware, however, that the pressure drop at the flood point can never be determined precisely since it increases infinitely at approaching the flood point. Hence, very small variations of the gas flow drastically increase the pressure drop.

Several attempts have been made to develop a general method for flood point prediction. [Sherwood et al. 1938] was the first to present an empirical flood point correlation based on a variety of experimental data. He managed to get a single flood point line by plotting a modified gas load versus the flow parameter, i.e., the square root of the ratio of the kinetic energy of liquid and gas. Later on [Leva 1954], completed the Sherwood diagram by implementing the pressure drop in form of parameter lines. Several modifications of this diagram have been made by, e.g., [Eckert 1970], see Fig. 8-36. Recently, Kister adapted this diagram to modern packings [Kister 1992]. Surprisingly, the liquid viscosity η_L is combined with the gas load in the ordinate of all theses diagrams.

A theoretically more profound flood point correlation has been developed by [Mersmann 1965]. In his diagram, see Fig. 8-37 the gas load (expressed in form of dry pressure drop) is plotted versus a modified liquid load. The term for the liquid load has been derived from the theory of film flow on a vertical plate. It includes the viscosity of the liquid. The diagram contains the flood point line and parameter lines for the pressure drop of an irrigated packing. Additionally, the loading point is marked by a dashed line.

The Mersmann diagram as well as the Sherwood diagram and its modifications

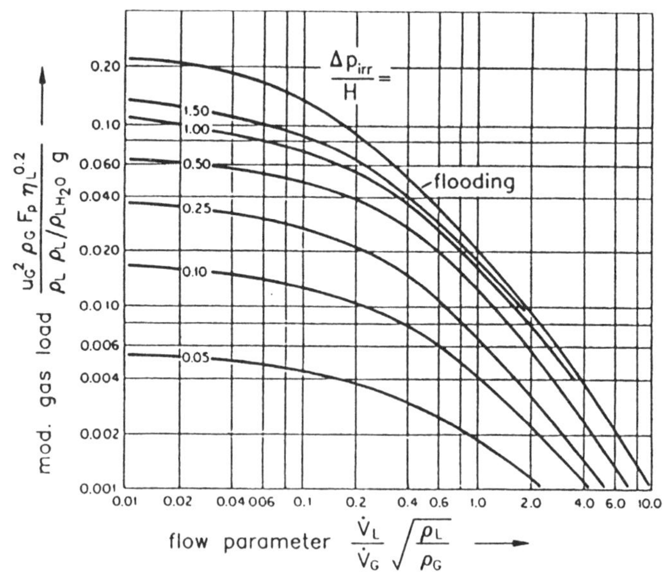

Fig. 8-36. Correlation of flood point and irrigated pressure drop according to [Eckert 1970]. The packing factor F_p depends on the type of packing. (The diagram is valid for u_G in ft/s, g in ft/s^2, η_L in cP (= g/(ms)), Δp_{irr} in inches of water, H in ft).

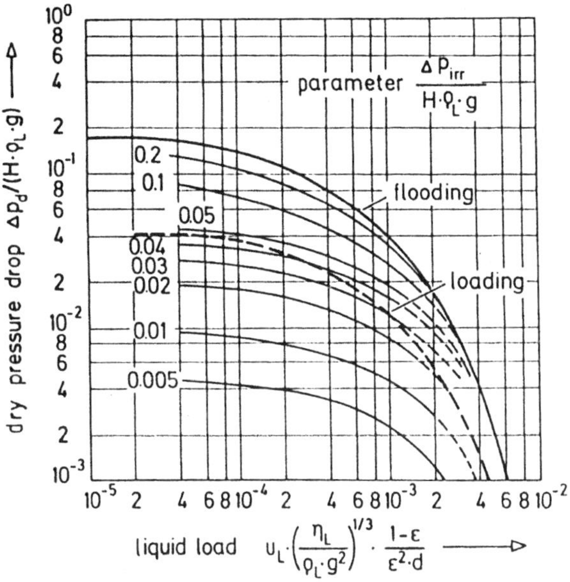

Fig. 8-37. Correlation of flood point and irrigated pressure drop according to [Mersmann 1965].

claim, in principle, to be valid for all types of packings and to enable a genuine prediction of the fluid-dynamic behavior of any packing. Hereby, the packing is characterized by the porosity, the specific area and the packing factor or the dry pressure drop. These parameters are, however, not sufficient to precisely describe the complex structure of a packing. Hence, these generalized correlations imply rather high inaccuricies.

Often used is a plot of experimental flood point data versus the flow parameter as is shown in Fig. 8-38. The flood point lines of different types of packings significantly differ. Such diagrams are often published by vendors, mostly on the basis of air/water data. These diagrams, however, are not able to predict the fluid-dynamic behavior of a new type of packing. They are just an adequate plot of experimental data. Even their applicability to other systems is ambiguous.

A more rigorous approach to flood point prediction should be based on an efficient model for the pressure drop of an irrigated packing. If such a model is valid even above the loading point (see Chapter 8.3.3.2) it will predict the sharp increase of pressure drop near the flood point. Such models based on a channel packing structure have been developed by [Mackowiak 1991 and Billet 1995]. A model that is based on the particle structure will be outlined in Chapter 8.3.3.2.

8.3.2.2 Minimum Liquid Load In contrast to tray columns, packed columns require a minimum liquid load to ensure sufficient mass transfer. Below this minimum value, only a very small part of the packing surface is wetted and liquid and gas are no longer in intimate contact. This results in a considerable drop of separation

Size d[mm]		Plastics dumped packing	D_C m	H m	z/V 1/m³
▽	50	NOR-PAC ring	0.30	1.50	7320
□	50	Hiflow ring	0.30	1.40	6280
○	45	Hackette	0.30	1.35	12850
⊘	50	INTALOX saddle	0.30	1.45	9422
⧄	50	INTALOX saddle	0.45	1.20	9875

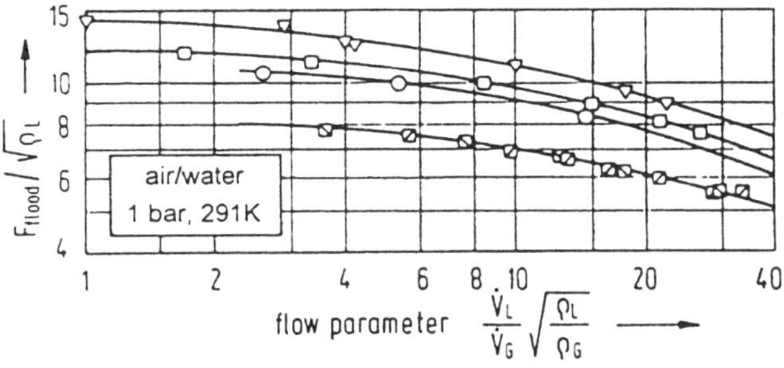

Fig. 8-38. *Presentation of flood point data [Billet 1995].*

efficiency. The lower capacity limit of liquid load depends, of course, on the type of the packing, the quality of packing supplements and the physical properties (e.g., surface tension) of the liquid. A rule of thumb gives the following values for dumped packings:

- $u_{Lmin} > 10\ m^3/(m^2h)$ in aqueous systems and
- $u_{Lmin} > 2\ m^3/(m^2h)$ in organic systems

High efficient structured packings with high quality liquid distributors allow an effective column operation with liquid loads as low as $u_{Lmin} = 0.2\ m^3/(m^2\ h)$ in organic systems, e.g., [Sulzer Chemtech 1994]. This value is comparable to the flow rate of a heavy rainfall.

A very useful equation for estimating minimum liquid load has been published by [Schmidt 1979]:

$$u_{L,min} = 7.7 \cdot 10^{-6} \left(\frac{\rho_L\ \sigma^3}{\eta_L^4\ g} \right)^{2/9} \left(\frac{g}{a} \right)^{1/2} \tag{8.65}$$

where ρ_L is the liquid density, σ the surface tension, g the acceleration due to gravity, η_L the liquid viscosity, and a the specific area of the packing. This equation accounts for the most important parameters and gives in most cases the right order of magnitude of the minimum liquid load.

8.3.3 Two-Phase Flow in Packed Columns

A packed column establishes a pure countercurrent flow of liquid and gas streams. The liquid trickles downward over the surface of the packing in the form of rivulets or films. Considerable drop formation can also occur within the bed if a large packing size is used.

8.3.3.1 Liquid Holdup The fluid-dynamic behavior of a packed column is governed by the amount of liquid that accumulates within the packing during operation. The accumulated liquid is called liquid holdup. The volume of the liquid V_L is conveniently referred to the volume V of the packing:

$$h_L = V_L/V \tag{8.66}$$

Hence, a packed column establishes a three-phase system with a solid phase, a liquid phase and a gas phase.

During operation the irrigated liquid flows through the bed countercurrently to the gas flow. Whenever a liquid element touches the packing surface it is slowed down due to the high viscosity of the liquid. Hence, most of the liquid trickles downward in the bed in the form of thin films or small rivulets. Only a small fraction of the liquid may move downwards in the form of falling droplets.

Present knowledge of liquid holdup is rather poor. This lack of knowledge may result from the fact that the liquid holdup is not of prime interest for column operation as, for instance, pressure drop, flooding or even interfacial area is. Liquid holdup is an intrinsic quantity that controls the fluid-dynamics of the column. Thus, knowledge of the liquid holdup is of great importance for a deeper understanding of the flow mechanism in the column. It facilitates the prediction of the fluid-dynamic operation parameters.

Many authors distinguish the liquid holdup into: a static holdup and a dynamic holdup. The static holdup results from the action of capillary forces that hold some liquid in narrow sections (pinches) of the packing. This fraction of the holdup has a very high residence time in the bed since it is hardly replaced by any fresh liquid during operation. When the irrigation has been stopped the static holdup remains within the packing. The static holdup is quite high in small packings with a plate structure, e.g., small Raschig rings with a high volumetric area a, see Fig. 8-39. The dynamic holdup strongly depends on liquid load. The higher the liquid load the larger the dynamic holdup. In modern packings with a lattice structure the dynamic holdup is much larger than the static holdup.

Figure 8-40 shows a typical result of holdup measurements [Billet 1995]. The holdup h_L is plotted versus the gas load u_G with the liquid load u_L as parameter. The holdup h_L is in the range of 5 to 20% of the bed volume. Hence, it can be higher than the volume of the solid material of the packing (in the tested packing approximately 6% of the bed volume). At low gas loads the holdup is a function of liquid load only. At the loading point the gas starts increasing the holdup. As is seen from the diagram the gas velocity at the loading point decreases with increasing liquid load.

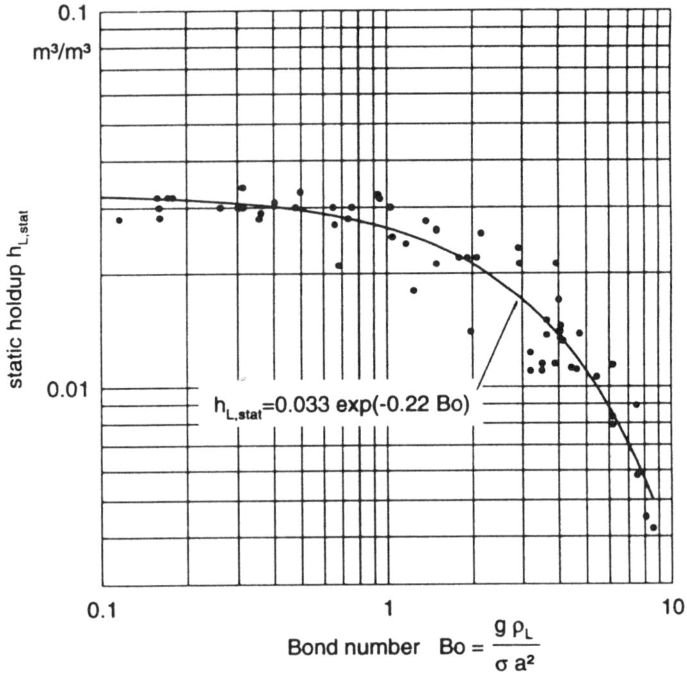

Fig. 8-39. *Plot of experimental data of static holdup [Engel et al. 1997].*

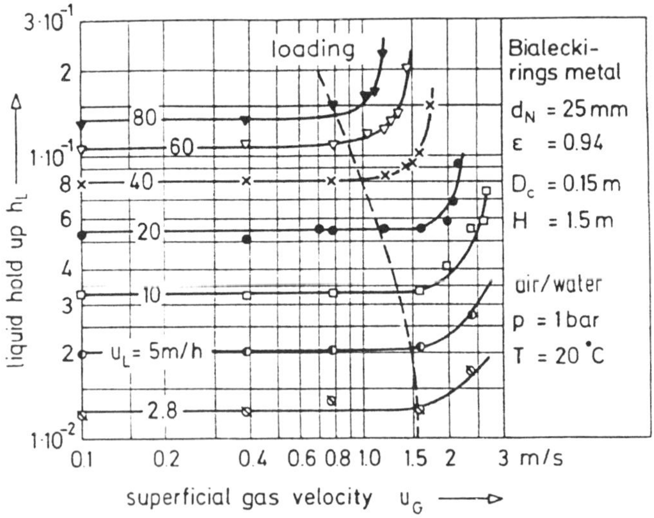

Fig. 8-40. *Liquid holdup data of a random metal packing of Bialecki rings [Billet 1984]. Figs. 8-40 and 8-44 establish a full set of fluid-dynamic data.*

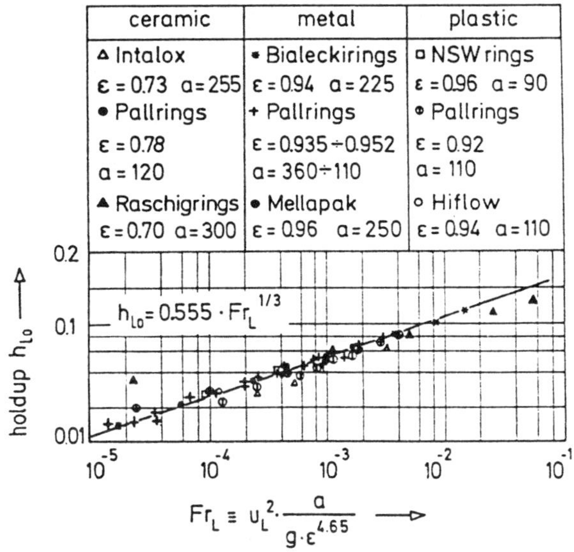

Fig. 8-41. *Correlation of the holdup below loading point [Stichlmair et al. 1989].*

8.3.3.1.1 Holdup Below the Loading Point Several holdup correlations have been published in literature. Most of them describe the holdup below the loading point only. Here, the holdup depends primarily on the Froude number of the liquid. However, different definitions of the Froude number have been used by various authors. Since most of the liquid holdup is attached to the surface of the packing there should exist a close relationship between liquid holdup h_L and specific area a of the packing. Hence, the Frounde number should be defined with the specific area a (and not with the particle diameter d_p). Further parameters that influence the liquid holdup are the surface tension, the viscosity of the liquid and the liquid load.

Figure 8-41 shows a plot of experimental holdup data h_{Lo} below the loading point versus the Froude number of the liquid. The data base contains differnt types of packings (random as well as ordered packings) and different packing materials. All data have been taken with the system air/water. The data are correlated by:

$$h_{Lo} = 0.555\ Fr_L^{1/3} \quad \text{with } Fr_L = \frac{u_L^2\ a}{g\ \epsilon^{4.65}} \tag{8.67}$$

The correlation does not account for any properties of the liquid since they have not been varied in the experiments. Thus, it is valid for water and similar liquids only. Further holdup correlations have been developed by [Mersmann and Deixler 1986, Buchanan 1969]. These correlations account for the influence of surface tension σ and liquid viscosity η_L but have been validated only by small scale packings. Most published data are correlated with a mean deviation lower than 16% by a correlation developed by [Engel et al. 1997]:

Table 8-8: Correlations for the Liquid Holdup h_{Lo} Below the Loading Point

	Exponent of					
	a	ϵ	σ	η_L	ρ_L	u_L
[Bemer and Kalis 1978]	0.333	—	—	—	—	0.666
[Billet and Bravo 1989]	—	−0.666	—	0.333	−0.666	0.666
[Blaß and Kurtz 1976]	0.16 to 0.22	1	0.14	0.74 to 0.8	−0.34 to −0.28	0.4 to 0.46
[Gelbe 1968]	0.73 to 0.85	−0.586	0.143	0.145 to 0.267	−0.41 to −0.28	0.33 to 0.45
[Mohunta et al. 1969]	0.75	—	—	0.25	−0.25	0.75
[Bornhuetter 1991]	0.57	−1.14	—	0.19	−0.19	0.57
[Mackowiak 1991]	0.5	−1	—	0.167	−0.167	0.5
[Billet and Schultes 1993]	0.566 to 0.633	—	—	0.083 to 0.183	−0.083 to 0.183	0.566 to 0.633
[Stichlmair et al. 1989]	0.333	1.55	—	—	—	0.666
[Engel et al. 1997]	0.717	—	0.125	0.2	−0.325	0.333

$$h_{Lo} = 0.93 \left(\frac{u_L^2\, a}{g}\right)^{1/6} \left(\frac{\eta_L^2\, a^3}{\rho_L^2 g}\right)^{1/10} \left(\frac{\sigma\, a^2}{\rho_L\, g}\right)^{1/8} \tag{8.68}$$

Similar correlations have been developed by several authors mostly in the form of potential equations. Table 8-8 lists the parameters accounted for in these correlations and the values of their exponents. The large variation of the involved parameters and of their exponents demonstrates the poor and ambigous knowledge in this field.

8.3.3.1.2 Holdup Above the Loading Point Above the loading point the liquid holdup is significantly influenced by the gas flow. The friction forces between gas and liquid retard the velocity of the surface of the liquid film and, in turn, increase both the thickness of the film and the pressure drop. Probably as important, however, may be the build up of a pressure gradient along the column height. The liquid has to move downward in the bed against an increasing static pressure.

The significance of the pressure gradient for the increase of liquid holdup is demonstrated in Fig. 8-42. Here, the holdup data of Fig. 8-40 are plotted versus the dimensionless pressure drop $\Delta p_{irr}/(H\ \rho_L\ g)$ (and not versus the superficial velocity of the gas u_G). The pressure drop data have been taken from Fig. 8-44. Both diagrams establish a consistent set of fluid-dynamic data because they have been collected during the same run of experiments. In the new plot the parameter lines are equidistant, and loading and flooding lines become independent on the parameter lines, i.e., on liquid load. From the plot in Fig. 8-42 follows a simple correlation for the holdup above the loading point:

$$h_L = h_{Lo} \cdot \left(1 + 20 \left(\frac{\Delta P_{irr}}{H\ \rho_L\ g}\right)^2\right) \tag{8.69}$$

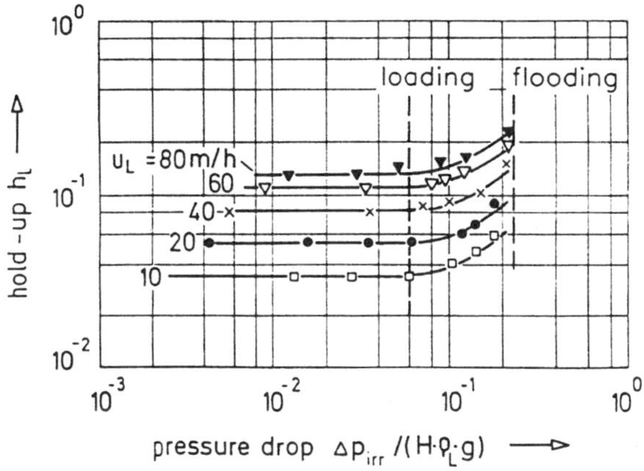

Fig. 8-42. *Modified plot of the holdup data of Fig. 8-40.*

The factor 20 and the exponent 2 have been adjusted to the experimental data. However, they should approximately be valid for all types of packings [Stichlmair et al. 1989]. The holdup below the loading point h_{Lo} is described by Equation (8.67) or (8.68). A similar approach was developed by [Buchanan 1987].

It is worth while to note that the holdup does not neccessarily become infinite at the flood point as is often assumed. This will be outlined in detail in Chapter 8.3.3.2.

8.3.3.2 Pressure Drop The most important operating parameter of two-phase flow in packed columns is the pressure drop of the gas. It is significantly influenced by the type of packing, the gas flow rate and the liquid flow rate. First, the pressure drop of the gas in a dry packing is considered.

8.3.3.2.1 Pressure Drop of a Dry Packing The pressure drop of the gas flow through a dry packed bed can be formulated either by the channel or by the particle model.

8.3.3.2.2 Channel Model Structure In the channel model structure the basis is the equation of the pressure drop in a pipe of length H:

$$\frac{\Delta p_d}{H} = f \frac{\rho_G}{2} u^2/d \tag{8.70}$$

Expressing the gas velocity in the channel by $u = u_G/\epsilon$ (u_G is the superficial gas velocity in the column) and the diameter d by 4 ϵ/a Equation (8.62) leads to:

$$\frac{\Delta_d}{H} = \frac{1}{4} f \frac{\rho_G}{2} u_G^2 \frac{a}{\epsilon^3} \tag{8.71}$$

Formulation of the equivalent diameter d_{eq} by Equation (8.64) yields:

$$\frac{\Delta p_d}{H} = \zeta_E \frac{1-\epsilon}{\epsilon^3} \rho_G u_G^2 \frac{1}{d_p} \tag{8.72}$$

This is the well known Ergun Equation [Ergun 1952] often used for calculating the pressure drop of a packed column. The friction factor ζ_E expresses the flow resistance of the bed. It has to be determined by experiments with a whole bed. Values for the friction factor ζ_E are, for instance, given in [Kister 1992].

It is important to be aware that the porosity term $(1 - \epsilon)/\epsilon^3$ in the Ergun Equation (8.72) is the result of a simplified model. This term has never been verified by comparison with experimental data since in a packed bed the porosity can not be varied. The value of the porosity is only a function of the shape of the particles and the structure of the bed. Hence, the published values of the friciton factor ζ_E are, in principle, only valid for the porosities of the tested beds.

8.3.3.2.3 Particle Model Structure Applying the particle model structure yields a different equation for the dry pressure drop. Basis for the development of a pressure drop equation is the pressure drop equation of a fluidized bed. The fundamental equation is derived in the Appendix to Chapter 8. The result is:

$$\frac{\Delta p_d}{H} = \frac{3}{4} \zeta_o \frac{1-\epsilon}{\epsilon^{4.65}} \rho_G u_G^2 \frac{1}{d_p} \tag{8.73}$$

This equation differs from the Ergun equation in the friction factor ζ_o and in the porosity term. Here, ζ_o denotes the friction factor of a single particle and not, as ζ_E does, of the whole bed. Thus, the pressure drop of a bed of spheres, for instance, can be predicted from a knowledge of the friction factor of a single sphere.

The porosity term $(1 - \epsilon)/\epsilon^{4.65}$ has been determined by experiments in fluidized beds under the variation of the porosity from $\epsilon = 0.4$ to $\epsilon = 1.0$. Hence, the porosity term in Equation (8.73) is well proven in contrast to the porosity term of the Ergun Equation (8.72).

Figure 8-43 shows a correlation of the friction factor ζ_o versus the Reynolds number for some important packings [Stichlmair et al. 1989]. The curves are correlated by using the following type of equation:

$$\zeta_o = \frac{C_1}{Re} + \frac{C_2}{Re^{1/2}} + C_3 \tag{8.74}$$

where C_1, C_2, and C_3 are constants that depend on the type of packing; for values, see Table 8-9. The correlation of Fig. 8-43 has been developed from published data of the dry pressure drop. Since even standard packings of different vendors are slightly different, it is recommended to use dry pressure drop data published by the vendors (as well as porosity and specific area data).

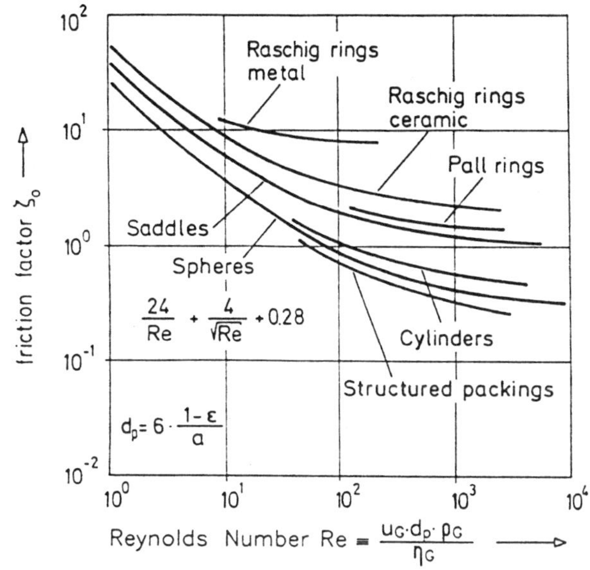

Fig. 8-43. Correlation of the friction factor of single particles of some packings [Stichlmair et al. 1989]. The lines are correlated by Equation (8.74).

8.3.3.2.4 Pressure Drop of an Irrigated Packing The pressure drop of an irrigated bed is always higher than that of a dry bed, see Fig. 8-44. This increase of pressure drop is caused by the liquid that accumulates within the bed during irrigation. The liquid holdup in the bed decreases, for instance, the voidage of the bed, and in turn, increases the velocity of the gas flow.

The pressure drop of an irrigated bed can be calculated either by the channel model or the particle model.

8.3.3.2.5 Channel Model Structure Most authors (e.g., Buchanan, Bravo, Billet, Bemer, and Kalis) use the channel model. They find:

$$\frac{\Delta p_{irr}}{H} = \frac{\Delta p_d}{H} \cdot \frac{1}{(1 - C_h \cdot h_L)^5} \tag{8.75}$$

Table 8-9: Approximate Factors for the Friction Factor Correlation Equation (8.74)

Packing type	C_1	C_2	C_3
Spheres	24	4	0.28
Raschig rings (metal)	48	8	2.0
Raschig rings (ceramic)	60	1	7.5
Pall rings	33	7	1.4
Bialecki rings	50	7	2.5
Saddles	32	7	1.0
Structured packings	18	4	0.2

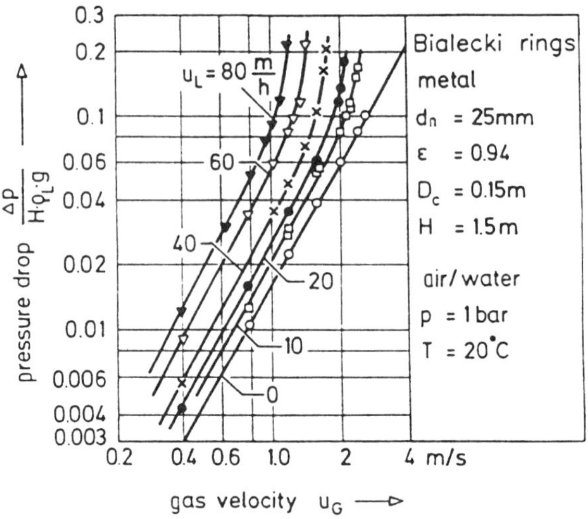

Fig. 8-44. *Dry and irrigated pressure drop of a random metal packing of Bialecki rings [Billet and Mackowiak 1984] (see also Fig. 8-40).*

The pressure drop of the dry bed is calculated by Equation (8.71) or the equivalent Equation (8.72). Hence:

$$\frac{\Delta p_{irr}}{H} = \frac{1}{4}\, f\, \frac{\rho_G}{2}\, u_G^2\, \frac{a}{\epsilon^3}\, \frac{1}{(1 - C_h\, h_L)^5} \tag{8.76}$$

Here, C_h is an empirical factor that depends on the type of packings. Some values of C_h have been published by [Bemer and Kalis 1978, Buchanan 1969, Billet and Mackowiak 1984].

Equation (8.76) has been successfully used by many authors for predicting the pressure drop of an irrigated packing. However, from a theoretical point of view, the equation looks inconsistent since the influence of the volumetric fraction of the solid material ϵ_s ($\epsilon_s = 1 - \epsilon$) differs from the influence of the liquid holdup h_L, i.e. $\Delta p \sim (1 - \epsilon_s)^{-3}$ and $\Delta p \sim (1 - C_h\, h_L)^{-5}$, respectively. In a consistent model the exponents in the two terms must not be different since the holdup of the liquid is considered in the model as additional solid material.

8.3.3.2.6 Particle Model Structure Application of the particle model to an irrigated packing leads to a more consistent formulation if all the changes caused by the liquid holdup in the bed are considered. The most significant changes are:

- Reduction of the voidage of the bed
- Increase of the particle diameter
- Change of the friction factor of the particle

- Increase of the velocity difference

All these changes of the bed structure can be expressed by the liquid holdup h_L if the holdup is treated as some additional solid material.

The porosity of the irrigated bed is:

$$\epsilon_{irr} = \epsilon - h_L \quad \text{or} \quad \epsilon_{irr} = \epsilon\,(1 - h_L/\epsilon) \tag{8.77}$$

The smaller voidage of the irrigated packing is caused by an increased particle diameter d_p. If the number of particles is considered to be the same in the irrigated and the dry bed a simple relation holds:

$$(1 - \epsilon_{irr})/d_{pirr}^3 = (1 - \epsilon)/d_p^3 \quad \text{or} \quad d_{pirr} = d_p\left(\frac{1 - \epsilon_{irr}}{1 - \epsilon}\right)^{1/3} \tag{8.78}$$

Substituting equation (8.77) yields:

$$d_{pirr} = d_p\left(\frac{1 - \epsilon\,(1 - h_L/\epsilon)}{1 - \epsilon}\right)^{1/3} \tag{8.79}$$

The liquid holdup in the bed has no significant influence on the shape of the particle. An irrigated Raschig ring, for instance, looks like a dry Raschig ring. Thus, the function of the friction factor (Equation 8.74) will be unchanged to a first approximation. But the particle diameter is increased by the liquid holdup. Hence, the Reynolds-Number of an irrigated particle is higher and, in turn, the friction factor of an irrigated particle is lower, see Fig. 8-43. The Equation (8.74) can be approximated for small variations of the Reynolds-Number by:

$$\zeta_o \propto Re^c \tag{8.80}$$

The exponent in (8.80) is determined by the derivative of Equation (8.74):

$$c = \frac{1}{\zeta_o}\left(-\frac{C_1}{Re} - \frac{C_2}{2Re^{1/2}}\right) \tag{8.81}$$

Typically, the exponent c lies between $c = -1$ (laminar flow) and $c = 0$ (turbulent flow). From Equation (8.80) follows:

$$\frac{\zeta_{oirr}}{\zeta_o} = \left(\frac{Re_{irr}}{Re}\right)^c = \left(\frac{d_{pir}}{d_p}\right)^c \tag{8.82}$$

Replacing d_{pirr} by Equation (8.79) yields:

$$\zeta_{oirr} = \zeta_o \left(\frac{1 - \epsilon (1 - h_L/\epsilon)}{1 - \epsilon} \right)^{c/3} \tag{8.83}$$

In cases with low gas but very high liquid flow rates the superficial gas velocity u_G must be corrected to account for the increased velocity difference between gas and liquid phases. The correction is well approximated by:

$$\frac{u_{Girr}}{u_G} = 1 + \frac{u_L}{u_G} \frac{\epsilon}{h_L} \tag{8.84}$$

This correction is only important at very high liquid loads seldom encountered in distillation. It is not accounted for in the following.

In the particle model structure, the pressure drop of the dry and the irrigated packing is described by the same equation (Equation 8.73). Hence, the pressure drop of an irrigated packing is:

$$\frac{\Delta p_{irr}}{H} = \frac{3}{4} \zeta_{oirr} \frac{1 - \epsilon_{irr}}{\epsilon_{irr}^{4.65}} \rho_G u_G^2 \frac{1}{d_{pirr}} \tag{8.85}$$

The quantities ϵ_{irr}, d_{pirr}, and ζ_{oirr} are formulated by Equations (8.77), (8.79) and (8.83), respectively.

Conveniently, the pressure drop of an irrigated packing is related to the pressure drop of a dry packing. Combining Equations (8.73) and (8.85) yields:

$$\frac{\Delta p_{irr}}{\Delta p_d} = \left(\frac{1 - \epsilon (1 - h_L/\epsilon)}{1 - \epsilon} \right)^{(2 + c)/3} (1 - h_L/\epsilon)^{-4.65} \tag{8.86}$$

The second term of Equation (8.86) is very similar to Equation (8.75) which has been derived by the channel model.

Equation (8.86) is the result of a simplified theoretical model and, therefore, needs to be validated by experimental data. However, experimental results with consistent data sets of both holdup and pressure drop are rather hard to find in the literature. Available data for Bialecki rings [Billet and Mackowiak 1984], ceramic Raschig rings [Buchanan 1967] and the structured packing Gempak 2A [Billet 1985] are plotted in Fig. 8-45. The lines represent Equation (8.86). The accordance is very good. The differences between the three different packings are caused by the first term of Equation (8.86) that is, in essence, a function of the bed porosity. This term replaces the empirical factor C_h in Equation (8.75).

Figures 8-46, 8-47, and 8-48 demonstrate the ability of Equation (8.86) to predict the pressure drop of different packings. The dry pressure drop has been adjusted to the experimental data. The parameter lines in the diagrams were determined by Equation (8.86) in connection with Equation (8.68) for the holdup. Below the loading point the parameter lines run parallel to the dry pressure drop line. Above the loading point the gradients of the parameter lines increase with increasing gas and liquid

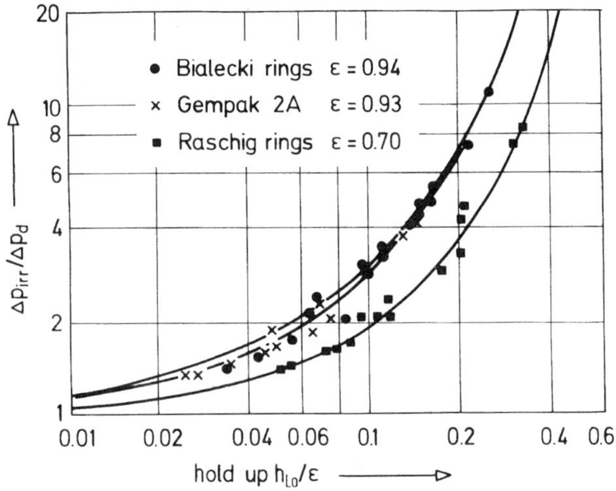

Fig. 8-45. *Ratio of the irrigated and dry pressure drop versus liquid holdup. The lines represent Equation (8.86).*

load. When the gradient becomes infinite, flooding occurs. In total, the accordance between model and experiment is very good. Existing smaller deviations can be attributed mainly to the holdup correlation.

Equation (8.86) describes the pressure drop of an irrigated bed in the whole range of operation, i.e., below and above the loading point. It also correctly describes the

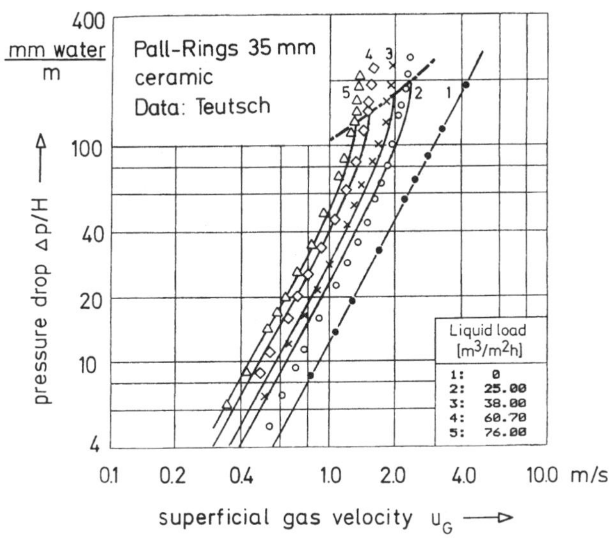

Fig. 8-46. *Comparison of Equation (8.86) with experimental data of a ceramic random packing.*

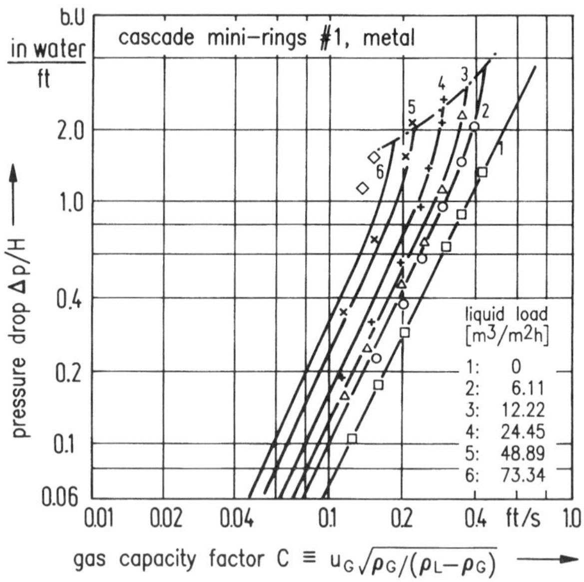

Fig. 8-47. *Comparison of Equation (8.86) with experimental data of a metal random packing.*

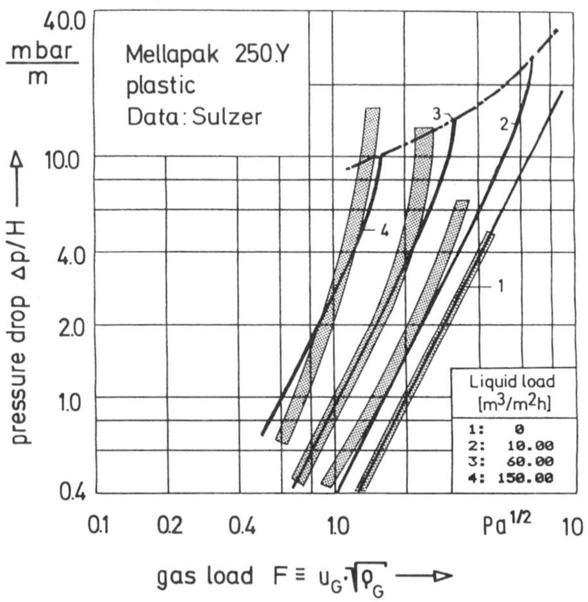

Fig. 8-48. *Comparison of Equation (8.86) with experimental data of a plastic structured packing.*

steep increase of the pressure drop near the flood point. Thus, it should be possible to determine the flood point from the pressure drop equation.

8.3.3.2.7 Determination of the Flood Point from the Pressure Drop Equation The flood point is reached when an infinitely steep increase of the pressure drop occurs with increasing gas load, e.g., [Buchanan and Dixon 1987]. Thus, if the dry pressure drop is used as a measure of the gas load, the flood point condition is:

$$\partial \Delta p_{irr} / \partial \Delta p_d = \infty \tag{8.87}$$

In order to perform the derivative $\partial \Delta p_{irr}/\partial \Delta p_d$ the relation between irrigated and dry pressure drop has to be formulated explicity. The holdup relation (8.69) has to be inserted into Equation (8.86) since it depends on the irrigated pressure drop. The result is:

$$\frac{\Delta p_{irr}}{H\rho_L g} = \frac{\Delta p_d}{H\rho_L g}\left(\frac{1-\epsilon\left(1-\frac{h_{Lo}}{\epsilon}\left(1+20\left(\frac{\Delta p_{irr}}{H\rho_L g}\right)^2\right)\right)}{1-\epsilon}\right)^{(2+c)/3} \times\left(1-\frac{h_{Lo}}{\epsilon}\left(1+20\left(\frac{\Delta p_{irr}}{H\rho_L g}\right)^2\right)\right)^{-4.65} \tag{8.88}$$

Obviously, the derivative $\partial \Delta p_{irr}/\partial \Delta p_d$ cannot be formed since the right-hand side of Equation (8.88) contains Δp_{irr}. Therefore, the flood point condition has to be expressed by the inverse of Equation (8.87):

$$\partial \Delta p_d / \partial \Delta p_{irr} = 0 \tag{8.89}$$

The derivative of Equation (8.88) is:

$$\left(\frac{\Delta p_{irr}}{H\rho_L g}\right)_f^{-2} - \frac{40\,\frac{2+c}{3}\,h_{Lo}}{1-\epsilon+h_{Lo}\left(1+20\left(\frac{\Delta p_{irr}}{H\,\rho_L\,g}\right)_f^2\right)} = 0 - \frac{186 h_{Lo}}{\epsilon-h_{Lo}\left(1+20\left(\frac{\Delta p_{irr}}{H\rho_L\,g}\right)_f^2\right)} \tag{8.90}$$

The subscript f denotes the pressure drop at the flood point. It has to be found from Equation (8.90) by iteration. In general, values of $(\Delta p_{irr}/(H\rho_L g))_f$ are in the range of 0.1 to 0.3, which agrees well with experimental results.

The procedure for flood point calculation from Equation (8.90) is as follows: First, a liquid load is chosen and the holdup h_{Lo} below loading point is evaluated from Equation (8.67) or a similar holdup relation. Then, the pressure drop at flood point is iteratively calculated from Equation (8.90). From a knowledge of $(\Delta p_{irr}/(H\rho_L g))_f$ the dry pressure drop $\Delta p_d/(H\rho_L g)$ is evaluated from Equation (8.88) and, finally, the gas load at flood point from Equation (8.73). A more straightforward procedure is the calculation of the pressure drop for constant liquid load as a function of gas load. The result should be plotted in form of the diagram in Fig. 8-46. Such a plot is very helpful for chosing the operation point that must have a sufficient distance to the flooding point. The two procedures give the same values of the flood point.

The method for flood point prediction from a sound pressure drop relation has

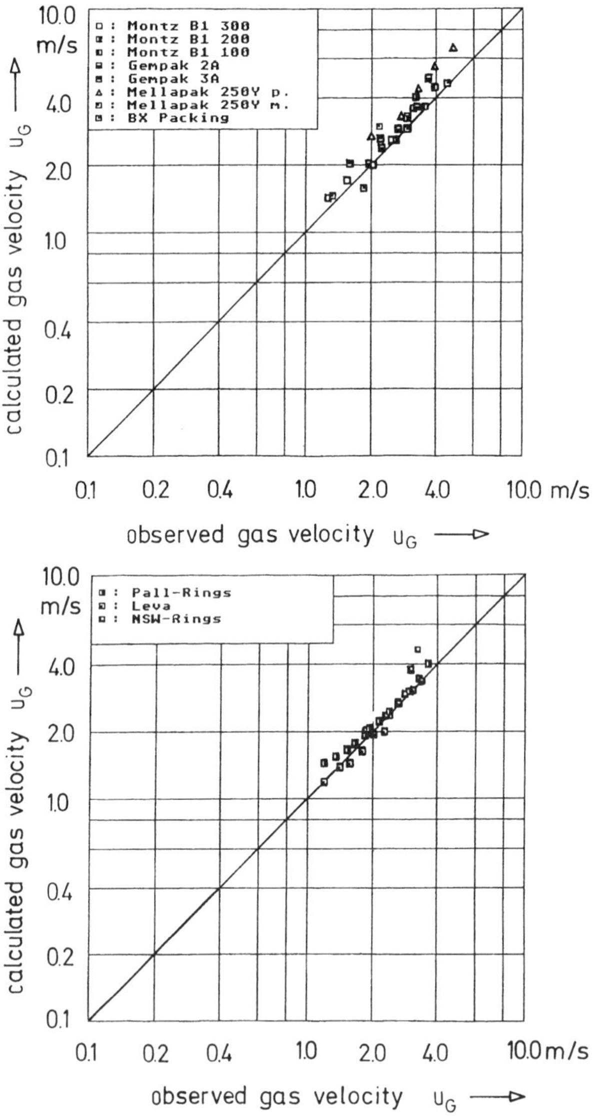

Fig. 8-49. *Parity plots of the modeled flood point and experimental data of structured and random packings.*

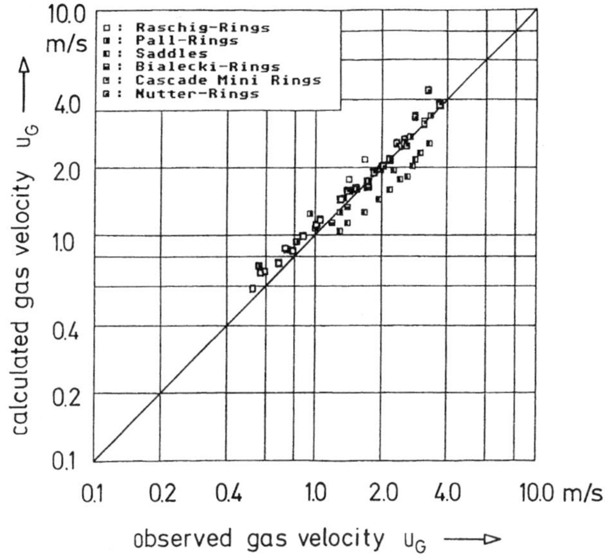

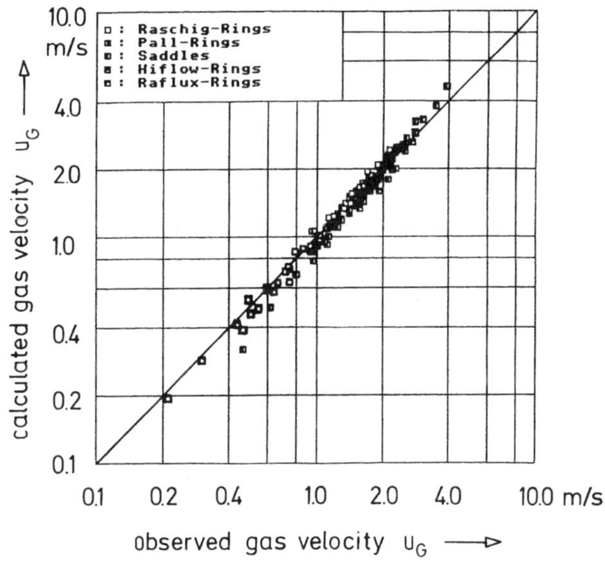

***Fig. 8-49.** (continued)*

been tested on a variety of published experimental data [Stichlmair et al. 1989]. The parity plots for structured packings and dumped metal, ceramic and plastic packings are shown in Fig. 8-49. The accordance is very good.

Figure 8-50 shows a very interesting plot of Equation (8.88). The abscissa is the liquid holdup h_{Lo} below the loading point. It is, in essence, a function of the liquid

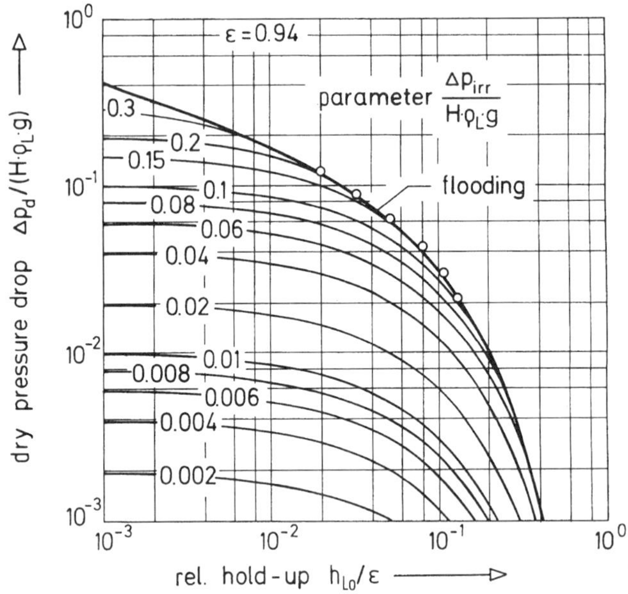

Fig. 8-50. Plot of Equation (8.87) in analogy to the [Mersmann 1965] correlation of Fig. 8-37. The data points of flooding have been taken from Fig. 8-44.

load only. The ordinate is the dimensionless dry pressure drop that depends on gas load only. Lines of constant irrigated pressure drop are included in the diagram as parameter. The flood point line has been determined from Equation (8.90). This diagram is very similar to the Mersmann diagram of Fig. 8-37. The only difference is the abscissa. However, the term used in the Mersmann diagram correlates the liquid holdup. The parameter lines in the Mersmann diagram have been developed by a large number of experimental data whereas the parameter lines in Fig. 8-50 have all been predicted by Equation (8.88).

The particle model developed by [Stichlmair et al. 1989], enables the prediction of the fluid-dynamic behavior of packed columns from a knowledge of bed porosity ϵ, specific area a and dry pressure drop (i.e., friction factor ξ_o) only. No other empirical factors depending on the type of packing are required. It is supposed in the model that the holdup increases the pressure drop in the same manner below and above the loading point. However, some experimental data, e.g., [Spiegel and Meier 1992] indicate a different behavior.

The decisive factor in calculating the fluid-dynamic behavior of an irrigated column is the liquid holdup. Published data (usually for air/water) are not sufficient to provide a reliable equation for this; especially lacking are data at low interfacial tensions and viscosities, which are typical in distillation.

8.3.3.3 Maldistribution of Liquid The aim of any packed column design is the realization of a countercurrent flow of gas and liquid in the form of a plug flow.

The term plug flow implicates no axial back mixing and uniform flow rate profiles across the column cross-section. However, this ideal flow regime is impossible to realize neither in random nor in structured packings.

Deviations from plug flow are called maldistribution. In principle, maldistribution can exist in the gas phase as well as in the liquid phase. However, the pressure drop of the gas flow flattens the flow rate profile of the gas and, consequently, the degree of maldistribution is practically quite low in the gas phase. In the liquid phase, however, the degree of maldistribution can be very high. It often causes serious problems since it adversely affects mass transfer efficiency of the column. The problem of maldistribution has attracted increased interest in recent years since packings are more and more installed in large diameter columns that increase the risk of severe maldistribution.

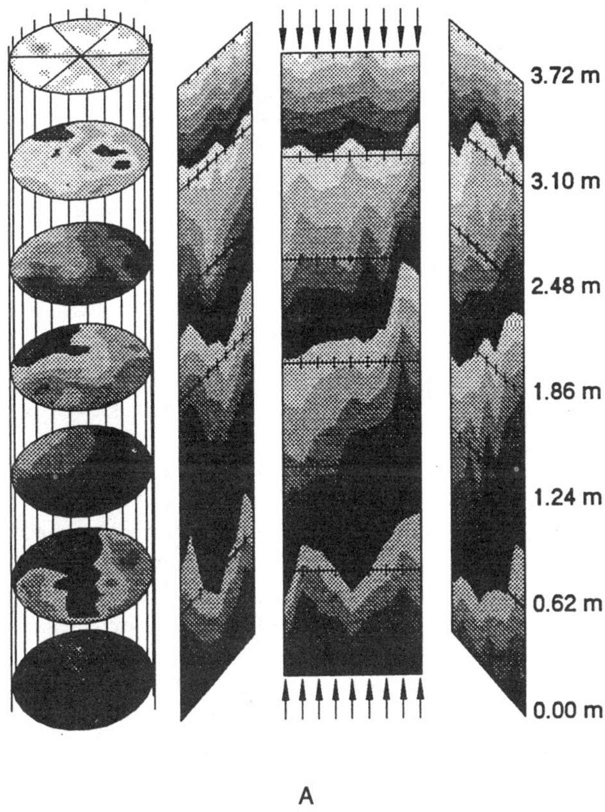

Fig. 8-51. *Maldistribution in a random packing of 35 mm metal Pall rings [Potthoff 1992]. Column diameter D_c = 0.63 m, packing height H = 3.72 m, liquid load u_L = 11.2 m/h, gas load F = 2.47 $Pa^{1/2}$. Plot of liquid temperatures at three longitudinal sections of the column. Deviations of the isotherms from horizontal lines indicate the degree of maldistribution: (A) Uniform liquid feed. Experimental value of H_{OG} = 0.36 m.*

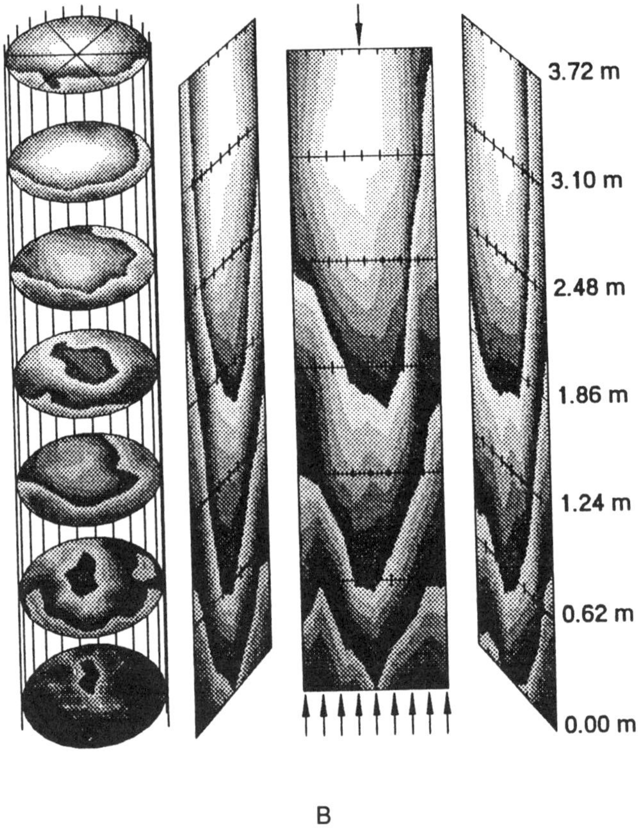

Fig. 8-51. *(B) Single point liquid feed. Experimental value of* H_{OG} *= 0.67 m.*

Present knowledge on maldistribution in packed columns is rather poor. This is true for both the degree of maldistribution that exists in a column of industrial size and the effect of maldistribution on mass transfer efficiency of the whole column. [Hurter 1893] was the first to perform maldistribution experiments on a large scale column ($1.2 \cdot 1.2\ m^2$). He used a single-liquid feed at the top of a column with a coke bed and collected the liquid at eight different locations below the bed. This technique has been used by most researchers since then, e.g., [Kirschbaum 1931, Barker et al. 1935, Scott 1935, Stanek and Kolar 1973 Dutkai, and Ruckenstein 1970, Onda et al. 1973, Hoek et al. 1986, Kourl and Sohlo 1987, Stoter et al. 1990]. Many attempts have also been made to study and model the influence of maldistribution on mass transfer efficiency of a column, e.g., Mullin 1957, Changez and Sawistowski 1963, Meier Huber 1969, Schlünder 1979, Kunesh 1987, Porter and Jones 1987, Stoter et al. 1990, Zuiderweg et al. 1993].

A simple experimental technique to answer both questions is to operate the test column with air and hot water (like a cooling tower) and to measure the liquid temperatures within the packing as well as the states of gas and liquid at both ends of the column [Stichlmair and Stemmer 1987]. Mass transfer from liquid to gas in this system results in a fall of liquid temperatures instead of liquid concentrations. Liquid temperatures can easily be measured in the packing by thermocouples. In the case of plug flow the liquid temperatures ought to be uniform in each cross section of the column. Deviations of the isotherms from horizontal lines indicate the degree of maldistribution.

Figures 8-51 to 8-53 show experimental results of experiments with a test column 0.63 m in diameter and 3.6 m in packing height [Potthoff and Stichlmair 1991, Potthoff 1992]. Approximately 450 thermo-couples have been installed among seven cross sections of the column. Lines of constant liquid temperatures (i.e., isotherms) were determined by linear interpolation of the experimental data. The isotherms are presented in three longitudinal sections of the column. The crosses mark the positions of the thermo-couples.

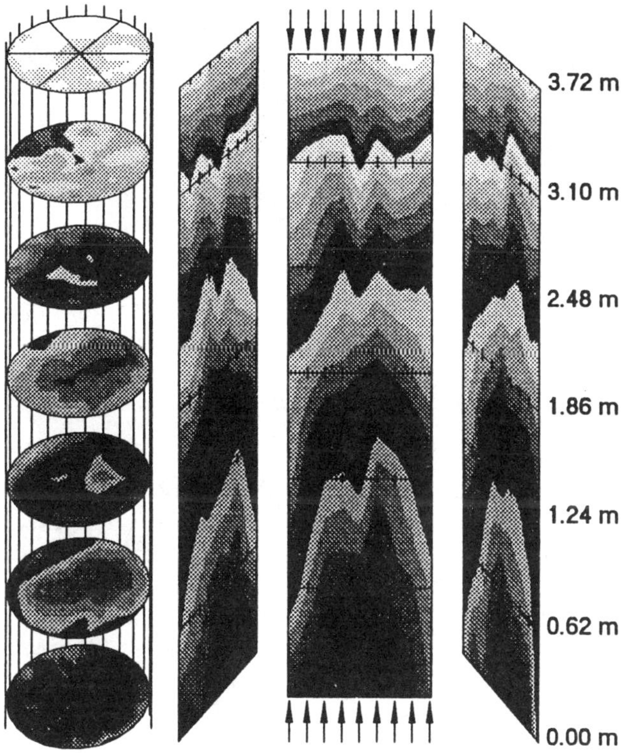

A

Fig. 8-52. Maldistribution in a random packing of ceramic torus saddles [Potthoff 1992] Column diameter D_c = 0.63 m, packing height H = 3.72 m, liquid load u_L = 6.42 m/h, gas load F = 1.39 $Pa^{1/2}$. Plot of liquid temperatures at three longitudinal sections of the column. Deviations of the isotherms from horizontal lines indicate the degree of maldistribution: (A) Uniform liquid feed. Experimental value of H_{OG} = 0.40 m.

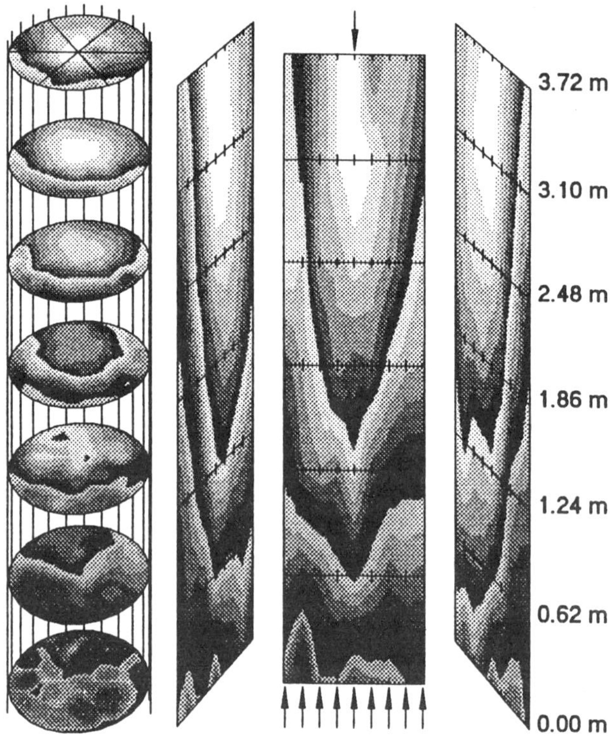

B

Fig. 8-52. *(B) single point liquid feed. Experimental value of* H_{OG} = *0.78 m.*

Figure 8-51A shows results of a metal random packing of 35 mm Pall rings. At uniform liquid distribution at the top the maldistribution develops after a liquid flow length of approximately two column diameters. Apparently, no wall flow exists, rather some lateral shifting of the liquid flow. The degree of maldistribution obviously increases as the liquid flow proceeds downwards in the bed. However, the isotherms flatten in the lowest section of the bed. This surprising behavior is caused by the mass transfer between the liquid and the gas phase. The entering gas has a state (i.e., concentration and temperature) that is uniform across the bed. Due to the non-uniform liquid temperatures the mass transfer rate is locally different, which causes the flattening of the isotherms. Thus, the highest deviations in the states of both liquid and gas exist in the middle section of the column [Zuiderweg and Hoek 1987].

An experimental result with the same packing, the same gas and liquid load but single-point liquid feed atop the packing is presented in Fig. 8-51B. The isotherms deviate strongly from horizontal lines, rather they run parallel to the column wall. Hence, most of the liquid proceeds downwards in the center of the column. Obviously, the packing has no, or a very limited, ability, for liquid distribution. These results prove the necessity of a uniform liquid distribution on top of the packing since the quality of liquid distribution on top of the bed is decisive for the flow profile within the whole bed.

Similar results of a random ceramic packing (35 mm Torus saddles) are presented in Fig. 8-52. At uniform liquid feed (Fig. 8-52A) a significant wall flow develops as the liquid proceeds downward in the packing. A totally adverse flow pattern exists at single-point liquid feed on top of the bed.

The results of tests with a structured metal packing (Ralupak) are depicted in Fig. 8-53A for uniform and in Fig. 8-53B for pointwise liquid feed. In both cases, a lateral shifting of the liquid flow profile is observed. However, at single-point liquid feed the flow pattern flattens after a flow path of approximately two column

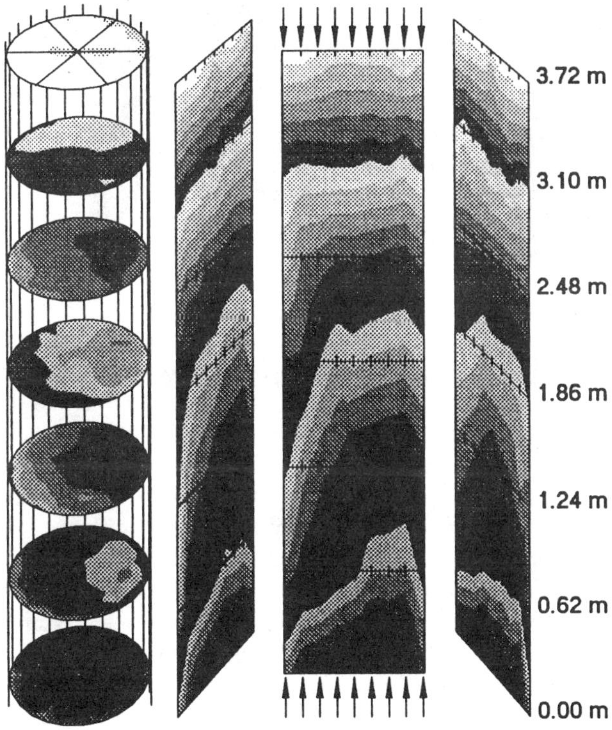

A

Fig. 8-53. Maldistribution in a structured packing [Potthoff 1992]. Column diameter D_c = 0.63 m, packing height H = 3.72 m, liquid load u_L = 11.2 m/h, gas load F = 2.43 $Pa^{1/2}$. Plot of liquid temperatures at three longitudinal sections of the column. Deviations of the isotherms from horizontal lines indicate the degree of maldistribution: (A) Uniform liquid feed. Experimental value of H_{OG} = 0.34 m.

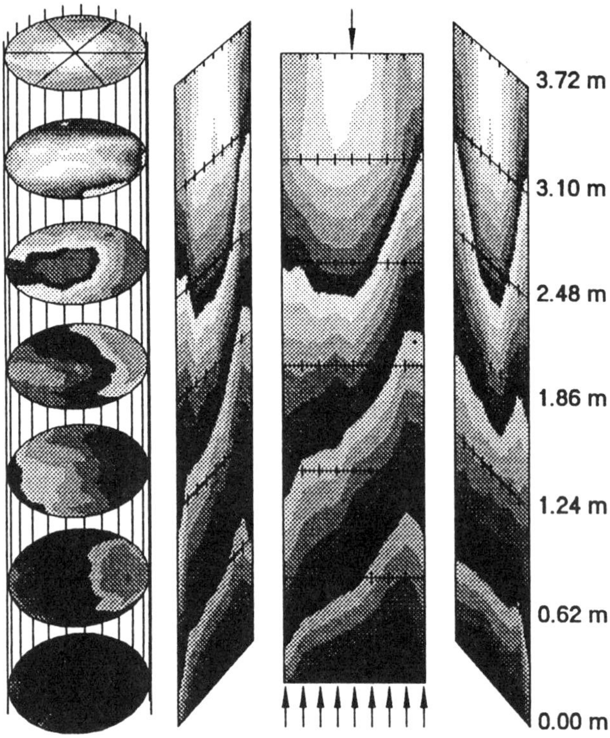

B

Fig. 8-53. *(B) Single point liquid feed. Experimental value of* $H_{OG} = 0.35$ *m.*

diameters, i.e., 1.2 m. This structured packing shows a good ability for liquid distribution. However, the flow patterns are far from that of a plug flow.

A large variety of different dumped and structured packings has been tested with this technique. In all experiments a significant degree of maldistribution was detected even at ideal liquid distribution on top of the packing. Thus, the assumption of plug flow always made in dimensioning packed columns is never correct.

In these experiments the height of a transfer unit has been determined from the measurements of gas and liquid states at both ends of the column. Plug flow has been assumed in these calculations, but this assumption is wrong. Hence, not genuine, but pseudo, values of the height of a transfer unit have been evaluated. However, for practical column dimensioning a knowledge of these pseudo transfer units is required. The values for uniform and pointwise liquid feed differ significantly. More results are presented in Chapter 8.3.4.2. These results prove the decisive influence of maldistribution on the height of a transfer unit, i.e., on the separation efficiency of a packed column.

8.3.3.4 Interfacial Area Mass transfer efficiency of a column is governed by the interfacial area within the packing and the concentration gradients between gas and liquid. A knowledge of the value of the interfacial area is an essential prerequisite for the prediction of the separation that can be achieved by the column.

Interfacial area is, first of all, a quantity of two-phase flow within the column. It is defined as the area where gas and liquid are in intimate contact. It is important to be aware that the interfacial area between gas and liquid is not identical with the specific area a of the packing. But a close relation exists between these two quantities. Furthermore, not all of the interfacial area existing in the bed is virtually effective for mass transfer. Of prime interest is the effective interfacial area a_{eff}.

Three groups of parameters that influence the effective interfacial area have to be distinguished [Schultes 1990]:

- Design parameters of the packing, i.e., specific area, porosity and shape, size and material of the packing
- Properties of the liquid, i.e., liquid density, liquid viscosity and surface tension.
- Operating parameters, i.e., liquid load, gas load and acceleration due to gravity.

Additional parameters, as liquid distributors or wall tighteners are only of influence if not properly designed.

The significance of the packing material and the liquid properties to the effective interfacial area of a vertical plate is demonstrated in Fig. 8-54. Water moves downwards on a metal plate in the form of small rivulets that might even meander on the plate. An organic liquid as heptane, however, forms a broad thin film on such a plate, see Fig. 8-54A. On a ceramic plate the differences of the flow pattern of water and heptane are much smaller, see Fig. 8-54B. Here, even water forms a rather broad film indicating a better wettability. A totally different behavior exists on a plastic plate (e.g., polypropylene), see Fig. 8-54C. The water rivulet breaks into a chain of droplets that moves downwards on the plate much faster than a rivulet or a film.

The wettability of a packing material by a given liquid is often quantified by the angle θ formed by a drop sitting on a horizontal plate. The smaller the angle θ the better the wettability. Figure 8-55 shows a plot of $\cos\theta$ versus the surface tension σ of the liquid [Zech 1978]. The plate is fully wettable if the surface tension σ of the liquid is smaller than the critical surface tension σ_c of the plate that is determined by extrapolating the parameter lines to $\cos\theta = 1$. In many empirical correlations for the effective interfacial area, the ratio σ/σ_c is included to account for the wettability of the packing material. It is, however, ambiguous as to whether or not such a behavior is observed after a long time column operation.

Several researchers have developed correlations for the effective interfacial area. Some of them are presented in Chapter 8.3.4.2. The parameters accounted for in the

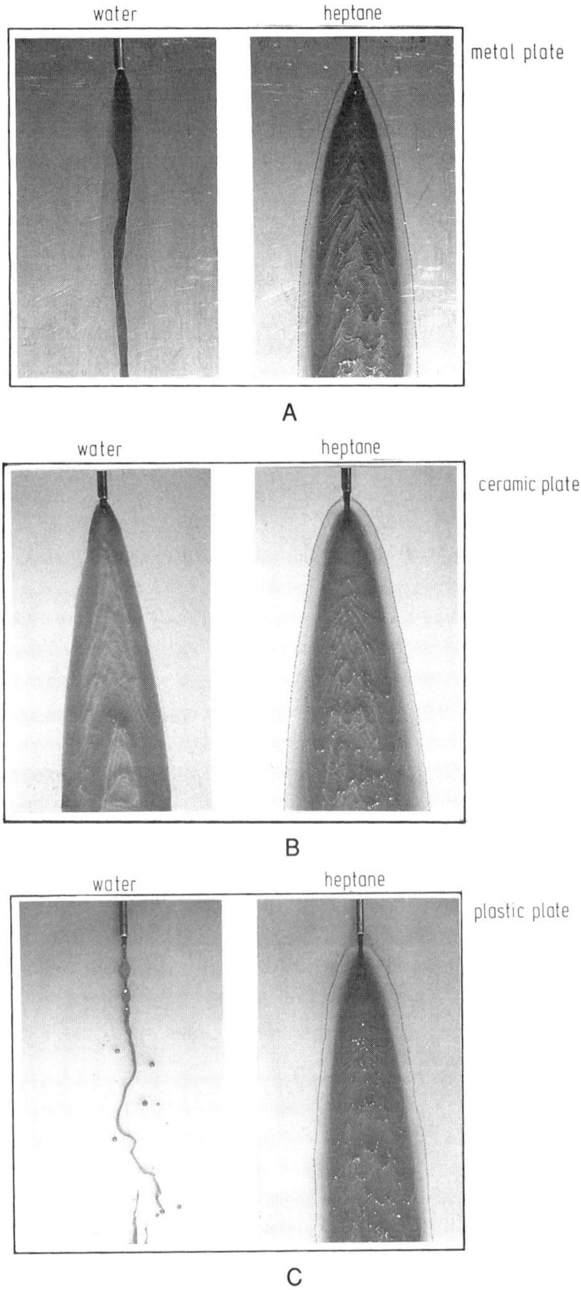

Fig. 8-54. *Flow pattern of water and heptane on a vertical plate: (A) Metal plate, (B) ceramic plate, (C) polypropylene plate.*

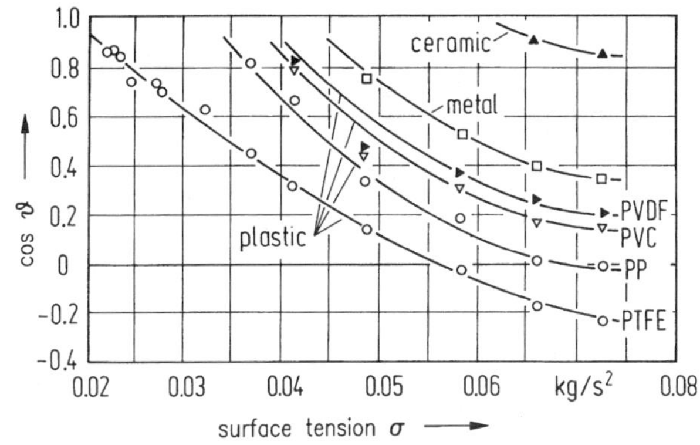

Fig. 8-55. *Wettability of several packing materials [Zech 1978].*

correlations and their exponents are listed in Table 8-10. All authors agree that the liquid load u_L is one of the major parameters with a dependence according to $a_{eff}/a \sim u_L^{0.4}$. The values of the other exponents differ rather strongly, e.g., from 0.174 to 0.95 for liquid density ρ_L and from -0.5 to $+0.392$ for liquid viscosity η_L.

Figure 8-56 presents a plot of the relative interfacial area a_{eff}/a versus the liquid load u_L [Schultes 1990]. The correlations of different researchers have been evaluated for air/water in a 35 mm metal Pall ring packing. The values of the correlations differ significantly, but the dependence on liquid load is quite uniform.

In principle, a close relationship should exist between the effective interfacial area and the liquid holdup in the bed since it is the holdup that forms the interfacial area. Recently, [Wagner et al. 1997] developed the following relation for a_{eff}/a from the particle model structure:

$$\frac{a_{eff}}{a} = \left(1 + \frac{h_L}{1 - \epsilon}\right)^{2/3} - 1 \tag{8.91}$$

Table 8-10: Exponents Used in Correlations for the Effective Interfacial Area a_{eff}/a by Various Investigators

	Exponent of					
	u_L	u_G	ρ_L	η_L	σ	g
[Bravo, Fair 1982]	0.392	0.392	—	0.392	−0.392	—
[Kolev 1973]	0.392	—	0.46	—	−0.46	0.264
[Puranik, Vogelpohl 1974]	0.307	—	0.174	−0.041	−0.182	—
[Shi, Mersmann 1985]	0.4	—	0.05	0.2	−0.15	−0.15
[Zech 1978]	0.5	—	0.95	−0.5	−0.45	0.45
[Billet, Schultes 1993]	0.4	—	0.55	0.2	−0.75	+0.45

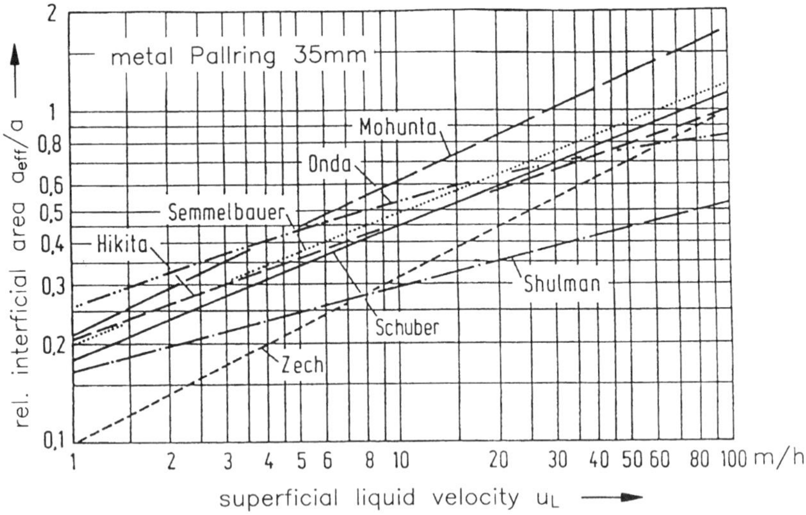

Fig. 8-56. Interfacial area of a metal Pall ring packing (d_n = 35 mm) predicted by different empirical correlations [Schultes 1990].

This relation agrees quite well with the empirical correlations of Fig. 8-56. The term $h_L/(1-\epsilon)$ describes the ratio of liquid and solid material within the bed. This ratio should be decisive for the interfacial area when the liquid holdup in the bed has the same geometric structure as the solid material.

It must be mentioned that present knowledge on interfacial area is rather poor in spite of all the research activities described here. It is not sufficient for dimensioning industrial columns with sufficient accuracy (see Chapter 8.3.4.3).

8.3.4 Mass Transfer in Packed Columns

The prediction of mass transfer in a packed column is based on the assumption that both gas and liquid move through the column bed in a type of plug flow. This means that back mixing in axial direction does not exist. Furthermore, both phases must be uniformly distributed over the mass transfer area and throughout the packed height. However, these assumptions will never be fully met in practice.

8.3.4.1 Height of a Transfer Unit From the transfer unit concept the following relationship is obtained (cf. Section 4.1.2):

$$dN_{OG} = K_{OG}/\dot{G}\ dA \tag{8.92}$$

With $dA = dH\ A_c\ a_{eff}$ (A_c is the cross-sectional area of the column and a_{eff} is the effective interfacial area of the packing) the differential height dH of the packing is:

$$dH = dN_{OG} \frac{\dot{G}}{K_{OG}\, A_c\, a_{eff}} \tag{8.93}$$

The second term of this equation is defined as the height H_{OG} of an overall gas side transfer unit:

$$H_{OG} = \frac{\dot{G}}{K_{OG}\, A_c\, a_{eff}} \tag{8.94}$$

The term $\dot{G}/A_c$ is proportional to the superficial velocity u_G of the gas. Simple transformation gives:

$$\frac{\dot{G}}{A_c} = u_G\, \rho_G/M_G \tag{8.95}$$

The overall gas phase mass transfer coefficient K_{OG} is calculated from the coefficients of the phases involved, i.e., gas phase k_G and liquid phase k_L:

$$\frac{1}{K_{OG}} = \frac{1}{k_G} + \frac{m}{k_L} \tag{8.96}$$

where m is the slope of the equilibrium curve on the y-x diagram. Furthermore:

$$k_G = \beta_G\, \rho_G/M_G \quad \text{and} \quad k_L = \beta_L\, \rho_L/M_L \tag{8.97}$$

From Equations (8.94) to (8.100) follows for the height of a transfer unit:

$$H_{OG} = \frac{u_G}{a_{eff}\, \beta_G}\left(1 + m\,\frac{\beta_G}{\beta_L}\,\frac{M_L}{M_G}\,\frac{\rho_G}{\rho_L}\right) \tag{8.98}$$

In distillation, the slope m of the equilibrium line is very small. Hence, the second term in the paranthesis of Equation (8.98) is very small indicating that most of the mass transfer resistance lies in the gas phase.

Equation (8.98) contains three unknown quantities specific to the apparatus:

- The effective interfacial area a_{eff}
- The mass transfer coefficient β_G in the gas
- The mass transfer coefficient β_L in the liquid

Relations for these three quantities have to be known for predicting the mass transfer efficiency of a packed column.

The height H of the packing required to meet the product specifications follows from:

$$H = H_{OG} \, N_{OG} \tag{8.99}$$

In this equation N_{OG} denotes the number of transfer units. It is a thermodynamic quantity that describes the difficulty of the separation from a knowledge of the system properties without any regard to the apparatus, see Chapter 4.1.3. The height of a transfer unit H_{OG} is that quantity that translates the thermodynamic quantity N_{OG} into packing height H.

Often, the concept of equilibrium stages is used instead of the concept of transfer units. Then the difficulty of the separation is expressed by the number n of equilibrium stages. The height of the packing required to perform a given separation is calculated from:

$$H = n \cdot HETP \tag{8.100}$$

The term HETP (= height equivalent to a theorectical plate) describes the mass transfer efficiency of the packing. The following relation exists, see Equation (4.11):

$$HETP = H_{OG} \cdot \frac{\ln(m/(\dot{L}/\dot{G}))}{m/(\dot{L}/\dot{G}) - 1} \tag{8.101}$$

In the case of parallel operating and equilibrium lines, i.e., $m/(\dot{L}/\dot{G}) = 1$, HETP and H_{OG} have the same value.

The key quantity for the determination of column height is the height of a transfer unit. It can be predicted by fundamental laws of mass transfer.

8.3.4.2 Mass Transfer Coefficients and Effective Interfacial Area Several correlations for the mass transfer coefficient in the gas, β_G, and in the liquid, β_L, are available in the literature. Most of these correlations have been developed from experimental mass transfer data in the form of the height of a transfer unit H_{OG}. From these data the volumetric mass transfer coefficient $a_{eff} \cdot K_{OG}$ can be determined rigorously under the assumption of plug flow within the bed. No rigorous methods are available for splitting the volumetric mass transfer coefficient into the interfacial area a_{eff} and the mass transfer coefficient K_{OG}. Hence, the researchers start either with a relation for the interfacial area and adjust the mass transfer coefficient, or with a model for the transfer coefficient and adjust the interfacial area. Thus, the correlations developed by different researchers for the mass transfer coefficients and the interfacial area establish a consistent set of relations that should never be divided. It is not allowed, for instance, to combine correlations of different authors for calculating the mass transfer in a packed column.

Most correlations for mass transfer coefficients and interfacial area have the structure of a potential equation, e.g., Equations (8.109) to (8.111). Table 8-11 lists both the parameters involved in such correlations and the values of their exponents for the gas side mass transfer coefficient β_G. Most authors agree on the exponent 0.66 of the diffusion coefficient, i.e., $\beta_G \sim D_G^{0.66}$. The exponents of the gas load u_G

Table 8-11: Exponents Used in Correlations for the Gas Side Mass Transfer Coefficient β_G by Various Investigators

	Exponent of			
	u_G	D_G	η_G	ρ_G
[Billet 1985]	0.75	0.66	−0.417	0.417
[Mersmann, Deixler 1986]	0.66	0.66	−0.33	0.33
[Onda et al. 1968]	0.7	0.66	−0.367	0.367
[Shulman, DeGouff 1952]	0.65	0.66	—	0.65
[van Krevelen 1948]	0.8	0.66	−0.467	0.467
[Zech, Mersmann 1978]	0.66	0.66	−0.33	0.33
[Billet, Shultes 1993]	0.75	0.66	−0.417	0.417

differ only in a small range from 0.66 to 0.8. A larger variation exists for the exponents of the viscosity η_G and density ρ_G. Here, the values lie in the range from 0.33 to 0.417. In spite of the good agreement of the values of the exponents the absolute values of the gas side mass transfer coefficient β_G differ significantly between the authors (see Example 8.2).

A similar listing of the liquid side mass transfer coefficient β_L is presented in Table 8-12. All authors agree on the exponent 0.5 of the diffusion coefficient, i.e., $\beta_L \sim D_L^{0.5}$. This results from the penetration model for mass transfer. The exponents on the liquid load u_L vary in a rather large range from 0.167 to 0.66. With respect to liquid density ρ_L and viscosity η_L the authors do not even agree on the sign of the exponents. Some researchers find a dependency of the transfer coefficient on surface tension σ and others do not. Hence, present knowledge on liquid side mass transfer coefficient is rather ambiguous and inconsistent. However, since mass transfer in distillation columns is dominated by the gas phase and not by the liquid phase, this poor knowledge does not count too much in most cases.

A few important correlations for the interfacial area a_{eff} and the mass transfer coefficients β_G and β_L are presented in detail in the following.

Table 8-12: Exponents Used in Correlations for the Liquid Side Mass Transfer Coefficient β_L by Various Investigators

	Exponent of					
	u_L	D_L	ρ_L	η_L	σ	g
[Billet 1985]	0.33	0.5	0.33	−0.33	—	0.33
[Schultes 1990]	0.33	0.5	0.167	−0.167	—	—
[Shi, Mersmann 1985]	0.19	0.5	0.23	−0.23	0.05	0.22
[Shulman et al. 1955]	0.45	0.5	−0.05	0.05	—	—
[Zech 1978]	0.167	0.5	−0.15	—	0.15	−0.017
[Onda et al. 1968]	0.66	0.5	0.167	−0.833	—	0.33

8.3.4.2.1 Onda Model In 1968 Onda published correlations for the effective interfacial area a_{eff} and the mass transfer coefficient in both phases [Onda et al. 1968]:

$$\frac{a_{eff}}{a} = 1 - \exp\left(-1.45\left(\frac{\sigma_c}{\sigma}\right)^{0.75} \cdot \underbrace{\left(\frac{u_L \rho_L}{a\ \eta_L}\right)^{0.1}}_{= Re_L} \cdot \underbrace{\left(\frac{u_L^2\ a}{g}\right)^{-0.05}}_{= Fr_L} \cdot \underbrace{\left(\frac{u_L^2\ \rho_L}{\sigma\ a}\right)^{0.2}}_{= We_L}\right) \tag{8.102}$$

with

Material:	Polyethylene	Polyvenylchlorid	Ceramic	Glas	Metal
σ_c:	33	40	61	73	75 mN/m

$$\beta_L = 0.0051\left(\frac{\rho_L}{g\ \eta_L}\right)^{-0.33} \cdot \underbrace{\left(\frac{\rho_L\ u_L}{a_{eff}\ \eta_L}\right)^{0.66}}_{= Re_L} \cdot \underbrace{\left(\frac{\eta_L}{\rho_L\ D_L}\right)^{-0.5}}_{= Sc_L} (a\ d_n)^{0.4} \tag{8.103}$$

$$\beta_G = C \underbrace{\left(\frac{\rho_G\ u_G}{a\ \eta_G}\right)^{0.7}}_{= Re_G} \cdot \underbrace{\left(\frac{\eta_G}{\rho_G\ D_G}\right)^{0.33}}_{= Sc_G} \cdot (a\ d_n)^{-2} \cdot (a\ D_G) \tag{8.104}$$

with $C = 2.0$ for $d_n < 15$ mm and $C = 5.23$ for $d_n > 15$ mm

Here, d_n denotes the nominal diameter of the particles as defined by the vendors. Onda's correlation is valid for dumped packings only. It is considered to give rather conservative results. Hence, the risk of undersizing the packing height is very low.

8.3.4.2.2 Bravo/Fair Correlation This correlation has been published in 1982 [Bravo and Fair 1982]. The authors developed a new equation for the effective interfacial area and used the mass transfer coefficients of the Onda correlation:

$$\frac{a_{eff}}{a} = 19.76\ \frac{\sigma^{0.5}}{H^{0.4}}\left(\frac{\eta_L\ u_L}{\sigma_L}\ \underbrace{\frac{u_G\ \rho_G}{a\ \eta_G}}_{= Re_G}\right)^{0.392} \tag{8.105}$$

The mass transfer coefficients are calculated by Ondas' correlations, i.e., Equations (8.103) and (8.104).

The Bravo-Fair correlation gives a dependence of the effective area on surface tension σ and packing height H in order to account for the influence of liquid maldistribution. These two quantities make the correlation dependent on the units. SI units have to be used, i.e., kg/s^2 for the interfacial tension σ and m for the packing height H.

Table 8-13: Selected values of the factors C_L and C_G of the [Billet and Schultes 1993] Correlation

Type	Material	C_L	C_G
Raschig ring	ceramic	1.325	0.298
Pall ring	metal	1.218	0.362
	plastic	1.000	0.398
	ceramic	1.227	0.415
Bialecki ring	metal	1.531	0.302
Berl saddle	ceramic	1.410	0.288
Intalox saddle	ceramic	1.677	0.488
Ralu pak	metal	1.334	0.385
Impuls pak	metal	0.983	0.270
Montz pak	metal	1.068	0.406

8.3.4.2.3 Billet/Schultes Correlation On a very large data base Billet developed several models for interfacial area and mass transfer coefficient. The actual version is [Billet and Schultes 1993]:

$$\frac{a_{eff}}{a} = 1.5\,(a\,d_h)^{-0.5}\underbrace{\left(\frac{u_L\,d_h\,\rho_L}{\eta_L}\right)^{-0.2}}_{=\,Re_L}\cdot\underbrace{\left(\frac{u_L^2}{g\,d_h}\right)^{-0.45}}_{=\,Fr_L}\cdot\underbrace{\left(\frac{u_L^2\rho_L\,d_h}{\sigma}\right)^{0.75}}_{=\,We_L} \tag{8.106}$$

$$\beta_L = C_L\,12^{1/6}\left(\frac{u_L\,D_L}{h_L\,d_h}\right)^{0.5} \tag{8.107}$$

$$\beta_G = C_G\,(\epsilon - h_L)^{-0.5}\left(\frac{a}{d_h}\right)^{0.5} D_G\underbrace{\left(\frac{u_G\,\rho_G}{a\,\eta_G}\right)^{0.75}}_{=\,Re_G}\underbrace{\left(\frac{\eta_G}{\rho_G\,D_G}\right)^{0.33}}_{=\,Sc_G} \tag{8.108}$$

$$\text{with } d_h = 4\,\epsilon/a \text{ and } h_L = \left(12\,\frac{\eta_L}{g\,\rho_L}\,u_L\,a^2\right)^{0.33}$$

The factors C_L and C_G depend on the type of packings, see Table 8-13.

8.3.4.2.4 Problems in Predicting Mass Transfer in Packed Columns At the present state of the art, all models for mass transfer in a packed column suppose ideal plug flow of gas and liquid within the packing. In a real column, however, the flow pattern of gas and, in particular, of liquid are much more complex. They deviate significantly from plug flow. Even in modern structured packings with a good initial liquid distribution, maldistribution cannot be completely avoided. The degree of maldistribution can be detected by the temperature measuring technique described in Chapter 8.3.3.3. All tests performed until now revealed more or less drastic deviations from plug flow. Maldistribution may become the dominant factor,

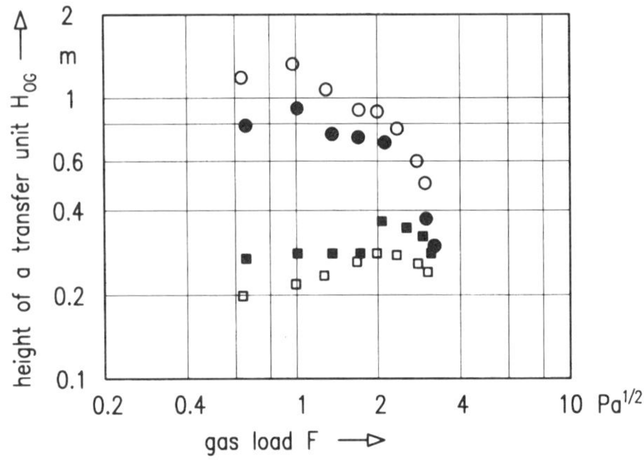

Fig. 8-57. Height of a transfer unit H_{OG} of a metal Pall ring packing at uniform (□, ■) and single point (○, ●) liquid feed [Potthoff 1992].

especially in large diameter columns operated with low liquid loads, conditions that often prevail in vacuum distillation columns.

It is a common understanding that maldistribution adversely affects mass transfer efficiency of a packed column, e.g., [Shariat and Kunesh 1995]. This statement is based on many theoretical studies and, for instance, on the experiments described in Chapter 8.3.3.3. The measurement of temperature profiles within the bed enables a deep insight into the mass transfer zone which is normally treated as a black box. This technique allows in the same experiments measurement of the maldistribution within the bed and determination of the mass transfer efficiency, expressed for instance, as height of a transfer unit H_{OG}, see Figs 8-51 to 8-53. The values of H_{OG} differ very much for uniform liquid feed (i.e., low maldistribution) and single-point liquid feed (i.e., high maldistribution). Figure 8-57 shows typical results of tests with a 35 mm metal Pall ring packing, see also Fig. 8.52. The ratio $\dot{V}_L/\dot{V}_G$ has been kept constant to hold the operation line in a parallel position to the equilibrium line. Hence, an increase of the gas load is accompanied by an increase of liquid load. The values of H_{OG} differ by a factor up to five between uniform and single-point liquid feed. At high gas and liquid loads the column has been operated above the loading point where the gas flow influences liquid flow. The gas flattens the liquid flow rate profiles, i.e., it reduces the degree of liquid maldistribution and, in turn, the differences of H_{OG} in single point and uniform liquid feed vanish, see Fig. 8-57.

Similar results of experiments with a structured packing (Ralupak) are shown in Fig. 8-58. Since this packing has a good ability for liquid distribution (see Fig. 8-53), the measured values of H_{OG} do not depend so significantly on the mode of liquid feed. However, not all structured packings have such a good ability for liquid distribution as further experiments proved [Potthoff 1992].

The influence of maldistribution on mass transfer efficiency of packed columns

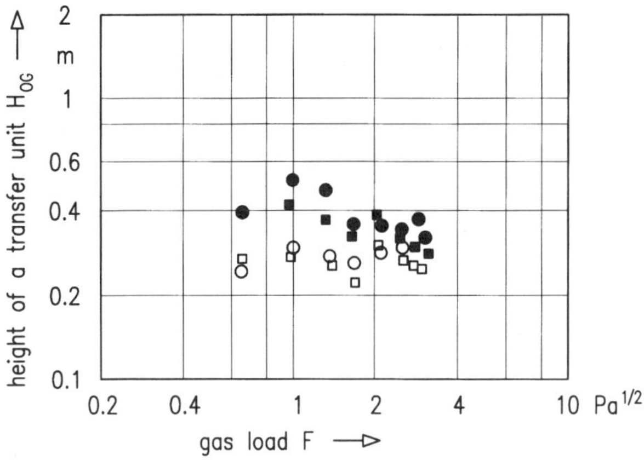

Fig. 8-58. *Height of a transfer unit H_{OG} of a structured packing (Ralupak) at uniform (□, ○) and single point (■, ●) liquid feed [Potthoff 1992].*

depends, moreover, on the distance between operating and equilibrium line in the McCabe-Thiele diagram. The smaller the distance, the higher is the adverse effect on mass transfer efficiency. A special effect exists at the column ends. The quality of the bottom fraction is deteriorated by liquid maldistribution as well as the quality of the overhead fraction by gas maldistribution. This unfavorable effect is especially strong when the required product purity is very high, since a local low purity is not compensated for by a high purity. This explains the empirical know-how that very pure bottom products are hard to achieve in packed columns. Tray columns are superior to packed columns in such separations.

It must be concluded that most published experimental data of mass transfer efficiency (*e.g.*, in the form of H_{OG} or a_{eff} K_{OG}) are not true values, but pseudo values, because they are falsified by maldistribution. Hence, the correlations for the effective interfacial area a_{eff} and the mass transfer coefficient β_G and β_L are also not true values since they include maldistribution effects. A sound mass transfer model should use true values of a_{eff}, β_G and β_L and separately account for the effect of maldistribution. However, such a model has not yet been developed.

8.3.4.3 Practical Determination of the Height of a Transfer Unit The fundamental laws of mass transfer are highly developed and their application to packed columns is, in principle, clear. But the most significant parameters, e.g., effective interfacial area and mass transfer coefficients, cannot be predicted from first principles at the present state of the art. Only empirical correlations are available, and most of these correlations have been developed by experiments in rather small-scale columns operated with air/water, a system that is not representative of distillation. Thus, it is not possible to predict mass transfer efficiency of a column with sufficient reliability.

Presently, practical dimensioning of packed columns is always based on empirical knowledge either from industrial columns operated with similar systems or from the know-how of the manufacturer of the packings. Even the methods applied to the scale up are ambigous. Hence, column dimensioning always involves some uncertainity and, in turn, rather high safety factors have to be implemented.

In dumped packings the height of a transfer unit H_{OG} strongly depends on the column size. A small-scale column is much more efficient than a large-scale column. This behavior is obviously caused by the fact that small-sized particles with a high specific area are used in small diameter columns. The larger the column diameter, the larger the packing elements and, in turn, the lower the specific area of the bed. A simple rule of thumb says that the height of a transfer unit is approximately half of the column diameter.

In modern structured packings the same geometric dimensions are applied to small-scale (approximately $D_c>0.4$ m) and large-scale columns. Hence, the dependency of mass transfer efficiency on column diameter is, if at all, quite small. A rule of thumb says that the height of a transfer unit is approximately 0.4 m to 0.6 m. In other words, a packing height of 1 m establishes a separation of 2 to 3 equilibrium stages. Separation efficiency is only slightly influenced by the gas and liquid load, provided the column is operated within the operation region, i.e., with a sufficient margin to the borders that are set by flooding and dewetting, respectively. When approaching the flood point, the height of a transfer unit increases drastically which is equivalent to a significant decrease of separation efficiency.

A large collection of published values of the height of a transfer unit H_{OG} or the height equivalent to a theoretical plate HETP is included in the books of [Kister 1992 and Billet 1995]. These collections contain data of packing manufacturers and of research institutes.

Empirical correlations for H_{OG} and HETP, respectively, that are not based on the fundamentals of mass transfer, have been developed by several authors, e.g, [Porter and Jenkins 1979, Strigle 1987]. The correlation of Strigle, based on proprietary data of Norton, Fractionation Research Inc. and published data from Billet, is:

$$\text{HETP} = \exp(\text{n} - 0.187 \ln \sigma + 0.213 \ln \eta_L) \tag{8.109}$$

This equation is the result of pure empirism and, moreover, it is not correct in the dimensions. Surface tension σ must be inserted in dyn/cm and liquid viscosity η_L in cP ($= 10^{-3}$ kg/(m s)), the values of HETP have the unit ft (feet). The only parameter specific to the packing is n. Its values lie in the range from 1.13 to 1.72 [Strigle 1987]. This correlation is valid for atmospheric distillations with surface tensions from 4 to 36 dyn/cm and liquid viscosities from 0.08 to 0.83 cP.

The confidence of all methods for predicting the mass transfer efficiency of packed columns is rather poor. It is not sufficient for a save column design. In industry there is a huge demand for more reliable mass transfer data and correlations. This led to the foundation of mass transfer research institutes, e.g., FRI (Fractionation Research Inc.) at Stillwater, Oklahoma, and SRP (Separation Research Program) at the University of Texas, Austin. Both institutions conduct mass transfer experiments with distillation systems in large diameter columns (0.4 m, 1.2 m and 2.4 m, respectively). Most of the data are confidential but some have been published, e.g., [Kunesh 1987, Shariat and Kunesh 1995, Fair, and Bravo 1990].

Example 8-2: DIMENSIONING A PACKED COLUMN

Determine the diameter D_c and the height of a transfer unit H_{OG} of a packed column for distilling the mixture ethylbenzene/styrene at a pressure p = 133.3 mbar. The internal flows are $\dot{G}$ = 0.01162 kmol/s and $\dot{L}$ = 0.01017 kmol/s.

The system properties are:

	Liquid Phase	Gas Phase
Molar mass	M_L = 105 kg/kmol	M_G = 105 kg/kmol
Density	ρ_L = 818 kg/m^3	ρ_G = 0.49 kg/m^3
Surface tension	σ = 0.025 kg/s^2	
Viscosity	η_L = $4.09 \cdot 10^{-4}$ kg/(m s)	η_G = $7.84 \cdot 10^{-6}$ kg/(m s)

Solution:

1. Column diameter

Decisive for the column diameter is in most cases the gas flow.

Assumption of the operational gas flow rate: F = 2.0 Pa$^{1/2}$ (this value needs to be checked against flooding). Superfical gas velocity: $u_G = F/\sqrt{\rho_G} = 2.0\text{Pa}^{1/2}/\sqrt{0.49\ \text{kg/m}^3} = 2.86$ m/s. Volumetric gas flow: $\dot{V}_G = \dot{G}\ M_G/\rho_G = 1.162 \cdot 10^{-2}$ kmol/s 105 kg/kmol/0.49 kg/m^3 = 2.49 m^3/s. Cross section: $A_c = \dot{V}_G/u_G$ = 2.49 m^3/s/2.86 m/s = 0.871 m^2. Column diameter:

$$D_c = \sqrt{4\ A_c/\pi} = \sqrt{4\ 0.871\ \text{m}^2/\pi} = 1.05\ \text{m}.$$

1.1. Selected column diameter: D_c = 1.1 m. Cross section: A_c = 0.95 m^2. Superficial gas velocity: u_G = 2.62 m/s. Volumetric liquid flow: $\dot{V}_L = \dot{L}\ M_L/\rho_L = 1.017 \cdot 10^{-2}$kmol/s105kg/kmol/818kg/m^3 = $1.31 \cdot 10^{-3}$ m^3/s. Superficial liquid velocity: $u_L = \dot{V}_L/A_c = 1.31 \cdot 10^{-3}$ m^3/s/ 0.95 m^2 = $1.37 \cdot 10^{-3}$ m/s.

1.2. Selection of the packing: Ceramic Pall rings d_n = 35 mm. From Table 8-6 follows: a = 165 m^2/m^3; ϵ = 0.76.

1.3. Minimum liquid load—Equation (8.65):

$$u_{L,min} = 7.7 \cdot 10^{-6} \left(\frac{\rho_L\ \sigma^3}{\eta_L^4\ g}\right)^{2/9} \left(\frac{g}{a}\right)^{1/2}$$

$$= 7.7 \cdot 10^{-6} \left(\frac{818\text{kg/m}^3(0.025\text{kg/s}^2)^3}{(4.09 \cdot 10^{-4}\text{kg/(m s)})^4 9.81\text{m/s}^2}\right)^{2/9} \left(\frac{9.81\text{m/s}^2}{165\text{m}^{-1}}\right)^{1/2} = 4.41 \cdot 10^{-4}\text{m/s}$$

This value is smaller than the actual liquid load $u_L = 1.37 \cdot 10^{-3}$m/s.

1.4. Pressure drop and holdup—Dry pressure drop from Equation (8.73):

$$\frac{\Delta p_d}{H} = \frac{3}{4}\zeta_o \frac{1-\epsilon}{\epsilon^{4.65}} \rho_G u_G^2 \frac{1}{d_p}$$

Friction factor ζ_o from equation (8.74):

$$\zeta_o = \frac{C_1}{Re} + \frac{C_2}{Re^{1/2}} + C_3.$$

From Table 8-9: $C_1 = 33$; $C_2 = 7$; $C_3 = 1.4$;

$$Re = \frac{u_G\, d_p\, \rho_G}{\eta_G} \quad \text{with } d_p = 6(1-\epsilon)/a;$$

$$d_p = 6(1-0.76)/165\text{m}^{-1} = 8.72\cdot 10^{-3}\ \text{m};$$

$$Re = \frac{2.62\text{m/s}\ 8.72\cdot 10^{-3}\text{m}\ 0.49\text{kg/m}^3}{7.84\cdot 10^{-6}\text{kg/(m s)}} = 1430;$$

$$\zeta_o = \frac{33}{1430} + \frac{7}{1430^{1/2}} + 1.4 = 1.60.$$

Slope c from equation (8.81):

$$c = \frac{1}{\zeta_o}\left(-\frac{C_1}{Re} - \frac{C_2}{2Re^{1/2}}\right) = \frac{1}{1.60}\left(-\frac{33}{1430} - \frac{7}{2\ 1430^{1/2}}\right) = -0.072;$$

$$\frac{\Delta p_d}{H\,\rho_L\, g} = \frac{3}{4}\,1.60\frac{1-0.76}{0.76^{4.65}}\ \frac{0.49\text{kg/m}^3}{818\text{kg/m}^3\ 9.81\text{m/s}^2}(2.62\text{m/s})^2\frac{1}{8.72\cdot 10^{-3}\text{m}} = 0.05$$

Holdup below loading point from Equation (8.68):

$$h_{Lo} = 0.93\left(\frac{u_L^2 a}{g}\right)^{1/6}\left(\frac{\eta_L^2 a^3}{\rho_L^2\, g}\right)^{1/10}\left(\frac{\sigma\, a^2}{\rho_L\, g}\right)^{1/8}$$

$$= 0.93\left(\frac{(1.37\cdot 10^{-3}\text{m/s})^2\ 165\text{m}^{-1}}{9.81\text{m/s}^2}\right)^{1/6}\left(\frac{(4.09\cdot 10^{-4}\text{kg/(m s)})^2\ (165\text{m}^{-1})^3}{(818\text{kg/m}^3)^2\ 9.81\text{m/s}^2}\right)^{1/10}$$

$$\cdot\left(\frac{0.025\text{kg/s}^2(165\text{m}^{-1})^2}{818\text{kg/m}^3\ 9.81\text{m/s}^2}\right)^{1/8} = 0.0246.$$

Holdup above loading point from Equation (8.69):

$$h_L = h_{Lo}\cdot(1 + 20(\Delta p_{irr}/H\,\rho_L\, g)^2).$$

Since the holdup h_L depends on the pressure drop it can only be determined iteratively.

1. Iteration: Assumption $\Delta p_{irr} = \Delta p_d$;

$$h_L = h_{Lo} \cdot \left(1 + 20\left(\frac{\Delta p_{irr}}{H\,\rho_L\,g}\right)^2\right) = 0.0246 \cdot (1 + 20\,(0.05)^2) = 0.0258.$$

Equation (8.86):

$$\frac{\Delta p_{irr}}{H\rho_L g} = \frac{\Delta p_d}{H\rho_L g}\left(\frac{1 - \epsilon(1 - h_L/\epsilon)}{1 - \epsilon}\right)^{(2 + c)/3} (1 - h_L/\epsilon)^{-4.65}$$

$$= 0.05\left(\frac{1 - 0.76\,(1 - 0.0258/0.76)}{1 - 0.76}\right)^{(2 - 0.072)/3} (1 - 0.0258/0.76)^{-4.65}$$

$$= 0.0627$$

2. Iteration:

$$h_L = h_{Lo}\cdot(1 + 20\left(\frac{\Delta p_{irr}}{H\rho_L g}\right)^2) = 0.0246\cdot(1 + 20\,(0.0627)^2) = 0.0265;$$

$$\frac{\Delta p_{irr}}{H\rho_L g} = 0.05\left(\frac{1 - 0.76\,(1 - 0.0265/0.76)}{1 - 0.76}\right)^{(2-0.072)/3} (1 - 0.0265/0.76)^{-4.65}$$

$$= 0.0631.$$

A third iteration gives the same result.

1.5. Check of flooding—Equation (8.90):

$$\left(\frac{\Delta p_{irr}}{H\rho_L g}\right)_f^{-2} = \frac{40\frac{2 + c}{3}h_{Lo}}{1 - \epsilon + h_{Lo}(1 + 20\left(\frac{\Delta p_{irr}}{H\rho_{Lg}}\right)_f^2)} + \frac{186\,h_{Lo}}{\epsilon - h_{Lo}(1 + 20\left(\frac{\Delta p_{irr}}{H\rho_{Lg}}\right)_f^2)}$$

Solution by iteration. Start at the right-hand side of Equation (8.90) with the actual pressure drop $\Delta p_{irr}/H\rho_L g = 0.0631$.

1. Iteration:

$$\left(\frac{\Delta p_{irr}}{H\rho_L g}\right)_f^{-2} = \frac{40\frac{2 - 0.072}{3}0.0246}{1 - 0.76 + 0.0246(1 + 20(0.0631)_f^2)}$$

$$+ \frac{186\;0.0246}{0.76 - 0.0246(1 + 20(0.0631)_f^2)} = 8.61$$

$$\left(\frac{\Delta p_{irr}}{H\,\rho_L g}\right)_f = \frac{1}{8.61^{1/2}} = 0.341$$

2. Iteration:

Start at the right hand side of Equation (8.90) with the result of the 1. iteration: The result of the second iteration is: $(\Delta p_{irr}/H\rho_L g)_f = 0.339$. A third iteration gives the same result. The pressure drop at the flooding point is much higher than the actual pressure drop. Hence, the operating point is far below flooding.

Alternative determination of the pressure drop:

Diagram of Mersmann in Fig. 8-37. Abscissa:

$$u_L\left(\frac{\eta_L}{\rho_L g^2}\right)^{1/3}\frac{1-\epsilon}{\epsilon^2 d_p}$$

$$= 1.37\cdot 10^{-3}\text{m/s}\left(\frac{4.09\cdot 10^{-4}\text{kg/(m s)}}{818\text{kg/m}^3(9{,}81\text{m/s}^2)^2}\right)^{1/3}\frac{1-0.76}{0.76^2 8.72\cdot 10^{-3}\text{m}} = 1.13\cdot 10^{-4}$$

Ordinate: $\dfrac{\Delta p_d}{H\rho_L g} = 0.05$

The value of the pressure drop follows from the parameter lines in the diagram: $\Delta p_{irr}/H\,\rho_L\,g = 0.065$. This value is in good accordance with the calculated one. The operating point lies very close by the loading point line in Fig. 8-37, i.e., far below flooding.

2. Mass transfer: Mass transfer is expressed by the height of a transfer unit H_{OG}. Equation (8.98):

$$H_{OG} = \frac{u_G}{a_{eff}\,\beta_G}\left(+ m\frac{\beta_G}{\beta_L}\frac{M_L}{M_G}\frac{\rho_G}{\rho_L}\right)$$

2.1. Interfacial area—Onda model, Equation (8.102):

$$\frac{a_{eff}}{a} = 1 - \exp(-1.45\left(\frac{\sigma_c}{\sigma}\right)^{0.75}\cdot\left(\frac{u_L\rho_L}{a\,\eta_L}\right)^{0.1}\cdot\left(\frac{u_L^2 a}{g}\right)^{-0.05}\cdot\left(\frac{u_L^2\rho_L}{\sigma_L a}\right)^{0.2})$$

$$= 1 - \exp\left(-1.45\left(\frac{0.061}{0.025}\right)^{0.75}\cdot\left(\frac{1.37\cdot 10^{-3}\text{m/s}\,818\text{kg/m}^3}{165\text{m}^{-1}\,4.09\cdot 10^{-4}\text{kg/(m s)}}\right)^{0.1}\right.$$

$$\left.\cdot\left(\frac{(1.37\cdot 10^{-3}\text{m/s})^2\,165\text{m}^{-1}}{9.81\text{m/s}^2}\right)^{-0.05}\cdot\left(\frac{(1.37\cdot 10^{-3}\text{m/s})^2\,818\text{kg/m}^3}{0.025\,\text{kg/s}^2\,165\text{m}^{-1}}\right)^{0.2}\right)$$

$$= 0.727;\ a_{eff} = 0.727\,a = 0.727\,165\,\text{m}^{-1} = 120\,\text{m}^{-1}.$$

2.2. Mass transfer coefficients—Equation (8.104):

$$\beta_G = C\left(\frac{\rho_G u_G}{a\,\eta_G}\right)^{0.7}\cdot\left(\frac{\eta_G}{\rho_G D_G}\right)^{0.33}\cdot(a\,d_n)^{-2}\cdot(a\,D_G);$$

$$D_G = 2.82{\cdot}10^{-5}\mathrm{m^2/s} \quad \text{(from equation of [Fuller et al. 1966]);}$$

$$\beta_G = 5.23\left(\frac{0.49\mathrm{kg/m^3}\,2.62\mathrm{m/s}}{165\mathrm{m^{-1}}\;7.84{\cdot}10^{-6}\mathrm{kg/(m\;s)}}\right)^{0.7}\cdot\left(\frac{7.84{\cdot}10^{-6}\mathrm{kg/(m\;s)}}{0.49\mathrm{kg/m^3}\;2.82{\cdot}10^{-5}\mathrm{m^2/s}}\right)^{0.33}$$

$$\cdot(165\mathrm{m^{-1}}\;0.035\mathrm{m})^{-2}\cdot(165\mathrm{m^{-1}}\;2.82{\cdot}10^{-5}\mathrm{m^2/s}) = 7.56{\cdot}10^{-2}\;\mathrm{m/s}$$

Equation (8.103):

$$\beta_L = 0.0051\left(\frac{\rho_L}{g\;\eta_L}\right)^{-0.33}\cdot\left(\frac{\rho_L\;u_L}{a^{eff}\;\eta_L}\right)^{0.66}\cdot\left(\frac{\eta_L}{\rho_L\;D_L}\right)^{-0.5}(a\;d_n)^{0.4}$$

$$D_L = 3.44{\cdot}10^{-9}\mathrm{m^2/s} \quad \text{(from equation of [Wilke, Chang 1955])}$$

$$\beta_L = 0.0051\left(\frac{818\mathrm{kg/m^3}}{9.81\mathrm{m/s^2}\;4.09{\cdot}10^{-4}\mathrm{kg/(m\;s)}}\right)^{-0.33}\cdot\left(\frac{818\mathrm{kg/m^3}\;1.37{\cdot}10^{-3}\mathrm{m/s}}{120\mathrm{m^{-1}}\;4.09{\cdot}10^{-4}\mathrm{kg/(m\;s)}}\right)^{0.66}$$

$$\cdot\left(\frac{4.09{\cdot}10^{-4}\mathrm{kg/(m\;s)}}{818\mathrm{kg/m^3}\;3.44{\cdot}10^{-9}\mathrm{m^2/s}}\right)^{-0.5}(165\;\mathrm{m^{-1}}\;0.035\;\mathrm{m})^{0.4} = 1.17{\cdot}10^{-4}\;\mathrm{m/s}$$

2.3. Height of a gas side transfer unit—Equation (8.98):

$$H_{OG} = \frac{u_G}{a_{eff}\,\beta_G}\left(1 + m\frac{\beta_G}{\beta_L}\frac{M_L}{M_G}\frac{\rho_G}{\rho_L}\right)$$

$$= \frac{2.62\mathrm{m/s}}{120\mathrm{m^{-1}}7.56{\cdot}10^{-2}\mathrm{m/s}}\left(1 + 1\frac{7.56{\cdot}10^{-2}\mathrm{m/s}}{1.17{\cdot}10^{-4}\mathrm{m/s}}\frac{105\mathrm{kg/kmol}}{105\mathrm{kg/kmol}}\frac{0.49\mathrm{kg/m^3}}{818\mathrm{kg/m^3}}\right)$$

$$= 0.289\mathrm{m}\,(1 + 0.387) = 0.40\,\mathrm{m}$$

Packing height: $H = H_{OG}\,N_{OG} = 0.40\mathrm{m}\;N_{OG}$. The number of transfer units N_{OG} follows from the thermodynamic calculation of the process.

Application of other correlations leads to following results:

model	a_{eff}/a	β_G	β_L	H_{OG}
[Onda]	0.727	0.0756 m/s	$1.17{\cdot}10^{-4}$ m/s	0.40 m
[Billet/Schultes]	0.49	0.19 m/s	$1.69{\cdot}10^{-4}$ m/s	0.285 m
[Bravo/Fair] (with H = 2 m)	0.55	0.0756 m/s	$1.17{\cdot}10^{-4}$ m/s	0.53 m

APPENDIX: PRESSURE DROP IN PACKED BEDS

Fluid flow through a collective (i.e., swarm) of particles is a fundamental process of two-phase flow. The pressure drop caused by the flow resistance of the bed depends significantly on the porosity of the bed. The higher the porosity the lower is the pressure drop.

Different from fixed beds, fluidized beds allow the variation of the bed porosity over a wide range, i.e, from 0.4 to 1. Hence, experiments in fluidized beds give a deep insight into the influence of bed porosity on pressure drop. Many researchers studied the fluid-dynamics of fluidized beds e.g., [Wilhelm, and Kwauk 1948, Richardson, and Zaki 1954, Loeffler, and Ruth 1959, Capes, and Mcllhinney 1968, Martin et al. 1981]. A prime objective of their work was the development of a relation between superficial velocity u_s (the subscript stands for swarm) and porosity ϵ of the bed. Well accepted is the relation of Richardson and Zaki, developed as early as in 1954:

$$u_s = u_o \epsilon^n \quad \text{with } n = f(Re) \tag{8A.1}$$

The term u_o denotes the terminal velocity of a single particle (in most experiments spheres). The exponent n depends on the Reynolds-Number as is shown in Fig. 8A-1. The value of the exponent in laminar flow, $n_l = 4.65$, is approximately twice the value of turbulent flow, $n_t = 2.39$.

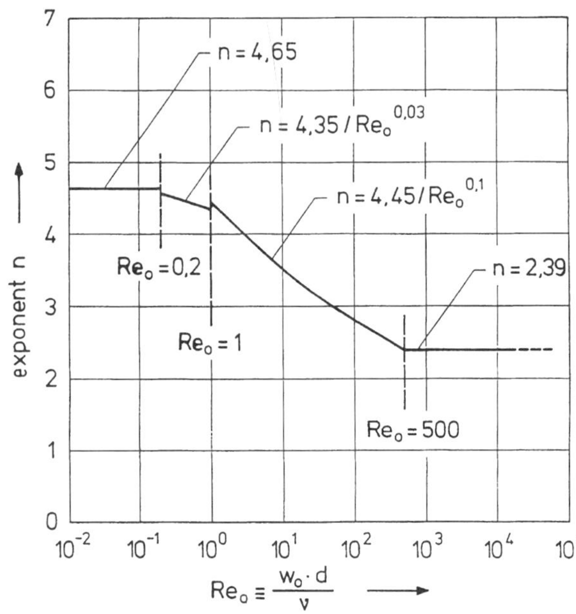

Fig. 8A-1. Dependence of the exponent n of Equation (8A.1) on the Reynolds-Number according to [Richardson and Zaki 1954].

A more straightforward relation is obtained when the influence of the bed porosity is correlated with the friction factor (and not with the velocity) since the velocity u depends on the friction factor ξ in a different manner ($u \sim \xi^{-1}$ in laminar flow and $u \sim \xi^{-1/2}$ in turbulent flow). Hence, Equation (8A.1) can be transferred into the following form:

$$\zeta_s = \zeta_o \, \epsilon^{-4.65} \quad \text{with } \zeta_o = 24/\mathrm{Re} + 4/\mathrm{Re}^{1/2} + 0.4 \tag{8A.2}$$

The friction factor ζ_s of a particle in a swarm depends on both the Reynolds-Number and the porosity. However, the dependency on the Reynolds-Number is attributed to the friction factor ζ_o of a single particle that is well known. The porosity term in Equation (8A.2) does not depend on the Reynolds-Number. The exponent -4.65 is valid for all flow regimes, i.e., for laminar flow, for turbulent flow and for the transition regime between them. Hence, Equation (8A.2) is more fundamental than Equation (8A.1).

Knowledge of the friction factor ζ_s of a particle in a swarm allows calculating the fluidization velocity u_s by equating the friction force and the weight minus buoyancy forces:

$$\frac{d^2 \, \pi}{4} \, \zeta_s \, \frac{\rho_c}{2} \, u_s^2 = \frac{d^3 \, \pi}{6} \, \Delta\rho \, g \tag{8A.3}$$

Here, ρ_c denotes the density of the continuous phase (i.e., the fluid) and $\Delta\rho$ the density difference of the dispersed (particles) and continuous phase.

Figure 8A-2 presents a comparison of experimental data (points) of several authors and calculated values, Equations (8A.2) and (8A.3) of the velocity ratio u_s/u_o under a wide variation of the Reynolds-Number ($\mathrm{Re}_o = 0.49$ to $\mathrm{Re}_o = 7150$). The porosity of the bed has been varied from $\epsilon = 0.4$ to $\epsilon = 1$. The accordance between the data of fluidized beds and Equation (8A.2) is excellent.

The fluidization of the whole bed of the height H is described by equating the pressure drop of the bed and the weight of the particles minus buoancy:

$$\Delta p \, A = A \, H \, (1 - \epsilon) \, \Delta\rho \, g \tag{8A.4}$$

Here, A denotes the cross-section of the bed and $(1 - \epsilon)$ the volume fraction of the particles. Eliminating $\Delta\rho \, g$ from Equations (8A.3) and (8A.4) yields:

$$\frac{\Delta p}{H} = \frac{3}{4} \, \zeta_s \, (1 - \epsilon) \, \rho_c \, u^2 \frac{1}{d} \tag{8A.5}$$

Replacing ζ_s by Equation (8A.2) leads to:

$$\frac{\Delta p}{H} = \frac{3}{4} \, \zeta_o \, \frac{1 - \epsilon}{\epsilon^{4.65}} \, \rho_c \, u^2 \, \frac{1}{d} \tag{8A.6}$$

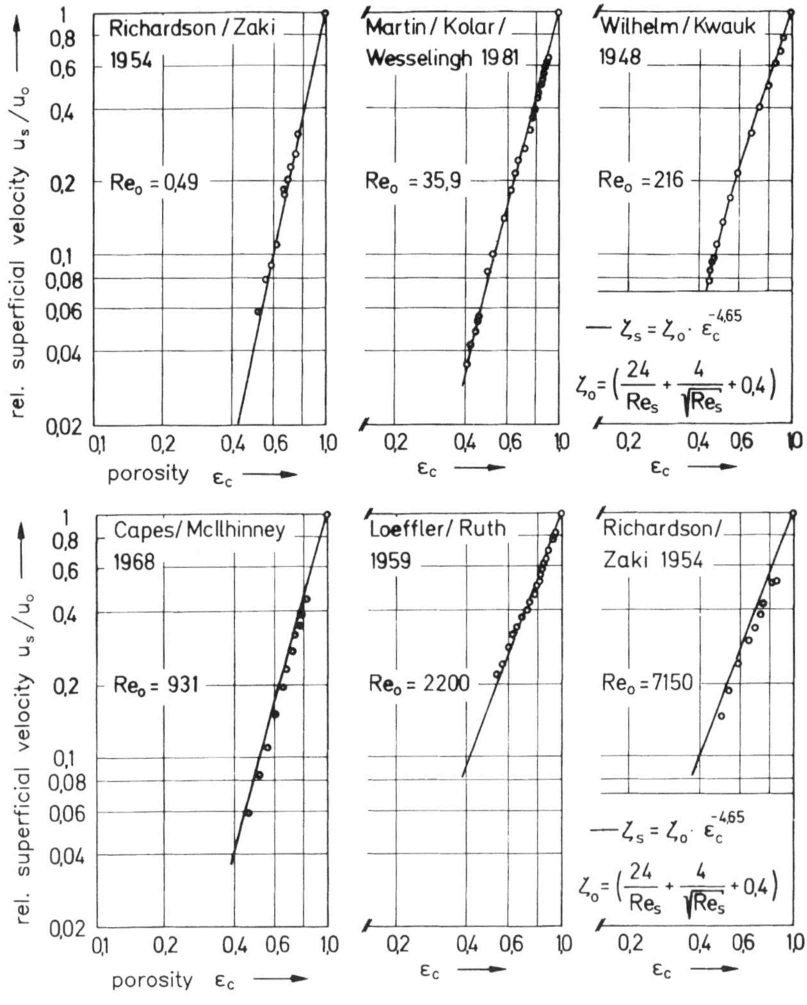

Fig. 8A-2. *Comparison of experimental fluidization data with Equation (8A.3).*

This equation has been derived for fluidized beds. The porosity term $(1 - \epsilon)/\epsilon^{4.65}$ is based on a large number of experimental data taken under a wide variation of the porosity. Hence, the porosity term is well proven.

However, the question of whether or not Equation (8A.6) derived for fluidized beds can be applied to fixed beds is still to be answered. Figure 8A-3 shows a large collection of pressure drop data of spheres and granulates in a fixed bed presented by [Coulson and Richardson 1976]. The line in the diagram represents Equation (8A.6), derived for fluidized beds. There is an excellent accordance. [Rumpf and Gupte 1971] published data of the pressure drop in a fixed bed of spheres (Fig. 8A-4). They managed to vary the bed porosity from $\epsilon = 0.366$ to $\epsilon = 0.640$ by shrinking

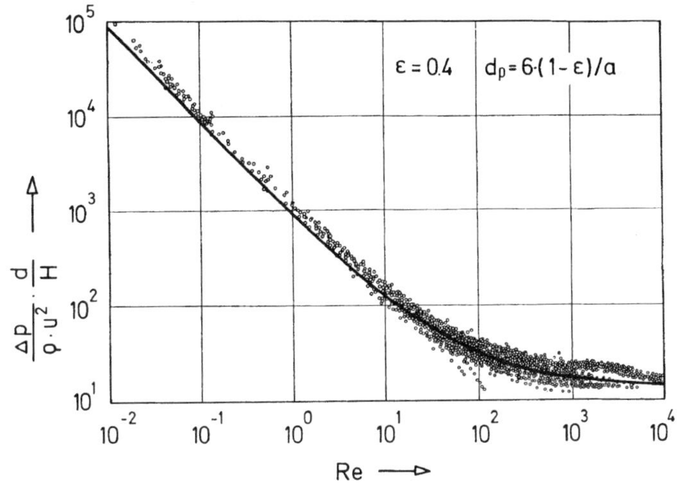

Fig. 8A-3. Pressure drop data of fixed beds of spheres and granulates according to [Coulson, and Richardson 1976]. The line represents Equation (8A.6).

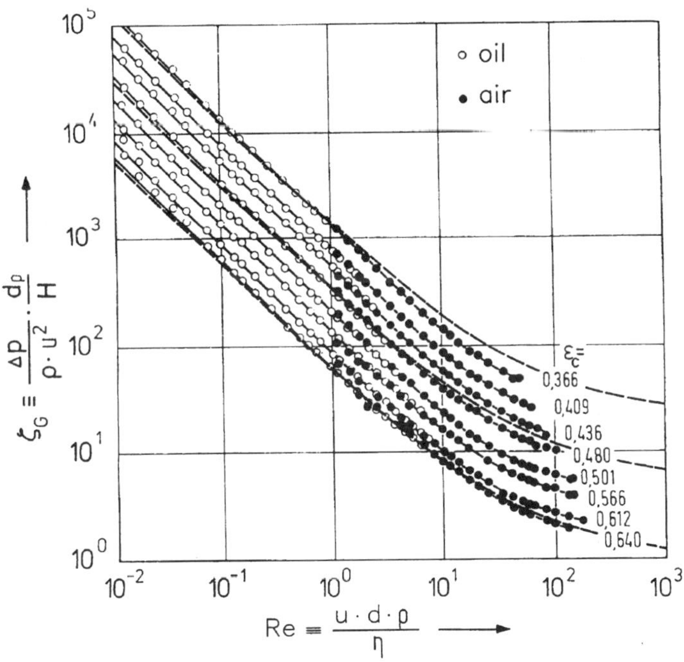

Fig. 8A-4. Pressure drop data of a fixed bed of spheres according to [Rumpf and Gupte 1971]. The porosity of the bed has been varied from $\epsilon = 0.366$ to $\epsilon = 0.640$. The dashed lines represent Equation (8A.6).

the plastic particles of the bed. The dashed lines in Fig. 8A-4 represent Equation (8A.6). Here too, the agreement between the experimental data and Equation (8.73) is excellent. Hence, Equation (8A.6) is valid for both fluidized beds and fixed beds and can consequently, be applied to column packings.

LITERATURE

Trays

Andrew, S.P.S.: ''Hydrodynamics of a Sieve Plate at High Liquid and Vapour Throughputs,'' *Ind. Chem. Eng.*, Symp. Series 32 (1969) No. 2, pp. 49–53.

Ashraf, F.A.; Cubbage, T. L.; and Huntington, R. L.: ''Entrainment in Oil Absorbers,'' *Ind. Eng. Chem* 26, (1934), pp. 1068–1072.

Barker, P. E. and Choudhury M. H.: ''Performance of Bubble-Cup Trays,'' *British Chem. Eng.* Vol. 4 (1959) pp. 348–350.

Bernard, J. and Sargent, R.: ''The Hydrodynamic Performance of a Sieve-Plate Distillation Column,'' *Trans. Inst. Chem. Eng.* Vol. 44 (1966) pp. T314–T327.

Bell, R. L.: ''Experimental Determination of Residence Time Distributions on Commercial Scale Distillation Trays Using a Fiber Optic Technique,'' *AIChE-J.* Vol. 18 (1972) No. 3, pp. 491–497.

Calcaterra, R. J.; Nicholls, C. W.; and Weber, J. H.: ''Free and Captured Entrainment and Plate Spacing in a Perforated Tray Column,'' *Br. Chem. Eng.* Vol. 13 (1968) No. 9, pp. 1294–1297.

Grovior, R. D. G., PhD Thesis University of Michigan Ann Arbor 1957

Fair, J. R. and Matthews, R. L.: ''Better Estimate of Entrainment from Bubble-Cup Trays,'' *Pet. Refin.* 37 (1958) No. 4, pp. 153–158.

Fair, J.: ''How to Predict Sieve Tray Entrainment and Flooding,'' *Petro/Chem Eng.* 33 (1961) No. 10, pp. 45–52.

''Final Report of the University of Delaware'': *Am. Inst. Chem. Eng.*, 1958.

''Final Report of the University of Michigan'': *Am. Inst. Chem. Eng.*, 1960.

Friend, L.; Lemieux, E. J.; and Schreiner, W. C.: ''New Data on Entrainment from Perforated Trays at Close Spacings,'' *Chem. Engineering* 31 (1960) pp. 101–108.

Fuller, E. N.; Schettler, P. D.; and Gieddings, J. C.: A New Method for Prediction of Binary gas-phase Diffusion Coefficients Ind. Eng. Chem. 58 (1966) No. 5, pp. 19–27

Gerster, J. A.; Hill, A.B.; Hochgraf, N.N.; and Robinson, D.G.: ''Tray Efficiencies in Distillation Columns,'' ''Final Report of the University of Delaware (1958),'' *Amer. Inst. Chem. Eng.*, 1958.

Gilliland, E.R.: Diffusion Cofficients in gaseous Systems, Ind. Eng. Chem. 26 (1934), No. 6, pp. 681–685.

Glitsch, Inc.: *Ballast Tray Design Manual.* Bulletin 4900, 4th Ed., Dallas, Texas, 1974.

Hunt, C. D. A.; Hanson, D. N.; and Wilke, C. R.: ''Capacity Factors in the Performance of Perforated-Plate Columns,'' *AIChE-J.* Vol. 1 (1955) No. 4, pp. 441–451.

Holbrook, G.E. and Baker, E. M.: ''A Study of Entrainment in a Bubble Cap Distillation Column,'' *AIChE-J.* Vol. 30 (1934) pp. 520–545.

Hoppe, K.; Krueger, G.; Bahr, A.; and Hentschel M.: ''Arbeitsbereich Druckverlust, Trennwirkung und Kosten des Hochleistungsstreckmetallbodens System Germania,'' *Chem. Techn.* Vol. 21 (1969) No. 9, pp. 536–542.

Hoppe, K. and Mittelstrass, M.: *Grundlagen der Dimensionierung von Kolonnenboeden.* Theodor Steinkopf, Dresden, 1967.

Huang, C.-J. and Hodson; J. R.: ''Perforated Trays,'' *Pet. Refin.* 37 (1958) No. 2, pp. 104–118.

Iyer, S. Rama and Murti, P. S.: ''Longitudinal Mixing in Sieve Plate Columns,'' *Indian J. Technol.* Vol. 3 (1965) No. 3, pp. 75–78.

Kastánek, F.: ''Efficiencies of Different Types of Distillation Plates,'' *Collection Czechoslov. Chem. Commun.* Vol. 35 (1970) pp. 1170–1187.

Kemp, H. S. and Pyle, C.: ''Hydraulic Gradient Across Various Bubble-Cap Plates,'' *Chem. Eng. Progr.* Vol. 45 (1949) No. 7, pp. 435–451.

Kirschbaum, E.: *Destillier- und Rektifiziertechnik.* Springer-Verlag, Berlin, 1950, p. 233.

Kirschbaum, E.: *Destillier- und Rektifiziertechnik.* Springer-Verlag, Berlin, 1960, p. 275.

Koch International Bulletin KT-5 (1968)

Kolmogoroff, A. N.: *Sammelband zur statistischen Theorie der Turbulenz.* Akademie Verlag, Berlin, 1958, pp. 151–155.

Kunesh, J. G.; Kister, H. Z.; Lockett, M. J.; and Fair, J. R.: ''Distillation: Still Towering over other Options,'' *Chemical Engineering Progress* No. 10 (1995) pp. 43–54.

Lewis, W. K.: ''Rectification of Binary Mixtures,'' *Ind. Eng. Chem.* 28 (1936) No. 4, pp. 399–402.

Lockett, M. J.; Spiller, G. T.; and Porter; K. E.: ''The Effect of the Operating Regime on Entrainment from Sieve Plates,'' *Trans. Inst. Chem. Eng.* Vol. 54 (1976) pp. 202–204.

Lockett, M.-J.: *Distillation Tray Fundamentals.* Cambridge University Press, Cambridge, England, 1986.

Loon, R. E.; Pinczewski, W. V.; and Fell, C. J. D.: ''Dependence of the Froth-to-Spray Transition on Sieve Tray Design Parameters,'' *Trans. Instn. Chem. Engrs*, Vol. 51 (1973) pp. 374–376.

Mashelkar R. A.: ''Bubble Columns'' *Br. Chem. Eng.* Vol. 15 (1970) No. 10, pp. 1297–1303.

Mersmann, A.: ''Wann werden alle Löcher einer Siebboden-Lochplatte durchströmt?'' *Chem. Ing. Tech.* Vol. 35 (1963) No. 2, pp. 103–107.

Mersmann, A.; Voit, H.; and Zeppenfeld, R.: ''Brauchen wir Stoffaustausch-Maschinen,'' *Chem. Ing. Tech.* 58 (1986) No. 2, pp. 87–96.

Molzahn, M. and Schmidt, R.: ''Belastungskennfelder von Trennkolonnen,'' *Vt Verfahrenstechnik* Vol. 9 (1975) No. 8, pp. 388–395.

Mukhlenov, I. P.: ''Investigation of a Suspended Layer of Mobile Foam in Sieve-Plate Equipment,'' *J. Appl. Chem. USSR* Vol. 31 (1958) No. 1, pp. 40–48.

Peavy, C. C. and Baker, E.M.: ''Efficiency and Capacity of a Bubble-Plate Fractionating Column,'' *Ind. Eng. Chem.* Vol. 29 (1937) No. 9, pp. 1056–1064.

Perry, R. H.; Green, D. W.; and Maloney, J. O.: *Perry's Chemical Engineer's Handbook.* Mc Graw-Hill, New York, 1984.

Pinczewski, W. V.; Benke, N. B.; and Fell, C. J. D.: ''Phase Inversion on Sieve Trays,'' *AIChE J.* Vol. 21 (1975) No. 6, pp. 1210–1213.

Pinczewski, W. V. and Fell, C. J. D.: ''The Transition from Froth-to-Spray Regime on Commercially Loaded Sieve Trays,'' *Trans. Inst. Chem. Eng.* Vol. 50 (1972) pp. 102–108.

Piqueur, H. and Verhoeye, L.: ''Research on Valve Trays, Hydraulic Performance in the Air-Water System,'' *The Canadian Journal of Chem Engineering* Vol. 54 (1976) pp. 177–184.

Porter, K. E. and Wong, P. F. Y.: ''The Transition from Spray to Bubbling on Sieve Plates,'' *Int. Distillation Symp.* (1969) pp. 17–28.

Ruff, K.; Pilhofer, Th. and Mersmann, A.: ''Vollständige Durchströmung von Lochböden bei der Fluid-Dispergierung,'' *Chem.-Ing.-Tech.* Vol. 48 (1976) No. 9, pp. 759–764.

Sherwood, T. K. and Jenny, F. J.: ''Entrainment in Plate Columns,'' *Industrial and Engineering Chemistry* Vol. 27 (1935) No. 3, pp. 265–272.

Smith, R. B.; Dresser, T.; and Ohlswager, S.: "Tower Capacity Rating Ignores Trays," *Hydrocarbon Processing & Petroleum Refiner* 40 (1963) pp. 183–184.

Smith, P. L. and van Winkle, M.: "Discharge Coefficients Through Perforated Plates at Reynolds-Numbers of 400 to 3000," *A.I.Ch:E. Journal* Vol. 4 (1958) No. 3, pp. 266–268.

Souders, M. and Brown, G. G.: "Design of Fractionating Columns," *Ind. Eng. Chem.* 26 (1934) No. 1, pp. 98–103.

Solomakha, G. P.; Chumakov, S. I.; and Galitskii, A. Ya.: "Gas Content of the Two-Phase Dynamic Layer on Bubble-Cup Trays," *Theor. Found. Chem. Eng.* (UdSSR) 6 (1972) No. 1, pp. 49–54.

Stichlmair, J.: "Untersuchungen zum statiönaren und dynamischen Verhalten einer adiabat betriebenen Absorptionskolonne." Thesis, Technical University of Munich, (1971).

Stichlmair, J. and Weisshuhn, E.: "Untersuchungen zum Bodenwirkungsgrad unter besonderer Berücksichtigung der Flüssigkeitsvermischung," *Chem.- Ing.- Techn.* Vol. 45 (1973) No. 5, pp. 242–247.

Stichlmair, J.: "Maximale Gasbelastung von Kolonnenböden," *Vt Verfahrenstechnik* Vol. 8 (1974) No. 11, pp. 323–326.

Stichlmair, J.: "Punktverstärkungsverhältnis bei Rektifikation und Absorption," *Chem.- Ing.- Techn.* Vol. 47 (1975) No. 23, p. 987.

Stichlmair, J.: *Grundlagen der Dimensionierung des Gas/Fluessigkeit-Kontaktapparates Bodenkolonne* Verlag Chemie, Weinheim, New York, 1978.

Stichlmair, J. and Hofer, H.: "Mitreißen von Flüssigkeit aus der Zweiphasenschicht von Bodenkolonnen," *Chem.- Ing.- Techn.* Vol. 50 (1978) No. 7, p. 553.

Stichlmair, J. and Ulbrich, S.: "Liquid Channeing on Large Trays and its Effect on Plate Efficiency," *Chem. Eng. Technol.* Vol. 10 (1987) No. 10, pp. 33–37.

Sulzen Chcmtech: *Prospect No. d/22.13.06-V.82-45*. Winterthur, Switzerland, 1982.

Takahashi, T. Miyahara, T. and Shimizu, K.: "Experimental Studies of Gas Void Fraction and Froth Height on a Perforated Plate," *J. Chem. Eng.* Vol. 7 (1974) No. 2, pp. 75–80.

Taylor, R. and Krishna, R.: *Multicomponent Mass Transfer*. John Wiley and Sons, New York, 1993.

Thomas, W. J. and Haq, M. A.: "Studies of the Performance of a Sieve Tray with 3/8-in. Diameter Perforations," *Ind. Eng. Chem., Process Des. Dev.* Vol. 15 (1976) No. 4, pp. 509–518.

Thomas, W. J. and Campbell, M.: "Hydraulic Studies in a Sieve Plate Downcomer System," *Trans. Inst. Chem. Eng.* 45 (1967) pp. T53–T63.

Volpert, M. and Stichlmair, J.: "Fluiddynamische Untersuchungen zur Ermittlung des Flüssigkeitsdurchsatzes durch die Ablaufschächte von Bodenkolonnen," *Chem. Ing. Techn.* Vol. 58 (1986) No. 7, pp. 606–607.

Wallis, G.: *One-Dimensional Two-Phase-Flow*. McGraw-Hill, New York, 1960, p. 377.

Weiss, S. and John, L.: "Zur Stoffübertragung bei der Destillation in Bodenkolonnen," *Chem. Tech.* Vol. 25 (1973) pp. 658–664.

Weisshuhn E.: "Untersuchungen zur Fluessigkeitsvermischung auf Siebboeden." Dissertation, TU, Muenchen, 1976.

Wenzel, L. A.: "Low Temperature Distillation," *Chem. Engineering Progress* Vol. 53 (1957) No. 6, pp. 272–277.

Wilke, G. R., Chang, P.: "Correlation of Diffusion Coefficients in Diluted Solutions," *AlChe Journal* 1 (1955) No 2, pp. 264–270

Packed Columns

Barker, T., Chilton, T. H., Vernon, H. C.: The Course of Liquor Flow in Packed Towers *AlChe* 31(1935) pp. 296–315.

Bemer, G. G. and Kalis, G. A. J.: ''A New Method to Predict Hold-Up and Pressure Drop in Packed Columns,'' *Trans. IChemE.* 56 (1978) pp. 200–204.

Billet, R.: *Packed Towers in Processing and Environmental Technology.* VCH-Verlagsgesellschaft, Weinheim, 1995.

Billet, R. and Mackowiak, J.: ''How to Use the Absorption Data for Design and Scale-Up of Packed Columns,'' *Fette Seifen Anstrichmittel* 86 (1984) No. 9, pp. 349–358.

Billet, R. and Schultes, M.: ''Modelling of Packed Tower Performance for Rectification, Absorption and Desorption.'' The Third Korea-Japan Symp. on Sep. Tech, Seoul, 1993.

Billet, R. and Schultes, M.: ''Predicting Mass Transfer in Packed Columns,'' *Chem. Eng. Technol.* 16 (1993) pp. 1–9.

Billet, R. and Schultes, M.: ''Fluid Dynamics and Mass Transfer in the Total Capacity Range od Packed Columns up to the Flood Point,'' *Chem. Eng. Technol.* 18 (1995) pp. 371–379.

Billet, R.: ''Packed Columns, Analysis and Design.'' Proceedings of the First Glitsch Packed Column Workshop Glitsch, Dallas, Texas, 1985.

Billet, R. and Bravo, J. L.: ''Modern Plastic Packings for More Efficient Separation Processes,'' *FAT Sci Technol.* 91 (1989) No. 8, pp. 329–334.

Billet, R.: ''Bewertung von Füllkörpern und die Grenzen ihrer Weiterentwicklung,'' *Chem.-Ing.-Tech.* 65 (1993) No. 2, pp. 157–166.

Blaß, E. and Kurz, R.: ''Der Einfluß grenzflächenenergetischer Größen auf den Zweiphasengegenstrom durch Raschigring-Füllkörpersäulen,'' *Vt Verfahrenstechnik* 10 (1976) No. 11, pp. 721–724. and *Vt Verfahrenstechnik* 11 (1977) No. 1, pp. 44–48.

Bornhuetter, K.: ''Stoffaustauschleistung von Füllkörperschüttungen unter Berücksichtigung der Flüssigkeitsströmungsformen.'' Thesis, Technical University of Munich, 1991.

Bravo, J. L. and Fair, J. R.: ''Generalized Correlation for Mass Transfer in Packed Distillation Columns,'' *Ind. Eng. Chem. Process Des. Dev.* 21 (1982) No 1, pp. 162–170.

Bravo, J. L.; Rocha, J. A.; and Fair, J. R.: ''Mass Transfer in Gauze Packings,'' *Hydrocarbon Processing* 64 (1985) No 1, pp. 91–95.

Bravo, J. L.; Rocha, J. A.; and Fair, J. R.: ''Pressure Drop in Structured Packings,'' *Hydrocarbon Processing* 65 (1986) No 3, pp. 45–49.

Buchanan, J. E: ''Holdup in Irrigated Ring-Packed Towers Below the Loading Point,'' *Ind. Eng. Chem. Fundam.* 6 (1967) No 3, pp. 401–407.

Buchanan, J. E.: ''Pressure Gradient and Liquid Holdup in Irrigated Packed Towers,'' *Ind. Eng. Chem. Fundam.* 8 (1969) No 3, pp. 502–511.

Buchanan, J. E. and Dixon, D. C.: ''Correlation of Flooding in Packed Columns and the Generalized Pressure Drop Relation.'' Department of Chemical Engineering, University of South Wales, Kensington, 1987.

Bühlmann, U.: Principles of Liquid Distribution and Scale-up Characteristics of Rhombopak Structured Packing Inst. Chem. Eng. Symposium Series 142 (1997) Vol 2, pp. 967–977

Capes, C. E. and McIlhinney, A.E.: ''The Pseudoparticulate Expansion of Screen-Packed Gas-Fluidized Beds,'' *AIChE Journal* Vol 14 (1968) No 6, pp. 917–922.

Changez, N. D. and Sawistowski, H.: ''Effect of Maldistribution on the Performance of Packed Columns,'' *Ind. Chem.* 39 (1963), pp. 181–185.

Coulson, J.M. and Richardson, J. F.: *Chemical Engineering*, Vol. 2: Unit Operations. Pergamon Press, Oxford UK, 1976.

Dutkai, E. and Ruckenstein, E.: ''New Experiments Concerning the Distribution of a Liquid in a Packed Column,'' *Chem. Eng. Sci.* 25 (1970) pp. 483–488.

Eckert, J. S.: ''Selecting the Proper Distillation Column Packing,'' *Chem. Eng. Prog.* 66 (1970) No. 3, pp. 39–44.

Engel, V.; Stichlmair, J. and Geipel, W.: ''A New Model to Predict Liquid Holdup in Packed Columns.'' Ind. Chem. Eng. Symp. Series, No. 142 (1997) pp. 939–947.

Ergun, S.: ''Fluid Flow Through Packed Columns,'' *Chem. Eng. Prog.* 48 (1952) pp. 89–94.

Fair, J. R.: ''A Half-Century Progress in Separation Technology,'' *Chem. Processing* (1988) No. 3, p. 58.

Fair, J. R.: ''Distillation: King in Separations,'' *Chem. Processing* (1990) No. 9, pp. 23–31.

Fair, J. R. and Bravo, J. L.: ''Distillation Columns Containing Structured Packing,'' *Chem. Eng. Prog.* 86 (1990) No. 1, pp. 19–28.

Fair, J. R. and Bravo, J. L.: ''Prediction of Mass Transfer Efficiencies and Pressure Drop in Structured Tower Packings in Vapor/Liquid Service,'' *Ind. Chem. Eng. Symp.* Ser. 104 (1987) p. A 183.

Fuller, E. N.; Schettler, P. D.; and Giddings J. C.: ''A New Method for Prediction of Binary Gas-Phase Diffusion Coefficients,'' Ind. Eng. Chem. 58 (1966) No 5, pp. 19–27.

Geipel, W. and Ulrich, H.-J.: *Fuellkoerper-Taschenbuch.* Vulkan-Verlag, Essen, 1991.

Gelbe, H.: ''A New Correlation for Liquid Holdup in Packed Beds,'' *Chem. Eng. Sci.* 23 (1968) pp. 1401–1404.

Hoek, P.J.; Wesselingh, J. A.; and Zuiderweg, F. J.: ''Small Scale and Large Scale Liquid Maldistribution in Packed Columns,'' *Chem. Eng. Res. Des.* 64 (1986), pp. 431–449.

Hurter, F.: ''Coke-Towers and Similar Apparatus,'' *J. Soc. Chem. Ind.* (1893) No. 3, pp. 227–232.

Kirschbaum, E.: ''Neue Fragen und Erkenntisse aus der Rektifiziertechnik,'' *Z. Ver. Dtsch. Ing.* 75 (1931) No. 39, pp. 1212–1216.

Kister, H. Z.: *Distillation Operation.* McGraw-Hill, New York, 1989.

Kister, H. Z.: *Distillation Design.* McGraw-Hill, New York, 1992.

Kolev, N,: ''Wirksame Austauschfläche von Füllkörperschüttungen,'' *Vt Verfahrenstechnik* 7 (1973) No. 7, pp. 71–75.

Kolev, N. and Semkov, K.: ''Über den Einfluß der Axialvermischung bei Rektifikationskolonnen mit ungeordneten Füllkörperschüttungen,'' *Chem. Eng. Process* 19 (1985) pp. 175–182.

Kouri, R. J. and Sohlo, J. J.: ''Liquid and Gas Flow Patterns in Random and Structured packings. *IChemE Symposium Series* 104 (1987) pp. B193–B212.

Kunesh, J. G.: ''Recent Developments in Packed Columns,'' *Can. J. Chem. Eng.* 65 (1987) No. 12, pp. 907–913.

Leva, M.: ''Flow Through Irrigated Dumped Packings-Pressure Drop, Loading, Flooding,'' *Chem. Eng. Prog. Symp. Ser.* 50 (1954) No. 10, p. 51.

Loeffler, A. L. and Ruth, B. F.: ''Particulate Fluidzation and Sedimentation of Spheres,'' *AIChE Journal* Vol 5 (1959) No. 3, pp. 310–313.

Mackowiak, J.: *Fluiddynamik von Kolonnen mit modernen Füllkörpern und Packungen für Gas/Flüssigkeitssysteme*. Sauerländer Verlag, Aarau, 1991.

Martin, B. L. A.; Kolar, Z.; and Wesselingh, J. A.: ''The Falling Velocity of a Sphere in a Swarm of Different Spheres,'' *Trans IChemE* Vol 59 (1981) pp. 100–104.

Meier, W. and Huber, M.: ''Measurements of the NEPT in Packed Columns with Artificial Maldistribution,'' *IChemE Symposium Series* 32 (1969) No. 4, pp. 31–37.

Mersmann A.: ''Zur Berechnung des Flutpunktes in Füllkörperschüttungen,'' *Chem. Ing. Techn.* Vol. 37 (1965) No. 3, pp. 218–226.

Mersmann A. and Deixler A.: ''Packed Columns,'' *Ger. Chem. Eng.* 9 (1986) No. 5, pp. 265–276.

Mersmann A.; Voit H., and Zeppenfeld R.: ''Brauchen wir Stoffaustauschmaschinen,'' *Chem. Ing. Tech.* 58 (1986) No. 2, pp. 87–96.

Mohunta, D. and Laddha, G.: ''Prediction of Liquid Phase Holdup in Random Packed Beds,'' *Chem. Eng. Sci.* 20 (1965) pp. 1069–1072.

Mohunta, D.; Vaidyanathan, A.; and Laddha, G.: ''Prediction of Liquid Phase Holdup in Random Packed Beds,'' *Indian Chemical Engineer* (1969) pp. 73–79.

Moore, F. and Rukovena, F.: ''Fluessigkeits- und Gasverteilung in handelsueblichen Fuellkoerperkolonnen,'' *CA-Chemie-Anlagen und Verfahren* 35 (1987) No. 5, pp. 33–41.

Mullin, J. W.: ''The Effect of Maldistribution on the Performance of Packed Columns,'' *Ind. Chem.* 33 (1957), S. 408–417.

Onda, K.; Takeuchi, H.; and Okumoto, Y.: ''Mass Transfer Coefficients Between Gas and Liquid Phases in Packed Columns,'' *Journal of Chemical Engineering of Japan* 1 (1968) pp. 56–62.

Onda, K.; Takeuchi, H.; and Maeda, Y.: ''Liquid Distribution in a Packed Column,'' *Chem. Eng. Sci.* 28 (1973), pp. 1677–1683.

Porter, K. E. and Jenkins, J.D.: ''The Interrelationship between Industrial Practice and Academic Research in Distribution and Absorption.'' *Institution of Chemical Engineers. Symp. Ser.* Vol 3, (1979) No. 56 p. 75–121.

Porter, K. E. and Jones, M. C.: ''Liquid and Gas Distribution in the Scale-up of Packed Columns,'' *IChemE Symposium Series* 104 (1987), pp. A245–A258.

Potthoff, R. and Stichlmair, J.: ''Maldistribution in Packungskolonnen,'' *Chem.-Ing.-Tech.* 63 (1991) No. 1, pp. 72.

Potthoff, R,: ''*Maldistribution in Fuellkoerperkolonnen.*'' *VDI Verlag, Reihe 3: Verfahrenstechnik* No. 294. Duesseldorf (1992).

Puranik, S. S. and Vogelpohl, A.: ''Effective Interfacial Area in Irrigated Packed Columns,'' *Chem. Eng. Sci.* 29 (1974) pp. 501–507.

Reichelt, W.: *Strömung und Stoffaustausch in Füllkörperapparaten bei Gegenstrom einer flüssigen und einer gasförmigen Phase* Verlag Chemie, Weinheim, 1974.

Richardson, J. F. and Zaki, W. N.: ''Sedimentation and Fluidization, Part I,'' *Trans. Inst. Chem. Eng.* 32 (1954) pp. 35–53.

Rocha, J. A.; Bravo, J. L.; and Fair, J. R.: ''Distillation Columns Containing Structured Packings: A Comprehensive Model for Their Performance 1. Hydraulic Models, *Ind. Eng. Chem. Res.* 32 (1993) No. 4, pp. 641–651.

Rumpf, H. and Gupte, A. R.: ''Einflüsse der Porosität und Korngrößenverteilung im Widerstandsgesetz der Porenströmung,'' *Chem.-Ing.-Tech.* 43 (1971) No. 6, pp. 367–375.

Schluender, E. U.: "Ueber den Einfluß von Phasenverteilungen auf den Wirkungsgrad von Trennstufen," *Ber. Bunsenges. Phys. Chem.* 83 (1979) pp. 1184–1193.

Scott, A. H.: "Liquid Distribution in Packed Towers," *Trans. Inst. Chem. Eng.* 13 (1935) pp. 211–217.

R. Schmidt: "The Lower Capacity Limits of Packed Columns," *Instit. Chem. Eng., Symp. Ser.* 56 (1979) Vol. 2, pp. 3.1/1–14.

Schultes, M.: *Einfluß der Phasengrenzfläche auf die Stoffübertragung in Füllkörperkolonnen* VDI Verlag, Reihe 3 No.230, Düsseldorf, 1990.

Shariat, A. and Kunesh, J. G.: "Packing Efficiency Testing on a Commercial Scale with Good (and not so good) Reflux Distribution," *Ind. Eng. Chem. Res.* 34 (1995) No. 4, pp. 1273–1279.

Sherwood, T. K.; Shipley, G. H.; and Holloway, F. A. L.: "Flooding Velocities in Packed Columns," *Ind. Eng. Chem.* 30 (1938) No. 7, pp. 765–769.

Shi, M. G. and Mersmann, A.: "Effective interfacial area in packed columns," *Ger. Chem. Eng.* 8 (1985) pp. 87–96.

Shulman, H. J. and DeGouff, J. J.: "Mass Transfer Coefficients and Interfacial Areas for 1-in Raschig Rings," *Ind. Eng. Chem.* 44 (1952) No. 8, pp. 1915–1922.

Shulman, H. J.; Ullrich, C. F.; Proulx, A. Z.; and Zimmerman, J. O.: "Wetted and Effective Interfacial Area, Gas and Liquid Phase Mass Transfer Rates," *AIChE J* 1 (1955) No. 2, pp. 253–258.

Shulman, H. J.; Ullrich, C. F.; and Wells, N.: "Total Static and Operating Holdups," *AIChE J* 1 (1955) No. 2, pp. 247–253.

Spiegel, L.: "Packungen in der Prozeßindustrie," *Process* 4 (1994) pp. 39–44.

Spiegel, L. and Meier, W.: "A Generalized Pressure Drop Model for Structured Packings," *Inst. Chem. Eng. Symp. Ser.* 128 (1992) pp. B85–B94.

Stanek, V. and Kolar, V.: "Distribution of Liquid over a Random Packing," *Coll. Czech. Chem. Commun.* 38 (1973) pp. 2865–2873.

Stichlmair, J. and Stemmer, A.: "Influence of Maldistribution on Mass Transfer in Packed Columns," *Inst. Chem. Eng. Symp. Ser.* 104 (1987) pp. B213–B224.

Stichlmair, J.; Bravo, L. J.; and Fair, J. R.: General Model for Prediction of Pressure Drop and Capacity of Coutercurrent Gas/Liquid Packed Columns, *Gas Separation & Purification* 3 (1989) No. 3, pp. 19–28.

Stoter, F.; Olujic, V.; and De Graauw, J.: "Measurements and Modelling of Liquid Distribution in Structured Packings." Vortrag Annual Meeting of AIChE, Chicago, 1990.

Strigle, R.F.: *Rondom Packungs and Packed Towers, Design and Applications.* Gulf Publishing Company, Houston, 1987.

Süess, P. and Spiegel, L.: "Holdup of mella pak structured packings," *Chem. Eng. Processing* 31 (1992) pp. 119–124.

Sulzer: *Prospect No. d/22.13.06-V.82-45.* Winterthur, Switzerland, 1982.

Sulzer Chemtech "Strukturierte Packungen für Destillation und Absorption," *Prospekt No 22.13.06.20 - IV.94–100.* Winterthur, Schweiz, 1994.

Sulzer Chemtech "Optiflow-Improved Efficiency in Distillation and Absorption Columns," *Prospect No 22.05.0640-VIII.96-30.* Wintherthur, Switzerland, 1996.

Teutsch, T.: "Druckverlust in Füllkörperschüttungen bei hohen Berieselungsdichten." Thesis, Technical University of Munich, 1962.

Ullrich, H.-J.; Deeke, W., and Geipel, W.: ''Fluessigkeitsverteilung mit offenen Verteilrinnen,'' *Vt-Verfahrenstechnik* 17 (1983) No. 4, pp. 223–230.

Van Krevelen, D.W. and Hoftijzer, P.J.: ''Kinetics of Simultaneous Absorption and Chemical Reaction,'' *Chem. Eng. Progr.* 44 (1948) No. 7, pp. 529–536.

Wagner, I.; Stichlmair, J.; and Fair, J. R.: ''Mass Transfer in Beds of Modern, High-Efficiency Random Packings,'' *Ind. Eng. Chem. Res.* 36 (1997) No. 1, pp. 227/237.

Wilhelm, R. H. and Kwauk, M.: ''Fluidization of Solid Particles,'' *Chemical Engineering Progress* Vol 44 (1948) No. 3, pp. 201–218.

Wu, K. Y. and Chen, G.K.: ''Large-Scale Pilot Columns and Packed Column Scale-Up,'' *Inst. Chem. Eng. Symp. Ser.* 104 (1987) B225.

Zech, J. B. and Mersmann, A.: ''Flüssigkeitsströmung und flüssigkeitsseitiger Stoffaustausch in berieselten Füllkörperschüttungen,'' *Chem.-Ing.-Tech.* 50 (1978) No. 7, pp. 549.

Zech, J. B.: ''Fluessigkeitsstroemung und Stoffaustausch in berieselten Fuellkoerperschuettungen.'' Thesis, Technical University of Munich, 1978.

Zuiderweg. F. and Hoek, P. J.: ''The Effect of Small Scale Liquid Maldistribution on the Separating Efficiency of Random Packings,'' *Ind. Chem. Eng. Symp. Ser.* 104 (1987) pp. B247.

Zuiderweg, F. J.; Hoek, P. J.; and Lahm, L.: ''The Effect of Liquid Distribution and Redistribution on the Separating Efficiency of Packed Columns,'' *Ind. Chem. Eng. Symp. Ser.* 104 (1987) pp. A217.

Zuiderweg, F. J.; Kunesh J. G.; and King, D. W.: ''A Model for the Calculation of the Effect of Maldistribution on the Efficiency of a Packed Column,'' *Trans IChemE* Vol 71 (1993) January, Part A, pp. 28–44.

9

Control of Distillation Processes

9.1 Control Loops
 9.1.1 Single-Loop Controller
 9.1.2 Ratio Control Loop
 9.1.3 Disturbance Feed-Forward Control
 9.1.4 Cascade Control Loop
9.2 Single Control Tasks
 9.2.1 Liquid Level Control
 9.2.2 Stream Partition Control
 9.2.3 Pressure Control
 9.2.4 Product Concentration Control
9.3 Design of Control Configurations for Distillation Columns
 9.3.1 Basic Control Systems without Composition Control
 9.3.2 One-Point Composition Control Systems
 9.3.3 Two-Point Composition Control Systems
 9.3.4 Range of Applications of the Control Configurations
  9.3.4.1 Range of Applicability of the Control Configurations
  9.3.4.2 Expanded Range of Application of the Control Configurations
9.4 Control Schemes for Distillation Processes
 9.4.1 Control Schemes for Continuously Operated Processes
  9.4.1.1 Separation of an Azeotropic Mixture by Changing the Pressure
  9.4.1.2 Control of an Air Separation Column
  9.4.1.3 Control of a Process with a Main and a Side Column
  9.4.1.4 Control of an Azeotropic Distillation Process
 9.4.2 Control Schemes for Batch Distillation Processes
Literature

The major objective of any control system for any process is the maintenance of the desired state of operation without any human action. This requires the ability to automatically prevent or suppress disturbances which can result from changes in the variables to which the operation is subject. In general, a process control system consists of a large number of control loops that may interact. A classical example of an interacting system is a distillation column with dual composition control. The sometimes formidable control problems of distillation columns are caused by loop and variable interactions, nonlinearities of the process, process and measurement delays and the large number of variables in such a unit.

A large variety of different control configurations for distillation columns has been developed. This variety has its origin partly in the large number of different separation tasks performed by distillation and partly in the complex structure of distillation columns. There is much written in the literature on the advantages and the disadvantages of various configurations, but case studies predominate over theoretically founded guidelines. This chapter is aimed at assisting the engineer in selecting the best control configuration for a given separation problem. Through the concept of stream partitioning, design strategies for control configurations, which are particularly helpful in conflicting situations, are developed in a systematical manner.

Computer controls and advanced controls are outside the scope of this text as their use in industry has been very limited. Neither are many other important issues not directly related to the control configuration problem discussed here. However, they are dealt with in the general literature about distillation process control, e.g., [Shinskey 1977, Luyben 1992, Kister 1989, Skogentad 1997].

Standard symbols for control systems are used in the following. The instrument is represented by a thin line circle. The purpose of the instrument is defined by a letter code contained within the circle. This letter code is constructed according to Table 9-1. The first letter denotes the measured or initiating variable and is used in accordance with Column 2 of the table. It is modified, if necessary, by the addition of a letter in accordance with Column 3. Succeding letters that characterize the display or output functions are in accordance with Column 4 of Table 9-1. If there are two or more succeeding letters they shall be placed one after the other in the sequence I R C T Q S Z A. The letter I may be omitted in the case of a self-indicating recorder.

9.1 CONTROL LOOPS

Different types of control loops occur in distillation processes. The most important ones are dealt with in the following.

9.1.1 Single-Loop Controller

A single-loop controller performs a regulation with a fixed set. Figure 9-1A gives the schematic presentation of the simplest case of a control loop, i.e., the maintenance of a constant liquid flow rate. The letter F stands for flow measurement and C for

Table 9-1. Letter code for Identification of Control Instrument Functions

	First Letter		Succeeding Letter
	Measured or Initiating Variable	Modifier	Display or Output Functions
A			Alarm
B			Display of state (for example, motor running)
C			Controlling
D	Density	Difference	
E	All electrical variables		Sensing Element
F	Flow rate	Ratio	
G	Gauging, position of length		
H	Hand (manually initiated) operated		
I			Indicating
J		Scan	
K	Time or time program		
L	Level		
M	Moisture of humidity		
N	User's choice		User's choice
O	User's choice		
P	Pressure or vacuum		Test point connection
Q	Quality: For example, analysis, concentration, conductivity	Integrate or totalize	Integrating or summating
R	Nuclear radiation		Recording
S	Speed or frequency		Switching
T	Temperature		Transmitting
U	Multivariable		Multifunction unit
V	Viscosity		Valve, damper, louvre, actuating element, unspecified correcting unit
W	Weight or force		
X	Unclassified variables		Unclassified functions (for example cathode-ray tube)
Y	User's choice		Computing relay, relay
Z			Emergency or safety acting

control. If the actual flow rate control point deviates from the set point, the controller varies the opening of the control valve to increase or decrease flow rate and, finally, to eliminate the deviation. The action of the controller on the valve is indicated by a thin dashed line.

In distillation processes a decentralized control system with single-loop controllers is preferred. This kind of control system is easier to understand and to tune, is more failure tolerant and is less sensitive to plant operation [Skogestad et al. 1990].

9.1.2 Ratio Control Loop

Sometimes the flow rate of one stream has to be kept proportional to that of a second one in distillation processes. This is expressed by the schematic of Fig. 9-1B. The second F (=fraction) in the controller symbol indicates that the controller keeps the flow rate of the second stream in a fraction of the first one. The disadvantage of ratio control is that two flow rate measurements are required. This increases the risk of failure of the control system. Hence, flow rate control is preferred to flow ratio control.

9.1.3 Disturbance Feed-Forward Control

In a feed-forward control loop the set point is adjusted to variations of an uncontrolled process variable. An example is presented in Fig. 9-1C. The objective of the control loop is a smooth flow of the lower stream. The set point of the flow controller depends on the temperature of the upper stream. The adjustment procedure in the controller is indicated by letter Q. Disturbance feed-forward control is seldom applied to distillation column control.

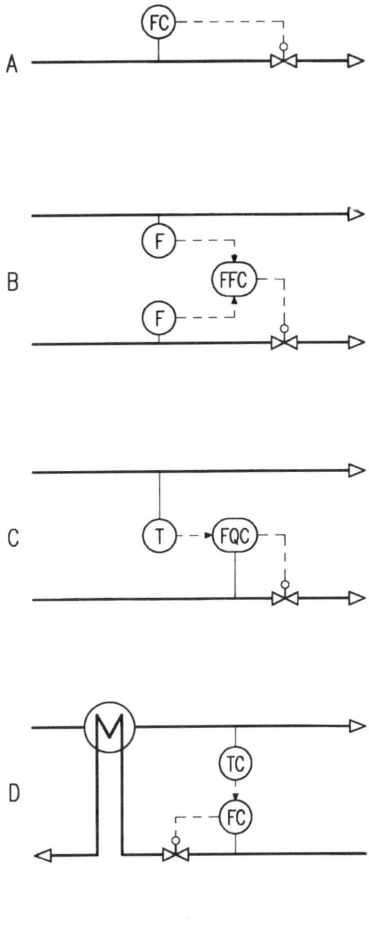

Fig. 9-1. Some examples of important control loops: (A) Single control loop, (B) ratio control loop, (C) disturbance feed forward control loop, (D) cascade control loop.

9.1.4 Cascade Control Loop

A cascade control loop consists of two loops, a master loop and a slave loop. The master control loop adjusts the set point of the slave control loop. For good operation, the master loop should be slow and the slave loop fast. An example of a cascade control loop is presented in Fig. 9-1D. The set point of the flow control loop is adjusted in a way that the temperature of the upper stream is kept constant. In contrast to disturbance feed-forward control, two variables (temperature and flow rate) are regulated by a cascade control loop. Cascade control loops are very important and, thus, often used in distillation process control.

9.2 SINGLE CONTROL TASKS

Typically, distillation processes contain control loops for flow rates, liquid levels, pressure and product concentrations. It is in this order that the complexity of the control problems rises. Especially important are liquid level control, stream partition control, pressure control and product quality control.

9.2.1 Liquid Level Control

Liquid level control is performed either by a single control loop, Fig. 9-2A, or by a cascade control loop, Fig. 9-2B. In order to ensure a smooth flow, the level controller must have a wide range of proportionality. The cascade control loop is preferred for large vessels since level variations are very slow compared to the quick response of flow rate variations.

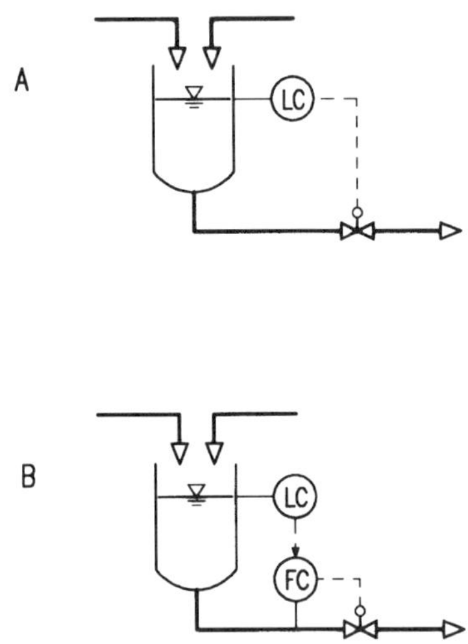

Fig. 9-2. Liquid level control: (A) Single control loop, (B) cascade control loop.

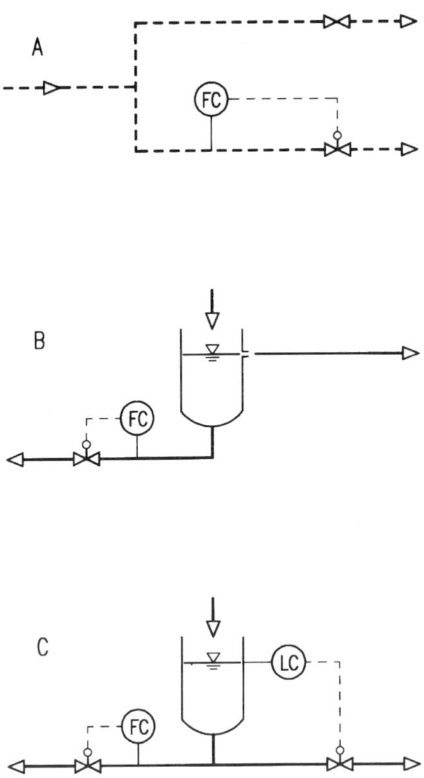

Fig. 9-3. *Stream partition control. Only one of the two partition streams is regulated by a fixed set, the other one is left free or is regulated by level control: (A) Partition of a gas stream, (B) partition of a liquid stream with overflow, (C) partition of a liquid stream with level control.*

9.2.2 Stream Partition Control

A very essential control task is the partition of a fluid stream into two or more smaller streams. Even when the flow rate of the feed stream is absolutely constant, it is not possible to control each of the partition streams with independent controllers because the measurement of the flow rates is never absolutely precise. It is impossible to meet the material balance by two independently acting flow rate controllers. Only one of the partition streams must be manipulated directly, the other one must be left free (Fig. 9-3). Preferably, the stream with the smaller flow rate is directly manipulated. Any partition control system performs best when the smaller stream is directly manipulated. If, for instance, a 100 m^3/h stream has to be split into two streams of 90 m^3/h and 10 m^3/h, respectively, a control system with $\pm 10\%$ accuracy keeps the streams in the range of 89–91m^3/h and 9–11 m^3/h, respectively, when the smaller stream is directly controlled. Control of the larger stream results in flow rates of 81–99 m^3/h and 1–19 m^3/h, respectively. The more the flow rates differ, the more important is direct control of the smaller stream.

In practical applications the partitioning of a liquid stream is performed by use of a vessel (e.g., reflux drum) because the flow rate of the directly manipulated stream can be maintained constant even when the feed stream is smaller for a short period of time (Fig. 9-3B). The resulting free stream is preferably regulated by a level controller (Fig. 9-3C).

Stream partition is one of the most essential control tasks in distillation processes. The above considerations can be summarized in two rules:

Rule 1: Only one stream can be directly manipulated by a fixed set point. The other one is left free or manipulated by a level controller.
Rule 2: The directly manipulated stream should be the smaller one and, in turn, the stream controlled by the level should be the larger one.

In cases where both of the rules cannot be met, Rule 1 takes precedence since it results from material balance that must never be violated. Rule 2, however, results from a consideration of flow rate measurement accuracy and flow rate fluctuations and, hence, is not as strict as Rule 1. If a well tuned level controller is used (e.g., a cascade control loop), the liquid level can be held even when the directly manipulated stream is larger than the stream used to control the level. A rule of thumb says that the directly manipulated stream can be up to five times larger than the stream used to control the level [Sieler 1970]. However, this violation of Rule 2 should only be made in conflicting situations. When one stream is much smaller than the other one,

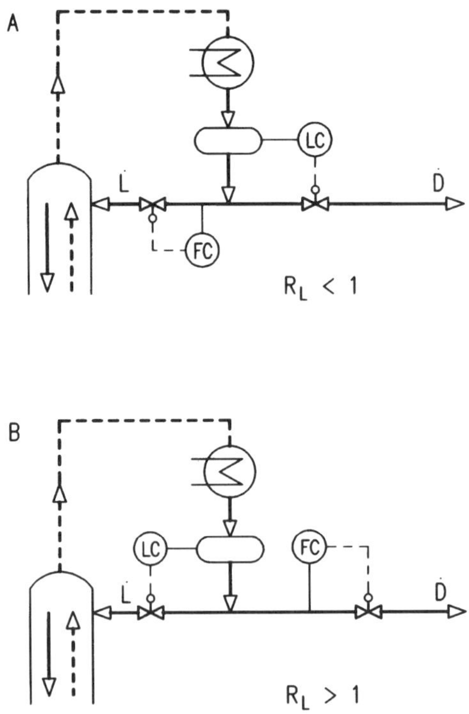

Fig. 9-4. Liquid partition of condensate at the top of the column: (A) Low reflux ratio R_L, (B) high reflux ratio R_L.

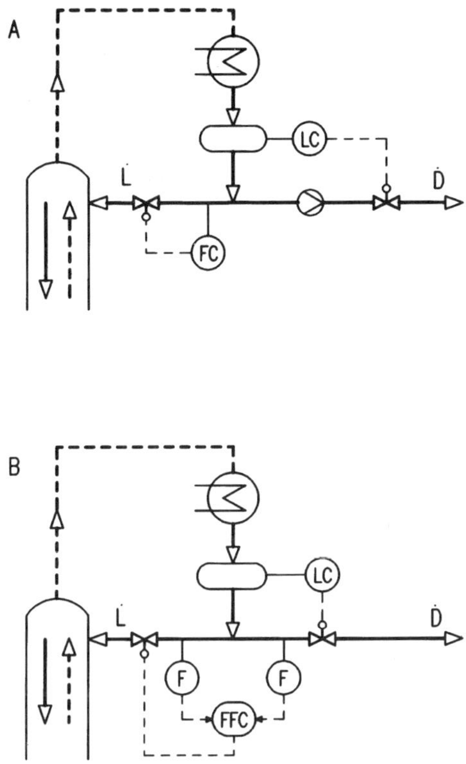

Fig. 9-5. *Liquid partition of condensate at the top of the column: (A) The pump must be installed before the control valve, (B) Control of reflux ratio $R_L = \dot{L}/\dot{D}$.*

e.g., by a factor of 10 or more, then Rule 2 must not be violated. Here, the smaller stream has to be directly manipulated.

Figures 9-4 and 9-5 show several control schemes at the top of a distillation column. All the vapor from the head of the column is condensed, collected in a drum and then partitioned into distillate $\dot{D}$ and reflux $\dot{L}$. When operated with a low reflux ratio ($R_L < 1$) the flow rate of the reflux is small and, thus directly manipulated. The flow rate of the distillate is controlled by the liquid level in the collector (Fig. 9-4A). In high reflux operation ($R_L > 1$) the distillate flow rate is small and, therefore, directly manipulated. The reflux is indirectly regulated by liquid level control (Fig. 9-4B). By such a control system an undisturbed operation is maintained even when the vapor arising in the column is, for a short period of time smaller than the directly manipulated stream. However, it is important to have enough holdup in the reflux drum to provide effective damping of disturbances in the column heating system. A common rule of thumb requires a residence time of the liquid in the condensate collector drum of five to ten minutes.

When the distillate has to be pumped (e.g., in a vacuum distillation column) the

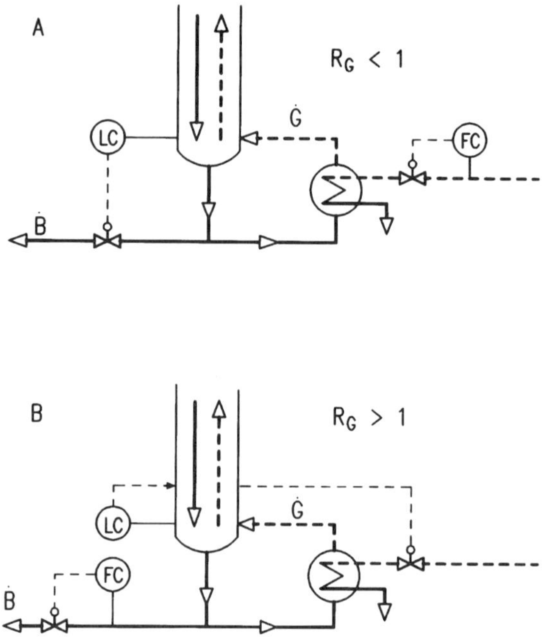

Fig. 9-6. Liquid partition at the column base: (A) Low reboil ratio R_G, (B) high reboil ratio R_G.

pump should be located before the control valve to avoid suction problems (Fig. 9-5A).

The schematic of a reflux ratio control loop is depicted in Fig. 9-5B. The disadvantages of such a control loop are the need of two flow rate measurements and the interaction with the level controller. Fluctuations of a poorly tuned level controller affect reflux flow rates. A cascade level control loop might be needed to overcome these difficulties.

Figure 9-6 presents two possible control schemes for stream partition at the bottom of a distillation column. The schematic in Fig. 9-6A is applied when the flow rate of the bottom product $\dot{B}$ is much higher than the boilup rate $\dot{G}$ (reboil ratio $R_G < 1$) because the smaller flow should be directly manipulated. If the reboil ratio is very high ($R_G > 1$), the control system of Fig. 9-6B is the better choice. Here, the flow rate of the heating stream is monitored by the liquid level.

9.2.3 Pressure Control

The simplest method of pressure control of an atmospheric column is a vent at the coldest point of the condenser. Another common method is to manipulate the draw-off of the inert gases from the collector drum as shown in Fig. 9-7A. An essential prerequisite is that there are enough noncondensable gases in the feed. This condition is always met with vapor feed or with a liquid feed that has not been exposed in a previous process to a temperature higher than the bottom temperature of the column.

The preferred control scheme for a vacuum column is presented in Fig. 9-7B. Addition of nitrogen avoids the need for a nozzle in the suction pipe of the vacuum pump. Thus, the vacuum pump is always in a safe range of operation.

Sometimes the overhead product is withdrawn as a vapor. In this case the flow rate of the overhead product is directly manipulated to maintain the pressure (Fig. 9-8A). Control of the rate of cooling water is necessary to hold the liquid level in the reflux drum.

If there are not enough inert gases in the feed, the column pressure might be manipulated via the temperature of the condensate. This is performed by manipulating the cooling water flow rate, see Fig. 9-8B. However, such a control system has a large dead time. This sometimes results in heavy interaction with temperature control loops for product quality (see Section 9.2.4).

In air-cooled condensers the control systems of Fig. 9-9 are often applied. The withdrawal of the condensate is manipulated in such a way that the liquid holdup

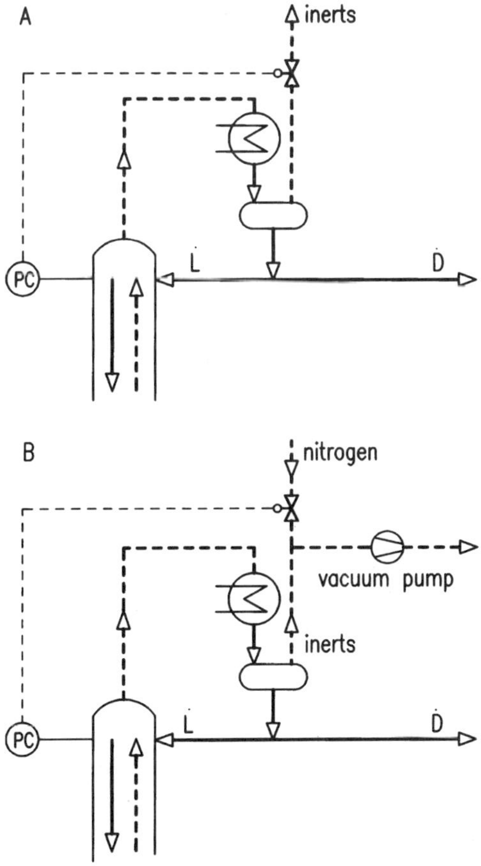

Fig. 9-7. *Pressure control: (A) Manipulating the flow rate of inerts; (B) vacuum column.*

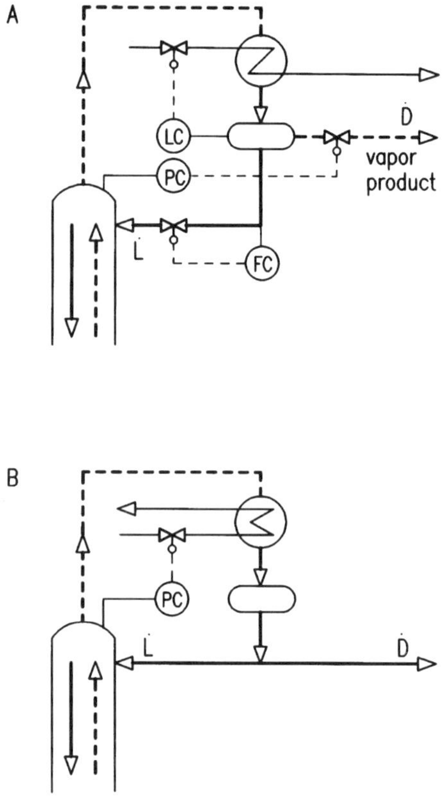

Fig. 9-8. *Pressure control: (A) Manipulating the flow rate of the vaporous overhead product, (B) manipulating condensate temperature.*

in the condenser partly floods the condenser that reduces the area effective for heat transfer (Fig. 9-9A). However, most dry air-cooled condensers are horizontal, with only two or four rows of tubes vertically oriented so that flooding does not give smooth control. Hence, bypassing vapors is common practice, see Fig. 9-9B [Shinskey 1977].

9.2.4 Product Concentration Control

A major problem of distillation processes is the control of product quality. The first requirement for product quality control is an adequate measurement. Product quality measurement is often nonspecific, i.e., the concentration measurement is replaced by a temperature measurement because it is more simple, much faster and more reliable. Since in a distillation column, liquid and vapor are at their bubble and dew points respectively, the temperature depends only on concentration, provided the pressure is kept constant. The temperature sensor should be located in the vapor in order to get a quick response to temperature variations.

A typical plot of a liquid temperature profile in a distillation column is presented in Fig. 9-10. Towards the column ends the temperature profile flattens. The sensivity

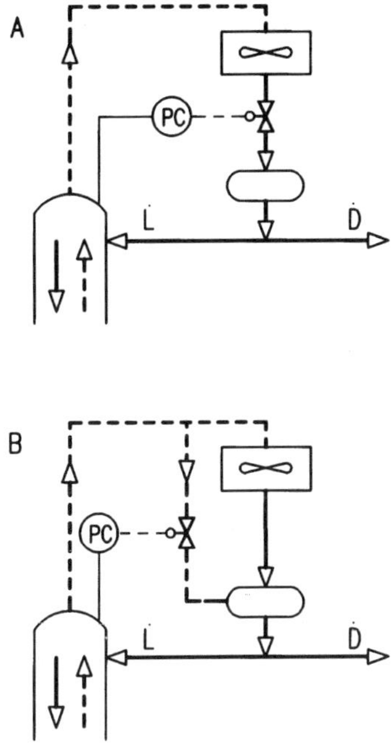

Fig. 9-9. *Pressure control of a column with air-cooled condenser: (A) Partial flooding of the condenser, (B) bypassing some vapor.*

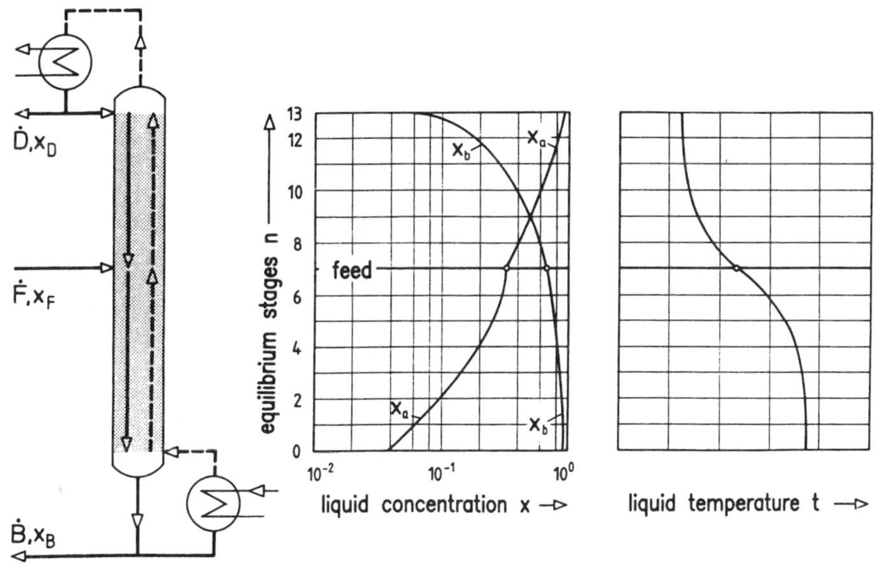

Fig. 9-10. *Temperature profile in a distillation column. The temperature profile flattens towards the column ends.*

of the temperature sensor may become limiting there. In this case the effects of small pressure changes can swamp the effects of composition changes on temperature. Hence, the sensor should be located a sufficient distance from the column ends, where a significant temperature variation is caused when product concentration changes.

9.3 DESIGN OF CONTROL CONFIGURATIONS FOR DISTILLATION COLUMNS

The schematic of a distillation column is depicted in Fig. 9-11A. It represents, in principle, a system of three interconnected stream partitions. Figure 9-11B shows the basic flow structure with two vessels and five streams in a more abstract manner. The column has been discarded to simplify the flow structure and to make the interconnections of the three stream partitions clearer. The stream partitions are:

- Partition of the feed $\dot{F}$ into overhead product $\dot{D}$ and bottom product $\dot{B}$. This partition is characterized by the product ratio $\dot{D}/\dot{B}$ or $\dot{D}/\dot{F}$ (since $\dot{B} = \dot{F} - \dot{D}$).

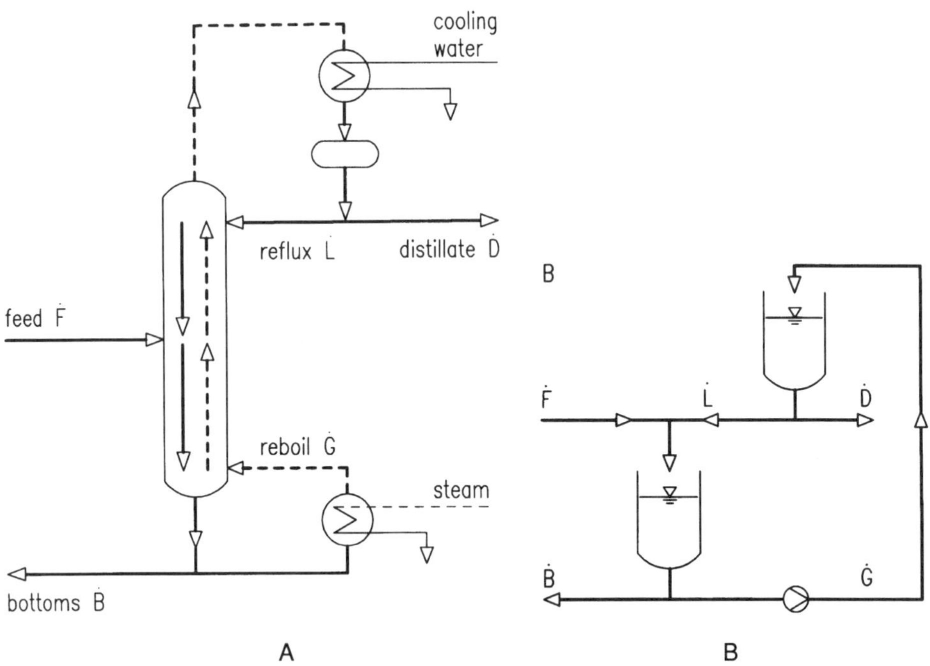

Fig. 9-11. Flow sheet of a single distillation column: (A) The column consists, in principle, of an interconnected system of three flow partitions, i.e., feed partition into products, condensate partition at the top and liquid partition at the bottom of the column, (B) basic flow structure of a distillation column from a control structure point of view.

- Partition of the condensate into reflux $\dot{L}$ and distillate $\dot{D}$. This partition is characterized by the reflux ratio $R_L = \dot{L}/\dot{D}$.
- Partition of the bottom liquid into boilup $\dot{G}$ and bottom product $\dot{B}$. Here, the parameter is the reboil ratio $R_G = \dot{G}/\dot{B}$.

The three stream partitions in the flow network of a distillation column are interconnected. The distillate $\dot{D}$ is, for instance, an element of the partition of the feed as well as of the partition of the condensate. The same holds for the partition of the bottoms since the bottom fraction $\dot{B}$ is an element of both the partition of the feed and the partition of the bottom liquid. These interconnections of the three stream partitions are expressed by the following equation that was derived in Chapter 4 (see Equation 4.21):

$$R_G = \frac{\dot{D}/\dot{F}\,(R_L + 1) - (1 - q_F)}{1 - \dot{D}/\dot{F}} \tag{9.1}$$

Here, q_F denotes the caloric factor defined in Chapter 4. In the two-phase region q_F represents the fraction of boiling liquid in the feed. In the case of a boiling liquid feed, the caloric factor is $q_F = 1$. In the case of a saturated vapor feed the caloric factor is $q_F = 0$. In these cases Equation (9.1) simplifies to:

$$R_G = \frac{\dot{D}/\dot{F}\,(R_L + 1)}{1 - \dot{D}/\dot{F}} \quad \text{for } q_F = 1 \qquad \text{or} \tag{9.2}$$

$$R_G = \frac{\dot{D}/\dot{F}\,R_L}{1 - \dot{D}/\dot{F}} - 1 \quad \text{for } q_F = 0$$

Due to the relation formulated by Equation (9.1) only two out of three partition parameters can be independently chosen. In other words, the flow network of a distillation column with three stream partitions has a degree of freedom of only two. At constant flow rate of the feed—as is supposed in this Chapter—the setting of the values of the partition parameters R_L, R_G or $\dot{D}/\dot{F}$ is equivalent to the setting of the flow rates of the partition streams. The ratio $\dot{D}/\dot{F}$ can be, for instance, replaced by the flow rate of the distillate $\dot{D}$.

In principle, four streams are available for manipulation in order to establish a column control system, i.e., distillate $\dot{D}$, reflux $\dot{L}$, boilup $\dot{G}$ and bottoms $\dot{B}$. However, only the flow rates of two of them can be independently manipulated. Depending on the manipulated streams, different control configurations exist. They are designated by the two directly manipulated streams. In a D-G configuration, for instance, the flow rates of distillate $\dot{D}$ and the boilup $\dot{G}$ are directly manipulated. From the laws of permutations and combinations it follows that six different pairs can be taken out of a collective of four. These different control configurations are listed in Table 9-2. However, not all of the six existing configurations can be applied to column control since, in addition, partition Rule 1 has to be considered. This rule demands that in each stream partition, one stream must be directly manipulated, and

Table 9-2: Check of the Existing Six Control Configurations for Consistency with Partition Rule 1

	Rule 1 for the Partition of			
Configuration	Feed into $\dot{D}$ and $\dot{B}$	Condensate into $\dot{D}$ and $\dot{L}$	Bottoms into $\dot{G}$ and $\dot{B}$	Judgment
D-L	met	violated	not met	not feasible
D-G	met	met	met	good
D-B	violated	met	met	not feasible
L-B	met	met	met	good
L-G	not met	met	met	feasible
B-G	met	not met	violated	not feasible

the other one left free. Table 9-2 indicates whether or not this rule is met in each partition.

The D-L configuration, for instance, meets Rule 1 in the partition of the feed $\dot{F}$ since just one ($\dot{D}$) of the two streams is directly manipulated and the second one ($\dot{B}$) is left free. In the condensate partition, however, the rule is violated since both partition streams are directly manipulated. Moreover, the partition of the bottoms does not meet Rule 1 since none of the two partition streams is directly manipulated. Thus, the D-L configuration must be discarded.

In the D-G configuration, the distillate $\dot{D}$ and the boilup $\dot{G}$ are the directly manipulated streams. Thus, partition Rule 1 is met in the partition of the feed into the products $\dot{D}$ and $\dot{B}$. The partition of the condensate into distillate $\dot{D}$ and reflux $\dot{L}$ follows Rule 1 as $\dot{D}$ is directly manipulated. The same holds for the partition of the bottoms since the boilup $\dot{G}$ is directly manipulated and the bottoms $\dot{B}$ is left free. Consequently, the D-G configuration meets all requirements. It is a favorable control configuration.

The D-B configuration, however, violates Rule 1 by directly manipulating both product streams $\dot{D}$ and $\dot{B}$ of the feed partition. This configuration is not feasible.

The L-B configuration meets all requirements of the partition Rule 1. This configuration is equivalent to the D-G configuration.

A special situation exists in the L-G configuration. The Rule 1 is met in the partition of the condensate into $\dot{D}$ and $\dot{L}$ as well as in the partition of the bottoms into $\dot{G}$ and $\dot{B}$. But none of the two streams of the feed partition into $\dot{D}$ and $\dot{B}$ is directly manipulated as required. Both product streams are left free. Such a configuration is feasible but not as favorable as the D-G and the L-B configurations [Wherry et al. 1984].

The last configuration listed in Table 9.2 is the B-G configuration. It violates the partition Rule 1 in the partition of the bottoms into $\dot{B}$ and $\dot{G}$. It is comparable to the D-L configuration and is, therefore, not feasible.

9.3.1 Basic Control Systems without Composition Control

The three feasible control configurations are illustrated in Figs. 9-12 to 9-14. In order to minimize disturbances caused by the feed, its flow rate and temperature

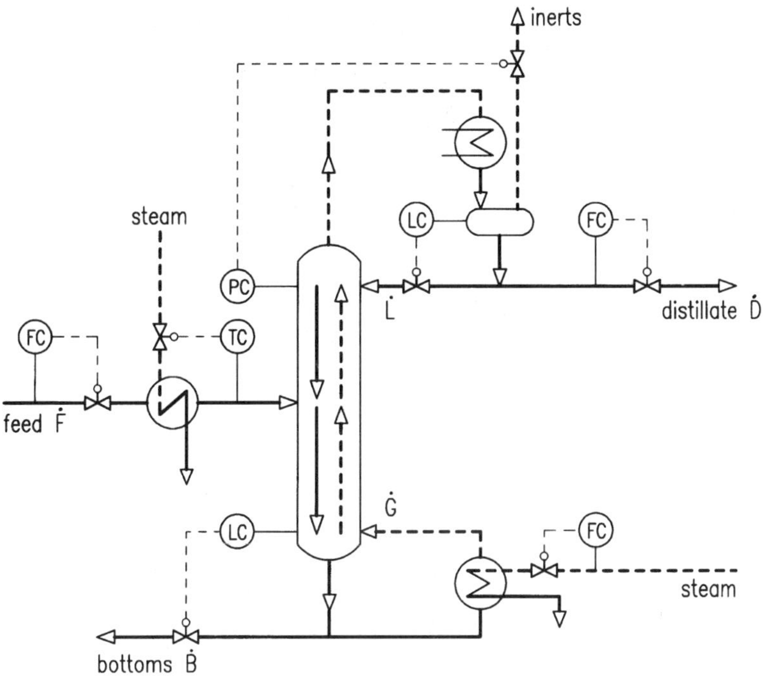

Fig. 9-12. Basic D-G control configuration.

are kept constant by separate controllers. The column pressure is controlled by manipulating the draw off of the inert gases from the condensate collector drum.

Figure 9-12 shows the D-G configuration where the distillate rate $\dot{D}$ and the boilup rate $\dot{G}$ are the directly controlled streams. Hence, the product withdrawal (i.e., material balance) and the energy input (i.e., energy balance) are manipulated independently, giving a very stable operation of the column.

The set point of the distillate flow controller has to meet the material balance:

$$\dot{D} < \dot{F} \cdot x_{Fa} \tag{9.3}$$

If this condition is violated, no pure overhead fraction can be gained regardless how high the separation efficiency of the column.

Equation (9.3) sets a necessary but insufficient condition for a pure overhead fraction. The set point of the boilup controller has to meet the condition:

$$\dot{G} > R_{Gmin} \cdot \dot{B} \tag{9.4}$$

The minimun reboil ratio R_{Gmin} is determined by the methods described in Section 4.2.4 and 4.3.3, respectively. The set point of the boilup controller is decisive for the separation efficiency of the column. If this set point is too low, no pure overhead

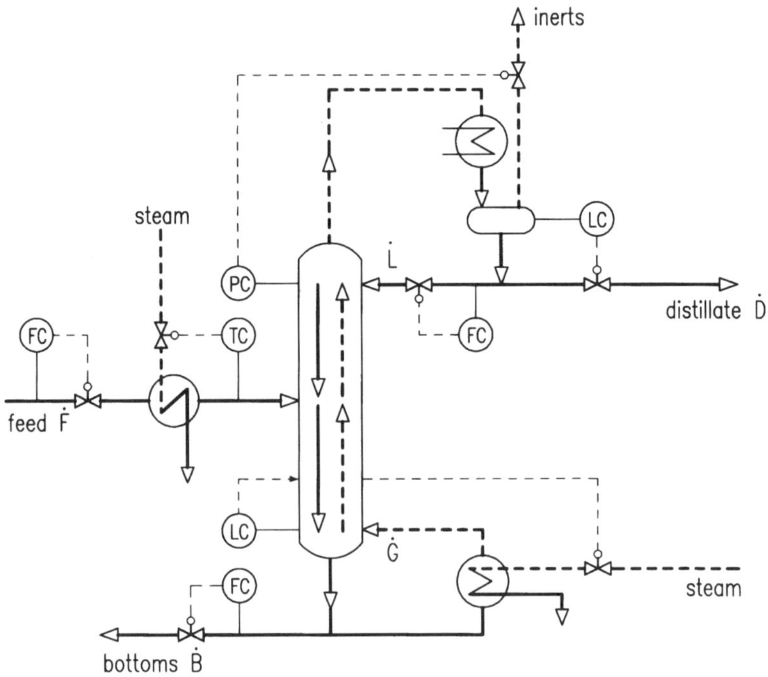

Fig. 9-13. Basic L-B control configuration.

fraction can be gained even in a column with a very large number of equilibrium stages.

The D-G configuration enables a very stable column operation since the material and energy balances are controlled by different control loops. The variation of the set point of the boilup controller does not, for instance, influence the material balance and vice versa. This control configuration should be applied when the overhead fraction is smaller than the bottom fraction (see Section 9.3.4).

The L-B configuration is depicted in Fig. 9-13. Here, the directly controlled streams are the reflux $\dot{L}$ and the bottom product $\dot{B}$. The set point of the bottom flow rate has to meet the material balance:

$$\dot{B} < \dot{F} \cdot x_{Fb} \tag{9.5}$$

This equation expresses that the flow rate of the bottom product must be lower than the amount of high boiler in the feed.

The set point of the reflux controller has to meet the following condition:

$$\dot{L} > R_{Lmin} \cdot \dot{D} \tag{9.6}$$

The minimum reflux ratio R_{Lmin} is determined by the equations derived in Section 4.2.4.

The L-B configuration is equivalent to the D-G configuration since the material and energy balances are manipulated by different controllers. The variation of the set point of the reflux controller does not affect the material balance and vice versa. This control configuration is applied to separations where the bottom fraction is smaller than the overhead fraction (see Section 9.3.4).

The basic L-G configuration is illustrated in Fig. 9-14. Here, the flow rates of the reflux $\dot{L}$ and the boilup $\dot{G}$ are directly manipulated. Neither of the two product streams $\dot{D}$ and $\dot{B}$ is directly controlled, i.e., the material balance around the column is left free.

The set point of the reflux controller must meet the condition:

$$\dot{L} > R_{Lmin} \cdot \dot{D} \tag{9.7}$$

Analogously, the condition for the set point of the boilup controller is:

$$\dot{G} > R_{Gmin} \cdot \dot{B} \tag{9.8}$$

As can be seen from Fig. 9-14, an increase of the set point of the boilup controller increases both the energy input into the bottom of the column and the flow rate $\dot{D}$ of the overhead product (since $\dot{D} = \dot{G} - \dot{L}$). Hence, the energy and material balances

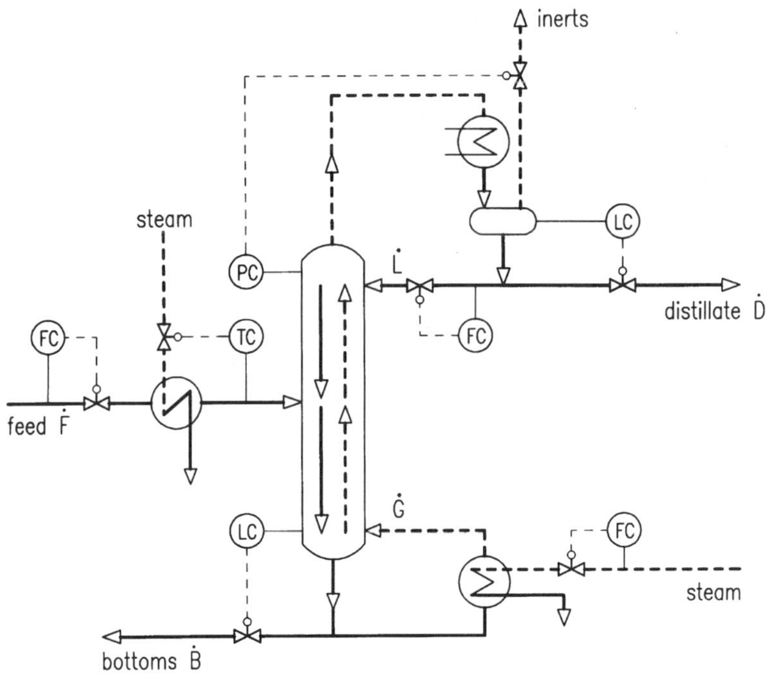

Fig. 9-14. Basic L-G control configuration.

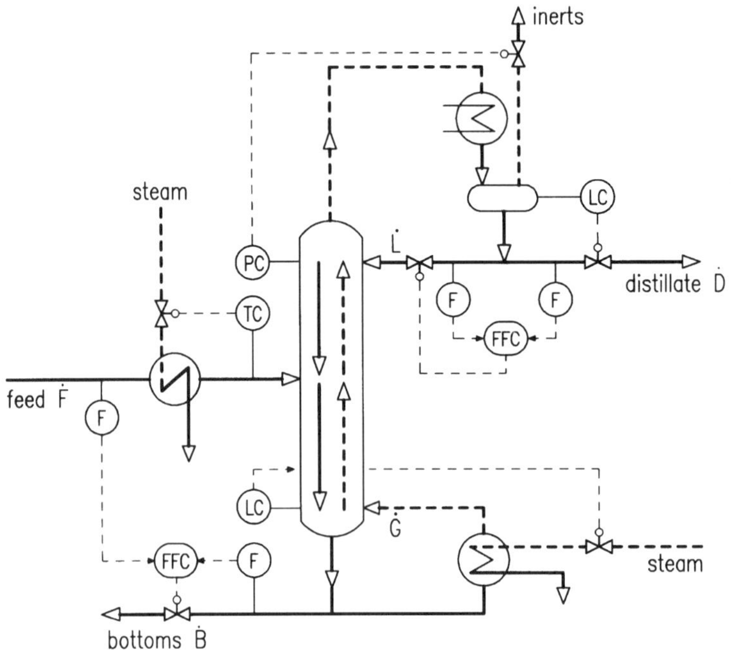

Fig. 9-15. *Ratio control applied to an L-B configuration.*

are not separately controlled. Manipulation at one end of the column causes unfavorable variations at the other end of the column.

Vice versa, an increase of the reflux flow rate $\dot{L}$ increases both the separation efficiency of the column and the flow rate of the bottom fraction. The increase of the reflux improves the separation efficiency of the column, but the increase of the bottom fraction may violate the condition that the bottom fraction must be less than the amount of high boiler in the feed. Consequently, the L-G configuration is not as favorable as the D-G and the L-B configurations.

Until now, column operations with constant feed flow rate and constant feed composition have been considered. However, these requirements are seldom met in industrial applications. Variations in feed flow rates can easily be compensated for when the two set points of the flow rates are replaced by the set points for the ratios, e.g., $\dot{B}/\dot{F}$ and $\dot{L}/\dot{D}$ (i.e., reflux ratio R_L). Such a control system is illustrated in Fig. 9-15. It corresponds completely with Fig. 9-13 with the exception that the two set points are not flow rates but flow ratios. This control system can be applied when the bottom product has to be recovered in pure form. When the low boiler has to be removed in pure form as distillate, an analogous ratio control structure can be developed from rate control structure (Fig. 9-12).

The main disadvantage of ratio control systems is the need for more flow rate measurements which makes them more difficult to implement and more sensitive to failure. Besides, the feed flow rate as well as the feed composition often vary in industrial applications. Hence, composition control is the better choice.

9.3.2 One-Point Composition Control Systems

The basic control configurations developed in the previous sections are only applicable when the composition of the feed is constant. If, for instance, the amount of the low boiler in the feed varies (and this is the normal situation in practice) the set point of the distillate flow controller must be permanently adjusted to the feed concentration. This is particularly important in separations with both high purity and high yield. If the distillate flow $\dot{D}$ is higher than the amount of low boiler in the feed (i.e., $\dot{D} > \dot{F}\ x_{Fa}$) some high boiler *b* is forced into the overhead fraction even when the separation efficiency of the column (i.e., number of stages and value of the reboil ratio) is extremly high.

Figure 9-16 shows the one-point composition control applied to the D-G configuration. The set point of the distillate flow controller is no longer externally set. It is manipulated by a temperature controller in the upper section of a column. If the distillate flow rate $\dot{D}$ is too high, some high boiler is forced into the overhead fraction that makes the temperature in the upper column section rise. This is detected by the temperature measurement device, and the temperature controller reduces the flow rate of the distillate. By proper location of the temperature measurement and proper selection of the set point of the temperature controller, the condition $\dot{D} < F \cdot x_{Fa}$ can be met provided the boilup $\dot{G}$ is high enough to establish the separation.

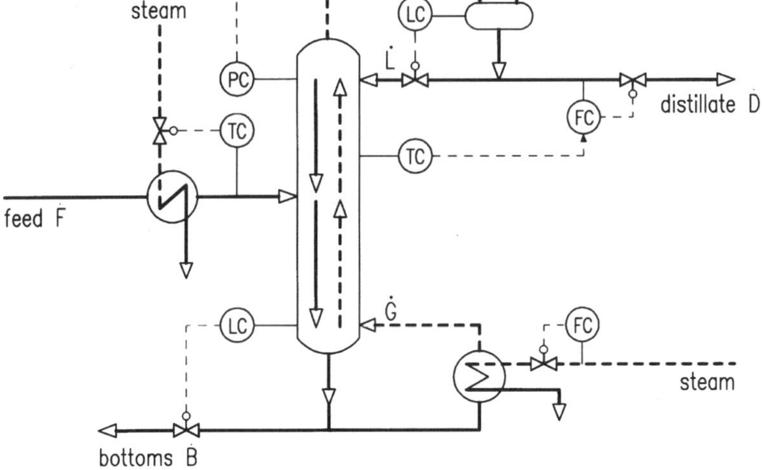

Fig. 9-16. *One-point composition control applied to an D-G configuration.*

In Fig. 9-17 the one-point composition control is applied to the L-B control configuration. Analogously to Fig. 9-16, a temperature controller in the stripping section of the column automatically adjusts the bottom flow rate $\dot{B}$ to the condition $\dot{B} < F \cdot x_{Fb}$ if the reflux $\dot{L}$ is high enough to establish the separation. The set point of the reflux controller must still be set externally in a proper way.

The configuration illustrated in Fig. 9-17 is the famous criss-cross configuration that, at first glance, is very astonishing since the temperature in the lower section of the column is manipulated by the flow rate of the bottom product $\dot{B}$ and not by the heating system. However, such a configuration performs very well when the bottom fraction $\dot{B}$ is very small [Bogenstaetter and Hengst 1959].

Application of one-point composition control to the L-G configuration can be established in two different manners, either to the reflux $\dot{L}$ or to the reboil $\dot{G}$. Figure 9-18 shows the application of the temperature control to the reflux. The set point of the reflux controller is automatically adjusted by the temperature controller in order to compensate for perturbations of feed flow rate and feed composition. A pure overhead product and a high yield of low boiler are possible if the set point of the boilup controller is high enough. An analogous control system is illustrated in Fig. 9-19. It should be applied when a high yield and a high purity of the bottom fraction is desired.

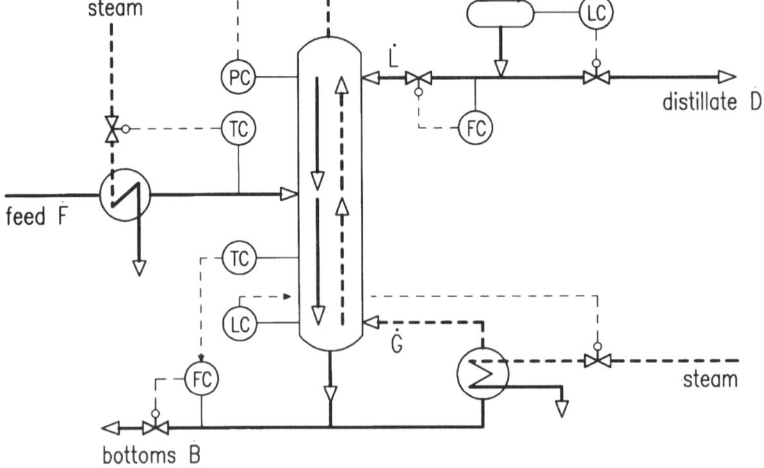

Fig. 9-17. One-point composition control applied to an L-B configuration.

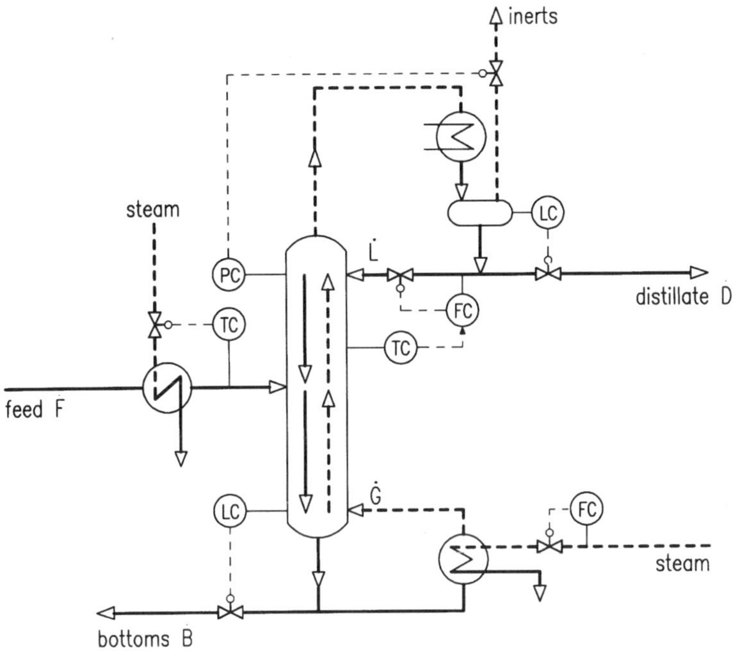

Fig. 9-18. *One-point composition control applied to an L-G configuration.*

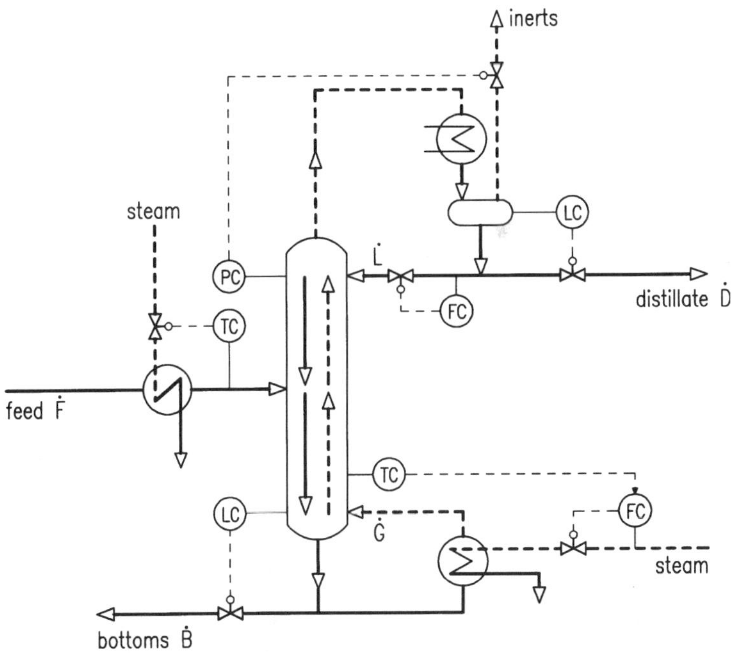

Fig. 9-19. *One-point composition control applied to an L-G configuration.*

One-point composition control is a very favorable modification of the basic control configurations. Most industrial columns are controlled by such control systems.

9.3.3 Two-Point Composition Control Systems

When the feed-flow rate as well as the feed composition are not constant and, furthermore, both pure distillate and pure bottoms have to be recovered, two-point composition control might be considered. Here, the flow rates of both manipulative streams are controlled by temperature controllers.

Figures 9-20 and 9-21 demonstrate the application of two-point composition control to the D-G configuration and to the L-G configuration, respectively. Several authors [Bogenstaetter, and Hengst 1959, Skogestad et al. 1990] report that such a control structure works well if the loops are carefully tuned and only PI-controllers are used. However, there is a high risk of severe control loop interaction. The two temperature controllers may act against each other. A common way to suppress interaction is to tune one loop very tight and the other loop loose. However, the performance of the slow loop is sacrificed to a certain extent.

The principle problem of two-point composition control lies in the proper selection of the set points of the two temperature controllers. Both set points have to lie

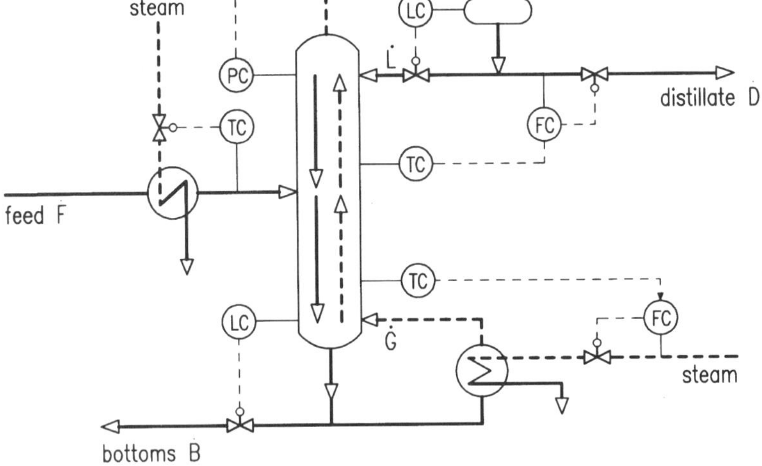

Fig. 9-20. Two-point composition control applied to a D-G configuration. There is a high risk of interaction of the two temperature controllers.

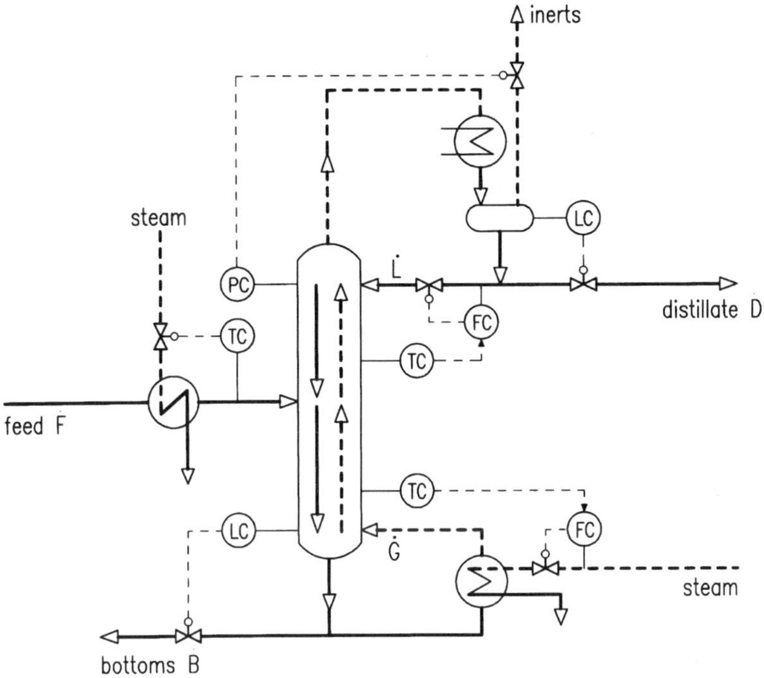

Fig. 9-21. *Two-point composition control applied to an L-G configuration. There is a high risk of interaction of the two temperature controllers.*

on the temperature profile developed within the column during operation. Thus, the two set points cannot be chosen independently. However, a pair of set points consistent with the initial state of the feed might not be consistent when the state of the feed has changed. Here, application of advanced control systems that offer the possibility of calculating consistent pairs of set points might be required [Wozny et al. 1991].

The D-B control configuration has been classified in Table 9-2 as not feasible since it violates the material balance of the column. In rare cases the D-B configuration can be made workable in form of a two point composition control. Both the bottoms and the distillate flow rates are manipulated by temperature controllers. If the bottoms flow rate is, for instance, higher than the amount of high boiler in the feed, some low boiler is entrained into the bottoms what decreases the temperature in the lower column section. If, vice versa, the flow rate of the distillate is too high some high boiler is entrained into the distillate what makes the temperature arise in the upper column section. Consequently, violations of the material balance around the column are detected by the temperature sensors in the upper and the lower column sections. The D-B control configuration can be realized in this form. However, the D-B configuration with two-point concentration control should be applied in very special cases only, e.g. very close boiling systems with both pure bottom and pure overhead fractions. It is a very risky control configuration but better than no control at all.

9.3.4 Range of Applications of the Control Configurations

As has been outlined in the previous sections, three different basic control configurations exist for the control of distillation columns. These basic control configurations are depicted in a rather simplified form in Fig. 9-22. A fourth configuration, the D-B configuration, is not feasible since it violates the material balance around a distillation column, see Table 9-2. Each of the three feasible control configurations can be modified in many respects.

Figure 9-23 shows some modifications of the D-G control configuration. If, for instance, the feed flow rate varies very much, ratio control (FFC) should be applied instead of flow control (FC). However, a lag compensation is required for the forward loop. If the feed composition varies, or if a high yield is required, temperature control replaces flow control of the overhead and bottom fraction, respectively. This one-point quality control is most widely used in industrial distillation processes. Two-

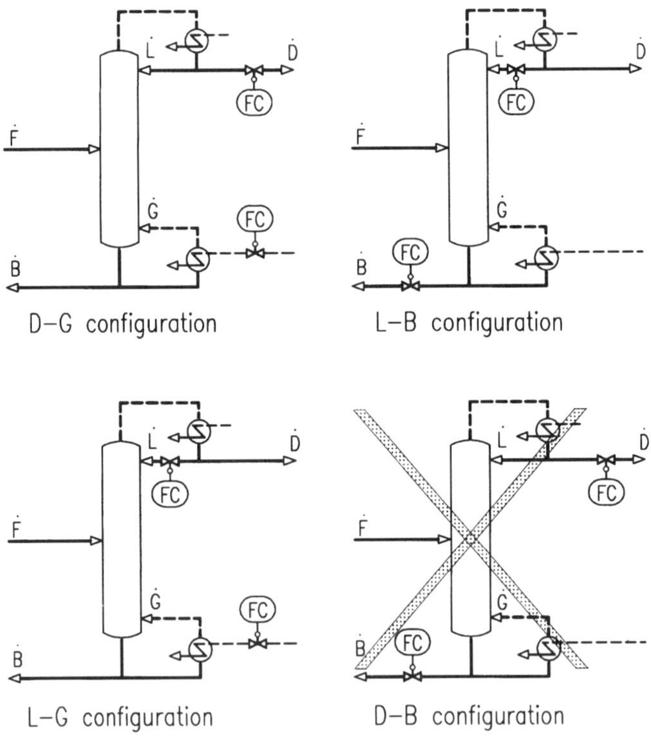

Fig. 9-22. Basic control configurations. The points of the directly manipulated flow rates are externally set.

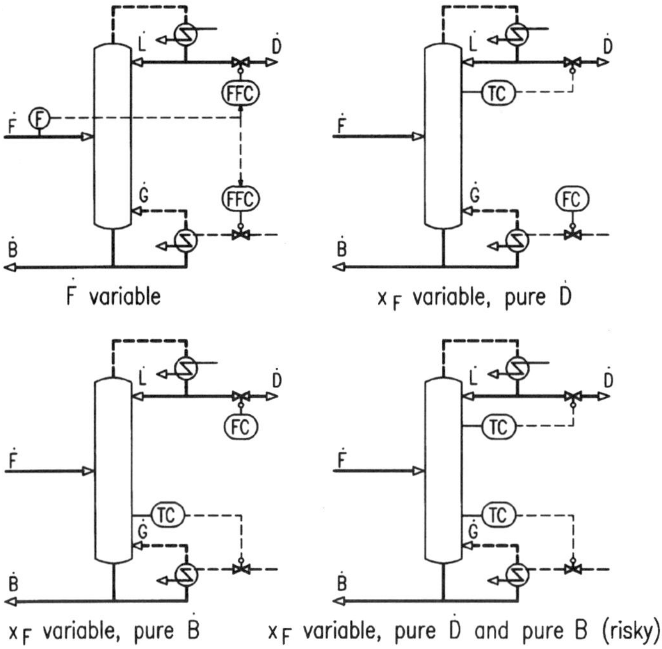

Fig. 9-23. *Some examples for the modification of the D-G configuration. The control system up right is very often used in industry.*

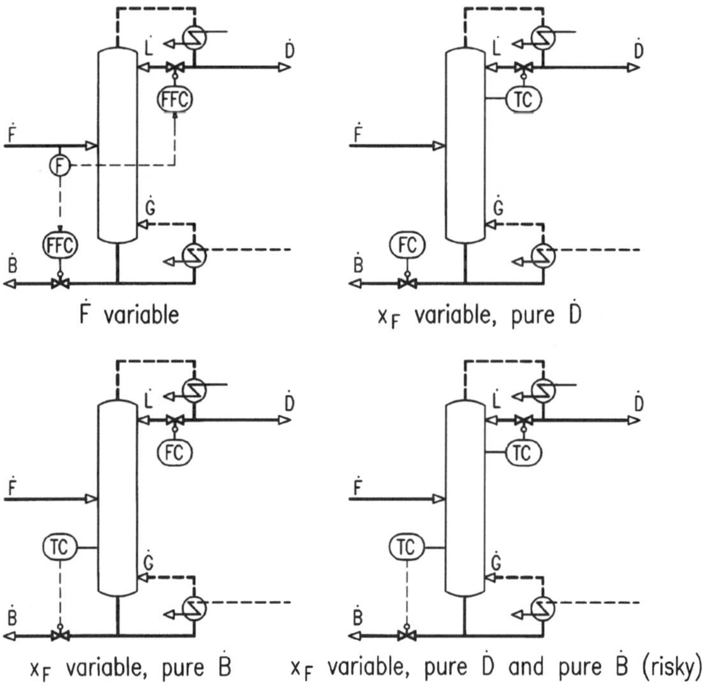

Fig. 9-24. *Some examples for the modification of the L-B configuration. The control system left down is the famous criss-cross control scheme.*

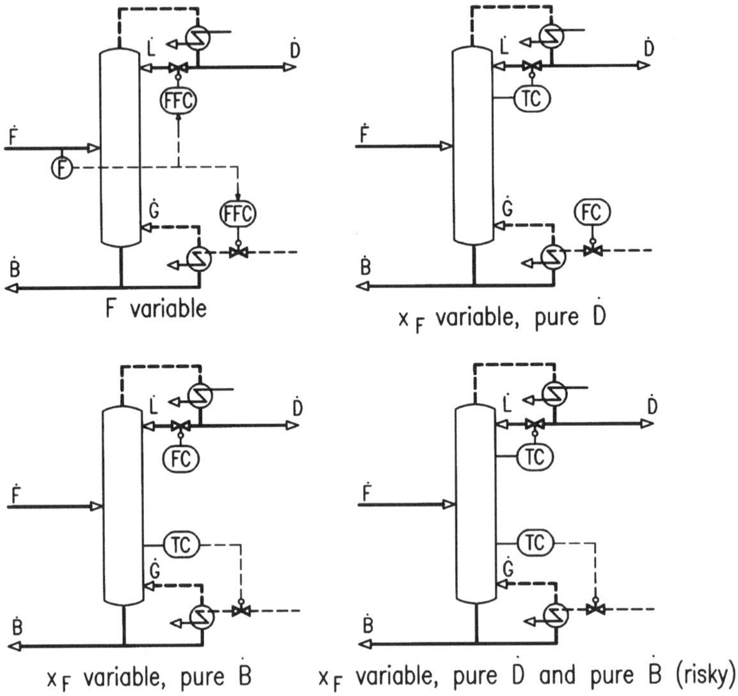

Fig. 9-25. *Some examples for the modification of the L-G configuration.*

point quality control is, although often proposed, very risky, e.g., [Heckle 1981]. The fundamental problem of two-point quality control is the thermodynamically consistent selection of the setpoints of the two temperature controllers.

Analogous modifications are presented in Figs. 9-24 and 9-25 for the L-B and the L-G control configuration, respectively. Much more control schemes are described in the literature and used in the industry. One must be aware, however, that all these control schemes are just modifications of the three feasible basic configurations. The principle difference, i.e., the choice of the directly manipulated streams, is not affected by these modifications.

A very important question that must be answered in developing a control system is which configuration should be applied to which type of separation. The appropriate control configuration for a given separation depends on the nature and state of the feed and on the yield and quality of the required products. Of great importance are, additionally, perturbances of flow rate and composition imposed on the feed stream. In this section the answer to this question will be given.

A distillation column represents, in essence, a system of three interconnected stream partitions as depicted in Fig. 9-11. Each stream partition is characterized by the value of the partition parameters reflux ratio R_L, reboil ratio R_G and product ratio $\dot{D}/\dot{B}$, respectively. The parameters are related by Equation (9.1). Hence, only two out of three partition parameters can be choosen independently, the value of the third one results from relation (9.1).

Figure 9-26A shows the graph of Equation (9.1). The reflux ratio R_L is plotted versus the distillate to feed ratio $\dot{D}/\dot{F}$ with the reboil ratio R_G as parameter. The values $R_L = 1$, $R_G = 1$ and $\dot{D}/\dot{B} = 1$ (i.e., $\dot{D}/\dot{F} = 0.5$), being of special importance, are marked by bold lines. The solid parameter lines refer to a boiling liquid feed, i.e., $q_F = 1$; the dashed ones to a saturated vapor feed, i.e., $q_F = 0$. This is the appropriate plot of Equation (9.1) if a pure overhead fraction has to be recovered from the feed.

For the recovery of a pure bottom fraction the plot of Fig. 9-26B is better. Here, the reboil ratio R_G and the bottom fraction $\dot{B}$ are of prime interest. Consequently, the reboil ratio R_G is plotted versus the bottom to feed ratio $\dot{B}/\dot{F}$. The course of the parameter lines is identical in both diagrams of Fig. 9-26. Just the lines for liquid feed transfer into the lines for vapor feed. From a thermodynamic point of view, the separation of a pure overhead fraction from a vapor feed is identical to the separation of a pure bottom fraction from a liquid feed, and vice versa. This can easily be verified by an upside down version of the McCabe-Thiele diagram.

For the selection of the proper control configuration the partition Rule 2 (see Section 9.1) has to be considered. This rule demands that in each stream partition the smaller of the two partion streams should be directly manipulated. The values of the partition parameters provide information as to which of the two partition streams is the smaller one. A value of the reflux ratio $R_L = 2$ (i.e., $\dot{L}/\dot{D} = 2$) expresses, for instance, that the flow rate of the distillate $\dot{D}$ is much smaller than that of the reflux $\dot{L}$ ($\dot{L} = 2\ \dot{D}$). Hence, the distillate flow rate $\dot{D}$ must be directly controlled. At a value $R_L = 0.5$, however, the reflux $\dot{L}$ is the smaller stream ($\dot{L} = 0.5\ \dot{D}$) and must be directly controlled. What is decisive for the control configuration of any stream partition is a value <1 or >1 of the relevant partition parameter.

In the D-G configuration, the distillate $\dot{D}$ and the boilup $\dot{G}$ are the directly controlled streams. Partition Rule 2 requires that the directly controlled streams are the smaller ones in each partition. Thus, the distillate $\dot{D}$ has to be smaller than the bottoms $\dot{B}$ to meet the Rule 2 for the partition of the feed $\dot{F}$, i.e., $\dot{D}/\dot{B} < 1$ (or $\dot{D}/\dot{F} < 0.5$). Applying this rule to the partition of the condensate demands that the distillate flow rate $\dot{D}$ is smaller than the reflux flow $\dot{L}$, i.e., the reflux ratio R_L ($=\dot{L}/\dot{D}$) is larger than one. Analogously, in the partition at the bottom the directly controlled stream $\dot{G}$ must be smaller than the bottoms flow rate $\dot{B}$, i.e., the reboil ratio R_G ($=\dot{G}/\dot{B}$) is smaller than one. Thus, the range of application of the D-G configurations is limited by the conditions: $\dot{D}/\dot{F} < 0.5$; $R_L > 1$ and $R_G < 1$.

These limitations define the upper left region in the diagram of Fig. 9-27. The D-G configuration should be applied to this region only. From the plot follows that the condition $R_G < 1$ is more strict than the condition $\dot{D}/\dot{F} < 0.5$ (for boiling liquid feed).

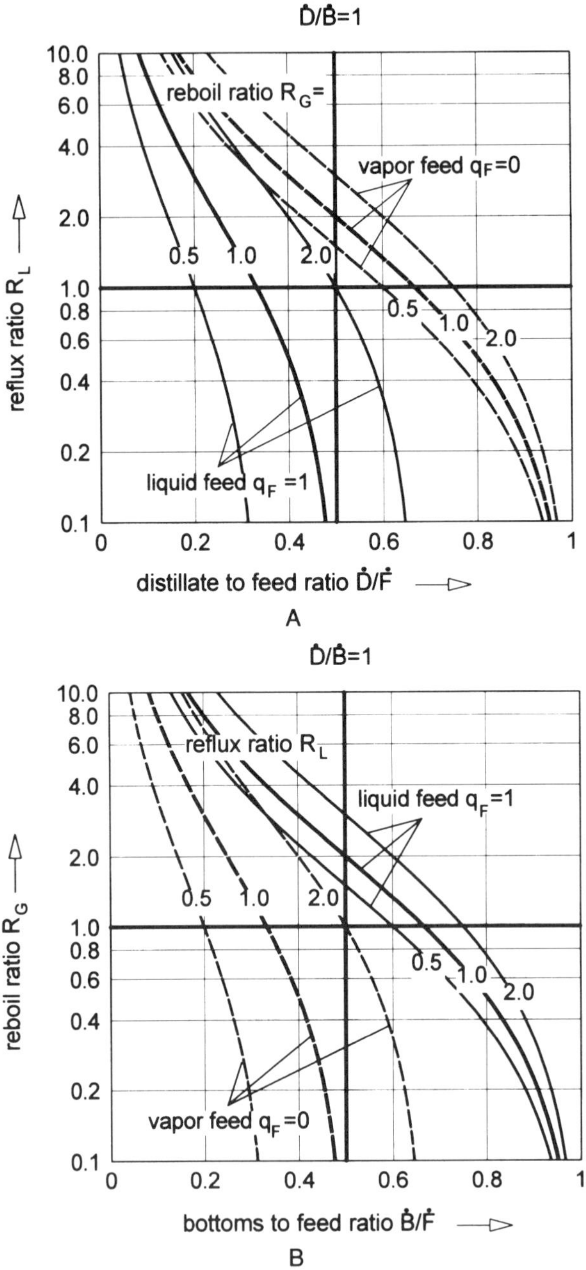

Fig. 9-26. *Graphical plot of the relation between the partition parameters R_L, R_G and $\dot{D}/\dot{B}$ according to Equation (9.2) for boiling liquid feed ($q_F = 1$) and saturated vapor feed ($q_F = 0$): (A) Pure overhead product, (B) pure bottom product.*

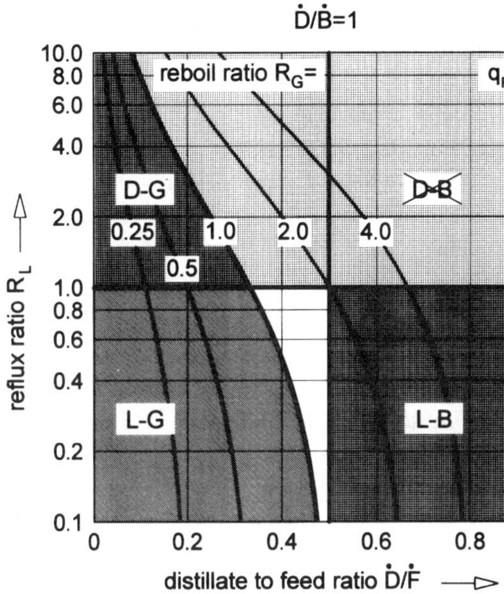

Fig. 9-27. Range of applicability of the different control configurations for separation of a pure overhead fraction from a boiling liquid feed.

In the L-B configuration the directly controlled streams are the reflux $\dot{L}$ and the bottom fraction $\dot{B}$. From Rule 2 it follows for the partition of the condensate that the flow rate of the reflux $\dot{L}$ is smaller than the flow rate of the distillate $\dot{D}$. Hence, the reflux ratio R_L ($=\dot{L}/\dot{D}$) must be lower than one. In the partition of the feed into distillate $\dot{D}$ and bottoms $\dot{B}$ the flow rate of the bottoms should be smaller than the distillate, i.e., $\dot{D}/\dot{B} > 1$ (or $\dot{D}/\dot{F} > 0.5$). Applying partition Rule 2 to the partition at the bottom results in the condition R_G ($=\dot{G}/\dot{B}$) > 1. Consequently, the limitations for the application of the L-B configuration are: $R_L < 1$, $\dot{D}/\dot{F} > 0.5$ and $R_G > 1$.

The region defined by these conditions is marked in the diagram of Fig. 9-27 (lower right). Here, the condition $\dot{D}/\dot{F} > 0.5$ is more strict than the condition $R_G > 1$ (for a boiling liquid feed).

Limitations for the application of the L-G configuration can be derived in an analogous manner. The reflux $\dot{L}$ being the directly controlled stream should be smaller than the distillate $\dot{D}$, i.e., the reflux ratio R_L ($=\dot{L}/\dot{D}$) must be smaller than 1. The second directly controlled stream is the boilup $\dot{G}$. According to Rule 2 it should be smaller than the bottoms product $\dot{B}$ which results in the condition R_G ($=\dot{G}/\dot{B}$) < 1. Consequently, the range of application of the L-G configuration is: $R_L < 1$ and $R_G < 1$. These limitations define the left down region in the diagram of Fig. 9-27.

The upper right region in Fig. 9-27 is described by the conditions $R_L > 1$ and $R_G > 1$. Hence, the distillate $\dot{D}$ and the bottoms $\dot{B}$ should be the directly controlled streams since both the reflux $\dot{L}$ and the boilup $\dot{G}$ are very large. However, such a D-B control configuration violates partition Rule 1 since both product streams of the feed partition are directly controlled. Such a control structure is forbidden as the material balance can never be met, see Table 9-2.

The diagram in Fig. 9-27 is valid for a boiling liquid feed and a pure overhead fraction. Analogous diagrams can easily be developed for a saturated vapor feed and for the separation of a pure bottom fraction. All these diagrams can be applied not only to binary mixtures but also to multicomponent mixtures, to nonideal or even azeotropic mixtures and to finite and infinite numbers of stages. The only restriction made is constant molar overflow.

9.3.4.1 Range of Applicability of the Control Configurations

The selection of an adequate control configuration depends, as has been demonstrated above, on the values of the partition parameters R_L, R_G and $\dot{D}/\dot{B}$. However, these values cannot be freely chosen, they are rather determined by the thermodynamics of the distillation process. The values of the partition parameters are the result of the thermodynamic calculation of the distillation process. Hence, their values are known when it comes to the design of the control configuration.

The question arises: Which values of the partition parameters R_L, R_G and $\dot{D}/\dot{B}$ are typically encountered in which systems to be separated? This question cannot be answered generally because the partition parameters depend in a very complex manner on the nature of the system. However, for ideal binary mixtures and an infinite number of stages, simple relations between the partition parameters and the system properties (i.e., relative volatility α and feed concentration x_{Fa}) exist, see Section 4.2.4 on minimum reflux and reboil. Since the practical operation of distillation columns is only a little bit over operation with minimum reflux and reboil, respectively, the results of the following considerations are relevant to practical column operation.

The distillate to feed ratio $\dot{D}/\dot{F}$ follows from a material balance when a pure overhead product is withdrawn:

$$\dot{D}/\dot{F} = x_{Fa} \quad \text{with } x_{Da} = 1 \tag{9.9}$$

The minimum reflux ratio follows from Section 4.2.4 (Equation 4.35):

$$R_{Lmin} = \frac{1}{(\alpha - 1)\, x_{Fa}} \quad \text{for } n \rightarrow \infty \text{ and } x_{Da} = 1 \tag{9.10}$$

This equation is valid for a boiling liquid feed, i.e., $q_F = 1$. With Equation (9.9) follows:

$$R_{Lmin} = \frac{1}{(\alpha - 1)\ \dot{D}/\dot{F}} \quad \text{for } n \rightarrow \infty \text{ and } x_{Da} = 1 \tag{9.11}$$

The analogous relations for the withdrawal of a pure bottom fraction are:

$$\dot{B}/\dot{F} = x_{Fb} \quad \text{with } x_{Bb} = 1 \tag{9.12}$$

The minimum reboil ratio is calculated from Equation (4.42) assuming boiling liquid feed, i.e., $q_F = 1$:

$$R_{Gmin} = \left(\frac{\alpha}{1 + (\alpha - 1)\ x_{Fa}} - 1\right)^{-1} \tag{9.13}$$

$$= \left(\frac{1}{1 - (1 - 1/\alpha)\ x_{Fb}} - 1\right)^{-1} \quad \text{for } n \rightarrow \infty \text{ and } x_{Bb} = 1$$

Combining Equations (9.12) and (9.13) gives:

$$R_{Gmin} = \left(\frac{1}{1 - (1 - 1/\alpha)\ \dot{B}/\dot{F}} - 1\right)^{-1} \quad \text{for } n \rightarrow \infty \text{ and } x_{Bb} = 1 \tag{9.14}$$

Similar relations exist in both cases for a saturated vapor feed. All these relations can be implicated in the plot of Equation (9.2). The resulting diagrams are presented in Fig. 9-28. The dashed parameter lines α = const. characterize the nature of the system to be separated (relative volatility α). The concentration of the feed is included in the abscissa according to Equations (9.9) and (9.12), respectively. The regions of the control configurations, as developed in the previous section, are also marked in Fig. 9-28.

From Fig. 9-28A follows, for instance, that the D-G configuration should be applied to wide boiling systems (with relative volatilities $\alpha > 2.5$) and low concentration of the feed ($x_{Fa} < 0.33$) when a pure overhead fraction has to be separated from a boiling feed. In the case of a saturated vapor feed (Fig. 9-28B) the range of application of the D-G configuration is larger. The feed concentration is as high as $x_{Fa} = 0.5$ when the relative volatility is larger than $\alpha = 3$. For the separation of a pure bottom fraction from a liquid feed (Fig. 9-28C) the D-G configuration can be applied if the volatility α is higher than 4, provided high concentrations of the high boiler *b* in the feed ($x_{Fb} > 0.66$). For a vaporous feed (Fig. 9-28D), the range of feed concentration is expanded to values as low as $x_{Fb} = 0.5$, for relative volatilities $\alpha > 2$.

The L-B configuration should be applied to wide boiling systems, too, but high feed concentrations of the low boiler ($x_{Fa} > 0.5$) and, consequently, low feed concentration x_{Fb} of the high boiler (see Fig. 9-28A). Here, the range of application is larger for separating a pure overhead product from a liquid feed than from a vapor feed (Fig. 9-28B).

The application of the L-G configuration is in all four cases restricted to intermedi-

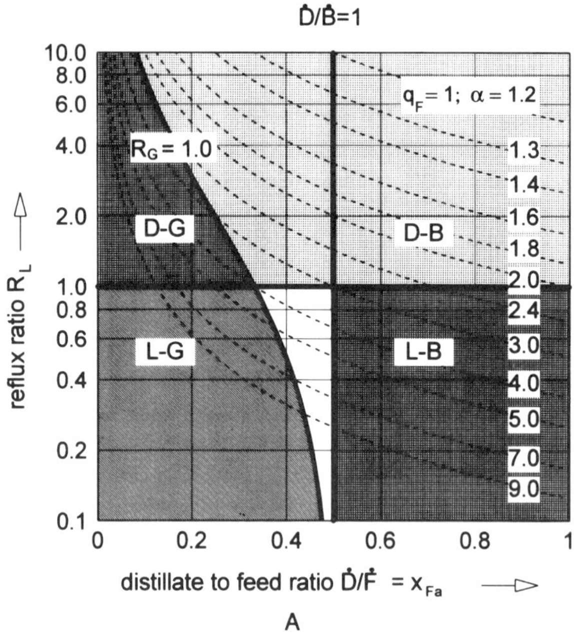

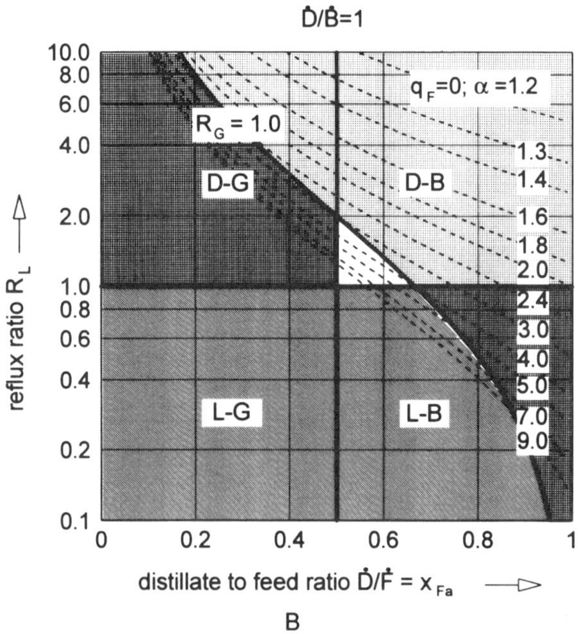

Fig. 9-28. *Range of applicability of the basic control configurations. The dashed parameter lines are valid for ideal binary mixtures and infinite number of stages (minimum reflux and reboil): (A) Pure overhead fraction and boiling liquid feed (caloric factor $q_F = 1$), (B) pure overhead fraction and saturated vapor feed (caloric factor $q_F = 0$).*

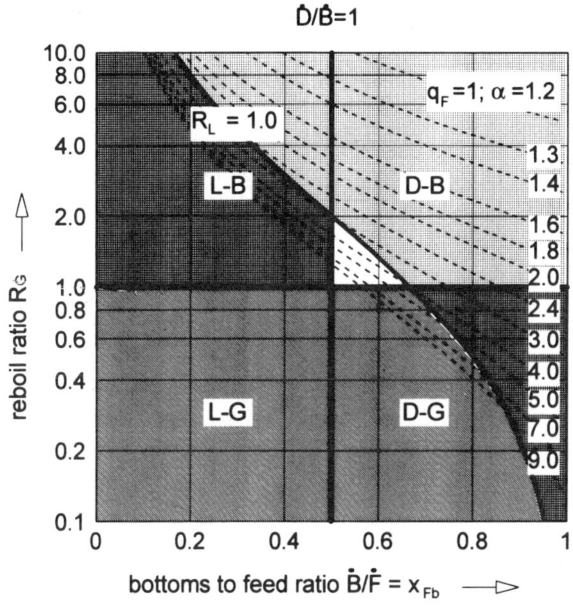

C

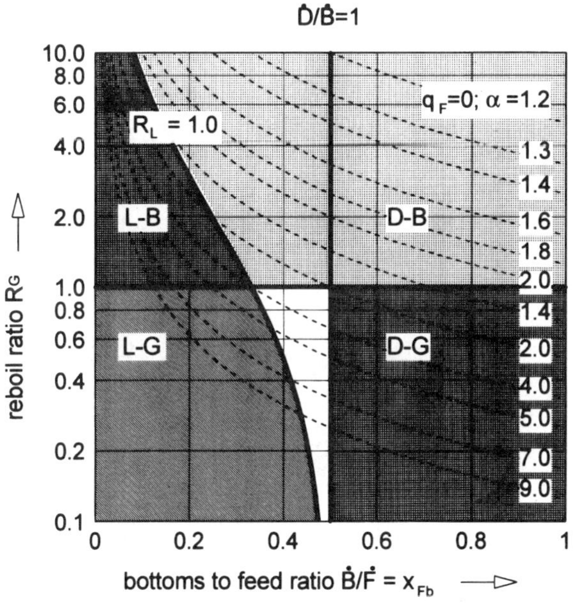

D

Fig. 9-28. *(continued) (C) Pure bottom fraction and boiling liquid feed (caloric factor q_F = 1), (D) pure bottom fraction and saturated vapor feed (caloric factor q_F = 0).*

ate feed concentrations x_{Fa} and x_{Fb}, respectively, and to systems with very high values of relative volatilities $\alpha > 4$. Thus, this control configuration, although often considered in literature as a standard configuration, has only a very limited range of applicability to practical distillation problems. The restriction to very wide boiling systems and the disadvantage of not decoupling the material balance and the enthalpy balance leads to the recommendation not to apply L-G control configuration. As will be shown in the next section, its range of applicability is also covered by the expanded ranges of the D-G and L-B configuration, respectively.

9.3.4.2 Expanded Range of Application of the Control Configurations Partition Rule 2 that requires direct control of the smaller partition stream is not a strict rule. A control system also performs well when the directly controlled stream is a little larger than the free stream. Hence, the range of applicability of the different control configurations can be somewhat expanded. However, if the directly controlled stream is much larger than the free one, the control system will fail at higher perturbances of flow rates. The more the partition parameter deviates from value 1, the more strictly must partition Rule 2 be obeyed.

The expanded range of applicability of all feasible control configurations is presented in this Section only for the case of a boiling liquid feed from which a pure overhead fraction has to be separated. Analogous diagrams for a vapor feed and pure bottom fractions are easily developed. In the expanded regions of applicability the critical values of the partition parameters are assumed to be 2 (moderately shaded area) and 4 (lightly shaded area).

Figure 9-29 shows the ranges of applicability of the D-G control configuration. In the heavily shaded region the smaller stream is directly controlled. The expanded regions, where the larger stream is directly controlled and the smaller stream is left free, are moderately or lightly shaded. This diagram shows that the D-G control figuration can be applied to systems with relative volatilities as low as $\alpha = 1.4$ (at $x_{Fa} < 0.3$) and with feed concentrations of the pure overhead or bottom fraction as high as $x_{Fa} = 0.75$ (at $\alpha = 6$).

The range of applicability of the L-B control configuration is shown in Fig. 9-30. It covers the region of high feed concentrations of the pure overhead product. Only very close boiling systems with relative volatilities $\alpha < 1.3$ are not capable of control by this configuration.

The expanded range of applicability of the L-G configuration is plotted in Fig. 9-31. It mainly covers the range of small feed concentrations x_{Fa} and very large relative volatilities $\alpha > 3.5$. Most of its range of applicability is also covered by the D-G and the L-B configurations. Whenever possible, these configurations should be given precedence over the L-D configuration since material and energy balances are not decoupled in L-D configuration.

In summary, the D-G configuration is the best control configuration for low-feed concentrations of the pure product and, in turn, the L-B configuration is the best one for high-feed concentrations of the pure product. Both configurations cover all feed concentrations and nearly all relative volatilities α. Only systems with very low volatilities $\alpha < 1.3$ are difficult to control by these configurations. Here, applica-

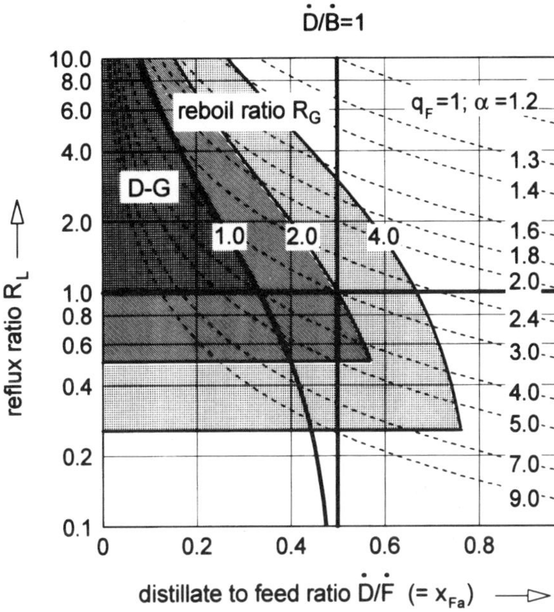

Fig. 9-29. *Expanded range of applicability of the D-G configuration, for separation of a pure overhead fraction from a boiling liquid feed.*

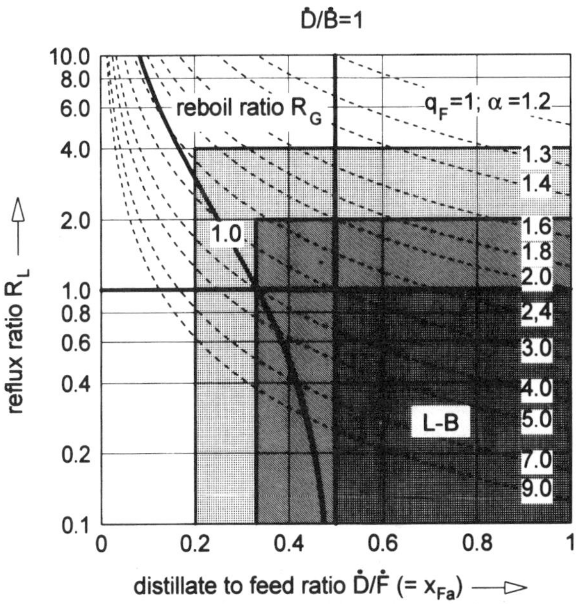

Fig. 9-30. *Expanded range of applicability of the L-B configuration, for separation of a pure overhead fraction from a boiling liquid feed.*

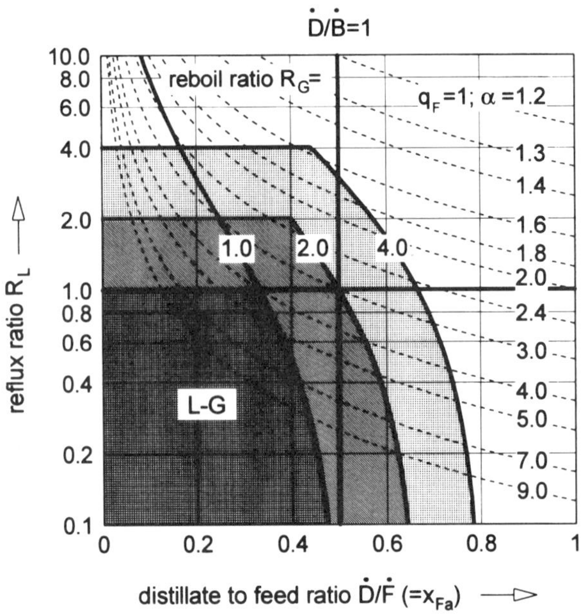

Fig. 9-31. Expanded range of applicability of the L-G configuration, for separation of a pure overhead fraction from a boiling liquid feed. This configuration is less advantageous than the D-G and the L-B configuration, since material balance and energy balance are not separately controlled.

tion of the D-B configuration with two composition control should be considered, see Section 9.3.3.

9.4 CONTROL SCHEMES FOR DISTILLATION PROCESSES

A distillation process consists, in general, of a sequence of several distillations columns and many auxiliary devices, e.g., vessels, heat exchangers, pumps, compressors etc. Typically, several streams are recycled within the process making the structure very complex. Hence, the control configuration of a distillation process is not a mere compilation of the control schemes of single columns. Some examples for process control are presented in the following section.

9.4.1 Control Schemes for Continuously Operated Processes

9.4.1.1 Separation of an Azeotropic Mixture by Changing the Pressure Binary mixtures with an azeotrope can be separated by changing system pressure as has been outlined in Chapter 5.5. Two columns are required for separating such a system. An important example is the mixture tetrahydrofuran/water. The basic process is depicted in Fig. 5-32. Since the mixture exhibits a minimum azeotrope, the pure products are drawn off as bottom fractions. In order to perform the separation the two columns have to be operated with different pressures.

In Chapter 7 it has been shown that thermal coupling of columns provides a very effective means for energy saving. The most essential prerequisite for column coupling is that the temperature level of the condenser of one column is higher than the temperature level of another column's reboiler. To meet this condition the columns have, in general, to be operated at different pressures. Thus, processes for separating azeotropic mixtures by changing system pressure are always candidates for thermal-column coupling. Details are presented in Fig. 7-14A for the system tetrahydrofuran/water.

The control system of this process is presented in Fig. 9-32 [Abu-Eishah and Luyben 1985]. In each column both the overhead and the bottoms products are manipulated by level controllers as in the basic control structure presented in Fig. 9-14. Such an L-G control structure is not optimal, but acceptable. Ratio control is applied in both columns to reflux. To avoid too many flow measurements, the flow rate of the distillate of the first column is also used for the ratio control of the reflux of the second column. The set points of the reboil ratios are manipulated by the temperatures in the stripping section of each column. Thus, one-point composition control, as described in Section 9.3.2, has been applied to the process.

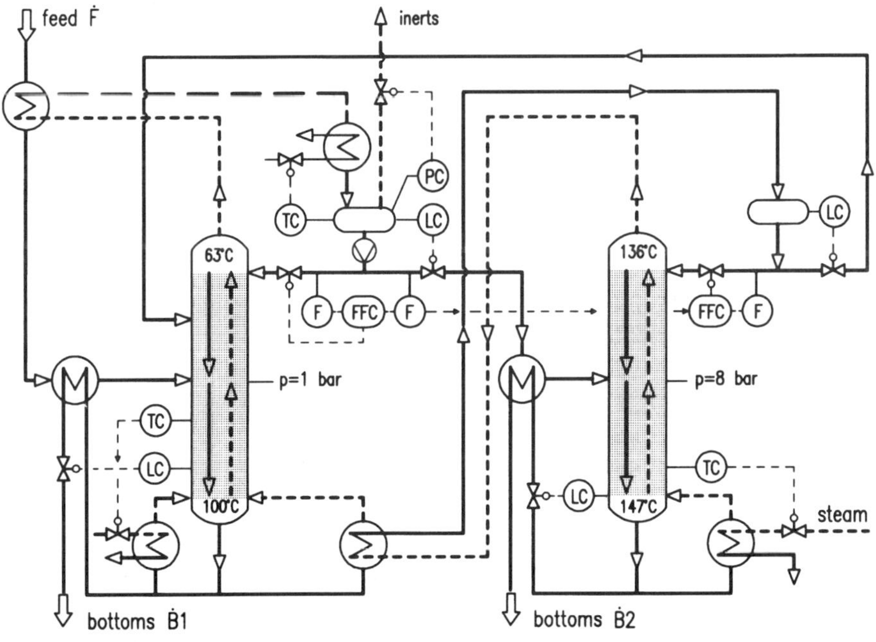

Fig. 9-32. *Control structure of a process for separating the binary azeotropic mixture tetrahydrofuran(a)/water(b) by changing system pressure. The columns are thermally coupled to save energy.*

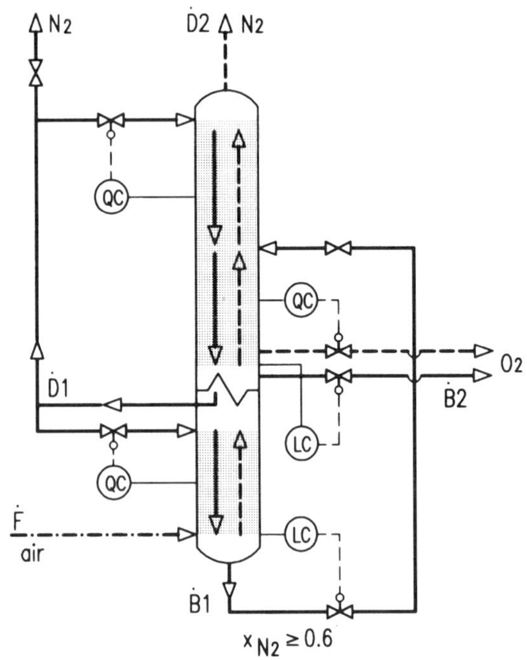

Fig. 9-33. *Control structure of an air separation double column. An L-B configuration is applied to the low pressure Column C-2 according to the rules developed in the previous Section. QC = quality control by oxygen sensors.*

9.4.1.2 Control of an Air Separation Column Air separation by distillation is performed in most cases by two thermally coupled columns, see Section 5.2.4. Some pure liquid nitrogen is removed from the partially condensed air in Column C-1 (Fig. 5-10) to serve as reflux in Column C-2. The bottom fraction $\dot{B}1$ of Column C-1, which consists of a rectifying section only, has an increased oxygen content of approximately 35 to 40 mol.-%. This fraction is fed into column C-2 to be separated into pure oxygen as bottoms $\dot{B}2$ and pure nitrogen as distillate $\dot{D}2$. Thermal coupling requires operation of Column C-1 at an increased pressure to increase the bubble point of nitrogen over the bubble point of oxygen in Column C-2. In most cases the high pressure Column C-1 is installed directly below the low pressure Column C-2 for combining the condenser and the reboiler in a single heat exchanger.

A well approved control configuration is shown in Fig. 9-33 [Zapp 1994]. The reflux of the high pressure Column C-1 is directly manipulated. The set point of the flow control is permanently adjusted by keeping the concentration in the middle of the column constant. Untypically for distillation column control, a concentration measurement is used instead of a temperature measurement. A cascade loop should be used to provide a quick response.

A slightly modified L-B configuration is installed in the low pressure Column C-2. This is in accordance with Fig. 9-28, because the nitrogen/oxygen mixture is a wide boiling one ($\alpha = 4$) and the content of the low boiler in the feed of Column C-2 is high ($x_{FN2} = 0.6 - 0.65$). Reflux and bottoms flow rates are directly manipulated by keeping the concentrations in the rectifying and stripping sections constant (two-point quality control).

Thermal coupling of the two columns makes the flow rates of the vaporos overhead fraction $\dot{D}2$ and the liquid bottom fraction $\dot{B}2$ unflexible. Hence, variation of the bottoms flow rate is established by withdrawal of some of the bottom product in vapor state.

Essential for effective control is the proper location of the concentration measurement devices. Relative gain analysis has proven an efficient tool for that [Zapp 1994].

9.4.1.3 Control of a Process with a Main and a Side Column Separation of a zeotropic ternary mixture into the pure substances *a, b* and *c* requires two distillation columns. If the intermediate boiling constituent *b* is present only in minor concentrations the use of a side column is advantageous. The principles of such processes have been dealt with in Section 5.3.2, see especially Fig. 5-24.

Control configurations for such processes have been studied by [Glinos and Malone 1985, Alatiqi and Luyben 1986, Raisch et al. 1993]. In ternary mixtures the bubble point is not a univocal function of concentration as it is (at constant pressure) in binary mixtures. Quality control via temperature control is only feasible if the temperature sensors are located in those parts of the column where mainly two components are present, i.e., towards the column ends.

An acceptable control structure is presented in Fig. 9-34. One-point quality control

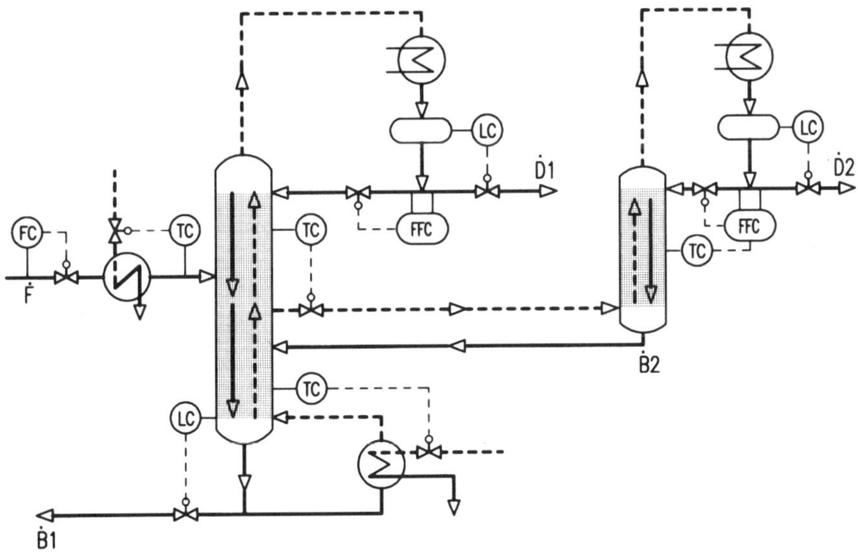

Fig. 9-34. Control structure of a process for separating a ternary mixture by a main and a side column.

is applied to both columns. In the main column the reboiler duty is controlled by the set point of a temperature in the stripping section of the column. The level in the reboiler is manipulated by the flow rate of the bottoms product. At the top of the column the level in the condensate collector is regulated by the flow rate of the distillate. The reflux is controlled by ratio control. The rate of the vaporos side stream is controlled by a proper temperature in the stripping section of the main column. In the side column the value of the reflux ratio is monitored by the set point of a temperature in the middle of the column.

This and other control configurations have been experimentally studied with the system methanol/ethanol/n-propanol by [Lang and Gilles 1989]. The controller managed to stabilize the plant but became unstable when larger disturbances of feed rate, and especially of feed concentrations, were imposed. Depending on the type of disturbances, only two out of three products meet the specifications at best. An improved control strategy suggested by [Alatiqi and Luyben 1986] uses the temperature difference below and above the point of side-stream withdrawal for manipulating the flow rate of the side stream. For a better control, advanced control strategies have to be applied. Processes with sidestream columns are prime candidates for multivariable robust control, e.g., [Levien and Morari 1987].

9.4.1.4 Control of an Azeotropic Distillation Process A very effective process for separating a binary mixture with a minimum azeotrope is azeotropic distillation, see Chapter 5.6.3.1. The separation of the feed into its constituents is established by an entrainer that is not completely miscible with the feed. The process consists of two distillation columns that deliver the products as bottom fractions and one decanter that regains the entrainer.

A classical example of an azeotropic distillation process is the separation of ethanol/water using toluene as entrainer, see Fig. 9-35. In principle, the whole process is a partition of the feed stream into the two product streams. In most cases, the ethanol content of the feed is low and, consequently, the ethanol product stream is smaller than the water product stream. Hence, the flow rate of the ethanol stream $\dot{B}2$ must be directly controlled and, in turn, the flow rate of the water product stream $\dot{B}1$ must be left free, i.e., it must be manipulated by a level controller. In order to account for a variable feed concentration the ethanol stream $\dot{B}2$ is manipulated by a temperature cascade control loop.

Application of partition Rule 1 to the Column C-1 requires direct flow control of the overhead fraction $\dot{D}1$ since the bottom fraction $\dot{B}1$ has been selected as the free partition stream. Hence, a D-G configuration is applied to Column C-1. Analogously, the decision of direct control of the bottom fraction $\dot{B}2$ of Column C-2 makes the overhead fraction $\dot{D}2$ a free partition stream that is manipulated by a level controller. Thus, Column C-2 is controlled by an L-B configuration. Additional control loops are required to maintain constant levels in the decanter and, most importantly, to circulate the proper amount of the entrainer within the process.

The proposed control scheme results from the partition of the feed into the products. This partition is superimposed on the partitions of the columns and the decanter. It must be checked whether or not the partition rules are met for all of these internal

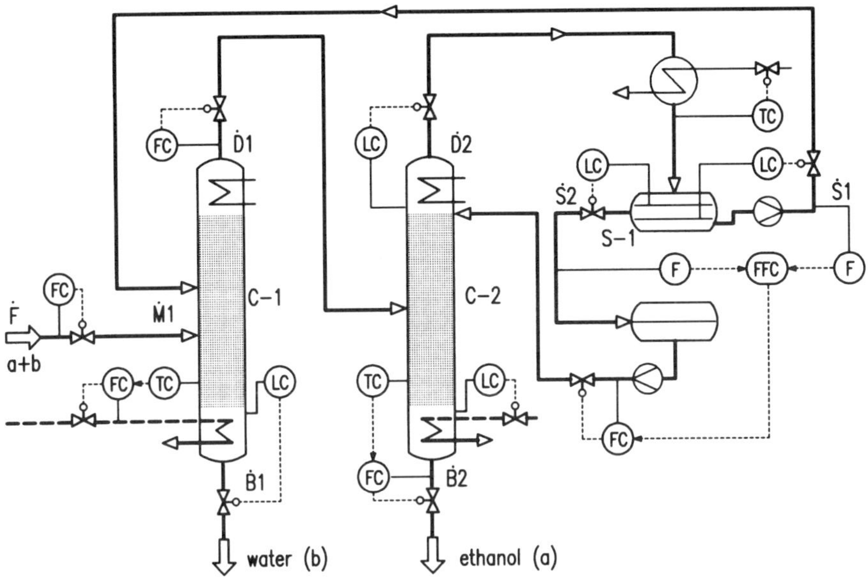

Fig. 9-35. Control structure of an azeotrope distillation process for separating ethanol-water with toluene as entrainer.

stream partitions. In conflicting situations, precedence should be given to the superimposed partition since it stabilizes the operation of the whole process.

9.4.2 Control Schemes for Batch Distillation Processes

In continuously operated distillation columns there exists a steady nonlinear temperature profile within the column. The temperature gradient $\partial T/\partial n$ (with n = number of stages) significantly varies along the column. The tray that exhibits the highest temperature gradient is the optimal sensor point for quality control since disturbances that make the temperature and concentration profiles move within the column are most easily detected there.

In batch distillation, however, the temperature profile is not constant over time. The temperatures steadily increase, i.e., the temperature profile moves upwards during operation. The point of the highest gradient of the temperature profile is no longer fixed within the column.

According to [Egly and Ruby 1980] the temperature sensors should be located so that they characterize product quality over the entire time of operation and that, as well, they exhibit the highest sensitivity to disturbances. The authors identified 4 trays that meet these requirements by rigorous simulation of the non-steady state behavior of a 50-tray column. The resulting control configuration is schematically depicted in Fig. 9-36. Four temperature sensors are implemented in the column for identification of unwanted variations of the process. The actual temperatures of these 4 trays are compared with the theoretical values by a computer. The desired state

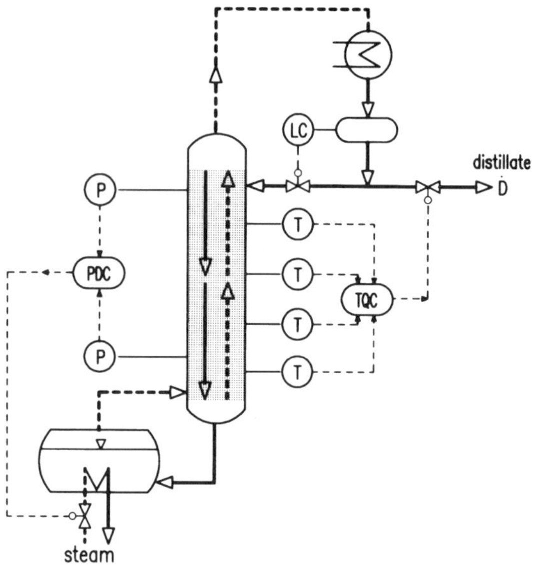

Fig. 9-36. Control structure of a batch distillation process operated with constant product quality.

of operation is maintained by manipulating the distillate rate and, thus, at constant heating, the reflux rate. The authors studied the effectiveness of this control configuration at the example of a binary mixture with constant product quality. The reflux ratio is automatically increased by the control system, and disturbances are compensated for very effectively.

LITERATURE

Abu-Eishah, S.I. and Luyben, W.-L.: "Design and Control of a Two-Column Azeotropic Distillation System," *Ind. Eng. Chem. Process Des. Dev.* Vol. 24 (1985) No. 1, pp. 132/140.

Alatiqi, I. M. and Luyben, W. L.: "Control of a Complex Sidestream Column/Stripper Distillation Configuration," *Ind. Eng. Chem. Process Des. Dev.* 25 (1986) No. 3, pp. 762/767.

Bogenstaetter, G. and Hengst, K.: "Regelung von Destillationskolonnen," *Chem.-Ing.-Techn.* 31 (1959) No. 7, pp. 425/431.

Deshpande, P. B.: *Distillation Dynamics and Control.* Instrument Society of America, Research Triangle Park, NC, 1985.

Egly, E. and Ruby, V.: "Betrieb und Regelung energieoptimaler Batch Prozesse," *Vt "Verfahrenstechnik"* 14 (1980) No. 5, pp. 351/355.

Glinos, K. and Malone, M. F.: "Steady-State Control of Sidestream Distillation Columns," *Ind. Eng. Chem. Process Des. Dev.* 24 (1985) No. 3, pp. 608/613.

Heckle, M.: *Destillationskolonnen, in Hengstenberg, J.* Springer Verlag, Berlin, 1981

Kister, H. Z.: *Distillation Operation.* McGraw-Hill Publishing Company, New York, 1989.

Lang, L. and Gilles, E. D.: *Multivariable Control of Two Coupled Distillation Columns for Multicomponent Separation on a Pilot Plant Scale.* Preprints Dycord, Maastricht, 1989, pp. 53/60.

Levien, K. L. and Morari, M.: "Internal Model Control of Coupled Distillation Columns," *AIChE Journal* 33 (1987) No. 1, pp. 83/98.

Luyben, W. L.: *Process Modelling, Simulation, and Control for Chemical Engineers* McGraw-Hill Kogakuska, Tokyo, 1973, pp. 332/351.

Luyben, W. L.: *Practical Distillation Control* Van Nostrand Reinhold, New York, 1992.

Patwardhan, A. A. and Edgar, T. F.: "Nonlinear Model Predictive Control of a Packed Distillation Column," *Ind. Eng. Chem. Res.* 32 (1993) pp. 2345/2356.

Raisch, J.; Lang, L.; and Gilles, E.-D.: "H_{oo}-Reglerentwurf für Zwei- und Dreistoffdestillationsprozesse," *at-Automatisierungstechnik* 41 (1993) No. 6, pp. 215/224.

Rijnsdorp, J. E.: "Interaction in Two Variable Control Systems for Distillation Columns—I *Automatica* Vol. 1 (1965) pp. 15/28.

Rijnsdorp, J. E.: "Interaction in Two Variable Control Systems for Distillation Columns—II *Automatica* Vol. 1 (1965) pp. 29/52.

Shinskey, F. G.: *Distillation Control.* McGraw-Hill, New York, 1977.

Sieler, W.: "Zur Auswahl von Regelsystemen bei Destillationskolonnen," *Regeltechnische Praxis und Prozeß-Rechentechnik* (1970) No. 4, pp. 142/148.

Skogestad, S. and Morari, M.: "Control Configuration Selection for Distillation Columns," *AIChE-Journal* Vol. 33 (1987) No. 10, pp. 1620/1635.

Skogestad, S.; Lundström, P.; and Jacobsen, E. W.: "Selecting the Best Distillation Control Configuration," *AIChE-Journal* Vol. 36 (1990) No. 5, pp. 753/764.

Skogestad, S.: "Dynamics and Control of Distillation Columns—A Critical Survey," *Preprint IFAC Symposium DYCORD,* Maryland, 1992, pp. 1/25.

Skogestad, S.: Dynamics and Controll of Distillation Columns—a Critical Survey, Modelling, Identification and Control 18 (1997) No. 3, pp. 177–217

Stichlmair, J.: "Conceptual Design of the Control Configuration of Distillation Columns," *Chemical Engineering and Processing* 34 (1995) pp. 61/69.

Sturm, B. and Winkler, O.: *Messen, Steuern und Regeln in der Chemischen Technik, Band III,* pp. 357/375. Springer Verlag, Berlin 1981.

Wany, S. J. and Yu, S. W.: "Robust Feedback Design for Nonlinear High Purity Distillation Column Control," *Computers Chem. Engng.* 17 (1993) No. 9, pp. 897/908.

Wherry, T. C.; Peebles, J. R.; McNesse, P. M.; Teter, P. O.; Worsham, R. E.; and Young, R. M.: Process control, in: Perry, R. H.; Green, D. W.; and Maloney, J. D.: *Perry's Chemical Engineers Handbook* Sixth Edition pp. 22–115/22–118. McGraw-Hill, New York, 1984.

Wozny, G.: Lüder, Th.; and Canavas, C.: "Expertensysteme zur Automatisierung thermischer Trennprozesse," *atp—Automatisierungstechnische Praxis* 33 (1991) No. 8, pp. 430–440.

Yu, C. C. and Luyben, W. L.: "Robustness with Respect to Integral Controllability," *Ind. Eng. Chem. Res.* 26 (1987) pp. 1043/1045.

Zapp, G.: "Die dynamische Simulation einer Luftzerlegungsanlage und die Anwendung der Relative Gain Analyse auf den Entwurf von Regelstrukturen," *Linde Berichte aus Technik und Wissenschaft* 71 (1994) pp. 13/18.

Zhu, Z.-X. and Jutan, A.: "Stability Robustness for Decentralized Control Systems," *Chemical Engineering Science* Vol. 48 (1993) No. 13, pp. 2337/2343.

Index

Acetone 227, 248, 306, 343
Activity coefficient
  binary mixtures 9
  definition 9
  heterogeneous mixtures 18
  multicomponent mixtures 54
  ternary mixtures 54
Air separation 204
Ammonia recovery from waste water 200
Antoine equation 10
  list of constants 10
Area
  active 368
  effective 445, 450
  interfacial 389, 445
Argon recovery from air 219
Atmospheric tower 220
Azeotrope
  data collection 67
  definition 21
  hetero 17, 21, 43
  list of mixtures 22
  maximum boiling 21, 44
  minimum boiling 21, 42, 43
  reactive 257–262
  saddle 26, 30, 44
Azeotropic distillation
  batch processes 321
  continuous processes 238
  control 514
  list of systems 240
Azeotropic mixtures 21, 67

Batch distillation 284–325
  control 515
  multistage 284
  singlestage 87
  steam distillation 212
Batch distillation, multistage 284–325
  azeotropic mixtures 305
  binary mixtures 286–300
  energy demand 290, 294, 300
  modes of operation 286
  multicomponent mixtures 308
  processes 312–325
  ternary mixtures 300–308
Benzene 227, 239, 248, 306
Billet-Schultes Correlation 453
Boiling point, bubble point 11
  binary mixtures 11
  mixtures with total immiscibility 18, 211
  non-ideal ternary mixtures 56
  surface 27–30
  ternary mixtures 24–31
Border distillation line 40, 57
  reactive 266
Bravo/Fair correlation 452
Bubble cap trays 364, 368
Bubble point, see boiling point
Butanol 222, 343

Caloric factor 119
Capacity factor 370
Carbondioxide outburst at lake Nyos 82
Cascade control loop 478
Channel model structure 414
Chemical reaction 18, 253
Chloroform 306
Closed condensation 72
Closed distillation 72
  binary mixture 73
  continuous 73
  multicomponent mixture 74
Column coupling
  direct 339
  indirect 342
  thermal 342
Composite curve
  cold 349
  grand 352
  hot 350
Condensate collector drum
  residence time 481
Condensation
  counter current 72
  semi-continuous 103
  single stage 72
Constant molar over flow
  criteria 126
Constraints for industrial distillation processes 193–197
Control 474–516
  configurations 486
  distillation processes 474, 510
  letter code 476
  loops 475–478
  single loops 475
  tasks 478–486
Control configurations 486
  basic 498
  expanded range of application 508
  modified 499–501
  range of application 504
Control loops 475–478
Control of
  liquid level 478
  reboil 482
  reflux 480
  stream partition 479
  pressure 483
  product quality 484
Control of an air separation column 512
Control of an azeotropic distillation process 514
Control of product quality 484
Control schemes for distillation processes 488–501
Convergence pressure
  curve 24
  diagram 35
Costs
  cooling media 195
  heating media 194
  steam 194
Countercurrent condensation 72
Countercurrent distillation 72
Criteria for entrainer selection
  processes in one distillation field 225
  processes in two distillation fields 230
Criteria for use of tray or packed columns 366
Crude oil distillation 220
Cyclohexane 239, 248

Dalton's law 8
Deacidification of palm oil 214
Decolorization of fats and oils 214
Deodorization of esters 213
Desalination 334
De Santis equation 61
Design of heat exchanger networks 348–359
Design parameters
  packed columns 408
  tray columns 367
Dew point 11–13
  surface 31
  ternary mixtures 14
Dichloromethane 247
Dimensioning
  packed columns 408–461
  tray columns 367–408
Direct coupling of columns 339–343
Distillation
  multistage, batchwise 284
  multistage, continuous 107
  semi-continuous 104
  single stage, batchwise 87
  single stage, continuous 73

Distillation, combined with
absorption 240
adsorption 250
decantation 241
desorption 244
extraction 246–249
membrane permeation 251
Distillation, continuous
binary mixtures 116–133
multicomponent mixtures 170–190
ternary mixtures 133–170
Distillation lines 36–44
border 42
calculation 38
charts 41–45,98–101
definition 37
reactive 262–267
Dividing wall column 341
Downcomer 373
Dumped packing 410

Energy demand, see heat requirement
Enthalpy-concentration diagram 124
Entrainer 224
criteria for selection 225, 230
list 240
processes 226, 231
Entrainment 382
Equations of state 59–67, 64
Equilibrium
chemical 18
curve 34–36
ratio 31–34
stage 111, 128
vapor-liquid 6–69
Equilibrium stage
concept 111
number 112
Ethanol 223, 233, 239, 240, 242, 250, 251, 304, 323, 515
Ethylene glycol 242, 277
Excess enthalpy 47
definition 45
equations 48
Exergy losses 331
Exhaust steam 194
Extractive distillation
batchwise 325
continuous 240
list of systems 241

Feasible bottom and top fractions 141–147
Feasible dimensions of columns 196
Feasible pressures 195
Feasible products
binary mixtures 122
pure products 149–151
ternary mixtures 141–147
Feasible temperatures 193
Feed forward control 477
Flash 74
adiabatic 81, 82
process 84
Flooding 419, 435–438
Flow regimes 376
Froth height 379
Fugacity 9
coefficient 60

Gas, see also vapor
Gas capacity factor 370
Gas distributor 418
Gas load of trays
maximum 369
minimum 370
Generalized process
batchwise 315–321
continuous 230–233
Gibbs-Duhem equation 45
Grand composite curve 352

Hamiltonian function 297
Heat capacity ratio 332
Heat exchangers networks 348, 354
dual flow 357
grid representation 355
Heat of vaporization 76
Heat pump 332, 333
Heat requirement, of columns
batch distillation 290, 294, 300
binary mixtures 131, 330
continuous distillation 336–348
ternary mixtures
high boiler separation 170
low boiler separation 170
preferred separation 169
Heat requirement, of processes
path a 335, 346

Heat requirement, of processes *(continued)*
  path a-c 338, 340, 347
  path c 336, 346
Height of a transfer unit 449
Henry's law 9
Heptane 227
HETP (= height equivalent to a theoretical plate) 450, 455
Historical 2–4
Hold down plate 418
Hybrid Processes
  batchwise 321–325
  continuous 237–252
Hydrogenchloride 322
  concentration of diluted acid 233, 237
  recovery from inert gases 202

Indirect Coupling 343–348
Industrial distillation processes 192–278
  reactive 272–278
Interfacial area
  packings 443, 450
  trays 389
Intermediate
  condenser 121
  reboiler 121
  vessel 285, 317
Iteration history 183–190

Jacobian matrix 178–181
  partial derivatives 182

Linde column
  double column 206
  control scheme 512
Linde process for air separation 204
Liquid collector 417
Liquid distributor 417
Liquid entrainment 382
Liquid holdup
  influence on product quality 304, 309
  of packings 423, 426
  on trays 378
Liquid level control 478
Liquid load of packings
  minimum 421
Liquid load of trays
  maximum 373
  minimum 374
Liquid maldistribution
  packing 438–445
  trays 386–389
Liquid mixing on trays 385
Liquid redistributor 417

Maldistribution of liquid
  in packed columns 438–445
  in trayed columns 386–389
Margules equation 48
Mass transfer
  efficiencies 397
  equipment 361
  packing 448
  trays 395
Mass transfer coefficients
  packing 450
  trays 401
Mass transfer efficiencies 397
Mass transfer equipment 361–466
  types of construction 363
Maximum gas load 369–371
Maximum liquid load 373
McCabe-Thiele method 118, 128
Membrane 251
MESH-Equations 171
  evaluation 174
Methanol 247, 275, 304
Middle vessel 285, 317
Mixing 385
Model structures of packing 414
MTBE synthesis 275
Multiple-stage-flash (MSF) process 334
Multistage continuous distillation 107–190

Natural gas 85
Newton-Raphson algorithm 176
Nitric acid 245
Nitromethane 221
NRTL-equation 48

Onda correlation 452
One point composition control 493
Open condensation 72, 90, 93
  batchwise 72
  multicomponent mixtures 102
Open distillation 72
  batchwise 87

multicomponent mixtures 97
Operating line
rectifying section 117
stripping section 118
Operating region
packed columns 418–422
trayed columns 369–376, 374
Orifice coefficient 394

Packed Columns 365, 408–461
Packing
dumped 409
model structures 413
ordered 412
structured 412
supplements 415
Palm oil raffination 214
Particle model structure 414
Peng-Robinson equation 61
Pinch
double 154
heat exchanger networks 350
single 157, 159
tangential 129
ternary distillation 151
Pinch technology 350
Point efficiency 397, 400
Pontryagin's maximum principle 296
Preferred separation 152, 154
Pressure drop of packings
dry 427, 462
irrigated 428–453
Pressure drop of trays 392
bubble cap 396
dry 392
sieve trays 391–395
valve 396
Principle of distillation separation 1
Processes
batch 314
continuous 197–279
in one distillation field 314
in two distillation fields 315
Propanol 239, 304
Pump around 220

Raffination
of crude oil 220
of palm oil 214
Range of applications of control configurations
binary mixtures 504
expanded range 508
Raoult's law 9
Ratio control loop 476
Rayleigh equation
binary mixtures 88, 286
ternary mixtures 92
Reactive distillation 252–278
distillation lines 256, 262–267
equilibrium 253
processes 267–278
systems 274
Reboil ratio 123
binary mixtures 130
c-path 167
definition 118
preferred separation 165
ternary mixtures 151–168
total 127
Rectification 4, 108
Redlich-Kwong equation 61
list of parameters 62, 63
Reflux policy 286, 298
Reflux ratio 123
a-path 166, 168
binary mixtures 129
definition 117
preferred separation 155
ternary mixtures 151
total 127
Reflux ratio control 492
Relative gain analysis 513
Residuum curves
liquid 92, 95, 98–101
vapor 95, 96, 98–101
Rigorous column simulation 171
Rules for stream partition 480

Schematic of a distillation column 486
Separation paths 214
Separation processes, azeotropic mixtures
control 510
heterogeneous mixtures 221
pressure change 223, 510
use of entrainer 224
one distillation field 225
two distillation fields 229

Separation processes, zeotropic mixtures 214
high boiler separation 121
low boiler separation 215
number of separation paths 216
preferred separation 169
Separation sequences 335
a-path 335
a-c-path 338
c-path 332
Side columns 215, 219, 220
Sieve trays 364
Single stage distillation 72
batchwise 87
continuous 72
semi-continuous 103
Slop cut 312
Sodium hydroxide (NaOH) 246
Steam distillation 209, 212
Stoichiometric lines 253
Stream partition
control 479
rules 480
Structured packings 412
Sugar cane 4
Sulfuric acid 197, 233, 237, 245,322
Support grid 418
Symbols for control systems 476

TAME synthesis 278
Taylor series 175
Temperature-enthalpy diagram 349
Tetrahydrofuran (THF) 224, 239, 246, 345, 511
Thermal coupling 343–348
Toluene 240, 323, 515
Total immiscibility 18
Transfer units 400, 455
concept 113
definition 114
number 115
total reflux 128
Tray columns 363, 367–408
Tray efficiency 398, 402
Tray to tray calculation 133, 137
Two phase flow
packings 423–448
trays 367–391
Two point composition control 496
Types of construction
packed columns 365
trays 363

Uniquac equation 49

Valve trays 364
Van der Waals equation 60, 61
Van Laar equation 48
Vapor-liquid equilibrium 6, 17
chemical reaction 17
data collections 67
empirical findings 19–30
formulation 31
from equations of state 59
from excess enthalpy 45
ideal mixtures 17
Vapor pressure, see also Antoine equation
mixtures 11, 75
pure substances 10
Vessel, intermediate 285, 317
Virial equation 60

Water purification 207–209
Wettability 447
Wilson equation 48
parameters 52–53
ternary mixtures 54

Zeotropic mixtures 214–221